I0066266

TRANSACTIONS

OF THE

AMERICAN PHILOSOPHICAL SOCIETY

HELD AT PHILADELPHIA

FOR PROMOTING USEFUL KNOWLEDGE

———

NEW SERIES—VOLUME XXVI

OCTOBER, 1935

———

Siwalik Mammals in the American Museum of Natural History
EDWIN H. COLBERT

————

PHILADELPHIA:

THE AMERICAN PHILOSOPHICAL SOCIETY

104 SOUTH FIFTH STREET

1935

LANCASTER PRESS, INC., LANCASTER, PA.

PREFACE

The American Museum collection of fossil vertebrates from the Siwalik beds of northern India was made during the year of 1922 by Dr. Barnum Brown, who at that time was carrying on extended explorations in India, Samos Island and Greece, in order that the Museum might have representative faunal collections from these areas, so elaborately studied by various European palaeontologists. As the result of Dr. Brown's work in India, the expenses of which were partially defrayed by funds provided by Mrs. Henry Clay Frick, the American Museum now houses one of the outstanding collections of fossil mammals from the Siwaliks; a collection that ranks not only with the classic assemblage of fossils gathered together early in the last century by Hugh Falconer and Proby T. Cautley, and now contained in the British Museum in London, but also with the Geological Survey of India collection in Calcutta, accumulated and described by Richard Lydekker and by Guy E. Pilgrim.

It might be well to say, in passing, that a recent expedition from Yale University has been working in the Siwalik beds of India, and has brought back to this country a fine collection of fossil mammals, which it is hoped will greatly supplement the collection now in the American Museum. American students of mammalian evolution will now be able to gain a comprehensive understanding of the Upper Tertiary faunas of India, and without the necessity of a long trip to London or to Calcutta, they will be able to compare these Asiatic faunas with North American faunas, a method of study that undoubtedly will throw a great deal of illumination on vexing problems of phylogeny, of mammalian migrations and of intercontinental stratigraphic correlations.

The American Museum Siwalik collection has not heretofore been thoroughly studied. It had been originally intended that Dr. William Diller Matthew, formerly Curator in Chief of the Department of Vertebrate Palaeontology of the American Museum, should conduct the investigation of the Siwalik collection, and with this purpose in mind he visited, in 1926, the type collections in London and in Calcutta, in order to obtain a broad background for his subsequent labors. As a result of his work on the above mentioned collections he published a preliminary paper, to serve as a foundation for his intended detailed studies of the Siwalik fossils in the American Museum. Unfortunately, Dr. Matthew's brilliant contributions to mammalian palaeontology were cut short by his most untimely death, and so occupied was he just previous to his passing with certain other palaeontological research problems, that he never had an opportunity of pursuing serious and protracted studies on the American Museum Siwalik collection.

The purpose of the present work is an attempt to make a thorough study of certain of the mammalian orders which comprise the bulk of the American Museum Siwalik collection, not only by means of my own researches but also by the correlation of my work with that which is being done or which has been done by other authors. Professor Henry Fairfield Osborn has critically examined the proboscideans from this collection and the results of his studies are embodied in his monograph on the Proboscidea. Dr. William King Gregory and Dr. Milo Hellman have made an exhaustive study of the three species of *Dryopithecus* in the collection.   Dr. Guy E. Pilgrim, formerly Superintendent of the Geological Survey of India, has recently completed a detailed study of the Bovidae.   Therefore, the present study, while directed at all of the other mammalian groups not studied by the above mentioned authors, will also present a summary of their work as far as is practicable.

Before I began the preparation of this memoir in the year 1932, Professor Osborn had devoted many years of study in cooperation with Dr. Guy E. Pilgrim, to the faunas characteristic of the successive horizons in the Siwaliks of India and downward into the older horizons and faunas of Baluchistan, with special emphasis on the correlation of these faunas with faunas of other parts of Eurasia.   These studies were based upon Dr. Pilgrim's faunistic analysis of the Siwalik horizons, and they were embodied in a special chapter of Osborn's Memoir on the Proboscidea, preliminary manuscript of which was placed at my disposal.   I am greatly indebted to Dr. Pilgrim and to Professor Osborn for the liberal use of their prior studies.

As a background for the present study, I was able to spend some time at the British Museum in London, during the fall and winter of 1931–1932, examining the Siwalik collection in that institution.   I owe many thanks to the officers of the Department of Geology in the British Museum, to Dr. W. D. Lang, Dr. W. E. Swinton and Mr. Harold Sealy particularly, for the opportunity of studying the fossils under their jurisdiction, and for the numerous courtesies extended to me by these gentlemen.

I wish to acknowledge the kindness of the President and the Trustees of the American Museum, for making possible my trip to London.   I must acknowledge the kindness of Professor Osborn and of Mr. Childs Frick for their generous assignment of this collection to myself as a subject of research.   My greatest thanks are due to Dr. Barnum Brown for the innumerable conferences and the valuable aid he has given to this study, especially as regards problems of stratigraphy and of the geologic occurrences of fossils.   In this connection I would like to express also my debt to Mr. G. Edward Lewis of Yale University and to Dr. Guy E. Pilgrim for their help, based on their first hand knowledge of Siwalik stratigraphy.   Dr. Pilgrim has also rendered valuable aid on questions of faunal associations.   I would like to express my gratitude to Dr. Richard S. Lull and to Dr. Malcolm R. Thorpe for permission to look over the Yale University Siwalik collection.

Acknowledgments are especially due to Dr. William King Gregory, who has constantly devoted much time to advice and help on this work.   His valuable suggestions have done much towards giving a clearer understanding of phylogenetic problems, and his exceeding interest has served as a constant inspiration towards carrying the study to completion. Dr. Gregory has devoted much time to a critical examination of the manuscript for this paper.

I wish further to acknowledge the work of the artists who have prepared the illustrations.   Margaret Matthew Colbert and John C. Germann made the line drawings.   Mr. Germann retouched the photographs, which were made by Mr. Hugh Rice.   The large infolded map of northern India was made by Dr. Erwin Raisz, and it was placed at my disposal by Professor Osborn.   The maps of the American Museum collecting localities were traced by D. F. Levett Bradley from originals furnished by the Indian Geological Survey.

I wish to thank Mrs. R. H. Nichols for her aid in the tedious task of sorting and completely cataloguing the collection, and for other help in the preparation of this memoir. Finally I wish to express my great indebtedness to the Indian Geological Survey, for the geological maps furnished by them, and for the permission to use such data as appear on these maps.

EDWIN H. COLBERT

AMERICAN MUSEUM OF NATURAL HISTORY
      January 7, 1935

*Published by permission of the
Trustees of the American Museum
of Natural History*

SIWALIK MAMMALS IN THE AMERICAN MUSEUM OF NATURAL HISTORY

By Edwin H. Colbert

CONTENTS

SIWALIK MAMMALS IN THE AMERICAN MUSEUM OF NATURAL HISTORY

By Edwin H. Colbert

PART I. INTRODUCTION

Previous Publications Dealing with American Museum Siwalik Vertebrates

Several of the more interesting fossils in the American Museum Siwalik collection have been the subjects of papers by various authors. Most of these papers have been preliminary notices, issued as descriptions of new species or as detailed accounts of especially important forms. One, the Anthropological Paper by Gregory and Hellman is a thorough monograph, based in part on the primate remains discovered by Dr. Brown. Three other papers are general in their scope. One of these is a popular descriptive article by Dr. Brown giving an account of his expedition in India, and another is a popular article by Dr. Matthew dealing with Siwalik fossils and the general aspects of the Siwalik problem. The third is Dr. Matthew's "Critical Observations upon Siwalik Mammals," mentioned in the preface, a long report of extraordinary value.

The papers published by the American Museum, concerning Siwalik vertebrates are listed below. These publications, together with the present work and with Dr. Pilgrim's contribution on the Bovidae, bring to a completion the studies on the mammals in the collection, and with the exception of a few more shorter contributions which will come out on the remaining undescribed reptilian remains, they bring to a close the studies on the Siwalik collection. Thus, some ten years or so since it was made, the collection has been virtually completely studied, and it will now remain in the Museum as important reference material for future studies on vertebrate evolution. In the accompanying list all of the Siwalik contributions are included for the sake of completeness.

1. **Matthew, W. D.,** 1924. Fossil Animals of India. Nat. Hist., XXIV, No. 4, pp. 208–214.
2. **Brown, Barnum, Gregory, William K., and Hellman, Milo,** 1924. On Three Incomplete Anthropoid Jaws from the Siwaliks, India. Amer. Mus. Novitates, No. 130.
3. **Brown, Barnum,** 1925. Glimpses of India. Nat. Hist., XXV, pp. 109–125.
4. **Osborn, Henry Fairfield,** 1926. Additional New Genera and Species of the Mastodontoid Proboscidea. Amer. Mus. Novitates, No. 238, pp. 4–6.
5. **Gregory, William K., and Hellman, Milo,** 1926. The Dentition of Dryopithecus and the Origin of Man. Amer. Mus. Anthropological Papers, XXVIII, Pt. 1.
6. **Brown, Barnum,** 1926. A New Deer from the Siwaliks. Amer. Mus. Novitates, No. 242.
7. **Matthew, W. D.,** 1929. Critical Observations upon Siwalik Mammals. Bull. Amer. Mus. Nat. Hist., LVI, Art. VII, pp. 437–560.

1

8. **Osborn, Henry Fairfield,** 1929. New Eurasiatic and American Proboscideans. Amer. Mus. Novitates, No. 393, pp. 2–6, 8–15, 18, 21–23.

9. **Mook, Charles C.,** 1932. A New Species of Fossil Gavial from the Siwalik Beds. Amer. Mus. Novitates, No. 514.

10. **Colbert, Edwin H.,** 1933. The Skull of *Dissopsalis carnifex* Pilgrim, a Miocene Creodont from India. Amer. Mus. Novitates, No. 603.

11. **Colbert, Edwin H.,** 1933. The Presence of Tubulidentates in the Middle Siwalik Beds of Northern India. Amer. Mus. Novitates, No. 604.

12. **Colbert, Edwin H.,** 1933. A New Mustelid from the Lower Siwalik Beds of India. Amer. Mus. Novitates, No. 605.

13. **Colbert, Edwin H.,** 1933. The Skull and Mandible of *Conohyus*, a Primitive Suid from the Siwalik Beds of India. Amer. Mus. Novitates, No. 621.

14. **Colbert, Edwin H.,** 1933. A Skull and Mandible of *Giraffokeryx punjabiensis* Pilgrim. Amer. Mus. Novitates, No. 632.

15. **Colbert, Edwin H.,** 1933. Two New Rodents from the Lower Siwalik Beds of India. Amer. Mus. Novitates, No. 633.

16. **Colbert, Edwin H.,** 1933. An Upper Tertiary Peccary from India. Amer. Mus. Novitates, No. 635.

17. **Mook, Charles C.,** 1933. A Skull with Jaws of *Crocodilus sivalensis* Lydekker. Amer. Mus. Novitates, No. 670.

18. **Pilgrim, G. E.,** 1934. Correlation of Ossiferous Sections in the Upper Cenozoic of India. Amer. Mus. Novitates, No. 704.

19. **Colbert, Edwin H.,** 1934. A New Rhinoceros from the Siwalik Beds of India. Amer. Mus. Novitates, No. 749.

20. **Colbert, Edwin H.,** 1935. "Distributional and Phylogenetic Studies on Indian Fossil Mammals. I. American Museum Collecting Localities in Northern India." Amer. Mus. Novitates, No. 796.

21. **Colbert, Edwin H.,** 1935. "Distributional and Phylogenetic Studies on Indian Fossil Mammals. II. The Correlation of the Siwaliks of India as Inferred by the Migrations of *Hipparion* and *Equus*." Amer. Mus. Novitates, No. 797.

22. **Colbert, Edwin H.,** 1935. "Distributional and Phylogenetic Studies on Indian Fossil Mammals. III. A Classification of the Chalicotherioidea." Amer. Mus. Novitates, No. 798.

23. **Colbert, Edwin H.,** 1935. "Distributional and Phylogenetic Studies on Indian Fossil Mammals. IV. The Phylogeny of the Indian Suidae and the Origin of the Hippopotamidae." Amer. Mus. Novitates, No. 799.

24. **Colbert, Edwin H.,** 1935. "Distributional and Phylogenetic Studies on Indian Fossil Mammals. V. The Classification and the Phylogeny of the Giraffidae." Amer. Mus. Novitates, No. 800.

HISTORICAL REVIEW

Fossil bones in India were certainly known to the ancients, who had various beliefs as to the nature of these objects. Perhaps the first recorded mention of fossils in India is to be found in a Compendium of the History of the Moghul and Pathan Emporers, by Ferishta, written in 1360, quoted by Falconer in 1868. He mentions that laborers, working on a water way under the direction of Feroz the third, encountered in the course of their diggings the bones of human giants.

"In course of the operations bones of elephants and men were discovered in the unbedded mound. Those of the human forearm measured three yards. Some of the bones were petrified, while others were still in the condition of bone." [1]

These fossils were discovered in the Siwalik Hills.

Perhaps the first serious efforts to recover the remains of Siwalik vertebrates were expended by Captain W. S. Webb, who obtained some fossil bones, in the course of his travels incident to a survey of the heights of the Himalaya Mountains. He procured them from natives, who had made a practice of collecting the fossils from "the plains of Tibet" for use as charms. The bones thus procured by Captain Webb were sent to England and were referred to by Dr. Buckland in his "Reliquiae Diluvianae."

"But in Central Asia the bones of horses and deer have been found at an elevation of 16,000 feet above the sea in the Himalaya Mountains. The bones I am now speaking of are at the Royal College of Surgeons in London, and were sent last year to Sir E. Home, by Captain W. S. Webb, who procured them from the Chinese Tartars of Daba, who assured him that they were found in the north face of the snowy ridge of Kylas, in lat. 32°, at a spot which Captain Webb calculates to be not less than 16,000 feet high: they are only obtained from the masses that fall with the avalanches from the regions of perpetual snow, and are therefore said by the natives to have fallen from the clouds, and to be the bones of genii." [2]

Then again, we find that Mr. John Crawford, serving in the embassy to Ava during 1826, mentions the discovery of fossil bone along the Irrawaddy River in Burma. His discoveries were described by W. Clift and later by Hugh Falconer.

At about this same time, or shortly thereafter, two army officers, Lieutenants W. E. Baker and H. M. Durand were contributing various short articles to the Asiatic Society of Bengal, in which they described certain fossil bones discovered by them during their work with the army engineers in northern India. Baker and Durand were content, however, in giving empirical description of their specimens, without assigning generic or specific names to them, or without setting forth any definite views as to their possible taxonomic or phylogenetic relationships.

The discoveries by Baker and Durand immediately aroused the interest of Dr. Hugh Falconer and Lieutenant Proby T. Cautley, two young men in the government service, and they proceeded to the localities where Baker and Durand had found their specimens, and there they succeeded in gathering together considerable quantities of fossil bones. Falconer's enthusiasm was great, and he immediately launched himself into the detailed study of the numerous specimens which he and Cautley had discovered.

Consequently the first really reliable and complete descriptions of Siwalik fossils (from India proper as distinguished from the scattered specimens from Burma) were begun early in the eighteen thirties, in the Transactions of the Asiatic Society of Bengal, and from that time on, over a period of twenty or thirty years, Dr. Falconer continued his vigorous and brilliant researches on Siwalik vertebrates—studies that were destined to become classic in the field of vertebrate palaeontology—and in this work he was aided by Captain

[1] Falconer, Hugh, 1868, Palaeontological Memoirs, Vol. I, p. 4.
[2] Buckland, W., 1823, "Reliquiae Diluvianae," p. 222.

Cautley.   Falconer and Cautley may therefore be considered as the pioneers in the study of Siwalik vertebrates.

Falconer's work was later supplemented by Richard Lydekker who, during the period from 1876 to 1886, published a series of exhaustive monographs in the Palaeontologica Indica, as well as numerous shorter contributions in the Records of the Geological Survey of India.   At the same time that Falconer and later that Lydekker were engrossed in their studies on Siwalik mammals, members of the Geological Survey of India were busily engaged in researches on the stratigraphy of the area from whence the vertebrates were obtained. These studies were naturally destined to aid greatly in an understanding of the fossil vertebrates.   Among the authorities who first prosecuted serious researches into the question of Siwalik stratigraphy, may be mentioned the name of Henry Benedict Medlicott,

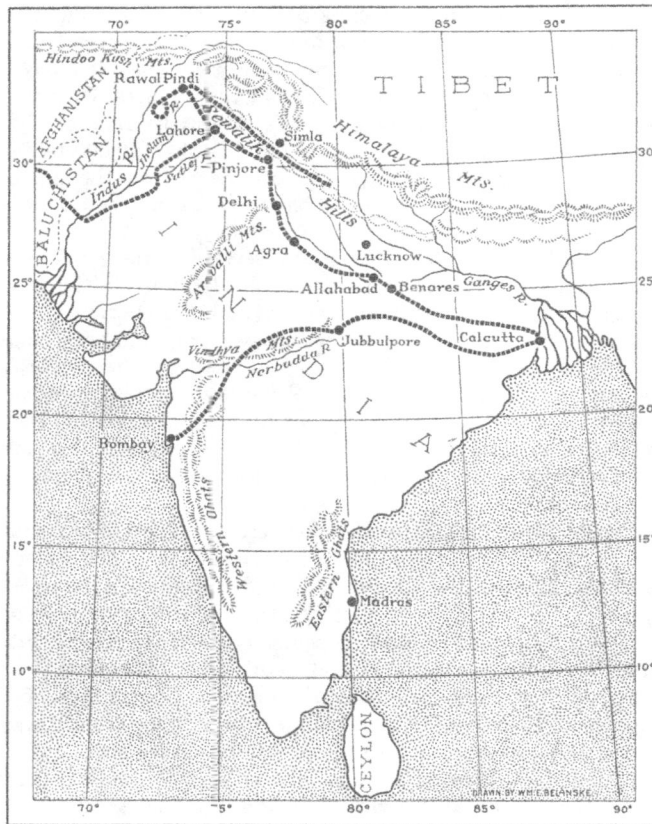

Fig. 1.   Map of India, showing the route of the Siwalik Hills Indian Expedition of the American Museum of Natural History.   From Brown, 1925.

whose work was especially valuable as it added supplementary knowledge to the problems dealt with by Falconer and by Lydekker.

Finally, during the past two decades, the status of our knowledge of Siwalik vertebrates has been greatly extended by the work of Dr. Guy E. Pilgrim, for many years a Superintendent in the Geological Survey of India. Dr. Pilgrim virtually opened the field in the discovery and study of Lower Siwalik vertebrates, and due to his efforts the succession of vertebrate faunas in the Siwalik series was definitely established.

Dr. Pilgrim's work has been recently supplemented by the studies of Dr. W. D. Matthew, mentioned above.

PART II. GEOLOGICAL CONSIDERATIONS

GEOGRAPHY OF THE SIWALIK HILLS AND THE SALT RANGE

The term "Sewalik Hills" was applied in a geographical sense by Dr. Hugh Falconer, who used it to denote the range of hills along the southern flanks of the Himalaya Mountains, stretching from the Indus River on the north to the Brahmaputra River on the south. Although many geographers have followed this usage of the term, there have been others that would apply it in a more restricted sense, confining it to the more northerly reaches of the "Sewaliks," as defined by Falconer, stretching over a linear distance of some four hundred miles immediately south of the headwaters of the Indus. This latter usage of the term "Siwalik" is the one generally followed by modern authorities. The name "Sub-Himalaya" has been used also, as a designation to include these Himalayan foothills.

The Siwalik Hills form a ridge between the great flood plain of the Ganges River and the Himalaya mountains. These hills are relatively low, ranging from three thousand to four thousand feet in height above sea level, and they have a general northwest-southeast trend parallel to the Himalayas. They are distinguished by their asymetric form, having in general a steep face or scarp on the southern side towards the valley, and gentler slopes on the northern sides. The hills actually have their expression as a series of parallel ridges, forming a belt some eight miles or so in width. Behind and between these ridges are narrow valleys or duns, running parallel to the general trend of the range, and the largest of these, the Dehra Dun, separates the Siwalik Hills from the greater reaches of the Himalaya range. Many rivers traverse the Siwalik Hills, at right angles to their northwest-southeast trend, cutting through the highlands by narrow gorges and debouching on the plain to the south in sandy flats.

It was from these Siwalik Hills that the upper Tertiary and Pleistocene fossil vertebrates were first discovered and studied by Hugh Falconer; hence the name "Siwalik" as applied to the beds containing the fossils. The studies of Lydekker and of Pilgrim opened up new vistas in the history of mammalian evolution in India, for these authors described fossils from the Salt Range of the Punjab States, fossils which proved to be immediately antecedent and in many cases directly ancestral to the typical Siwalik vertebrates. Thus the term "Siwaliks" in a stratigraphic sense was extended down to include these lower beds. Consequently it is necessary to include the Salt Range area in a geographic consideration of the Siwalik strata.

The Salt Range forms a separate line of low flat-topped hills, genetically quite distinct from the Siwalik Hills overthrust fault-scarp, and trending in a slight southwesterly direction, along the northern side of the Jhelum River. It is flanked by a fairly wide belt of parallel ridges and badlands. Intermittent streams traverse these ridges, at right angles to their general east-west trend, and join the main drainage lines of the Indus, notably the Jhelum and the minor branches parallel to it.

Immediately to the north of the Salt Range is the Potwar Plateau, an elevated plain situated between the northern slopes of the Salt Range and the Rawalpindi district. The level reaches of the Potwar Plateau, consisting of Siwalik and related beds, merge into the gently dipping strata that form the top of the Salt Range.

6

GENERAL GEOLOGIC CONSIDERATIONS OF THE SIWALIK BEDS

Historical Review

Dr. Falconer, one of the first students of Siwalik stratigraphy, placed the Siwalik beds as he knew them, namely those beds now constituting the Upper Siwalik division, in the middle portion of the Tertiary system, making them of Miocene age. He recognized the fact, however, that the Siwalik beds might be representative of a considerable period of time, thereby being the equivalent of more than one Tertiary period as recognized elsewhere, especially in Europe. The following statement outlines his views on this question (it speaks well for his sagacity and understanding), and for its general truth it is as fully applicable today, in the light of our greatly extended knowledge of Siwalik faunas, as it was when he made it.

"It would be, perhaps, not unphilosophical to conceive that the epoch of the Siwalik fauna may have lasted through a period corresponding to more than one of the Tertiary periods of Europe." [1]

Of course Falconer was speaking of the Upper Siwalik fauna.

Falconer had struck to the heart of the Siwalik problem when he stated that the "epoch of the Siwalik fauna" lasted through a considerable portion of the Tertiary. He erred, however (and when one considers the material on which his views were founded such an error is quite to be expected), in thinking that a single Siwalik fauna stretched over a considerable time period. Perhaps it may be well to quote here Dr. Falconer's conclusions on the "Geological and Climatal Bearings of the Sewalik Fauna" as the best summation of his views.

"Besides the mere zoological interest of the subject, the Sewalik inquiries involve these conclusions.

"1. The upheavement of a narrow belt of the plains of India at the foot of the Himalayahs into hills 3,500 ft. high along 11° of longitude, or about 800 miles, after the long establishment on the continent of such modern forms as quadrumana, camel, giraffe, and existing species of reptilia.

"2. A great upheavement of the Himalayahs, extending to many thousand feet, and equal to the elevation of a tract which formerly bore a tropical fauna, up to a height which now causes a climate of nearly arctic severity. Remains of rhinoceros, antelope, hyæna, horse, large ruminants, etc., found at 16,000 feet above the sea.

"3. Conditions in India during the tertiary period precisely the reverse of what have held in Europe. Instead of a succession of periods with successive decrease of temperature, India has now as high a temperature, if not higher, than it ever had during the tertiary period. The upheavements have operated to increase the heat. In lat. 30°, at 7,000 feet above the sea, the mean temperature, making the compensation for the elevation, and reducing it to the level of the sea, is 81.2° Fahr., or equal to that of the equator. The same excess of temperature holds generally over the continent, as contrasted with the eastern side of the continent of Asia.

"4. Instead of numerous subdivisions of the tertiary period with successive fauna, facts tend to the conclusion that India had one long term, and one protracted fauna, which lived through a period corresponding to several terms of the tertiary period in Europe." [2]

[1] Falconer, H., 1868, Palaeontological Memoirs, I, p. 28.
[2] Falconer, H., 1868, Palaeontological Memoirs, I, pp. 28–29.

Richard Lydekker, on the basis of his very extended studies on the Siwalik collections in the British Museum and in the Geological Survey of India Museum, recognized that the remains did not represent a unit fauna. Therefore he divided the Siwaliks into two horizons, and these he regarded as of Pliocene age.

Falconer had opened the study of Siwalik vertebrates with his researches on the Upper Siwalik fauna. Lydekker discovered the fauna of the Middle Siwaliks—the fauna to which he referred as "Lower Siwaliks." It remained for Dr. Guy E. Pilgrim, however, to complete the sequence of the Siwalik faunas. As early as 1864 Medlicott had recognized the Siwalik series to be made up of three horizons, which he termed upper, middle and lower. Dr. Pilgrim, collecting in regions where searches had not been made before, and working in beds much lower than those which usually yielded fossil mammals, showed that there are at least three distinct and successive faunas in the Siwalik series, namely the two known to Falconer and Lydekker, and in addition a new lower Siwalik fauna, quite unknown to the older authorities. Pilgrim, on the basis of his comparisons of the Siwalik faunas with similar European faunas, regarded the Upper Siwalik fauna as of upper Pliocene and lower Pleistocene age, the middle one as of lower Pliocene or Pontian age, and the lower one as of middle Miocene or Tortonian age.

The views of recent authors on the correlation of the Siwaliks, together with an appraisal of these views and an exposition of recent evidence on this perplexing question, will be presented below.

The Sequence and Lithology of the Siwalik Deposits

The following remarks concerning the divisions and the lithology of the Siwalik deposits are offered merely as a brief resumé of the descriptions published by Pilgrim and other authors who have dealt with the problem of the upper portion of the Tertiary sediments in northern India.   They are not intended to be original, either in scope or in outlook.

The Yale North India Expedition, recently returned from a collecting trip in the Punjab and adjacent regions, has brought back a considerable number of lithologic samples from the various zones of the Siwalik deposits.   It is the purpose of Mr. G. Edward Lewis, who made the palaeontological collections for Yale University, to contribute a detailed report on the lithology of the Siwalik series, with particular emphasis on its relationship to the stratigraphic sequence of those beds.   Since Mr. Lewis has the proper materials for making such a study, it is not advisable to enter at this point into a detailed consideration of Siwalik lithology.

The Divisions of the Siwaliks

Hugh Falconer, as has been pointed out above, considered the Siwalik beds as he knew them, to represent a single, continuous series of continental deposits. Lydekker, recognizing two faunas among the Siwalik fossils under his scrutiny, divided the series into an upper and a lower division. Pilgrim showed the necessity of a tripartite division of the Siwalik beds, as the result of his extended studies on the earlier faunas from the Punjab region. Moreover, he recognized certain faunal zones within the Siwalik beds, and these he named as follows:

Upper Siwaliks	{	Boulder Conglomerate zone
		Pinjor zone
		Tatrot zone
Middle Siwaliks	{	Dhok Pathan zone
		Nagri zone
Lower Siwaliks	{	Chinji zone
		Kamlial zone [3]

These zones were based by Dr. Pilgrim mainly on the evidence of faunal associations, and for the most part they cannot be definitely and lithologically distinguished in the field. For the purposes of field work, four distinct divisions of the Siwaliks are recognized by the Indian Survey. These are as follows:

Upper Siwaliks
Middle Siwaliks

Lower Siwaliks	{	Chinji
		Kamlial

There is a definite and distinct lithologic break between the Kamlial and the Chinji, and this is the only constant dividing line in the entire Siwalik series. The other beds, that is, the Chinji, Middle Siwaliks and Upper Siwaliks, grade into each other by such imperceptible degrees that it is quite impossible to establish very definite boundaries between these divisions. The Middle Siwaliks, for instance, are lithologically more or less distinct in their typical exposures, but as one approaches the Chinji beds below and the Upper Siwalik beds above, there are gradual transitions of lithology so that the passage from one level to the next is made without the expression of any definite stratigraphic break. This is a condition naturally to be expected in a series of beds deposited by continual and uninterrupted continental agencies. Consequently it becomes necessary to rely more or less on faunal evidences in order to determine the minor divisions in any such series.

Therefore Dr. Pilgrim's divisions of the Siwaliks are truly faunal horizons or zones, as used in the American sense of the word. They are contained within the four mappable units, as defined above, which latter might be regarded either in the light of formations or groups. The divisions of the Siwalik beds are thus seen to rest on the following bases, according to the work of Dr. Pilgrim.

Upper Siwaliks approximately 6,000 feet.

Boulder Conglomerate zone.

A division recognized chiefly by its lithologic peculiarities, namely its large heavy boulders, derived from beds of earlier ages and secondarily indurated by infiltrations of siliceous material. This zone was thought originally to contain the uppermost Siwalik fauna. Pilgrim now considers the Upper Siwalik fauna to be typically developed in the zone beneath, a conclusion substantiated by the work of Mr. Brown in the Punjab. Conformably underlying the Boulder Conglomerate is the Pinjor zone.

Pinjor zone.

Defined as the horizon containing the typical Upper Siwalik fauna. It is composed of Pleistocene sands and variegated clays. This zone grades downward into the Tatrot zone.

---

[3] The Kamlial was named by E. S. Pinfold and the name was subsequently adopted by Dr. Pilgrim.

Tatrot zone.

A zone in which fossils are scanty. It would seem to be characterized by forerunners of the Pinjor fauna and survivors of the Dhok Pathan fauna. The Tatrot is composed of hard brown sandstones, forming at the type locality protecting or capping beds over the Dhok Pathan sediments beneath.

Middle Siwaliks approximately 6,000 feet.

Dhok Pathan zone.

An abundantly fossiliferous horizon, containing the typical Middle Siwalik fauna. It is made up of light colored sands, containing considerable amounts of unweathered igneous minerals, notably feldspar. The Bhandar beds, originally distinguished by Pilgrim as a separate zone above the Dhok Pathan, but later combined by him with the latter, are shown by the American Museum collection to be a local facies development of the Dhok Pathan, carrying numerous and typical fossils. The sediments containing the Dhok Pathan grade down through a considerable thickness of unfossiliferous beds into the Nagri beds.

Nagri zone.

The lower division of the Middle Siwaliks, composed of red clays with included nodules. Typified by forerunners of the Dhok Pathan fauna, and by numerous holdovers from the Chinji fauna.

Lower Siwaliks approximately 4,000 feet.

Chinji zone.

A characteristic phase of about 2,300 feet of bright red clays, carrying beds of what Pilgrim has termed "pseudo-conglomerates." This zone contains the typical Lower Siwalik fauna. Pilgrim originally recognized two divisions of the Chinji, namely a lower and an upper division. Later work would seem to show that the fossils range pretty well throughout the thickness of the Chinji deposits, and that the upper and lower horizons recognized by Pilgrim really represent levels of unusual abundance of fossils, rather than zones of faunal differences. An unconformity, the only definite break in the Siwalik series, separates the Chinji from the lowest of the Siwalik zones.

Kamlial zone.

This zone is lithologically distinct from the overlying Chinji zone. It consists of about 1,700 feet of river sediments, containing numerous beds of conglomerates. Fossils are scarce, but those present seem to be definitely more primitive than the Chinji forms.

The relation of the several Siwalik horizons to each other are given in the accompanying chart, which is based in part on the correlation chart recently published by Dr. Pilgrim.

The Succession of Siwalik Strata

The Lithology of the Siwaliks and its Bearing on Sedimentation

D. N. Wadia in his textbook on the geology of India has summed up very well in a few words the sedimentary processes that were active in forming the Siwalik beds.

"The composition of the Siwalik deposits shows that they are nothing else than the

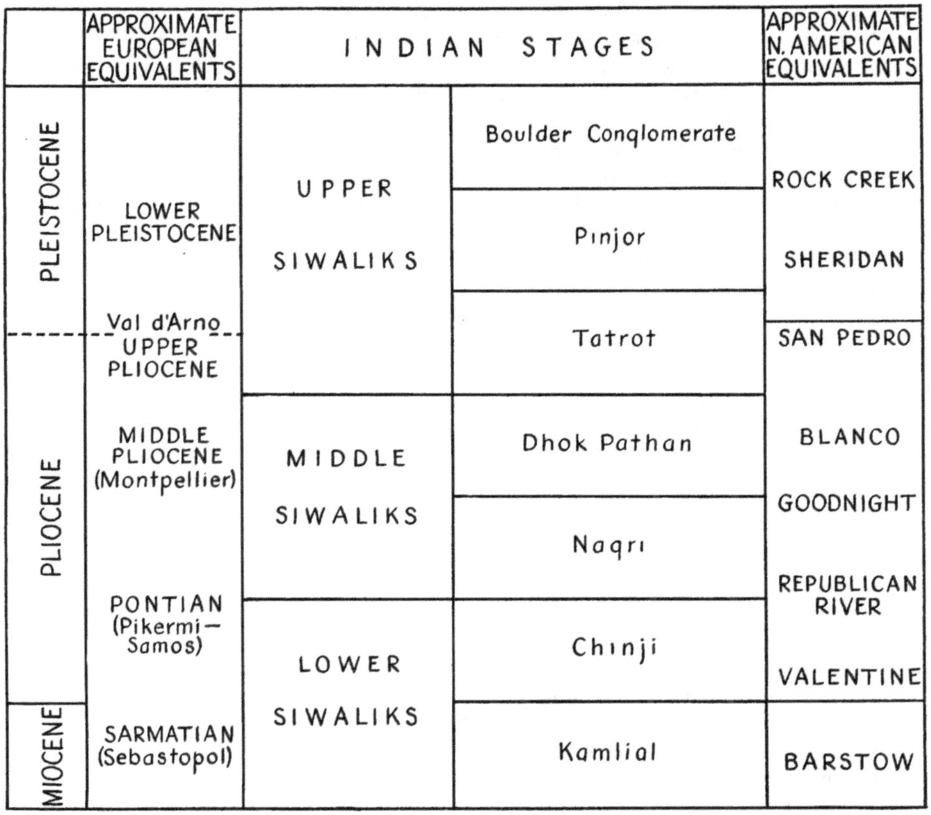

	APPROXIMATE EUROPEAN EQUIVALENTS	INDIAN STAGES		APPROXIMATE N. AMERICAN EQUIVALENTS
PLEISTOCENE	LOWER PLEISTOCENE	UPPER SIWALIKS	Boulder Conglomerate	ROCK CREEK
PLEISTOCENE	LOWER PLEISTOCENE	UPPER SIWALIKS	Pinjor	SHERIDAN
PLIOCENE	Val d'Arno UPPER PLIOCENE	UPPER SIWALIKS	Tatrot	SAN PEDRO
PLIOCENE	MIDDLE PLIOCENE (Montpellier)	MIDDLE SIWALIKS	Dhok Pathan	BLANCO
PLIOCENE	MIDDLE PLIOCENE (Montpellier)	MIDDLE SIWALIKS	Naqri	GOODNIGHT
PLIOCENE	PONTIAN (Pikermi— Samos)	LOWER SIWALIKS	Chinji	REPUBLICAN RIVER
PLIOCENE	PONTIAN (Pikermi— Samos)	LOWER SIWALIKS	Chinji	VALENTINE
MIOCENE	SARMATIAN (Sebastopol)	LOWER SIWALIKS	Kamlial	BARSTOW

FIG. 2. The divisions and stratigraphic position of the Siwalik Series.

alluvial detritus derived from the subaerial waste of the mountains, swept down by their numerous rivers and streams and deposited at their foot. This process was very much like what the existing river systems of the Himalayas are doing at the present day on their emerging to the plains of the Punjab and Bengal." [4]

Looking at the lithology of the Siwalik series it is to be noted that these rocks are the result of extremely rapid and perhaps almost continuous stream erosion and deposition. Lithologically the origin of the series is probably simple, representing increasingly coarse river detritus brought down from a rapidly rising mountain mass. Moreover there are no great secondary changes to be found in the Siwalik rocks, evidences of automorphism and glaciation being absent. The factor of predominant importance is that of erosion and deposition by rapidly flowing rivers.

Pilgrim has postulated that the lower Siwalik beds, composed of red and gray sands

[4] Wadia, D. N., 1919, "Geology of India," p. 231.

and clays, have within them numerous disconformities of minor extent, representing short breaks in sedimentation and consequently time intervals of more or less importance. According to his views, the lower Siwalik beds represent a considerable period of deposition.[5] Moreover, from the nature of these deposits he supposes that they were formed in estuaries and in numerous lake basins—the last remnants of former shallow sea ways.

On the other hand the Middle Siwalik deposits, made up of fairly coarse unweathered materials, derived directly from the adjacent mountain masses show that they were deposited at a rapid rate, probably by stream channels on flood plains, and in small basins.

In a like manner, the Upper Siwalik sediments, composed of very coarse materials, indicate extremely rapid deposition by heavily laden streams, carrying these materials from a rapidly rising source of supply.

Perhaps the conditions of sedimentation during the latter part of the Tertiary period and the early portion of the Pleistocene in India, were not greatly different from what they are today. Seasonal changes were probably marked. There was a dry season, during which relatively little material was carried by the streams, and this was followed by a wet season, a time when the heavily laden rivers brought great quantities of rocks, sands and silts from the high mountains and deposited this material on the flat plain. Perhaps the climate was not quite so moist or tropical as it is at the present time in India. In short, being isolated by the rising mountains to the north, and favored by warm moisture bearing winds from the south, India has enjoyed an increase in the precipitation and temperature during the later part of the Tertiary period, a course of climatic development somewhat the opposite to that which has taken place in Europe, Asia and North America.

To conclude these remarks concerning the sequence and lithology of the Siwalik series, it may be said that the upper Tertiary and Pleistocene strata of northern India represent an almost continual period of deposition, from the upper Miocene into the lower phases of the Pleistocene. Although there is a change in the sedimentation as we pass from the lower beds to the higher ones, from fine to coarse sediments, this change is gradual and intergraded. Thus all of the various Siwalik levels merge into one another and nowhere, except at the base of the Chinji zone, is it easy to establish a definite interformational boundary.

At this point I should like to insert some observations made by Mr. G. Edward Lewis of the Yale University North India Expedition, in a letter sent to me from India, where he was engaged in the collection of Siwalik vertebrates. Of course, this is a qualified view, based on field observations, and is not to be taken as a final and definitive opinion.

"It is of paramount importance to bear in mind that we are dealing with some 20,000 feet of strata representing sedimentation during the relatively short period between mid-Miocene (?) and mid-Pleistocene (?)—(everyone will agree that the Pliocene is included in toto, and the time overlap above, and more specifically below is subject to discussion), and that conditions of sedimentation resulted in cross-bedding throughout, dove-tailing, and an unbelievable amount of lateral variation (this applies to the major and subordinate units as well). Two parallel sections 100 meters apart usually give utterly different results as to lithology and fauna; suids predominate in one, proboscideans in the other. At a given level or group of levels, a massive stratum of sandstone up to 20 or 30 meters thick may lens out on either side within a distance of 80 meters.

"Constant areal dividing lines and definite boundary lines are absolutely impossible.

[5] Pilgrim, G. E., 1913, Rec. Geol. Surv. India, XLIII, p. 273.

Continuous gradual sedimentation seems to occur from basal Kamlial to uppermost Boulder Conglomerate, and while changing conditions of sedimentation enable the observer to recognize the very broad divisions of (1) Kamlial, (2) Chinji, (3) Middle, and (4) Upper Siwalik by a gradual variation in general lithology, he cannot do so in subordinate units except in a highly restricted area, usually the type locality. There are no abrupt faunal variations but a constant, orderly, slow, and gradual evolution from below upwards. There is only one fairly clear and definite boundary: that between Kamlial and Chinji. The former is characterized by hard, dark, ridge-forming sandstones (with dull reddish clays subordinate) which loom up above the more easily eroded soft, light grey, non-ferruginous Chinji sandstones and are very distinct from the predominant nodular clays of bright cinnabar red, so characteristic of the Chinji. Even here there is a transitional change, sands and clays are gradually replaced by sands and clays of a different type. In Siwalik stratigraphy gradual transition is the rule and if a boundary be drawn, lithological and faunal infringement must be expected both above and below it.

"Critical examination of the fauna collected within wide lateral limits of a given zone furnish the only accurate means of locating this zone stratigraphically.

"The Siwaliks are unique in the world of geology. If one could make a lithological hash of the Lance, Cloverly, Chugwater (turned clay), and the Tertiary continentals of our own West, spiced with lenticularity, cross-bedding, faults and folds, and served up as a 20,000 foot layer cake iced with everything from recent loess to travertine, not to mention alluvial deposits, with the Himalayas looming up over the lot, he would have a fine small-scale model of the situation as it struck me when I embarked into this field work."

FIG. 3. General view of the Lower Siwaliks, Chinji zone, near Chinji Rest House, Salt Range, Attock District, Punjab.

Nomenclature of Siwalik Rock Units

The question of the nomenclature of Siwalik rock units is a very perplexing problem, indeed, and though a solution cannot be effected at this time (being dependent upon more detailed field studies in India), a discussion is desirable, in order that the usage followed in this work will be clear.

The difficulty arises in the fact that India was more or less separated from Europe and Asia during the Tertiary, either by water or by mountain barriers, so that the sequence of

FIG. 4.   Dipping Chinji beds, near Chinji Rest House, Salt Range, Attock District, Punjab.
The dip is approximately 10°.

FIG. 5.   Middle Siwaliks, Nagri zone, at Hari Talyangar, Belaspur, Simla Hill States.

deposits is quite different in that region to what it is in other parts of the world.   That is, periods of deposition in India are not necessarily correlative with the standard periods recognized elsewhere.   They transcend the regularly recognized boundary lines of geologic time, thus increasing the difficulty (or even making impossible the propriety) of applying to them the standard names indicative of time and depositional sequence in other parts of the world.   The difficulties of Siwalik stratigraphic nomenclature have been but partially solved by the authorities who have worked on the problem.

FIG. 6.   Middle Siwaliks, Nagri zone, near Hari Talyangar, Belaspur, Simla Hill States.

FIG. 7.   General view of the Middle Siwaliks, Dhok Pathan zone, above the Dhok Pathan Rest House,
Attock District, Punjab.

The Indian Geological Survey has adopted a system of nomenclature for the Siwalik lithologic and faunal units somewhat at variance with the accepted usage of European and North American geologists.   The Siwalik beds in their entirety are designated by the

Indian Survey as the Siwalik *System*.   In Europe and North America the term *System* is reserved for a standard world-wide division of rocks.   Thus the Tertiary *period* was a time when the Tertiary *System* of rocks was deposited, or to follow the usage of some geologists, the Eocene *period* was a time when the Eocene *System* was formed.   Therefore the usage of the term *System* as followed by the Indian Survey is out of keeping with that followed by

FIG. 8.   Middle Siwaliks, Dhok Pathan zone, above Dhok Pathan, Attock District, Punjab.

FIG. 9.   Middle Siwaliks, Dhok Pathan zone, near Dhok Pathan, Attock District, Punjab.

geologists elsewhere, and according to the definitions adopted by the various official congresses, surveys and associations, the Indian Survey usage is incorrect.   Suppose then, we postulate that the Siwalik rocks do not constitute a system.   How should they be classified?

"*System*, a standard world wide division." [6]

[6] Committee on Stratigraphic Nomenclature, 1933.   "Classification and Nomenclature of Rock Units," Bull. Geol. Soc. America, XLIV, p. 429.

Obviously, since the term *system* is not applicable to the Siwalik rocks, a designation of lesser scope must be applied, and as such either *series* or *group* may be considered. Accord-

FIG. 10. Middle and Upper Siwaliks, near Hari Talyangar, Belaspur, Simla Hill States. The heavy capping bed in the upper right hand portion of the picture is presumably the Tatrot, overlying the Dhok Pathan.

FIG. 11. Upper Siwaliks, near Siswan, Siwalik Hills, Ambala District, Punjab. Alternating variegated clays and conglomerates near the top of the Pinjor and just below the Boulder Conglomerate.

ing to the recent report of the Committee on Stratigraphic Nomenclature published in the Bulletin of the Geological Society of America, a group may be defined as

"a local or provincial subdivision of a system, based on lithologic features," whereas a series is defined as "a major subdivision of a system"

or

"In part, the series is a convenient unit, of size approaching that of the comparable European subdivision but not necessarily equivalent to it.   In this second usage, a provincial name may be applied." [7]

The term *group*, however, has been applied by the International Geological Congress to designate the sequence of rocks deposited during an entire geologic era, and although this usage was never widely applied, still it seem advisable for the purpose of eliminating confusion, to disregard *group* as a term indicative of the Siwalik sequence of deposits.   Consequently, the second definition of series, given above, remains to be considered.   Since this definition states that the term *series* may be used to designate a local unit, not necessarily an equivalent of the standard European subdivision, it would seem to be a logical term as applied to the Siwalik beds.   Therefore it is here proposed to designate the Upper Tertiary

Fig. 12.   Upper Siwaliks, near Siswan, Siwalik Hills, Ambala District, Punjab.   Alternating variegated clays and conglomerates near the top of the Pinjor, just below the Boulder Conglomerate.

and Lower Pleistocene beds of northern India as the *Siwalik Series*, recognizing the fact that this series transcends the boundary between Tertiary and Quaternary deposition.

When our attention is directed to the lesser units of the Siwalik Series we are confronted with difficulties that would seem to defy a logical solution.   The Indian Survey has divided the Siwalik Series into a Lower, a Middle and an Upper division without applying any definitive names to each of these portions.   Then again, these divisions have been split into several "zones" namely the Kamlial and the Chinji comprising the Lower Siwaliks, the Nagri and the Dhok Pathan making up the Middle Siwaliks and the Tatrot, Pinjor and Boulder Conglomerate comprising the Upper Siwaliks.

Now the Indian Survey utilizes only four units for the Siwaliks in the maps of the Punjab region, these being the Kamlial, Chinji, Middle Siwaliks and Upper Siwaliks.   According to American usage these four divisions would be recognized as formations.   But is it

[7] *Op. cit.*, p. 429.

logical to designate certain "formations" by place names and others by positional names (middle, upper, etc.)? And if the Chinji and Kamlial are formations, what term should be applied to the Lower Siwaliks, which includes them? Moreover, since there is a continual gradation in sedimentation throughout the Siwalik Series with only one definite stratigraphic break, that between the Kamlial and the Chinji, what basis have we for defining formations either of greater or of lesser extent?

Since these various and at present unsolved problems are encountered in a consideration of the Siwalik Series, it seems best on the basis of our present knowledge, to disregard the term *formation* entirely. Thus the divisions, Lower Siwaliks, Middle Siwaliks and Upper Siwaliks are used in the above fashion, without any defining terms. This avoids the necessity of using *formation*, *group* or any other stratigraphic term in an improper sense, and it is logical in that it follows the usage already adopted by the Indian Geological Survey.

According to the definition of *zone* in the report of the Committee on Nomenclature of Rock Units, it is

"a subordinate unit containing the rocks deposited during the time of existence of a particular faunal or floral assemblage. It may be of the magnitude of a bed, a member, a formation or even a group." [8]

The above definition fits very well the smallest units employed by the Indian Survey. The various zones as defined by the Survey are units characterized by definite faunal assemblages. Moreover their stratigraphic value is as yet unknown. They may be either greater or less than a formation, a bed or a member. Therefore it is here proposed to use the term *zone* as a designation for the smallest Siwalik divisions—this term being the most appropriate, not only because of its applicability as defined, but also because of the fact that it has been widely used in the literature of the Indian Geological Survey.

The use of terms for the various Siwalik units, suggested above, is taken as the best and the most conservative solution for a difficult problem. It is not a perfect solution—that will remain for future stratigraphic work of a detailed nature in northern India. Since no logical or uniform system of stratigraphic nomenclature can at present be adopted, the above usages are taken as offering the best compromise in the light of our existing knowledge of Indian stratigraphy.

To sum up the foregoing arguments in a graphic way, the following chart is reproduced to show the method of nomenclature adopted for the Siwalik beds in the present work.

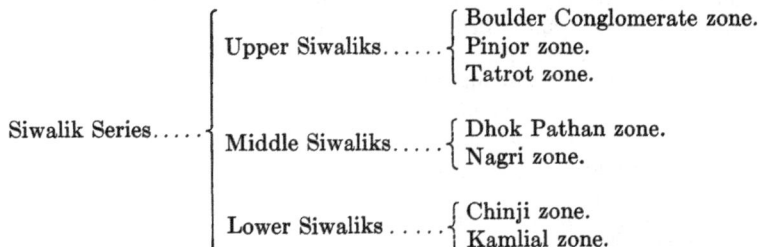

Siwalik Series.....	Upper Siwaliks......	Boulder Conglomerate zone.
		Pinjor zone.
		Tatrot zone.
	Middle Siwaliks.....	Dhok Pathan zone.
		Nagri zone.
	Lower Siwaliks.....	Chinji zone.
		Kamlial zone.

[8] *Op. cit.*, p. 429.

Geologic History of the Siwalik Deposits

The geologic history of the Siwalik beds in the Salt Range area, and of the Siwalik Hills themselves, is closely linked with the train of events that resulted as the consequence of the disappearance of ancient Gondwanaland and the Tethys Sea, and of the uplift of the Himalaya Mountains.   The story has been set forth in recent years by E. H. Pascoe and by G. E. Pilgrim, and for its details the reader should refer to those authors.[9]

At the beginning of the Tertiary period the peninsular portion of India was partially separated from the extra peninsular part to the north, by two marine transgressions; one from the southwest and the other from the southeast.   The former followed the valley of the Indus and extended eastwardly to the Himalaya region; the latter occupied a portion of what is now the Bay of Bengal and reached northward to the present Brahmaputra Valley.   These were areas of marine deposition, and in them accumulated great deposits of nummulitic limestones of Eocene age.

The Tertiary was a period of uplift in the Indian region.   During the early Miocene the two marine transgressions retreated and the peninsula of India became connected to the extra-peninsular region.   Sedimentation changed from marine to continental—in place of salt water limestones there were deposited in successive order brackish water estuarine beds, fresh water lake beds, and finally fluviatile flood plain deposits.   The deposition of the Siwalik beds was from its beginning a history of river sedimentation, a record of increasingly clastic deposits brought down by torrential waters from an extremely rapidly rising mountain mass.

As the sedimentation progressed from the Miocene through the Pliocene and into the Pleistocene, the deposits became increasingly coarser, denoting the increasing uplift of the Himalayas to the north and east, the main source of supply for the Siwalik beds.   Fine grained sandstones gave way to coarser grained sandstones and shales, and these in turn to arkosic deposits and finally to heavy conglomerates.

Pascoe and Pilgrim have each shown how, during the period of Siwalik deposition the drainage of northern India was opposite to its present configuration.   A great "Indobrahm" or "Siwalik" River flowed northward into the remnants of the Tethys Sea.   As the sea retreated into the present confines of the Indian Ocean, the Siwalik River followed it, turning at right angles from its northward course, and following the present Indus Valley to the sea. It was along this ancient Siwalik River that the Siwalik deposits accumulated.   The process of accumulation was mainly that of fluviatile deposition, quite similar to the processes that are typical of the present day Indus and Ganges and their northern tributaries.   Pilgrim has shown that the heavy boulder beds of the Upper Siwaliks were formed by a damming of the valley of the ancient Siwalik River, due to earth movements, and that owing to this damming of the river and its consequent formation of lake beds, there was a concentration of large boulders and pebbles, brought in by the steeply graded tributary rivers.

Contemporaneous with the development of the northward flowing Siwalik River, there

[9] Pascoe, E. H., 1920, "Petroleum in the Punjab and North-West Frontier Province," Mem. Geol. Surv. India, XL, pp. 450–473.

Pilgrim, G. E., 1913, "Correlation of the Siwaliks with Mammal Horizons of Europe," Rec. Geol Surv. India, XLIII, pp. 264–326.

Pilgrim, G. E., 1919, "Suggestions Concerning the History of the Drainage of Northern India, Arising out of a Study of the Siwalik Boulder Conglomerate," Jour. As. Soc. Bengal, N.S., XV, pp. 81–99, Pls. I, II.

was a southward flowing drainage system, paralleling the present Irrawaddy River system of Burma.

In the late phases of the Pleistocene the southward flowing drainage, emptying into the Bay of Bengal, was accentuated, probably by a certain amount of tilting, and it cut northward along its headwaters and beheaded the Siwalik River, thereby establishing the drainage system as it is now expressed in India.

Changes of Climate

It is quite possible that the progressive development of the Siwalik faunas was dependent upon a change from a relatively dry flood plain environment to a more moist flood plain and forest environment. At the beginning of the Siwalik deposition the Himalayan mountain mass was comparatively low, and was therefore less effective as a trap for moisture laden winds than are the present mountains.

Consequently the rising of the Himalayas may have caused a progressive increase in moisture and in temperature in the northern portion of India, due to the fact that the successively higher mountains caused an increase in precipitation along their southern flanks. Thus the Himalayas have acted as an increasingly effective barrier in isolating India from the rest of Asia, cutting out the cold temperatures sweeping down from the north and east, and retaining the warm moisture laden winds blowing in from the Indian Ocean.

The Correlation of the Siwaliks

As has been pointed out in the Introduction, the Siwalik beds were first supposed by Hugh Falconer to contain a unit fauna which he regarded as of Miocene age. Richard Lydekker came to the conclusion that the two faunas known to him were contained within the Pliocene period. Dr. Pilgrim, who established the succession of three Siwalik faunas, regarded the lower one to be of middle Miocene or Tortonian age, the middle one to be of lower Pliocene age, equivalent to the Pontian of Europe, and the upper one he placed at the top of the Pliocene and the base of the Pleistocene, more or less as an equivalent to the Val d'Arno fauna of Italy.

Dr. Matthew, in his 1929 review of Siwalik mammals, placed the lower, middle and upper Siwalik faunas somewhat higher in the geological time scale than did Dr. Pilgrim, making them of Pontian, of middle Pliocene and of lower Pleistocene age respectively.

Various lines of evidence tend to make the problem of Siwalik correlation a confusing one, and naturally such a situation has led to differences of opinion among students of the Siwalik faunal successions. Two methods of attack have been followed in the attempted solution of the perplexing question regarding the age of the Siwalik beds.

1. The several Siwalik faunas have been considered on the merits of their general aspects, and have accordingly been compared directly with the similar faunas of eastern Europe and of Asia. This method is the one that has been followed by Dr. Pilgrim.

2. The Siwalik faunas have been compared not only to the faunas of Europe and of Asia, but also to those of North America, and particular attention has been paid to the appearances of the various kinds of fossil horses and of the other definitely invading types into the region under consideration. This was the method followed by Dr. Matthew.

Now when the Siwalik faunas are compared directly to similar European faunas, the evidence seems to be greatly in favor of Dr. Pilgrim's views of correlation. The lower

Siwalik fauna is rather distinctly Miocene in its aspect, for it contains various species of *Dryopithecus*, an hyaenodont, primitive felids, mastodonts, listriodonts and other primitive pigs, rather primitive anthracotheres, *Macrotherium* and primitive bovids.

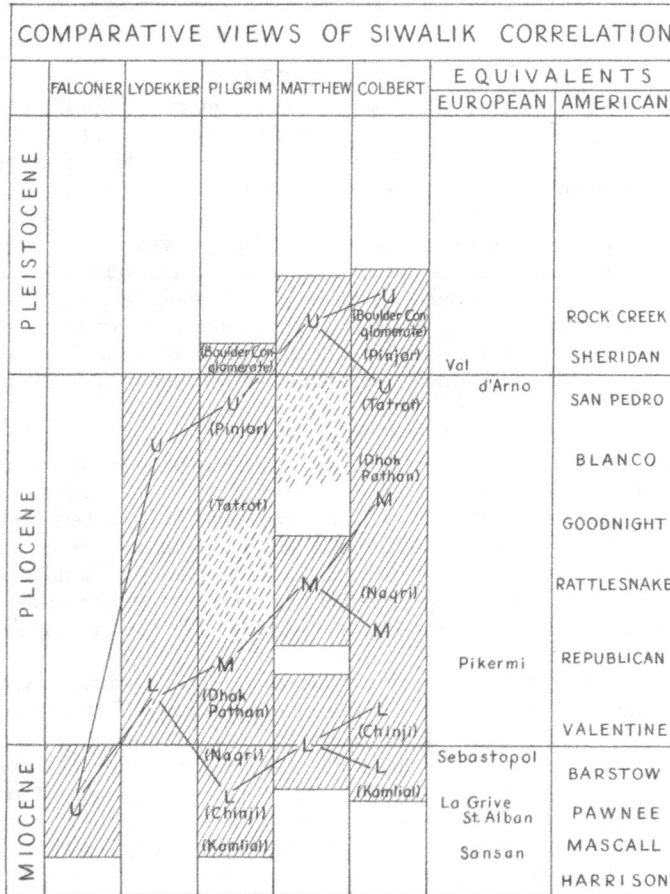

FIG. 13.   Comparative views of Siwalik correlation.

In a like manner the typical Middle Siwalik fauna from the Dhok Pathan area shows Pontian carnivores, *Orycteropus* like the species *O. gaudryi* from Samos, advanced mastodonts, and Pontian perissodactyls and artiodactyls.   And again, the Upper Siwalik fauna is composed of typical lower Pleistocene carnivores, mammoths and ungulates.   Naturally the above evidence seems strongly in favor of the correlations advocated by Pilgrim.

An examination of the fossil horses, however, throws a somewhat different light on the

question. It is now definitely established that the genus *Hipparion* is present at the base of
the Chinji beds, a fact that heretofore had been more or less in question. The material
collected by Mr. Brown, and in addition new evidence gathered by Mr. G. Edward Lewis
of the Yale North India Expedition, shows conclusively that *Hipparion* lived contem-
poraneously in India with the Chinji fauna.

Hipparion is a genus of North American origin. In the Valentine beds of northwestern
Nebraska, which represent the very top of the Miocene or the beginning of the Pliocene in
North America, we find the first appearance of the genus in the very primitive form *Hipparion gratum*. Every gradational transition may be found from *Hipparion gratum* down
into the advanced species of *Merychippus* which occur in beds slightly antecedent to or
equivalent to the Valentine. Thus, we have in Nebraska the material that shows the
actual transition from *Merychippus* to *Hipparion*, and thus our evidence for the North
American origin of *Hipparion* seems to be incontrovertible. *Hipparion* appears in Europe
in the Pontian, or at the earliest in the upper Sarmatian, which forces us to conclude that the
genus migrated from America to the Old World (probably by way of Asia) at a time immediately preceding or possibly correlative with the beginnings of Pontian deposition.
(Borissiak correlates a fauna containing *Hipparion* from Sebastopol, as of Sarmatian age.)
Therefore any Asiatic or European fauna containing *Hipparion* must be equal to or later
than the Valentine of North America. The argument then follows that the Chinji fauna,
instead of being middle Miocene in age, as is advocated by Dr. Pilgrim, is approximately
Pontian in age, and that its more primitive faunal character is due to the fact that it
represents a hold-over of primitive forms into a later period.

In a like manner, the Middle Siwalik fauna represents the hold-over of a typically
Pontian group of mammals into the middle of the Pliocene, or at least into post-Pontian
times.

The Upper Siwalik fauna contains *Equus*. As is the case with *Hipparion*, we have
definite evidence for the evolution of *Equus* in North America from the upper Pliocene
genus, *Plesippus*. The appearance of *Equus* in North America is approximately equivalent
with the beginning of the Pleistocene in the New World, and since this genus undoubtedly
migrated from North America to Asia and Europe, its appearance in the Old World must
also be indicative of the advent of the Pleistocene in that continent. Therefore, the Upper
Siwalik beds, containing *Equus*, may be regarded as definitely representing the Pleistocene
in India, and they are very probably of lower Pleistocene age.

These arguments have been previously put forward by Dr. Matthew. They are re-
peated here for the sake of emphasis, and they will be found, developed still further in the
sections of this work dealing with the Siwalik Equidae.[10]

Account is here taken of Dr. Pilgrim's argument that the Old World *Hipparion* may be
of a separate origin from the North American *Hipparion*, and of his alternative argument
that the genus *Hipparion*, in toto, may be of Old World origin from some as yet unknown
Miocene ancestor, which arguments permit the placing of the Lower Siwaliks in a geologic
position antedating the evolution of *Hipparion* in North America. The first argument, that
of the Old World *Hipparion* coming from a separate Old World ancestral stock, calls for too
great a degree of parallel evolution to be compatible with our wonderfully complete knowledge of the actual material demonstrating the evolution of the horse. The second argument,

[10] Matthew, W. D., 1929, Bull. Amer. Mus. Nat. Hist., LVI, Art. VII, pp. 528–530.

that of *Hipparion* being derived from some as yet unknown Eurasiatic ancestor and migrating into the New World seems a beggaring of the question, because our North American material actually demonstrates the transition of the genus *Merychippus* into the genus *Hipparion*. Consequently, Pilgrim's arguments run so contrary to the known facts regarding the origin and evolution of the genus *Hipparion*, and of the Equidae in general, that they lack weight, as opposed to the conclusions set forth by Osborn, Matthew, Gidley and other students of equine evolution.

Should the evidence of one or two genera, namely *Hipparion* and *Equus*, be given weight over the evidence of an entire fauna? In the case of the Siwalik problem, it should, because the horses in the Siwaliks represent invading elements from North America, while the large portion of the remaining members of the several faunas constitutes indigenous elements. Dr. Matthew has advocated the importance of invading forms in a fauna; he has pointed out that new forms suddenly appearing among an assemblage of animals, are much safer guides to the age of the fauna than are the indigenous members judged on their general aspects. Animals developed in a certain area may persist on past the period of their typical expression, thereby extending the time range of their faunal association and thus introducing doubts as to their true age, but as to the appearance of invading forms there can be no doubt. Invading animals link up a fauna with other definitely known faunas, thereby giving us the true clues to correlation.

Such really seems to be the case with the several Siwalik faunas. They actually are primitive hold-overs into later periods, a fact set forward by Dr. Matthew as follows:

"The appearance of new invading elements in a fauna is a safer guide to its correlation than the disappearance of old elements or the average composition of the fauna as a whole. The appearance of these new elements must be interpreted in the light of what is known of their origin and dispersal. When this is as directly recorded and fully documented as it is in the case of Tertiary Equidae or Camelidae, the evidence appears not open to any effective challenge. But more often the appearance of new elements in a fauna may be explained in several ways, the relative probability of which is not easy to test.

"India and the Oriental region generally are today characterized by the survival of many primitive types of mammals as well as by the absence, scarcity or recent appearance of some of the most progressive and specialized mammals. It compares in these respects with West Africa and tropical America. While it does not necessarily follow that this was true during the later Tertiary, yet it should be so considered until evidence proves the contrary; and so far from proving the contrary I believe that all of the evidence conforms with this assumption and much of it is difficult to explain in any other way. It should be added that the faunas of the Siwalik hills, of Burma, and of Java, should on this assumption contain progressively more and more of these relict elements or primitive survivors from the earlier northern faunas, and that the indications that these fossil faunas were archaic, not ancient, should in the same sequence be progressively more marked." [11]

Reference should be made to the accompanying charts (Figs. 12, 13) for the graphic representation of the views on Siwalik correlations set forth in the above pages.[12]

[11] Matthew, W. D., *op. cit.*, pp. 442–443.

[12] The above remarks regarding the correlation of the Siwaliks as inferred by the intercontinental migrations of *Hipparion* and *Equus* have been taken in part from Colbert, Edwin H., 1935, Amer. Mus. Novitates, No. 797. "Additional Remarks on the Correlation of the Siwaliks" also comes from this source.

While the fossil Equidae offer the strongest and most clear cut evidences for our conclusions as to the correlation of the Siwaliks, certain other mammals afford supplementary arguments in support of the facts demonstrated so clearly by the horses.   Among the other mammalian groups, the Giraffidae furnish us with strong arguments as to the Pontian and the post Pontian age of the Siwalik series.   Dr. Matthew has demonstrated in a most

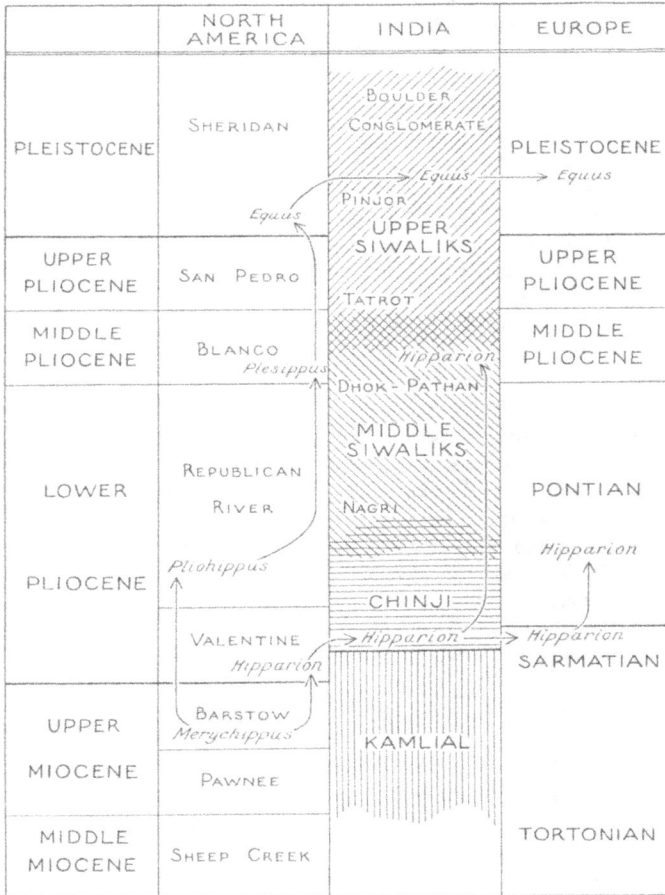

FIG. 14.   The correlation of the Siwaliks as based on the intercontinental migrations of
Upper Tertiary and Quaternary Equidae.

admirable fashion, that the oldest Siwalik giraffes are no older than the Pontian forms of Pikermi, Samos and other localities.   He has pointed out the following facts in regard to the Siwalik Giraffidae:

1. The Chinji giraffes are as progressive, if not more so, than the typical Pontian

giraffes of Pikermi.   That is, *Giraffa punjabiensis* from the Chinji beds, is more progressive than *Giraffa attica* of Pikermi, and *Giraffokeryx* from the Chinji is a close relative of *Palaeotragus* of the Pontian.   *Helladotherium duvernoyi* of the Pontian is closely comparable to *Vishnutherium* which occurs in the Chinji and in the Middle Siwaliks.

2. The Middle Siwalik giraffes are definitely more advanced and consequently later in age than the Pontian forms.   Thus the large, complexly horned forms, *Hydaspitherium* and *Bramatherium* are definitely post Pontian in their development.[13]

Therefore the remains of the Giraffidae from the Siwaliks would seem to indicate that the Chinji is about Pontian in age, while the Middle Siwaliks are definitely post-Pontian.

Additional Remarks on the Correlation of the Siwaliks

Since the foregoing remarks were written a paper by Teilhard and Stirton has appeared, which has some bearing on the problem of the correlation of the Siwaliks.[14]   In the above-mentioned paper (correlation table, p. 284) the Lower Siwaliks are indicated as of uppermost Vindobonian age, the Middle Siwaliks are correlated with the Pontian and the Upper Siwaliks are placed in the Astian, or uppermost Pliocene   This correlation, though in a way somewhat intermediate between Pilgrim's and Matthew's correlations, does tend to favor Pilgrim's views as to the age of the Siwaliks

On pages 281 and 282 of the above cited work the following statement appears:

"The argument has been advanced that, because *Hipparion* is derived from the North American Miocene genus *Merychippus*, hipparions should appear in North America in the Upper Miocene or prior to their appearance in Europe.   There is good evidence, however, which indicates that *Neohipparion* and *Nannippus* are derived from different species of *Merychippus;* accordingly, the Old World hipparions are probably descendants of an unknown Asiatic *Merychippus* or a *Merychippus* in this country older than the Niobrara River fauna, which contains advanced species of *Merychippus* showing intergradation with *Neohipparion* and *Pliohippus*.   At least, the species of *Merychippus* which show intergradation with the American genera did not give rise to *Hipparion* of the Old World."

The Niobrara River fauna, referred to in the foregoing quotation, is equivalent to the Valentine fauna contained in the Valentine formation, as used by Matthew, Simpson and others.

If the Old World *Hipparion* was independently derived from an Asiatic species of *Merychippus*, the arguments set forth above are materially weakened, and Pilgrim's views as to the correlation of the Lower and Middle Siwaliks are strengthened.   On the other hand, there is no real evidence for supposing that the Old World *Hipparion* was derived from an Old World *Merychippus*.   In the first place, an Old World Merychippus has never been found.   Pilgrim speaks of some of the Chinji *Hipparion* as being "slightly smaller and more brachyodont" than the Middle Siwalik *Hipparion*.   The material in the American Museum collection indicates, however, that the Chinji *Hipparion* was but little different if at all separate from the Middle Siwalik *Hipparion*.   Thus neither an Old World *Merychippus* nor a primitive Old World *Hipparion* is known from the material extant.

[13] Matthew, W. D., 1929, *op. cit.*, p. 553.

[14] Teilhard de Chardin, P., and Stirton, R. A., 1934, "A Correlation of some Miocene and Pliocene Mammalian Assemblages in North America and Asia with a Discussion of the Mio-Pliocene Boundary," Univ. of Calif. Publ., Bull. Dept. Geol. Sci., XXIII, No. 6, pp. 277–290.

In the second place, it seems rather unnecessary to postulate an Old World *Merychippus* or a primitive Old World *Hipparion* as the ancestors of the typical Old World *Hipparion*, since perfectly good structural ancestors of the proper kind are to be found in North America. *Hipparion gratum, Hipparion gratum tehonense* and related species would seem to be adequate as ancestral types for the Old World *Hipparion*. These species have, as do the Old World forms, a round oval protocone, a moderately high molar crown, a rather deep lacrymal fossa, and other features in common. Moreover, the North American form is primitive, as we might expect an ancestral species to be.

In Teilhard and Stirton there is a statement to the effect that the "species of *Merychippus* which show intergradation with the American genera [of *Hipparion*] did not give rise to *Hipparion* of the Old World." This may be true, but it does not argue against the strong probability that *Hipparion gratum* and its related species are the direct ancestors of the Old World *Hipparion*. Moreover, the primitive characters of *Hipparion gratum* as contrasted with the advanced and specialized characters of all of the Old World members of the genus, even the earliest forms, show that a certain time element was involved in the migration of the genus from North America to Eurasia.

Since there is no definite proof for the separate origin of the Old World *Hipparion* in Eurasia, it seems reasonable to look for their ancestors in North America, typified by such species as *Hipparion gratum*. Since the earliest of the Eurasiatic *Hipparion* are relatively advanced and specialized species, it is reasonable to think that a certain amount of geologic time passed between the appearance of the genus in North America and its migration to Asia and Europe. Therefore the correlation of the Siwaliks as advocated by Matthew would seem to be justified on the basis of all of the available evidence now known.

In the paper by Teilhard and Stirton the Valentine formation of northwestern Nebraska is divided. The lower phases supposedly transitional between the Miocene and the Pliocene are named the Niobrara River, while the upper portion, said to be definitely of Pliocene age, retains the name Valentine. The name Valentine is well established in the literature to indicate the formation and the time transitional between the Miocene and the Pliocene, and the abandonment of this usage of the term will lead to some confusion. Lacking more detailed and conclusive evidence on the question of the proposed division, the name Valentine, indicative of the transition from the Miocene to the Pliocene, is retained in this present work in its original meaning.

Regions Adjacent to the Siwalik Hills and the Salt Range

Beds equivalent or related to the Siwalik series are found in regions adjacent to northern India. The Irrawaddy series of Burma would seem to be approximately correlative with the Middle and part of the Upper Siwalik beds of India. Like the Siwaliks, the Irrawaddy series is characterized by intergrading deposits, laid down through a period of continuous sedimentation during upper Tertiary times. These beds contain numerous fossils of Tertiary mammals, and the series is especially noteworthy for the great quantities of fossil wood within it. Below the Irrawaddy series are marine beds, the Pegu series, and below these are the Pondaung beds carrying characteristic Eocene mammals.

To the west, in Baluchistan, and Sind, are the Bugti beds, constituting the Fatehjang stage, which contain a distinctive Oligocene fauna. The Bugtis would appear to be just below the lowest Siwalik horizons. The fauna is closely related to that of the Lower Siwalik fauna, pointing to a close genetic relationship between these two areas.   Above the Bugti beds in Sind, are the Lower and Upper Manchar series.   The Lower Manchar ranges through a considerable period of time and is seemingly equivalent to the Lower and Middle Siwaliks.   The Upper Manchar is equivalent to the Boulder Conglomerate of the Upper Siwaliks.

Again, to the southwest, in the Bay of Cambay, is Perim Island, on which are beds comparable to the Middle Siwalik horizons.   An interesting fauna is known from this locality, and its closest affinities are with the Dhok Pathan fauna of the Salt Range area.

PART III. MAMMALIAN FAUNAS OF THE SIWALIK SERIES

MAMMALIAN FAUNAS OF THE SIWALIK SERIES

The faunal lists presented on the following pages have been carefully compiled from various sources, and they represent to the best of our knowledge at the present date the Siwalik faunas as they are now known. These lists include all published names of genera and species and their stratigraphic occurrences. They are not intended as a critical review of the Siwalik faunas—questions of synonymy will be considered in the systematic portions of this work.

The proboscideans are listed according to the most recent works of Professor Henry Fairfield Osborn, published in various American Museum papers. The bovids have been arranged by Dr. G. E. Pilgrim, who has recently completed a survey of the Siwalik Bovidae in the American Museum. The new genera and species of Bovidae which will appear in a forthcoming Bulletin of the American Museum of Natural History, are not included in these lists.

MAMMALIAN FAUNAS OF THE SIWALIK SERIES OF INDIA

	Lower		Middle		Upper		
	Kamlial	Chinji	Nagri	Dhok Pathan	Tatrot	Pinjor	Boulder Conglomerate
Primates:							
Lorisidae:							
Indraloris lulli Lewis, 1933			×				
Cercopithecidae:							
Papio falconeri (Lydekker), 1886						×	
Papio subhimalayanus (v. Meyer), 1848						×	
Cercopithecus hasnoti (Pilgrim), 1910				×			
Semnopithecus palaeindicus Lydekker, 1884						×	
Macacus sivalensis Lydekker, 1878				×			
Pongidae:							
Simia cf. *satyrus* Linnaeus, 1766						×	
Dryopithecus punjabicus Pilgrim, 1910		×	×				
Dryopithecus pilgrimi Brown, Gregory. Hellman, 1924		×					
Dryopithecus cautleyi Brown, Gregory, Hellman, 1924			×				
Dryopithecus frickae Brown, Gregory, Hellman, 1924				×			
Dryopithecus chinjiensis Pilgrim, 1915		×					
Dryopithecus giganteus Pilgrim, 1915			×				
Dryopithecus sivalensis Lewis, 1934			×				
Sivapithecus orientalis Pilgrim, 1927			×				
Sivapithecus himalayanus Pilgrim, 1927			×				
Sivapithecus middlemissi Pilgrim, 1927		×					
Sivapithecus indicus Pilgrim, 1910		×	×				
Hylopithecus hysudricus Pilgrim, 1927			×				

MAMMALIAN FAUNAS OF THE SIWALIK SERIES OF INDIA (*Continued*)

	Lower		Middle		Upper		
	Kamlial	Chinji	Nagri	Dhok Pathan	Tatrot	Pinjor	Boulder Con- glom- erate
Primates: (*continued*)							
Pongidae:							
Palaeopithecus (?) *sylvaticus* Pilgrim, 1927			✕				
Palaeopithecus sivalensis Lydekker, 1879				✕			
Palaeopithecus sp. Pilgrim, 1913				✕			
Palaeosimia rugosidens Pilgrim, 1915		✕					
Ramapithecus brevirostris Lewis, 1934					✕		
Sugrivapithecus salmontanus Lewis, 1934			✕				
Sugrivapithecus hariensis Lewis, 1934			✕				
Bramapithecus thorpei Lewis, 1934		✕					
Incertae Sedis:							
Adaetontherium incognitum Lewis, 1934		✕					
Rodentia:							
Spalacidae:							
Rhizomys sivalensis Lydekker, 1878				✕			
Rhizomys punjabiensis Colbert, 1933		✕					
Rhizomys sp. Lydekker, 1884				✕		✕	
Muridae:							
Nesokia hardwickii (Gray), 1837						✕	
Hystricidae:							
Hystrix sivalensis Lydekker, 1878				✕			
Hystrix cf. *leucurus* Sykes, 1831						✕	
Sivacanthion complicatus Colbert, 1933		✕					
Lagomorpha:							
Leporidae:							
Caprolagus sivalensis Major, 1899						✕	
Carnivora (Creodonta):							
Hyaenodontidae:							
Dissopsalis carnifex Pilgrim, 1910		✕					
Dissopsalis ruber Pilgrim, 1910		✕					
Carnivora (Fissipedia):							
Canidae:							
Amphicyon sindiensis Pilgrim, 1932	✕						
Amphicyon palaeindicus Lydekker, 1876		✕	✕				
Amphicyon pithecophilus Pilgrim, 1932		✕					
Arctamphicyon lydekkeri (Pilgrim), 1910					✕		
Vishnucyon chinjiensis Pilgrim, 1932		✕					
Canis cautleyi Bose, 1880						✕	
Sivacyon curvipalatus (Bose), 1880						✕	
Procyonidae:							
Sivanasua palaeindica Pilgrim, 1932		✕					
Sivanasua himalayensis Pilgrim, 1932			✕				
Ursidae:							
Agriotherium sivalense (Falconer and Cautley), 1836						✕	
Agriotherium palaeindicum (Lydekker), 1878				✕			
Indarctos salmontanus Pilgrim, 1913				✕			

MAMMALIAN FAUNAS OF THE SIWALIK SERIES OF INDIA (*Continued*)

	Lower		Middle		Upper		
	Kamlial	Chinji	Nagri	Dhok Pathan	Tatrot	Pinjor	Boulder Conglomerate
Carnivora (Fissipedia): (*continued*)							
Ursidae:							
Indarctos punjabiensis (Lydekker), 1884				×			
Melursus (?) *theobaldi* (Lydekker), 1884						×	
Mustelidae:							
Sinictis lydekkeri Pilgrim, 1932						×	
Martes lydekkeri (Colbert), 1933		×					
Mellivora sivalensis (Falconer and Cautley), 1868						×	
Promellivora punjabiensis (Lydekker), 1884				×			
Eomellivora necrophila Pilgrim, 1932		×					
Eomellivora (?) *tenebrarum* Pilgrim, 1932				×			
Lutra palaeindica Falconer and Cautley, 1868						×	
Sivalictis natans Pilgrim, 1932		×					
Enhydriodon sivalensis Falconer, 1868						×	
Enhydriodon falconeri Pilgrim, 1931				×			
Sivaonyx bathygnathus (Lydekker), 1884				×			
Vishnuonyx chinjiensis Pilgrim, 1932		×					
Viverridae:							
Viverra chinjiensis Pilgrim, 1932		×					
Viverra bakeri Bose, 1880						×	
Vishnuictis salmontanus Pilgrim, 1932				×			
Vishnuictis durandi (Lydekker), 1884						×	
Hyaenidae:							
Ictitherium sivalense Lydekker, 1877				×			
Ictitherium indicum Pilgrim, 1910				×			
Hyaenictis bosei Matthew, 1929						×	
Lycyaena macrostoma (Lydekker), 1884				×			
Lycyaena macrostoma vinayaki Pilgrim, 1932				×			
Lycyaena proava (Pilgrim), 1910		×					
Lycyaena chinjiensis Pilgrim, 1932		×					
Crocuta carnifex (Pilgrim), 1913		×		×			
Crocuta gigantea (Schlosser), 1903			×	×			
Crocuta gigantea latro Pilgrim, 1932			×	×			
Crocuta mordax Pilgrim, 1932				×			
Crocuta sivalensis (Falconer and Cautley), 1868						×	
Crocuta felina (Bose), 1880						×	
Crocuta colvini (Lydekker), 1884						×	
Felidae:							
Mellivorodon palaeindicus Lydekker, 1884				×			
Vinayakia sarcophaga Pilgrim, 1932		×					
Vinayakia nocturna Pilgrim, 1932			×				
Aeluropsis annectens Lydekker, 1884				×			
Hyaenaelurus lahirii Pilgrim, 1932	×						
Megantereon palaeindicus (Bose), 1880						×	
Megantereon praecox Pilgrim, 1932			×				
Megantereon falconeri Pomel, 1853						×	

MAMMALIAN FAUNAS OF THE SIWALIK SERIES OF INDIA (*Continued*)

	Lower		Middle		Upper		
	Kamlial	Chinji	Nagri	Dhok Pathan	Tatrot	Pinjor	Boulder Conglomerate
Carnivora (Fissipedia): (*continued*)							
Felidae:							
Sansanosmilus serratus Pilgrim, 1932		×					
Sansanosmilus rhomboidalis Pilgrim, 1932		×					
Paramachaerodus pilgrimi Kretzoi, 1929				×			
Paramachaerodus indicus (Kretzoi), 1929				×			
Propontosmilus sivalensis (Lydekker), 1877				×			
Sivasmilus copei Kretzoi, 1929		×					
Felis subhimalayana Bronn, 1848						×	
Felis sp. Lydekker, 1884				×			
Panthera cristata (Falconer and Cautley), 1868						×	
Sivafelis potens Pilgrim, 1932						×	
Sivafelis brachygnathus (Lydekker), 1884						×	
Sivaelurus chinjiensis (Pilgrim), 1910		×					
Vishnufelis laticeps Pilgrim, 1932		×					
Tubulidentata:							
Orycteropodidae:							
Orycteropus browni Colbert, 1933			×				
Orycteropus pilgrimi Colbert, 1933			×				
Proboscidea:							
Dinotheriidae:							
Dinotherium sindiense Lydekker, 1880	×	×					
Dinotherium indicum Falconer, 1845.		×		×			
Dinotherium pentapotamiae Lydekker, 1876		×					
Dinotherium angustidens (?) Koch, 1845				×			
Trilophodontidae:							
Trilophodon pandionis (Falconer), 1868	×						
Trilophodon angustidens palaeindicus (Lydekker), 1884		×					
Trilophodon macrognathus (Pilgrim), 1913	×	×					
Trilophodon chinjiensis (Pilgrim), 1913		×					
Trilophodon hasnotensis Osborn, 1935 *				×			
Tetralophodon falconeri (Lydekker), 1880		×		×			
Tetralophodon punjabiensis (Lydekker), 1886				×			
Serridentinus hasnotensis Osborn, 1929		×					
Serridentinus metachinjiensis Osborn, 1929		×					
Serridentinus browni Osborn, 1926		×					
Serridentinus chinjiensis Osborn, 1929		×					
Serridentinus prochinjiensis Osborn, 1929		×					
Rhynchotherium chinjiensis Osborn, 1929				×			
Synconolophus dhokpathanensis Osborn, 1929				×			
Synconolophus propathanensis Osborn, 1929				×			
Synconolophus corrugatus (Pilgrim), 1913				×			
Synconolophus ptychodus Osborn, 1929		×		×			
Synconolophus hasnoti (Pilgrim), 1913				×			
Anancus perimensis (Falconer and Cautley), 1847				×			
Anancus properimensis Osborn, 1935 *		×					

MAMMALIAN FAUNAS OF THE SIWALIK SERIES OF INDIA (*Continued*)

	Lower		Middle		Upper		
	Kamlial	Chinji	Nagri	Dhok Pathan	Tatrot	Pinjor	Boulder Conglomerate
Proboscidea: (*continued*)							
Trilophodontidae:							
Pentalophodon sivalensis (Cautley), 1836						X	
Pentalophodon falconeri Osborn, 1935 *					X		
Elephantidae:							
Stegolophodon latidens (Clift), 1828				X			
Stegolophodon cautleyi (Lydekker), 1886				X			
Stegolophodon cautleyi progressus Osborn, 1929		X					
Stegolophodon nathotensis Osborn, 1929		X					
Stegolophodon stegodontoides (?) Pilgrim, 1913						X	
Stegodon bombifrons (Falconer and Cautley), 1847				X	X		
Stegodon cliftii (Falconer and Cautley), 1847				X			
Stegodon elephantoides (Clift), 1828				X			
Stegodon ganesa (Falconer and Cautley), 1845						X	
Stegodon insignis (Falconer and Cautley), 1845						X	
Stegodon pinjorensis Osborn, 1929						X	
Archidiskodon planifrons (Falconer and Cautley), 1845						X	
Hypselephas hysudricus (Falconer and Cautley), 1845						X	
Platelephas platycephalus (Osborn), 1929						X	
Perissodactyla:							
Equidae:							
Hipparion antelopinum (Falconer and Cautley), 1849				X			
Hipparion theobaldi (Lydekker), 1877		X	X	X			
Hipparion chisholmi (Pilgrim), 1910				X			
Hipparion punjabiense Lydekker, 1886				X			
Hipparion perimense (Pilgrim), 1910				X			
Hipparion sp. Lydekker, 1885		X					
Equus sivalensis Falconer and Cautley, 1849						X	
Equus namadicus Falconer and Cautley, 1849						X	
Chalicotheriidae:							
Nestoritherium sivalense Falconer and Cautley, 1843						X	
Nestoritherium (?) *sindiense* (Lydekker), 1876		X					
Macrotherium salinum Forster-Cooper, 1922		X	X				
Rhinocerotidae:							
Coelodonta platyrhinus (Falconer and Cautley), 1847						X	
Rhinoceros sivalensis (Falconer and Cautley), 1847						X	
Rhinoceros palaeindicus Falconer and Cautley, 1847						X	
Gaindatherium browni Colbert, 1934		X	X				
Aceratherium perimense Falconer and Cautley, 1847		X	X	X			
Rhinoceros planidens Lydekker, 1876				X			
Rhinoceros iravadicus Lydekker, 1876				X			
Aceratherium lydekkeri Pilgrim, 1910				X			
Aceratherium blanfordi Lydekker, 1884		X	X	X			
Chilotherium intermedium (Lydekker), 1884		X	X	X			

* From manuscript.

MAMMALIAN FAUNAS OF THE SIWALIK SERIES OF INDIA (*Continued*)

	Lower		Middle		Upper		
	Kamlial	Chinji	Nagri	Dhok Pathan	Tatrot	Pinjor	Boulder Conglomerate
Artiodactyla:							
Tayassuidae:							
Pecarichoerus orientalis Colbert, 1933		×					
Suidae:							
Palaeochoerus perimensis (Lydekker), 1887		×	×				
Palaeochoerus lahirii Pilgrim, 1926	×						
Conohyus sindiense (Lydekker), 1884	×	×	×				
Conohyus chinjiensis Pilgrim, 1926		×					
Conohyus indicus (Lydekker), 1884			×				
Sivachoerus prior Pilgrim, 1926				×			
Sivachoerus giganteus (Falconer and Cautley), 1847					×	×	
Tetraconodon magnus Falconer, 1868				×		×	
Tetraconodon mirabilis Pilgrim, 1926				×		×	
Tetraconodon minor Pilgrim, 1926			×				
Listriodon pentapotamiae (Falconer), 1868		×	×	×			
Listriodon theobaldi Lydekker, 1878		×					
Listriodon guptai Pilgrim, 1926	×						
Lophochoerus nagrii Pilgrim, 1926			×				
Lophochoerus himalayensis Pilgrim, 1926			×				
Lophochoerus exiguus Pilgrim, 1926		×					
Propotamochoerus salinus Pilgrim, 1926		×	×				
Propotamochoerus uliginosus Pilgrim, 1926		×	×	×			
Propotamochoerus hysudricus (Stehlin), 1899				×			
Propotamochoerus ingens Pilgrim, 1926				×			
Potamochoerus theobaldi Pilgrim, 1926						×	
Potamochoerus palaeindicus Pilgrim, 1926						×	
Dicoryphochoerus titan (Lydekker), 1884				×			
Dicoryphochoerus titanoides Pilgrim, 1926				×	×		
Dicoryphochoerus vagus Pilgrim, 1926				×	×	×	
Dicoryphochoerus chisholmi Pilgrim, 1926		×					
Dicoryphochoerus robustus Pilgrim, 1926			×				
Dicoryphochoerus haydeni Pilgrim, 1926		×					
Dicoryphochoerus instabilis Pilgrim, 1926		×					
Dicoryphochoerus durandi Pilgrim, 1926						×	
Dicoryphochoerus vinayaki Pilgrim, 1926				×			
Hyosus punjabiensis (Lydekker), 1878				×			
Hyosus tenuis Pilgrim, 1926				×			
Sivahyus hollandi Pilgrim, 1926				×			
Sanitherium schlagentweitii v. Meyer, 1866		×					
Sanitherium cingulatum Pilgrim, 1926		×					
Hippohyus sivalensis Falconer and Cautley, 1840–45						×	
Hippohyus lydekkeri Pilgrim, 1910				×			
Hippohyus tatroti Pilgrim, 1926					×		
Hippohyus grandis Pilgrim, 1926				×	×		
Hippohyus deterrai Lewis, 1934			×				
Sus hysudricus Falconer and Cautley, 1847						×	

MAMMALIAN FAUNAS OF THE SIWALIK SERIES OF INDIA (*Continued*)

	Lower		Middle		Upper		
	Kamlial	Chinji	Nagri	Dhok Pathan	Tatrot	Pinjor	Boulder Conglomerate
Artiodactyla: (*continued*)							
Suidae:							
Sus advena Pilgrim, 1926			×				
Sus comes Pilgrim, 1926				×			
Sus adolescens Pilgrim, 1926				×			
Sus praecox Pilgrim, 1926				×			
Sus peregrinus Pilgrim, 1926					×		
Sus bakeri Pilgrim, 1926						×	
Sus falconeri Lydekker, 1884						×	
Sus cautleyi Pilgrim, 1926						×	
Anthracotheriidae:							
Choeromeryx silistrense (Pentland), 1828		×		×			
Rhagatherium sindiense Lydekker, 1877		×					
Anthracotherium punjabiense Lydekker, 1877		×					
Hyoboops palæindicus (Lydekker), 1877		×					
Hemimeryx blanfordi Lydekker, 1883		×					
Hemimeryx pusillus (Lydekker), 1885		×	×				
Merycopotamus dissimilis Falconer and Cautley, 1836				×		×	
Merycopotamus nanus Falconer and Cautley, 1847						×	
Telmatodon sp. Pilgrim, 1910		×					
Hippopotamidae:							
Hippopotamus sivalensis Falconer and Cautley, 1836					×	×	
Hippopotamus iravaticus Falconer and Cautley, 1847				×			
Camelidae:							
Camelus sivalensis Falconer and Cautley, 1849						×	
Tragulidae:							
Dorcabune anthracotherioides Pilgrim, 1910		×					
Dorcabune hyæmoschoides Pilgrim, 1915		×					
Dorcabune sindiense Pilgrim, 1915		×					
Dorcabune nagrii Pilgrim, 1915			×				
Dorcabune latidens Pilgrim, 1915				×			
Dorcatherium majus Lydekker, 1876		×		×			
Dorcatherium minus Lydekker, 1876		×	×	×			
Dorcatherium sp. Pilgrim, 1913					×		
Tragulus sivalensis Lydekker, 1882				×			
Moschus sp. Lydekker, 1884				×			
Cervidae:							
Cervus sivalensis Lydekker, 1880						×	
Cervus punjabiensis Brown, 1926						×	
Cervus simplicidens Lydekker, 1876				×		×	
Cervus triplidens Lydekker, 1876				×		×	
Dicrocerus sp. Pilgrim, 1913		×					
Giraffidae:							
Giraffokeryx punjabiensis Pilgrim, 1910		×	×				
Propalaeomeryx sivalensis Lydekker, 1882		×					
Sivatherium giganteum Falconer and Cautley, 1836						×	

MAMMALIAN FAUNAS OF THE SIWALIK SERIES OF INDIA (Continued)

	Lower		Middle		Upper		
	Kamlial	Chinji	Nagri	Dhok Pathan	Tatrot	Pinjor	Boulder Conglomerate
Artiodactyla: (Continued)							
Giraffidae:							
Indratherium majori Pilgrim, 1910						×	
Vishnutherium iravaticum Lydekker, 1876				×			
Bramatherium perimense Falconer, 1845				×			
Hydaspitherium megacephalum Lydekker, 1876				×			
Hydaspitherium grande Lydekker, 1878				×			
Hydaspitherium magnum Pilgrim, 1910				×			
Hydaspitherium birmanicum Pilgrim, 1910				×			
Giraffa sivalensis (Falconer and Cautley), 1843						×	
Camelopardalis affinis Falconer and Cautley, 1843						×	
Giraffa punjabiensis Pilgrim, 1910				×			
Giraffa priscilla Matthew, 1929		×					
Giraffa sp. (Falconer), 1863		×					
Bovidae:							
Gazella (?) porrecticornis (Lydekker), 1878				×			
Damaliscus palaeindicus (Falconer), 1859						×	
Taurotragus latidens (Lydekker), 1884				×		×	
Perimia falconeri (Lydekker), 1886				×			
Boselaphus lydekkeri Pilgrim, 1910				×			
Tragocerus punjabicus Pilgrim, 1910				×			
Tragocerus perimensis (Lydekker), 1878				×			
Hippotragus (?) sivalensis (Lydekker), 1878						×	
Cobus (?) patulicornis (Lydekker), 1878						×	
Cobus gyricornis Falconer, 1868						×	
Cobus palaeindicus Lydekker, 1886						×	
Cobus sp. (Lydekker), 1886						×	
Capra sivalensis (Lydekker), 1886						×	
Bucapra daviesi Rütimeyer, 1877						×	
Proleptobos birmanicus Pilgrim, 1913				×			
Leptobos falconeri Rütimeyer, 1877						×	
Hemibos occipitalis (Lydekker), 1878						×	
Probubalus triquetricornis Rütimeyer, 1877						×	
Probubalus acuticornis (Falconer), 1868						×	
Probubalus antelopinus Rütimeyer, 1877						×	
Bubalus platycerus Lydekker, 1878						×	
Bubalus palaeindicus (Falconer), 1868						×	
Bison sivalensis (Falconer), 1868						×	
Bos acutifrons Lydekker, 1878						×	
Bos planifrons Lydekker, 1878						×	
Bos platyrhinus Lydekker, 1878						×	

PART IV. AMERICAN MUSEUM SIWALIK FOSSIL LOCALITIES

AMERICAN MUSEUM SIWALIK FOSSIL LOCALITIES

The Siwalik fossils in the American Museum collection were obtained from localities near the villages and rest houses listed below. In the succeeding pages of this work the names of villages or rest houses are often indicated without a designation of the District or State within which they are located. In the list now presented a complete designation is given of the villages, rest houses, Districts and States near which or within which American Museum Siwalik fossils were collected.

Chinji Rest House.... Nagri............... Dhok Pathan........ Dhulian.............	......Salt Range, Attock District, Punjab.
Hasnot............. Bhandar........... Kotal Kund........ Padhri............. Tatrot............. Nathot............ Phadial............	......Salt Range, Jhelum District, Punjab.
Hari Talyangar....... Bilaspur............	......Simla Hills, Bilaspur State, Punjab.
Pinjor.............. Mirzapur........... Siswan............. Chandigarh......... Moginand.......... Kalka............. Charnian...........	......Siwalik Hills, Ambala District, Punjab.
Ramnagar..........	......Jammu State, Kashmir.

EXPLANATION OF MAPS

The maps figured on the immediately following pages were drawn up from Indian Geological Survey topographic sheets, in which the geological formations had been drawn and colored by members of the Indian Survey. They are here reproduced by the kind permission of Dr. Leigh Fermor, Director of the Geological Survey of India, and all credit for the geographic and the geologic data on the maps should go to the Indian Survey.

The first map (Fig. 15) is a general outline map of India, showing on it two areas, marked A and B respectively, in which Mr. Brown made the collections of Siwalik mammals for the American Museum.

The two succeeding maps (Figs. 16 and 17) are enlargements of the areas A and B of the smaller map. On these maps are marked the numbers and the areas of certain Indian

Survey topographic sheets from which the succeeding maps were made. Furthermore, on the two maps under discussion there are certain areas marked by a stippled design, which represent the exact locations and boundaries of the nine detailed maps on the following pages.

The nine detailed maps show the locations of Mr. Brown's collecting grounds and the geographic and the geological features of these areas. These maps are on a scale of two miles to the inch. The boundaries of geological formations are marked by heavy lines, and the several stratigraphic units enclosed within these lines are indicated by certain letters in the following manner.

Ral	Recent.........	Alluvium
Qus	Quaternary.....	Upper Siwaliks
Tms	Tertiary........	Middle Siwaliks
Tch	Tertiary........	Chinji zone
Tka	Tertiary........	Kamlial zone

Chinji zone and Kamlial zone .... Lower Siwaliks

The stratigraphic classification used on these maps is that followed by the Indian Geological Survey, according to the official usage of that organization.

The localities at which Mr. Brown obtained the fossils in the American Museum Siwalik Collection are indicated by numbers and circles. These are the field numbers which were used by Mr. Brown in his field note book. Numbers 1 to 102 inclusive were located by Mr. Brown on topographic sheets when he was in the field, and they represent the exact localities at which fossils were found. Numbers 103 to 164 inclusive represent localities that were noted by Mr. Brown in his field records but which were not placed on the map at the time he was in the field. They have been subsequently located, according to the data contained in his field note book, and naturally they represent approximate locations of fossils. However, their positions may for all practical purposes be regarded as indicative of the exact locations of the fossils which they represent.

Thus, with these detailed maps showing the locations of the boundaries of geological formations, and the placing of the localities where fossils were discovered, it is possible to fix every specimen in the American Museum Collection, not only as regards its geographic position, but also as regards its approximate stratigraphic level.

Key to American Museum Siwalik Fossil Localities

The following list gives the locations and the geologic levels of the one hundred and sixty four fossil localities which appear on the nine accompanying maps of the northern Punjab (Figs. 18–26). These localities and their stratigraphic levels are given as they are set down in Mr. Brown's field notes. In a few cases there are certain discrepancies between the locality as it is recorded and as it is actually placed on the map, and likewise between the stratigraphic level of the locality and its stratigraphic level as recorded on the map. These few discrepancies are due to the fact that Mr. Brown estimated his distances, when he recorded them in his field note book, and naturally these estimates are sometimes at variance with the actual distances as shown on the maps. As explained above, the field numbers 1 to 102 inclusive were plotted by Mr. Brown in the field and are exact locations of the fossils recorded under those numbers. The field numbers 103 to 164 inclusive are approximate locations, made subsequently in the laboratory. The levels recorded by Mr. Brown are estimated, but are based on careful observations of the sequence of strata.

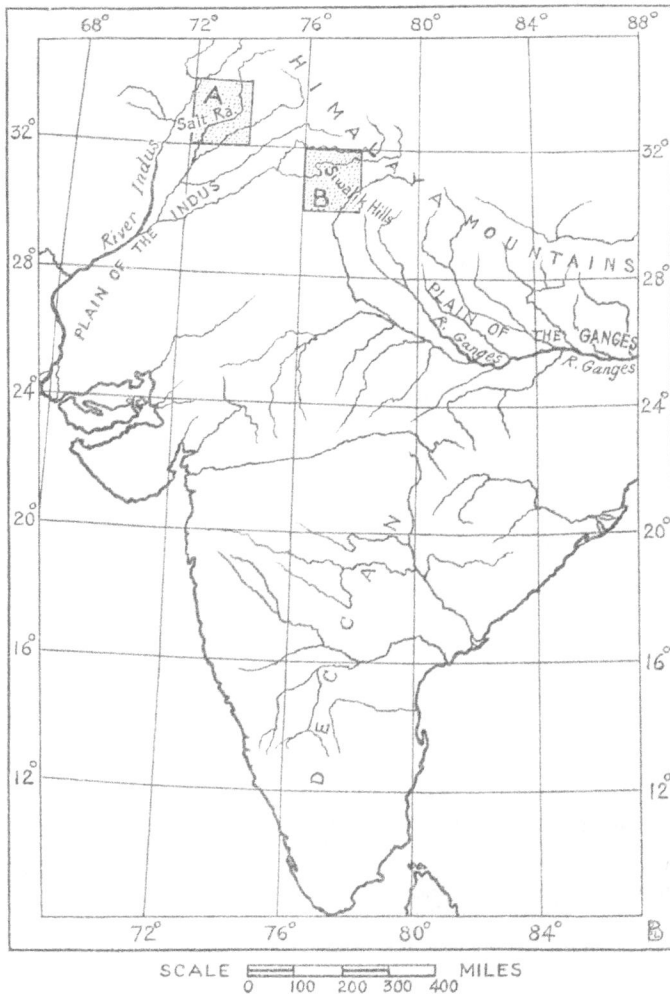

FIG. 15. Map of India, showing location of two key maps A and B, which include the localities whence the American Museum Siwalik collection was obtained.

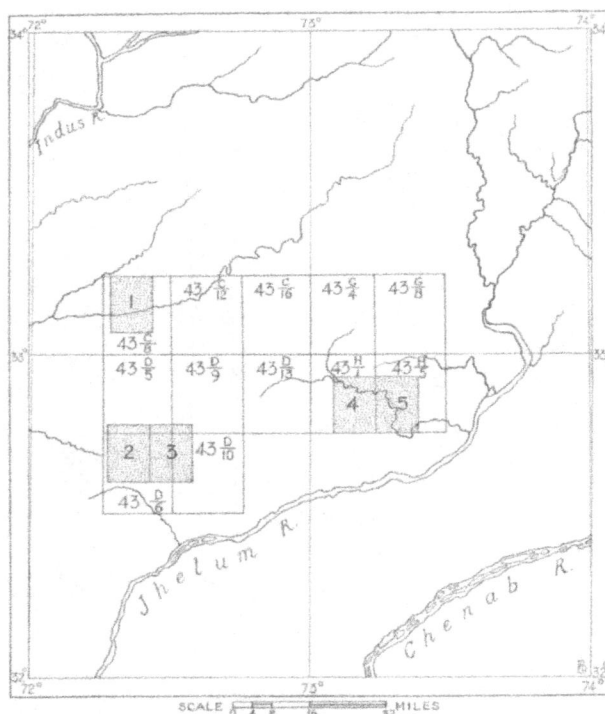

FIG. 16.   Key map A (see Fig. 15).   This is in the Salt Range area, near the headwaters of the Indus, the Jhelum and the Chenab Rivers.   On this map certain Indian Geological Survey sheets are indicated by rectangles and by the numbers 43 C/8, 43 C/12, etc.   The stippled rectangles, numbered 1 to 5 inclusive, are the specific areas from which fossils in the American Museum Siwalik collection were obtained.   These rectangles (1 to 5) are reproduced in detail in Figs. 18 to 22, inclusive.   Scale, 1 inch equals 32 miles.

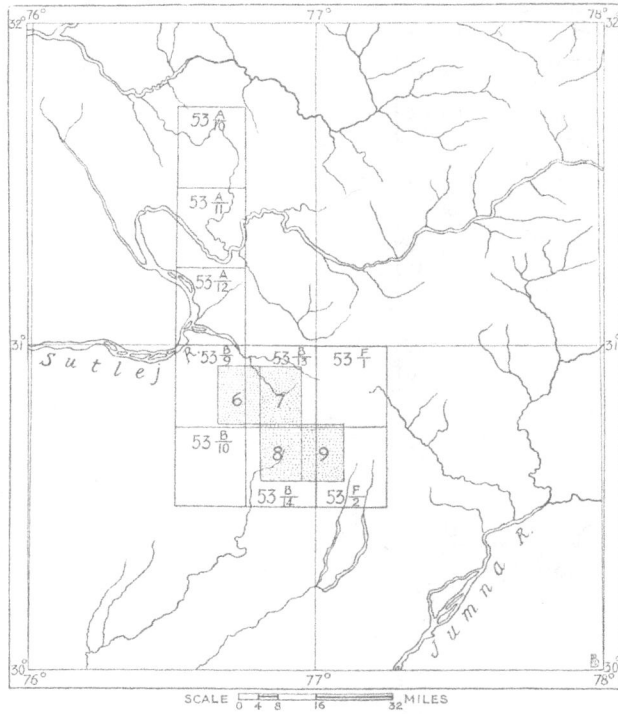

Fig. 17. Key map B (see Fig. 15). This is in the Siwalik Hill region, in the upper reaches of the Sutlej and the Jumna Rivers. Indian Geological Survey sheets and stippled rectangles are indicated as in Key map A. The stippled rectangles (6 to 9) are reproduced in detail in Figs. 23 to 26, inclusive. Scale, 1 inch equals 32 miles.

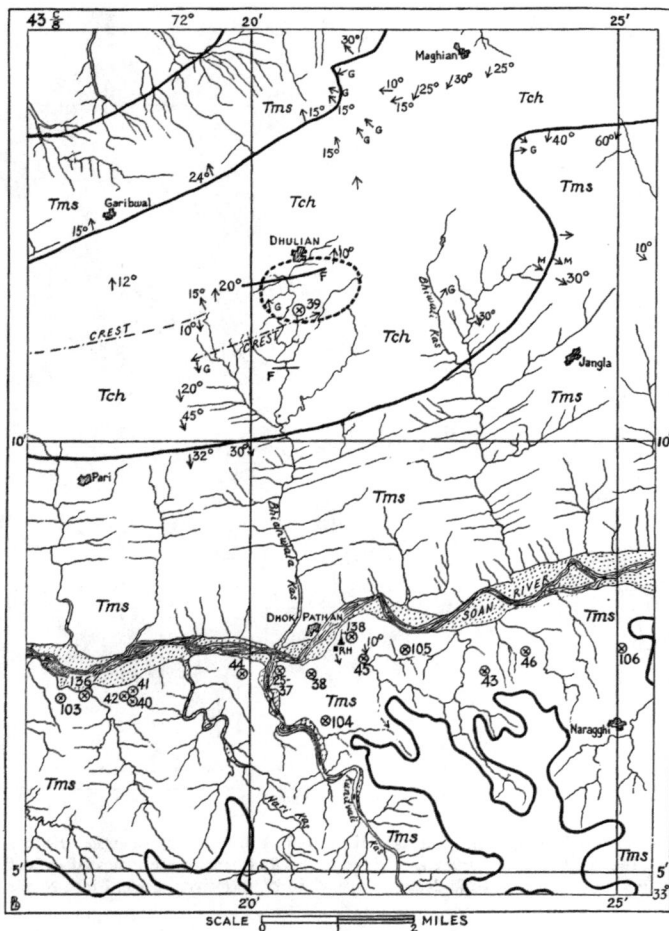

FIG. 18.   Rectangle No. 1 of key map A.   The region around Dhok Pathan and Dhulian.   American Museum fossil localities shown by x's enclosed in circles with accompanying numbers.   Scale, one inch equals two miles.   Legend: Tms = Tertiary, Middle Siwaliks; Tch = Tertiary, Chinji (Lower Siwaliks).

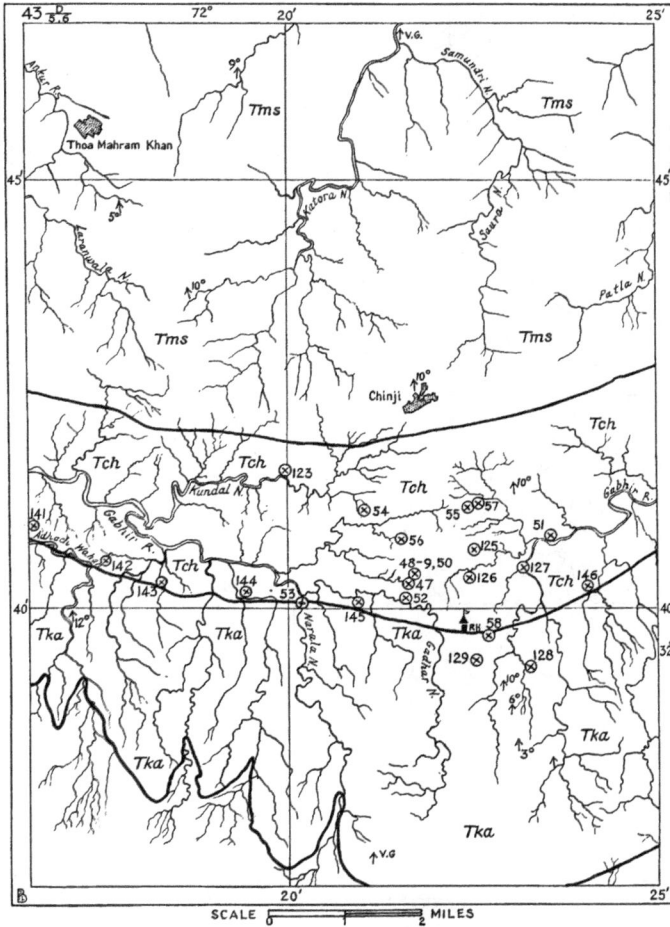

FIG. 19. Rectangle No. 2 of key map A. The region south of Chinji. Scale, one inch equals two miles. Legend: Tms = Tertiary, Middle Siwaliks; Tch = Tertiary, Chinji (Lower Siwaliks); Tka = Tertiary, Kamlial (Lower Siwaliks).

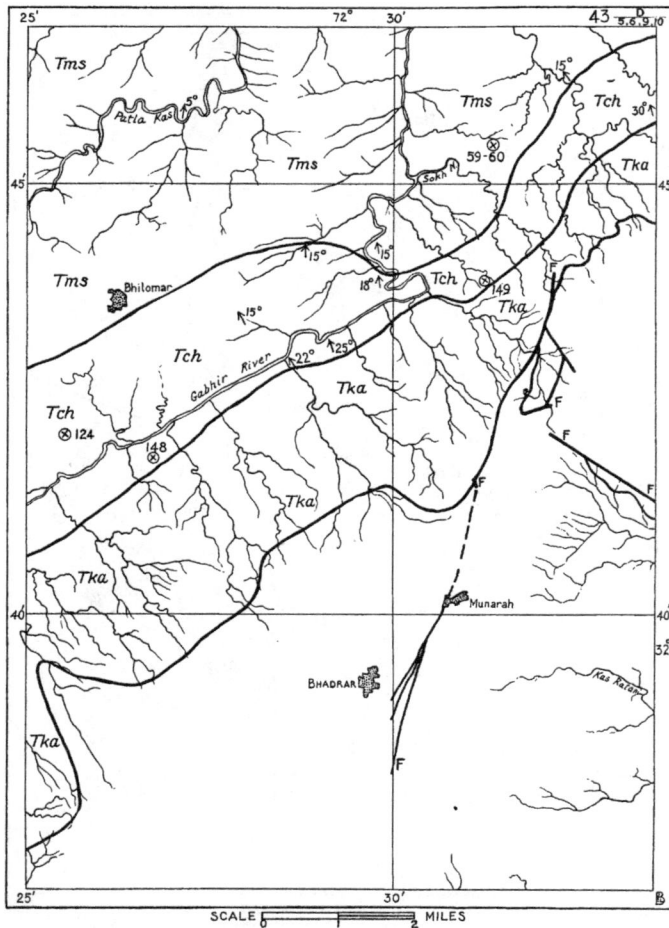

FIG. 20. Rectangle No. 3 of key map A. The region east of Chinji. Scale, one inch equals two miles. Legend: Tms = Tertiary, Middle Siwaliks; Tch = Tertiary, Chinji (Lower Siwaliks); Tka = Tertiary, Kamlial (Lower Siwaliks).

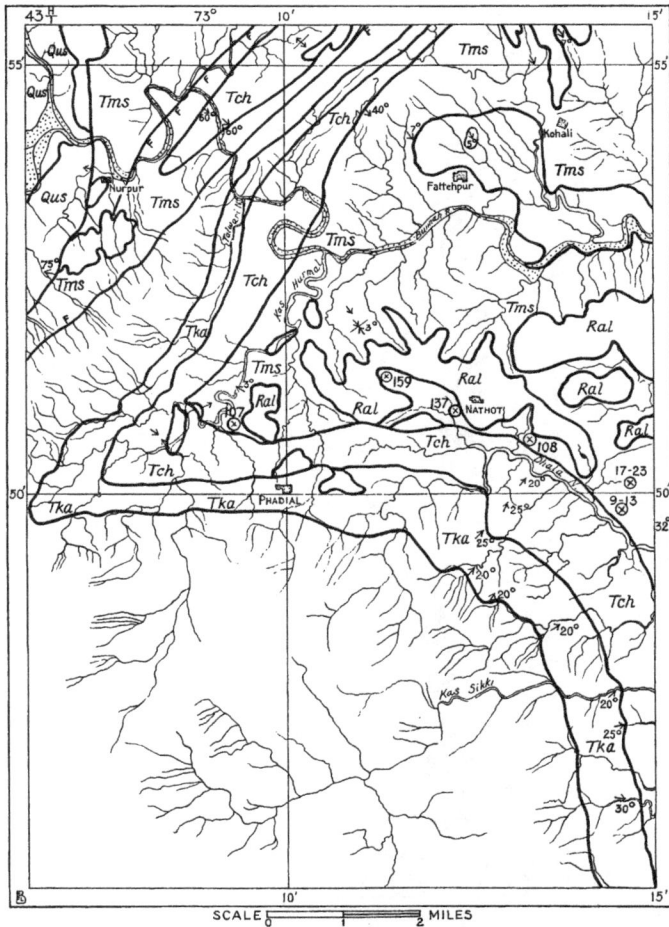

FIG. 21. Rectangle No. 4 of key map A. The region around Nathot. Scale, one inch equals two miles. Legend: Ral = Recent, alluvium; Qus = Quaternary, Upper Siwaliks; Tms = Tertiary, Middle Siwaliks; Tch = Tertiary, Chinji (Lower Siwaliks); Tka = Tertiary, Kamlial (Lower Siwaliks).

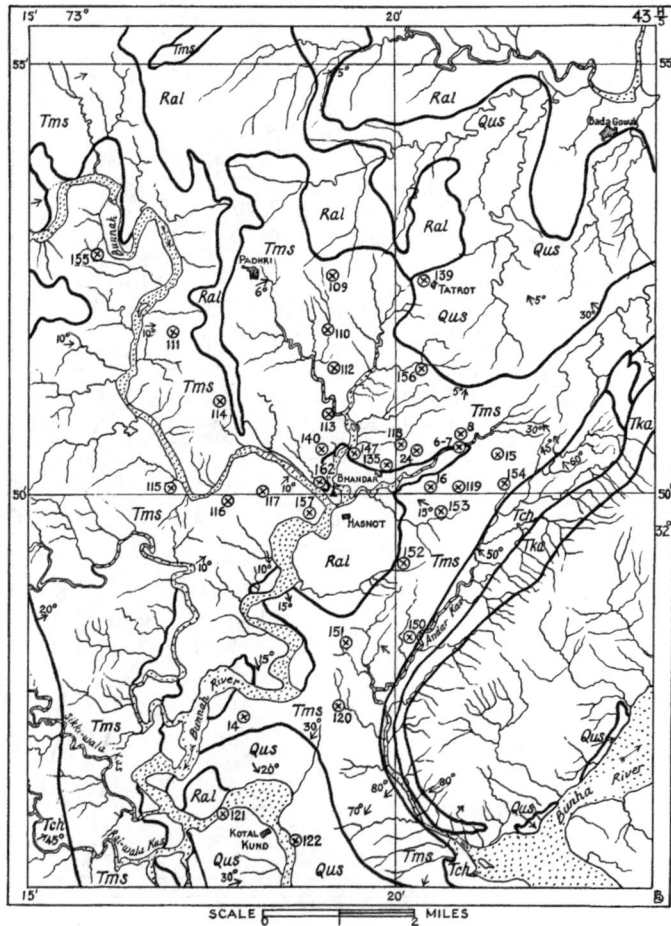

FIG. 22.   Rectangle No. 5 of key map A.   The region around Hasnot.   Scale, one inch equals two miles.   Legend: Ral = Recent, alluvium; Qus = Quaternary, Upper Siwaliks; Tms = Tertiary, Middle Siwaliks; Tch = Tertiary, Chinji (Lower Siwaliks); Tka = Tertiary, Kamlial (Lower Siwaliks).

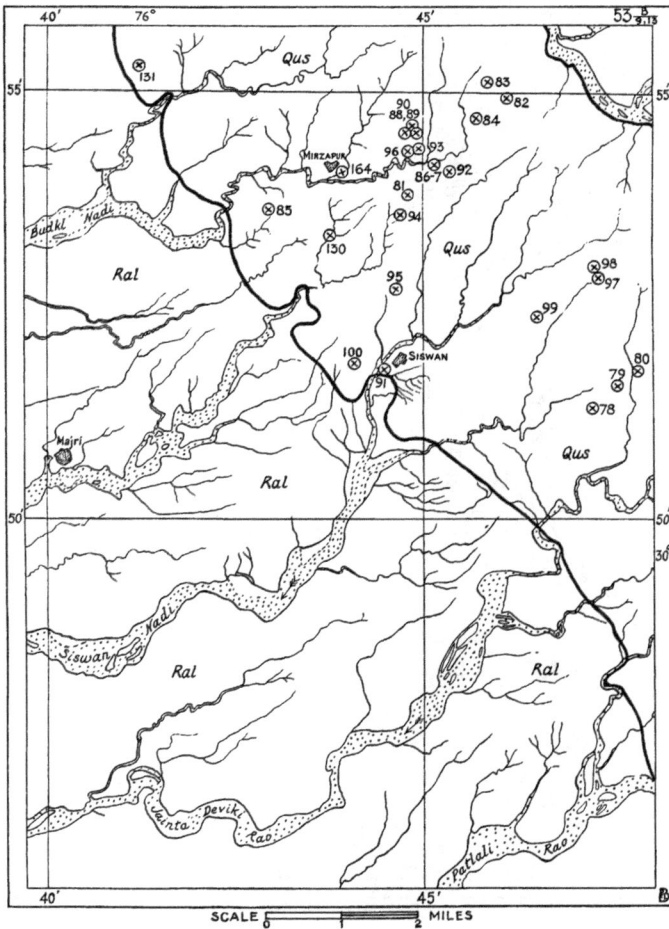

FIG. 23. Rectangle No. 6 of key map B. The region around Siswan. Scale, one inch equals two miles. Legend: Ral = Recent, alluvium; Qus = Quaternary, Upper Siwaliks.

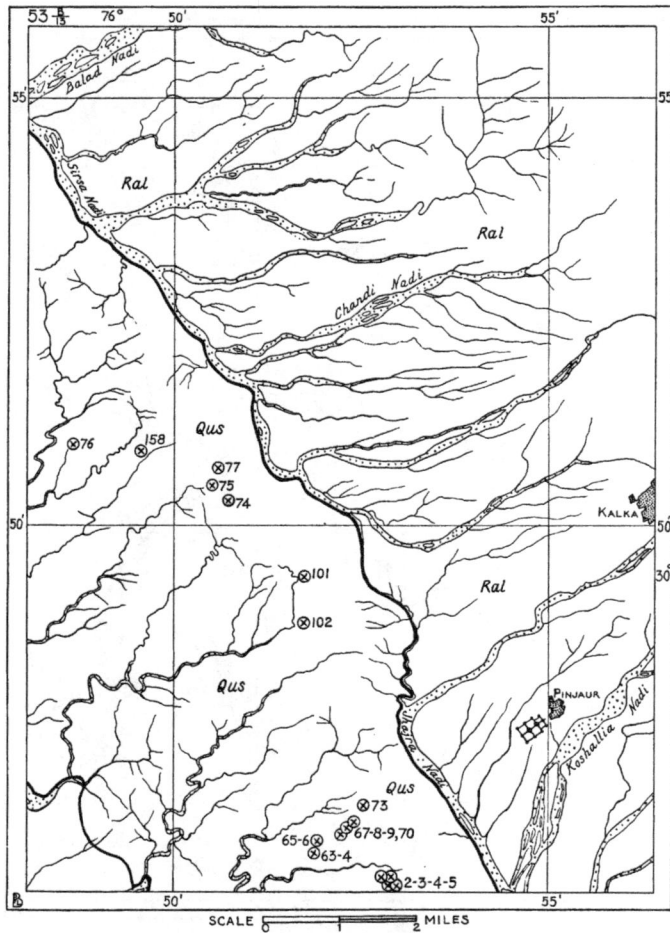

FIG. 24.   Rectangle No. 7 of key map B.   The region around Pinjaur.   Scale, one inch equals two miles.   Legend: Ral = Recent, alluvium; Qus = Quaternary, Upper Siwaliks.

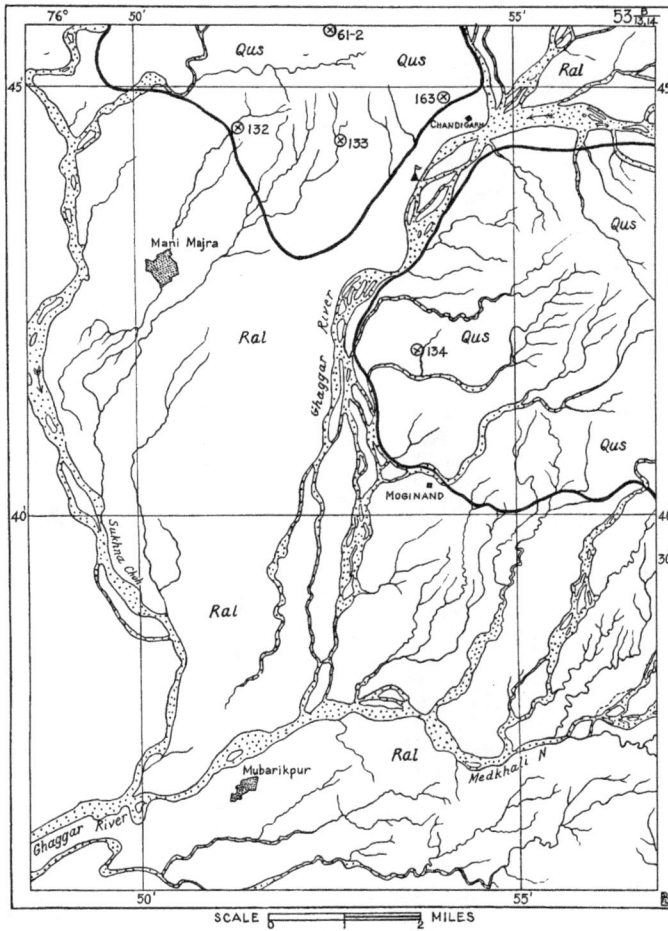

FIG. 25. Rectangle No. 8 of key map B. The region around Chandigarh. Scale, one inch equals two miles. Legend: Ral = Recent, alluvium; Qus = Quaternary, Upper Siwaliks.

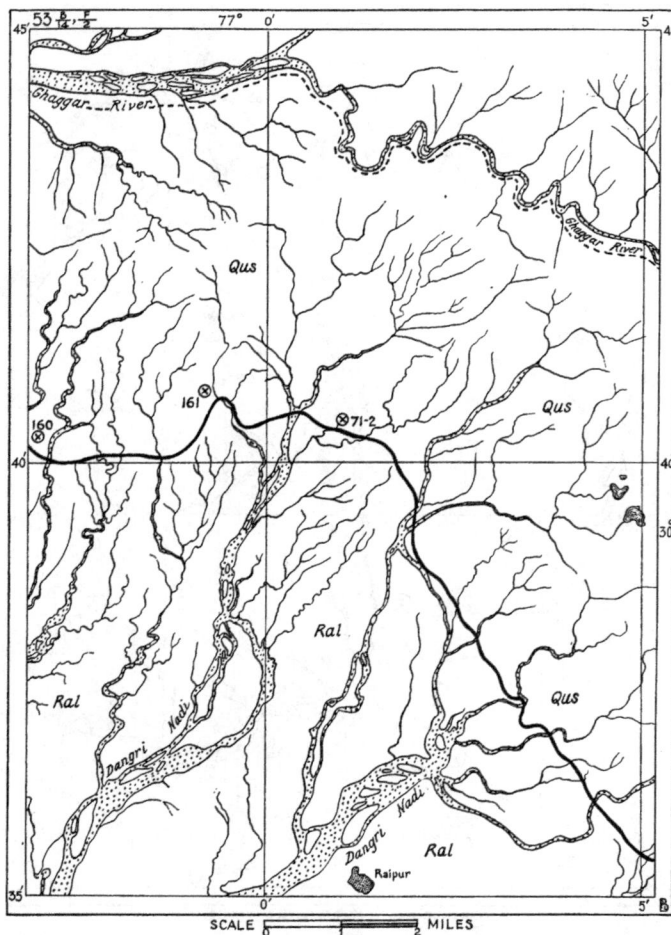

FIG. 26. Rectangle No. 9 of key map B.   The region east of Chandigarh.   Scale, one inch equals two miles.   Legend:
Ral = Recent, alluvium; Qus = Quaternary, Upper Siwaliks.

Field Number	Level	Location
1.	Middle Siwaliks	Near Hari Talyangar
2, 3, 4, 5.	Upper Siwaliks, top of variegated beds below conglomerate	3 miles northwest of Chandigarh
6, 7.	Upper portion of Middle Siwaliks	1½ miles northeast of Hasnot
8.	Middle Siwaliks, 100 feet above Bhandar bone bed	1½ miles northeast of Hasnot
9–13.	Middle Siwaliks, 1,000 feet below Bhandar bone bed	4½ miles west of Hasnot
14.	Upper Siwaliks, lower part	3½ miles northwest of Kotal Kund
15.	Middle Siwaliks, upper part	2 miles northeast of Hasnot
16.	Middle Siwaliks, upper part	1 mile northeast of Hasnot
17–23.	Middle Siwaliks, 1,000 feet below Bhandar bone bed	4½ miles west of Hasnot
24.	Middle Siwaliks, upper part	½ mile northeast of Bhandar
25–37.	Middle Siwaliks, upper part	½ mile southwest of Dhok Pathan
38.	Middle Siwaliks, upper part	1 mile south of Dhok Pathan
39.	Lower Siwaliks, 3,000 feet below Dhok Pathan quarry	Dhulian Dome, 6 miles north of Dhok Pathan
40.	Middle Siwaliks, upper part	3 miles west of Dhok Pathan
41, 42.	Middle Siwaliks, upper part	3 miles west of Dhok Pathan
43.	Middle Siwaliks, upper part	2 miles east of Dhok Pathan
44.	Middle Siwaliks, upper part	1 mile west of Dhok Pathan
45.	Middle Siwaliks, upper part	½ mile east of Dhok Pathan
46.	Middle Siwaliks, upper part	3 miles east of Dhok Pathan
47.	Lower Siwaliks, 1,100 feet above Chinji R.H.	1 mile northwest of Chinji Rest House
48–50.	Lower Siwaliks, 1,600 feet above Chinji R.H.	1 mile northwest of Chinji R.H.
51.	Lower Siwaliks, 400 feet above Chinji R.H.	1½ miles northeast of Chinji R.H.
52.	Lower Siwaliks, 400 feet above Chinji R.H.	1 mile west of Chinji R.H.
53, 54.	Lower Siwaliks	2 miles west of Chinji R.H.
55.	Lower Siwaliks, 1,600 feet above Chinji R.H.	1½ miles north of Chinji R.H.
56.	Lower Siwaliks, 1,600 feet above Chinji R.H.	1½ miles northwest of Chinji R.H.
57.	Lower Siwaliks, 1,600 feet above Chinji R.H.	1½ miles north of Chinji R.H.
58.	Lower Siwaliks, 100 feet below Chinji R.H.	At Chinji R.H.
59, 60.	Lower Siwaliks, 1,600 feet above Chinji R.H.	12 miles east of Chinji R.H.
61, 62.	Upper Siwaliks, below conglomerate	2 miles west of Chandigarh
63–70.	Upper Siwaliks, below conglomerate	3 miles west of Chandigarh
71, 72.	Upper Siwaliks, below conglomerate	15 miles east of Chandigarh
73.	Upper Siwaliks, below conglomerate	3 miles west of Chandigarh
74, 75.	Upper Siwaliks, below conglomerate	6 miles west of Kalka
76.	Upper Siwaliks, below conglomerate	8 miles west of Kalka
77.	Upper Siwaliks, below conglomerate	6 miles west of Kalka
78–80.	Upper Siwaliks, below conglomerate	9 miles west of Kalka
81.	Upper Siwaliks, upper clays below conglomerate	1 mile east of Mirzapur
82–84.	Upper Siwaliks, upper clays below conglomerate	3 miles northeast of Mirzapur
85.	Upper Siwaliks, upper clays below conglomerate	1 mile southwest of Mirzapur
86–87.	Upper Siwaliks, upper clays below conglomerate	3 miles north of Siswan
88.	Upper Siwaliks, upper clays below conglomerate	3 miles north of Siswan
89.	Upper Siwaliks, upper clays below conglomerate	3 miles north of Siswan
90.	Upper Siwaliks, upper clays below conglomerate	3 miles north of Siswan
91.	Upper Siwaliks, base of conglomerate	Siswan
92.	Upper Siwaliks, below conglomerate	3 miles north of Siswan
93.	Upper Siwaliks, below conglomerate	3 miles north of Siswan
94.	Upper Siwaliks, below conglomerate	2 miles north of Siswan
95.	Upper Siwaliks, below conglomerate	1 mile north of Siswan
96.	Upper Siwaliks, below conglomerate	1 mile east of Mirzapur

Field Number	*Level*	*Location*
97, 98.	Upper Siwaliks, below conglomerate	3 miles northeast of Siswan
99.	Upper Siwaliks, below conglomerate	2 miles northeast of Siswan
100.	Upper Siwaliks, near top of conglomerate	½ mile west of Siswan
101.	Upper Siwaliks, below conglomerate	2 miles south of Charnian
102.	Upper Siwaliks, below conglomerate	2½ miles south of Charnian
103.	Middle Siwaliks, upper part	4 miles west of Dhok Pathan
104.	Middle Siwaliks, upper part	1 mile south of Dhok Pathan
105.	Middle Siwaliks, upper part	1 mile east of Dhok Pathan
106.	Middle Siwaliks, upper part	4 miles east of Dhok Pathan
107.	Middle Siwaliks, lower part	½ mile west of Phadial
108.	Middle Siwaliks, 200 feet lower than Nos. 17–23	1 mile south of Nathot
109.	Middle Siwaliks (?) upper part, or Upper Siwaliks (?) lower part	3 miles north of Hasnot
110.	Middle Siwaliks, upper part	2 miles north of Hasnot
111.	Middle Siwaliks, upper part	3 miles northwest of Hasnot
112.	Middle Siwaliks, upper part	1½ miles north of Hasnot
113.	Middle Siwaliks, upper part	1 mile north of Hasnot
114.	Middle Siwaliks, upper part	2 miles northwest of Hasnot
115.	Middle Siwaliks, 500 feet below Bhandar bone bed	2 miles west of Hasnot
116.	Middle Siwaliks, 500 feet below Bhandar bone bed	1½ miles west of Hasnot
117.	Middle Siwaliks, upper part	1 mile west of Hasnot
118.	Middle Siwaliks, upper part	1 mile northeast of Hasnot
119.	Middle Siwaliks, upper part	1½ miles east of Hasnot
120.	Middle Siwaliks, upper part or Upper Siwaliks, lower part	3 miles south of Hasnot
121.	Upper Siwaliks	½ mile northwest of Kotal Kund
122.	Upper Siwaliks	½ mile east of Kotal Kund
123.	Lower Siwaliks, 1,600 feet above Chinji R.H.	3 miles northwest of Chinji R.H.
124.	Lower Siwaliks, level of Chinji R.H.	4 miles northeast of Chinji R.H.
125.	Lower Siwaliks, 600 feet above Chinji R.H.	1 mile north of Chinji R.H.
126.	Lower Siwaliks, 200 feet above Chinji R.H.	½ mile north of Chinji R.H.
127.	Lower Siwaliks, 100 feet above Chinji R.H.	1 mile northeast of Chinji R.H.
128.	Base of Lower Siwaliks	1 mile southeast of Chinji R.H.
129.	Lower Siwaliks, 100 feet below Chinji R.H.	½ mile south of Chinji R.H.
130.	Upper Siwaliks, middle of conglomerate	1 mile south of Mirzapur
131.	Upper Siwaliks, below conglomerate	4 miles west of Mirzapur
132.	Upper Siwaliks, below conglomerate	3 miles west of Chandigarh
133.	Upper Siwaliks, below conglomerate	1 mile west of Chandigarh
134.	Upper Siwaliks, below conglomerate	2½ miles south of Chandigarh
135.	Middle Siwaliks, 100 feet above bone bed	At Bhandar
136.	Middle Siwaliks, same level as No. 40	3½ miles west of Dhok Pathan
137.	Middle Siwaliks, near base	Near Nathot
138.	Middle Siwaliks, upper part	Near Dhok Pathan
139.	Middle Siwaliks, upper part or Upper Siwaliks, lower part	At Tatrot
140.	Middle Siwaliks, upper part	½ mile north of Hasnot
141.	Lower Siwaliks	6 miles west of Chinji R.H.
142.	Lower Siwaliks	5 miles west of Chinji R.H.
143.	Lower Siwaliks, 100 feet above Chinji R.H.	4 miles west of Chinji R.H.
144.	Lower Siwaliks, 400 feet above Chinji R.H.	3 miles west of Chinji R.H.
145.	Lower Siwaliks, 500 feet above Chinji R.H.	1½ miles west of Chinji R.H.
146.	Lower Siwaliks	1½ miles east of Chinji R.H.
147.	Middle Siwaliks, upper part	½ mile northeast of Hasnot

Field Number	Level	Location
148.	Lower Siwaliks	5 miles east of Chinji R.H.
149.	Lower Siwaliks	10 miles east of Chinji R.H.
150.	Middle Siwaliks, near base	2 miles south of Hasnot
151.	Middle Siwaliks, upper part	2 miles south of Hasnot
152.	Middle Siwaliks	½ mile southeast of Hasnot
153.	Middle Siwaliks, 200 feet below Bhandar bone bed	1 mile east of Hasnot
154.	Middle Siwaliks, upper part	2 miles east of Hasnot
155.	Middle Siwaliks, 100 feet below Bhandar bone bed	4½ miles northwest of Hasnot
156.	Middle Siwaliks, upper part	2½ miles northeast of Hasnot
157.	Middle Siwaliks, upper part	½ mile southwest of Hasnot
158.	Upper Siwaliks	7 miles west of Kalka
159.	Middle Siwaliks, lower part	2 miles northeast of Phadial
160.	Upper Siwaliks, below conglomerate	6 miles east of Chandigarh
161.	Upper Siwaliks, below conglomerate	12 miles east of Chandigarh
162.	Middle Siwaliks, upper part	At Hasnot
163.	Upper Siwaliks	At Chandigarh
164.	Upper Siwaliks, below conglomerate	At Mirzapur

Note.—Numbers 14, 59 and 60 are located on the accompanying maps in the positions recorded for them on the field maps. Evidently they are misplaced. Number 14 should be in the Upper Siwaliks, numbers 59 and 60 in the Lower Siwaliks.

INDEX TO AMERICAN MUSEUM FOSSIL LOCALITIES

The preceding list of fossil localities was arranged in a numerical fashion, according to the field numbers assigned to the specific areas at which American Museum Siwalik fossils were found.

The index now presented is arranged according to the towns, villages or rest houses near which American Museum Siwalik fossils were discovered. The names of the villages and rest houses are placed in an alphabetical order, and the numbers of the fossil localities near each village are listed with that village, together with the direction and distance of each locality from the village in question. For instance, if a fossil is designated as coming from a locality six miles east of Chandigarh, this index will show that its field locality number is 160. The number 160 may then be located on one of the large scale maps of American Museum fossil localities, and thus the location of the fossil both geographically and stratigraphically may easily be determined.

By using this index the reader may locate nearly all of the fossils listed in succeeding pages of the text of this work. A few isolated localities from which scattered fossils were obtained have not been included among the maps accompanying this report.

INDEX TO AMERICAN MUSEUM FOSSIL LOCALITIES

Name	Location		Locality Numbers
Bhandar	Near	Bhandar	135
	½ mile northeast of	"	24
Chandigarh	Near	Chandigarh	163
	6 miles east of	"	160
	12 miles east of	"	161

INDEX TO AMERICAN MUSEUM FOSSIL LOCALITIES (*Continued*)

Name	*Location*		*Locality Numbers*
	15 miles east of	Chandigarh	71, 72
	2½ miles south of	"	134
	1 mile west of	"	133
	2 miles west of	"	61, 62
	3 miles west of	"	63–70, 73, 132
	3 miles northwest of	"	2, 3, 4, 5
Charnian	2 miles south of	Charnian	101
	2½ miles south of	"	102
Chinji (Rest House)	Near	Chinji R. H.	58
	1 mile northeast of	"	127
	1½ miles northeast of	"	51
	4 miles northeast of	"	124
	1½ miles east of	"	146
	5 miles east of	"	148
	10 miles east of	"	149
	12 miles east of	"	59, 60
	1 mile southeast of	"	128
	½ mile south of	"	129
	1 mile west of	"	52
	1½ miles west of	"	145
	2 miles west of	"	53, 54
	3 miles west of	"	144
	4 miles west of	"	143
	5 miles west of	"	142
	6 miles west of	"	141
	1 mile northwest of	"	47, 48, 49, 50
	1½ miles northwest of	"	56
	3 miles northwest of	"	123
	½ mile north of	"	126
	1 mile north of	"	125
	1½ miles north of	"	55, 57
Dhok Pathan	Near	Dhok Pathan	138
	½ mile east of	"	45
	1 mile east of	"	105
	2 miles east of	"	43
	3 miles east of	"	46
	4 miles east of	"	106
	1 mile south of	"	38, 104
	½ mile southwest of	"	25–37
	1 mile west of	"	44
	3 miles west of	"	40, 41, 42
	3½ miles west of	"	136
	4 miles west of	"	103
	6 miles north of	"	39
Hasnot	Near	Hasnot	162
	½ mile northeast of	"	147
	1 mile northeast of	"	16, 118
	1½ miles northeast of	"	6, 7, 8

INDEX TO AMERICAN MUSEUM FOSSIL LOCALITIES (*Continued*)

Name	Location		Locality Numbers
	2 miles northeast of	Hasnot	15
	2½ miles northeast of	"	156
	1 mile east of	"	153
	1½ miles east of	"	119
	2 miles east of	"	154
	½ mile southeast of	"	152
	2 miles south of	"	150, 151
	3 miles south of	"	120
	½ mile southwest of	"	157
	1 mile west of	"	117
	1½ miles west of	"	116
	2 miles west of	"	115
	4½ miles west of	"	9–13, 17–23
	2 miles northwest of	"	114
	3 miles northwest of	"	111
	4½ miles northwest of	"	155
	½ mile north of	"	140
	1 mile north of	"	113
	1½ miles north of	"	112
	2 miles north of	"	110
	3 miles north of	"	109
Hari Talyangar	Near	Hari Talyangar	1
Kalka	6 miles west of	Kalka	74, 75, 77
	7 miles west of	"	158
	8 miles west of	"	76
	9 miles west of	"	78, 79, 80
Kotal Kund	½ mile east of	Kotal Kund	122
	½ mile northwest of	"	121
	3½ miles northwest of	"	14
Mirzapur	Near	Mirzapur	164
	3 miles northeast of	"	82, 83, 84
	1 mile east of	"	81, 96
	4 miles west of	"	131
	1 mile south of	"	130
	1 mile southwest of	"	85
Nathot	Near	Nathot	137
	1 mile south of	"	108
Phadial	2 miles northeast of	Phadial	159
	½ mile west of	"	107
Siswan	Near	Siswan	91
	3 miles northeast of	"	97, 98
	½ mile west of	"	100
	1 mile north of	"	95
	2 miles north of	"	94
	2 miles northeast of	"	99
	3 miles north of	"	86–90, 92, 93
Tatrot	Near	Tatrot	139

PART V. SYSTEMATIC DESCRIPTIONS AND DISCUSSIONS

PRIMATES

The primates in the American Museum Siwalik collection have been described by Brown, Gregory and Hellman (1924), and they have been most thoroughly monographed by Gregory and Hellman (1926). Consequently there is no point in making additional remarks about the American Museum Siwalik primates.

A great many genera and species of primates have been described by various authors who have studied the Siwalik faunas, and there can be no doubt but that a considerable amount of synonymy exists among the forms described. An excessive amount of splitting has been inevitable in this group, in view of the fragmentary nature of the fossils, and of their importance in relation to the origin of man. Indeed, this splitting may be desirable rather than otherwise, because it has furnished convenient names and designations for detailed considerations of the various specimens. It will be a problem of the future, when more complete material is known, for some student who has an intimate knowledge of all of the Siwalik primates, to correlate the specimens and reduce the synonymy to such a point that the nomenclature will express the true relations of the genera and species to each other and to other primates.

Since it is impossible to present here a critical review of the Siwalik primates, the described genera and species will be listed and the references to them cited. No attempts at discussions will be made.

LEMUROIDEA

LORISIDAE

Indraloris Lewis, 1933

Generic type, *Indraloris lulli* Lewis

Indraloris lulli Lewis

Indraloris lulli, Lewis, 1933, Amer. Jour. Sci., XXVI, pp. 134–138, figs. 1–2.
    Type.—Y.P.M. No. 13802, a left M_2.
    Paratypes.—None.
    Horizon.—From the lower portion of the Middle Siwaliks, Nagri zone.
    Locality.—One quarter of a mile east of Hari Talyangar village, northeastern Bilaspur State, Simla Hills.
    Diagnosis.—A relatively large lorisid, with high molar cusps. Lower second molar subrectangular, with a well developed fovea anterior.

ANTHROPOIDEA

CERCOPITHECOIDEA

CERCOPITHECIDAE

Presbytis Escholtz, 1821

Generic type, *Presbytis mitrata* Escholtz

Presbytis palaeindicus (Lydekker)

Semnopithecus palaeindicus, Lydekker, 1884, Pal. Indica (X), III, p. 123.

The name is merely listed in this reference. Lydekker refers back to the original descriptions of Falconer and Cautley, in which the type specimens of this species were described but not named.

Presbytis palaeindicus, Pilgrim, 1915, Rec. Geol. Surv. India, XLV, p. 2.

Additional References.—

Falconer, H., and Cautley, P.T. 1837, p. 354.   Specimens not named.

Falconer, H., 1868A, Pl. XXIV, figs. 56–58.   Specimens not named.

Lydekker, R., 1885C, pp. 2–3; 1886C, p. 5.

Pilgrim, G. E., 1910B, p. 198; 1913B, p. 325.

Matthew, W. D., 1929, p. 443.

Type.—B.M. No. 15710, a fragment of a right mandibular ramus, with P_4–M_3.

Paratypes.—B.M. No. 15711, a fragment of a right mandibular ramus with M_3.

Horizon.—Lower portion of the Upper Siwaliks, Tatrot zone.

Locality.—From the Siwalik Hills.

Diagnosis.—Teeth closely comparable to those of *Semnopithecus entellus*, but the ramus is not so heavy.   Talonid of M_3 large.

Cercopithecus Brünnich, 1722

Generic type, *Cercopithecus mona* Schreber

Cercopithecus asnoti (Pilgrim)

Semnopithecus asnoti, Pilgrim, 1910, Rec. Geol. Surv. India, XL, p. 64.

Cercopithecus asnoti, Pilgrim, 1915, Rec. Geol. Surv. India, XLV, pp. 3–6, Pl. I, figs. 1–3.

Additional Reference.—

Matthew, W. D., 1929, p. 448.

Type.—(Lectotype).—G.S.I. No. D 120, a right maxilla of an immature individual.

Cotype.—G.S.I. No. D 121, a left maxilla containing DM^{3-4}.

Horizon.—Middle Siwaliks, Dhok Pathan zone.

Locality.—From the neighborhood of Hasnot, Salt Range, Jhelum District, northern Punjab.

Diagnosis.—Characterized by a broad anterior shelf on the external wall of the first upper molar, and by the forward position of the transverse crest in the last premolar. These characters are used by Pilgrim as evidence that this species belongs to the genus *Cercopithecus*.

Macacus Lacépède, 1799

Generic type, *Simia inuus* Linnaeus

Macacus sivalensis Lydekker

Macacus sivalensis, Lydekker 1878, Rec. Geol. Surv. India, XI, p. 66.
  Additional References.—
      Lydekker, R., 1880B, p. 30; 1883C, pp. 82, 91; 1884D, p. 123; 1885B, p. 1; 1886C,
        p. 5, Pl. I, figs. 9–10.
      Pilgrim, G. E., 1910B, p. 198; 1913B, pp. 280, 288; 1915A, p. 2.
      Matthew, W. D., 1929, p. 448.
  *Type.—*G.S.I. No. D 2, two fragments of a maxilla.
  *Paratypes.—*None.
  *Horizon.—*Middle Siwaliks, Dhok Pathan zone.
  *Locality.—*From near Hasnot, Jhelum District, Punjab.
  *Diagnosis.—*Teeth small, molar series curved.   Posterior cusps of M³ relatively small.

Papio Erxleben, 1777

Generic type, *Papio sphinx* Erxleben

Papio subhimalayanus (von Meyer)

Semnopithecus subhimalayanus, von Meyer, 1848, Index Palaeontologicus, Bronn, p. 1133.
Cynocephalus subhimalayanus, Lydekker, 1885, Cat. Siw. Vert. Ind. Mus., p. 2.
Papio subhimalayanus, Pilgrim, 1910, Rec. Geol. Surv. India, p. 198.
  Additional References.—
      Baker, W. E., and Durand, H. M., 1836B, pp. 739–741.   (No specific name.)
      Falconer, H., 1868A, Pl. XXIV, figs. 1, 2.
      Lydekker, R., 1880B, p. 30; 1883C, pp. 82, 92; 1884D, p. 123; 1885C, pp. 4–5;
        1886C, pp. 6–7, Pl. I, figs. 3, 3a.
      Pilgrim, G. E., 1913B, p. 324; 1915A, p. 2.
  *Type.—*B.M. No. 31157, the right half of a palate.
  *Paratypes.—*None.
  *Horizon.—*Upper Siwaliks, presumably from the Pinjor zone.
  *Locality.—*From the Siwalik Hills.
  *Diagnosis.—*Comparable to the living baboons, but distinguished by its large size.

Papio falconeri (Lydekker)

Cynocephalus falconeri, Lydekker, 1886, Pal. Indica (X), IV, p. 7.
Papio falconeri, Pilgrim, 1910, Rec. Geol. Surv. India, XL, p. 198.
      The specimen listed above was described but not named in previous publications by
Falconer and Cautley, and by Lydekker.
  Additional References.—
      Falconer, H., and Cautley, P. T., 1837, p. 354.
      Falconer, H., 1868A, Pl. XXIV, figs. 3, 4.
      Lydekker, R., 1885C, p. 6.
      Pilgrim, G. E., 1913B, p. 325; 1915A, p. 2.
      Matthew, W. D., 1929, p. 443.

Type.—B.M. No. 15709, a mandible.
Paratypes.—None.
Horizon.—Upper Siwaliks.
Locality.—From the Siwalik Hills.
Diagnosis.—Symphysis elongated, with a flat oval surface. Cheek teeth relatively broad, and talonid of M₃ small.

HOMINOIDEA
PONGIDAE
Simia Linnaeus, 1758

Generic type, *Simia satyrus* Linnaeus

Simia cf. satyrus Linnaeus

This species was listed by Pilgrim as coming from the Upper Siwalik beds.
References.—

Pilgrim, G. E., 1910B, p. 198; 1913B, p. 325; 1915A, p. 2.
Matthew, W. D., 1929, p. 443.

Ramapithecus Lewis, 1934

Generic type, *Ramapithecus brevirostris* Lewis

Ramapithecus brevirostris Lewis

Ramapithecus brevirostris, Lewis, 1934, Amer. Jour. Sci., XXVII, pp. 162–166, Pl. I, figs. 1a–1b.
Type.—Y.P.M. No. 13799, a right maxilla and premaxilla with the dentition.
Horizon.—Upper Middle Siwaliks or lower Upper Siwaliks.
Locality.—One fourth mile east of Chakrana, 4 miles east of Hari Talyangar northwest of Bilaspur, Simla Hills.
Diagnosis.—Dentition parallels the hominoid type. The dental arcade of the upper jaw is parabolic. Face slightly prognathous. No diastema in the dental series, canine small, dentition compressed anteroposteriorly. Incisors approximately equal in size. Canine ellipsoid, with long axis normal to the curve of the dental arcade. Third premolar and succeeding teeth progressive.

Ramapithecus hariensis Lewis

Ramapithecus hariensis, Lewis, 1934, Amer. Jour. Sci., XXVII, pp. 166–167, Pl. I, figs. 2a–2b.
Type.—Y.P.M. No. 13807, maxillary fragment with the first and second right molars.
Horizon.—Lower Middle Siwaliks, from the Nagri zone.
Locality.—One fourth mile east of Hari Talyangar village, Bilaspur State, Simla Hills.
Diagnosis.—Like *Ramapithecus brevirostris* but more primitive. Cheek teeth not compressed anteroposteriorly to the degree typical of the foregoing species.

Sugrivapithecus Lewis

Generic type, *Sugrivapithecus salmontanus* Lewis

Sugrivapithecus salmontanus Lewis

Sugrivapithecus salmontanus, Lewis, 1934, Amer. Jour. Sci., XXVII, pp. 168–170, Pl. II,
    figs. 1a–1b.

Type.—Y.P.M. No. 13811, a left ramus with the roots of I_2–P_3 and with P_4–M_2 well
preserved.

Paratypes.—None.

Horizon.—Lower Middle Siwaliks.

Locality.—Three and three fourths miles west-southwest of Hasnot, Jhelum District,
Punjab.

Diagnosis.—Mandible with divergent rami and well developed chin.   Cheek teeth
elongate and laterally compressed.   Incisors and canine small, and the last premolar
hominid.

Dryopithecus Lartet, 1856

Generic type, *Dryopithecus fontani* Lartet

Dryopithecus punjabicus Pilgrim

Dryopithecus punjabicus, Pilgrim, 1910, Rec. Geol. Surv. India, XL, p. 63.

Additional References.—

    Pilgrim, G. E., 1910B, p. 198; 1913B, pp. 311, 320; 1915A, pp. 9–25, Pl. I, figs. 5,
        6, Pl. II, figs. 4, 5, Pl. III.

    Gregory, W. K., and Hellman, Milo, 1926, Various pages.

Type.—G.S.I. No. D 118, D 119, the right and left mandibular rami of the same
mandible.

Cotype.—G.S.I. No. D 185, a right maxilla.

Horizon.—Lower Siwaliks, Chinji zone.   Also from the lower portion of the Middle
Siwaliks, Nagri Zone, at Hari Talyangar.

Locality.—Chinji, Punjab.   Also from Hari Talyangar, Bilaspur state, Simla Hills.

Diagnosis.—"Maxillae and mandibles of this species occur in the Lower Siwaliks of
Chinji.   It is very near to *D. rhenanus*, but differs by the mesoconid being more in a line
with the protoconid and metaconid."   Pilgrim, G. E., 1910A, p. 63.

Dryopithecus chinjiensis Pilgrim

Dryopithecus chinjiensis, Pilgrim, 1915, Rec. Geol. Surv. India, XLV, pp. 25–27, Pl. II,
    figs. 6, 7.

Additional References.—

    Gregory, W. K., and Hellman, Milo, 1926, various pages.

    Pilgrim, G. E., 1927B, p. 10, fig. 4 of plate.

Type.—G.S.I. No. D 179, a left third lower molar.

Paratype.—G.S.I. No. D 180, a left first lower molar.

Horizon.—Lower Siwaliks, Chinji zone.

Locality.—Near Chinji, Attock District, Punjab.

Diagnosis.—Characterized by its high cusps and its large size.   Otherwise the tooth
characters connect this species with *Dryopithecus* although the height of the cusps makes it
comparable to *Sivapithecus*.

Dryopithecus giganteus Pilgrim

Dryopithecus giganteus, Pilgrim, 1915, Rec. Geol. Surv. India, XLV, pp. 27–29, Pl. II, fig. 8.

Additional Reference.—

Gregory, W. K., and Hellman, Milo, 1926, various pages.

Type.—G.S.I. No. D 175, a right lower third molar.

Paratypes.—None.

Horizon.—Lower portion of the Middle Siwaliks, Nagri zone.

Locality.—From a point north of the village of Alipur, in the eastern part of the Salt Range, Punjab.

Diagnosis.—Extremely large, which character alone distinguishes it from all other species of the genus.

Dryopithecus pilgrimi Brown, Gregory, Hellman

Dryopithecus pilgrimi, Brown, Gregory and Hellman, 1924, Amer. Mus. Novitates, No. 134, pp. 1–3, figs. 1–3.

Additional Reference.—

Gregory, W. K., and Hellman, Milo, 1926, various pages.

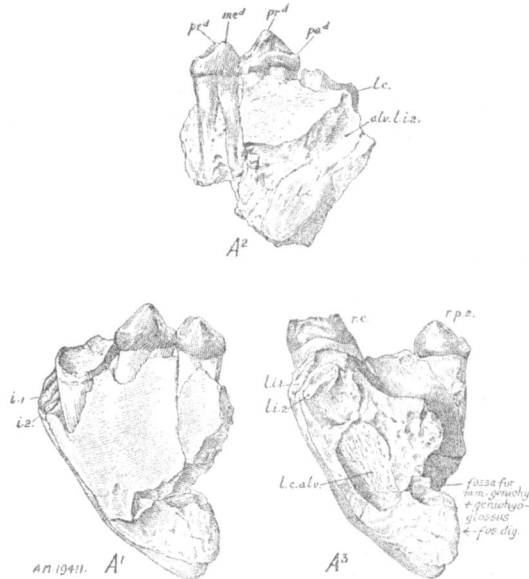

FIG. 27. *Dryopithecus pilgrimi* Brown, Gregory, Hellman. Type, Amer. Mus. No. 19411, mandibular symphysis, with incisor alveoli, canines and premolars of both sides. Left ramus as seen from the inside, above; the same as seen from the outside, below, left; right half of symphysis as seen from the inside, below, right. Natural size. From Brown, Gregory and Hellman, 1924.

Type.—Amer. Mus. No. 19411, the anterior part of a lower jaw, including the symphysis.

Paratypes.—None.

Horizon.—Lower Siwaliks, lower portion of the Chinji zone.

Locality.—Two miles east of Rammagar, Jammu State, Kashmir.

Specimens in the American Museum.—The type, listed above.

Diagnosis.—Like *Dryopithecus fontani*, but more progressive in the lesser crown height and the greater convexity of the cusps. P_3 compressed, P_4 wide, with well developed fovea anterior. External cingulum not present. Mandible moderately heavy, without symphyseal "simian shelf."

"*Relationships.*—This and the following species appear to be referable to the genus *Dryopithecus*, and they agree with the type species *D. fontani* from the Middle and Upper Miocene of Europe in all fundamental characters of the jaw and teeth. The breadth index of p_3 in the present species is not materially different from that in *D. fontani*. The lesser crown height is associated with the greater convexity of the cusps in the Siwalik specimens, which are thus in this respect somewhat more progressive (*i.e.*, more like the later anthropoids) than is *D. fontani*.

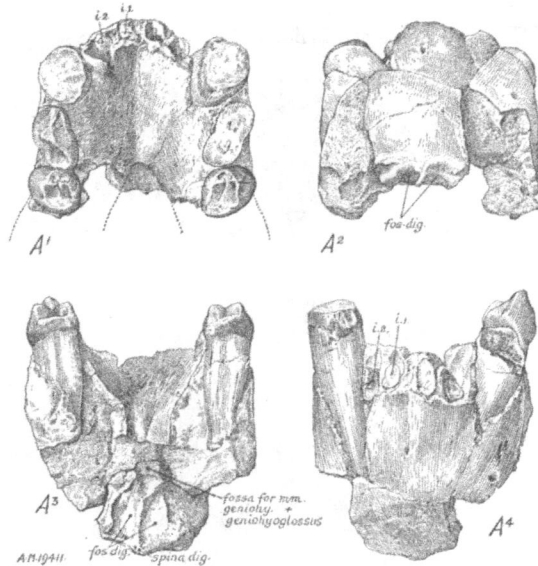

Fig. 28.   *Dryopithecus pilgrimi* [Brown, Gregory, Hellman. Type, Amer. Mus. No. 19411, mandibular symphysis, with incisor alveoli, canines and premolars of both sides. Crown view, above, left; ventral view, above, right; posterior view, below, left; anterior view, below, right. Natural size. From Brown, Gregory and Hellman, 1924.

"As compared with the p_3, which was referred by Pilgrim (1915, Pl. I, fig. 9) to *Sivapithecus*, that of *D. pilgrimi* is much more primitive and less bicuspidlike, *i.e.*, more compressed and with the anterior part less expanded.

"P_4 is relatively wider than in *D. fontani*, especially in the talonid, the anteroposterior sulcus and fovea anterior (trigonid fossa) are more pronounced; the external cingulum, which is distinctly suggested in p_4 of *D. fontani*, is wanting. The front part of the jaw

agrees generically with *D. fontani* as well as the longitudinal section of the symphysis."—
Brown, Barnum, Gregory, William K., and Hellman, Milo, 1924, p. 3.

Dryopithecus cautleyi Brown, Gregory, Hellman

Dryopithecus cautleyi, Brown, Gregory and Hellman, 1924, Amer. Mus. Novitates, No. 134,
    p. 5, fig. 4.
    Additional References.—
        Gregory, W. K., and Hellman, Milo, 1926, various pages.
        Lewis, G. E., 1934A, p. 171, Pl. II, fig. 2.
    Type.—Amer. Mus. No. 19412, the left half of a mandible, lacking the lower border
and the symphysis.
    Paratypes.—None.
    Horizon.—Lower portion of the Middle Siwaliks, Nagri zone.   About 1,000 feet below
the Bhandar bone bed.

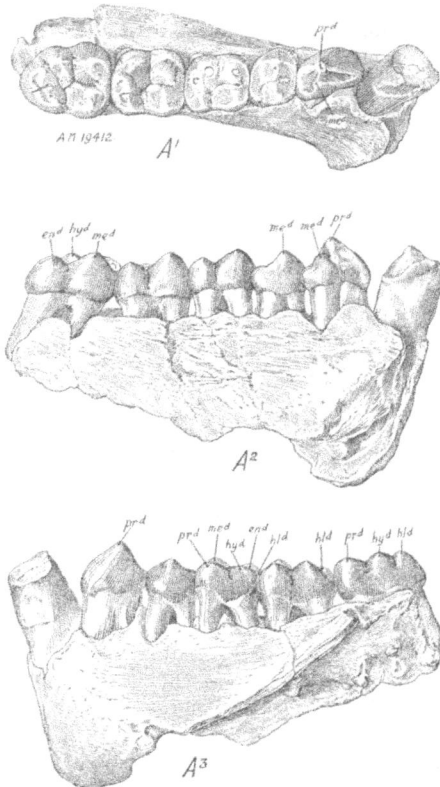

Fig. 29. *Dryopithecus cautleyi* Brown, Gregory, Hellman. Type, Amer. Mus. No. 19412, left ramus of mandible, with canine, premolars and molars. Crown view above, inner view in middle, outer view below. Natural size.
From Brown, Gregory and Hellman, 1924.

Locality.—Four and a half miles west of Hasnot, Salt Range, Jhelum District, Punjab.

Specimen in the American Museum.—The type, designated above.

Diagnosis.—Closely comparable to *Dryopithecus chinjiensis*, but smaller and without external cingulum on molars. Premolars more advanced than in the above mentioned species, being characterized by their relative width.

"*Relationships.*—The nearest relative of *D. cautleyi* seems to be its contemporary *D. chinjiensis* Pilgrim, known chiefly from a third lower molar (Pilgrim, *op. cit.*, Pl. II, fig. 6). The most conspicuous differences in *D. cautleyi* are the lack of an external cingulum, the smaller size of the tooth, the lesser distinctness of cusp 6 and the absence of cusp 7 behind the metaconid.

"*D. cautleyi* may well be a descendant of *D. pilgrimi*, from which it differs in the distinctly more advanced stage of evolution of the premolars."—Brown, Barnum, Gregory, William K., and Hellman, Milo, 1924, p. 5.

Dryopithecus (?) frickae Brown, Gregory, Hellman

Dryopithecus (?) *frickae*, Brown, Gregory and Hellman, 1924, Amer. Mus. Novitates, No. 134, pp. 5–8, fig. 5.

Additional Reference.—

Gregory, W. K., and Hellman, Milo, 1926, various pages.

MEASUREMENTS [15]

	D. pilgrimi Type A.M. 19411	*D. cautleyi* Type A.M. 19412	*D. (?) frickae* Type A.M. 19413
Ant. surface p_3 to post. convexity m_3	. . . .	55.6 mm.	. . . .
Ant. surface p_3 to post. surface p_4	18.2 mm.	18.2	. . . .
Ant. end m_1 to post. convexity m_3	. . . .	37.2	38.8 mm.
P_3, ant. post. (mesiodist.)	11.3	. . . .	. . . .
P_3, transv.	7.2	. . . .	. . . .
P_3, breadth index $\dfrac{tr. \times 100}{a.p.}$	63.7	. . . .	. . . .
P_4, ant. post	7.8	7.8	9.0
P_4, transv	8.7	9.2	10.2
P_4, breadth index $\dfrac{tr. \times 100}{a.p.}$	111	118	113
M_1, ant. post	. . . .	10.6	11.5
M_1, transv.	. . . .	9.5	11.0
M_1, breadth index $\dfrac{tr. \times 100}{a.p.}$	. . . .	90	95
M_2, ant. post	. . . .	12.6	13.5
M_2, transv.	. . . .	11.3	12.4
M_2, breadth index $\dfrac{tr. \times 100}{a.p.}$	. . . .	89.7	91.8
M_3, ant. post	. . . .	13.7	14.0
M_3, transv.	. . . .	12.3	12.7
M_3, breadth index $\dfrac{tr. \times 100}{a.p.}$	. . . .	89.8	90.7

[15] Adapted in part from Brown, Barnum, Gregory, William K., and Hellman, Milo, 1924, Amer. Mus. Novitates, No. 130, p. 9.

Type.—Amer. Mus. No. 19413, the left half of a mandible, lacking the front portion.

Paratypes.—None.

Horizon.—Middle Siwaliks, about the level of the Bhandar bone bed.

Locality.—Locality not given.

Specimen in American Museum.—The type, designated above.

Diagnosis.—Characterized by its relatively large size, M_2 being especially large. P_3 and P_4 more advanced than in the typical species of *Dryopithecus*.

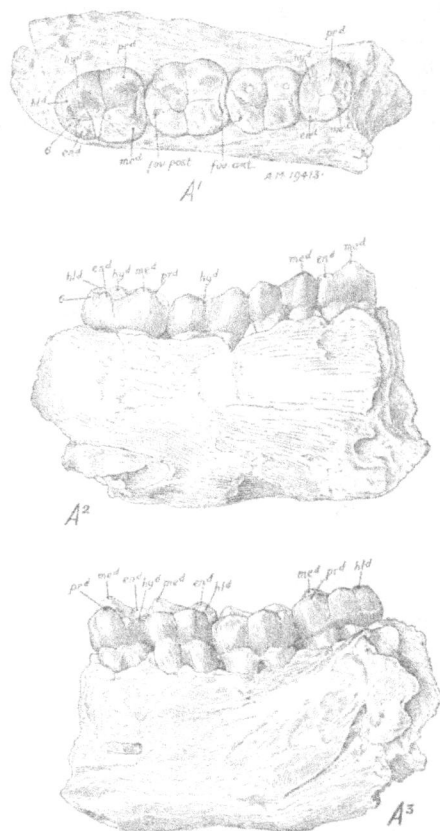

FIG. 30. *Dryopithecus frickae* Brown, Gregory, Hellman. Type, Amer. Mus. No. 19413, left ramus of mandible with last premolar and molars. Crown view above, inner view in middle, outer view below. Natural size. From Brown, Gregory and Hellman, 1924.

"*Relationships.*—*Dryopithecus* (?) *frickae* may well prove to be a descendant of *D. cautleyi*, from which it differs in the more advanced stage of p_4 and especially in the much larger size of m_2. In this feature it differs also from *D. fontani*, *D. rhenanus* and *D. punjabicus*. In fact, this difference is so marked that it leads to the suspicion that *D.* (?)

frickae represents a higher generic phase of the *Dryopithecus* line.   Mr. Brown in the field referred it to *Palaeopithecus*, a form from a slightly later horizon, known hitherto only from a palate figured by Lydekker and by Dubois.   We have considered the possibility of approximate occlusion in the lower jaw of *D. frickae* with the palate of *Palaeopithecus sivalensis* and have tried to project the lower cusps into the appropriate loci on the upper teeth.   But the wide differences in the state of wear of the upper and lower teeth in the two specimens prevent our obtaining positive correspondence between them, so that in default of further evidence we regard it as more conservative to refer the species *frickae* to *Dryopithecus*, especially as all the Indian 'genera' of anthropoids exhibit the fundamental '*Dryopithecus*' pattern of the molars with minor modifications."—Brown, Barnum, Gregory, William K., and Hellman, Milo, 1924, pp. 7–8.

As mentioned above (p. 56) the three specimens of *Dryopithecus* in the American Museum collection, namely the types of *D. pilgrimi*, *D. cautleyi* and *D. frickae*, have been so thoroughly and exhaustively studied by Dr. Gregory and Dr. Hellman (1926) that nothing additional need be said concerning them.

Dryopithecus sivalensis Lewis

Dryopithecus sivalensis, Lewis 1934, Amer. Jour. Sci., XXVII, pp. 171–172, Pl. I, fig. 5.
   Type.—Y.P.M. No. 13806, a fragment of a right mandibular ramus with M_2–M_3.
   Paratypes.—None.
   Horizon.—Lower Middle Siwaliks from the Nagri zone.
   Locality.—One fourth mile east of Hari Talyangar village, Bilaspur State, Simla Hills.
   Diagnosis.—Ramus broad and massive in comparison to the molars.   Molars small as compared to most species of *Dryopithecus*.   Cusps of molars low, hypoconid central, no cingulum present.   Molars quadrate in outline.

Hylopithecus Pilgrim, 1927

Generic type, *Hylopithecus hysudricus* Pilgrim

Hylopithecus hysudricus Pilgrim

Hylopithecus hysudricus, Pilgrim, 1927, Mem. Geol. Surv. India, Pal. Indica, N.S., XIV, pp. 11–12, fig. 6 of plate.
   Type.—G.S.I. No. D 200, an isolated lower molar, probably M_2.
   Paratypes.—None.
   Horizon.—Middle Siwaliks, Nagri zone.
   Locality.—Hari Talyangar, northwest of Bilaspur, Bilaspur State, Simla Hills.
   Diagnosis.—Distinguished by its very small size.   Cusps sharply defined and more or less isolated from each other.   No folds in crown of tooth.   A small fovea anterior is present.

Bramapithecus Lewis, 1934

Generic type, *Bramapithecus thorpei* Lewis

Bramapithecus thorpei Lewis

Bramapithecus thorpei, Lewis, 1934, Amer. Jour. Sci., XXVII, pp. 173–174, Pl. I, fig. 4.

Type.—Y.P.M. No. 13814, a left ramus with M_{2-3}.

Horizon.—Lower Siwaliks, upper Chinji zone.

Locality.—One and one half mile southeast of Hasnot, Salt Range, Jhelum District, Punjab.

Diagnosis.—Lower molars much compressed antero-posteriorly, with high crowns. Relief of crown complicated by numerous coarse grooves.

Palaeosimia Pilgrim, 1915

Generic type, *Palaeosimia rugosidens* Pilgrim

Palaeosimia rugosidens Pilgrim

Palaeosimia rugosidens, Pilgrim, 1915, Rec. Geol. Surv. India, XLV, pp. 29–31, Pl. II, fig. 9.

Additional Reference.—

Gregory, W. K., and Hellman, Milo 1926. Various pages.

Type.—G.S.I. No. D 188, a right M^3.

Paratypes.—None.

Horizon.—Lower Siwaliks, Chinji zone.

Locality.—From a locality near Chinji, Salt Range, Punjab. Exact locality unknown.

Diagnosis.—Molars characterized by the intensity of the wrinkling of the enamel surface. Tooth pattern similar to that of *Dryopithecus*.

Palaeopithecus Voigt, 1835

Generic type not mentioned

Palaeopithecus sivalensis Lydekker

Palaeopithecus sivalensis, Lydekker, 1879, Rec. Geol. Surv. India, XII, p. 33.

Troglodytes sivalensis, Lydekker, 1886, Pal. Indica (X), IV, pp. 2–4, Pl. I, figs. 1, 1a.

Anthropopithecus sivalensis, Pilgrim, 1910, Rec. Geol. Surv. India, XL, p. 198.

Additional References.—

Lydekker, R., 1885B, p. 1.

Dubois, E., 1897, p. 84.

Pilgrim, G. E., 1915A, p. 198.

Gregory, W. K., and Hellman, Milo, 1926. Various pages.

Matthew, W. D., 1929, p. 448.

Type.—G.S.I. No. D 1, a palate.

Paratypes.—None.

Horizon.—Middle Siwaliks.

Locality.—From Jabi, Punjab.

Diagnosis.—Molars with *Dryopithecus* pattern but distinguished by their breadth. Enamel thick.

Palaeopithecus (?) sylvaticus Pilgrim

Palaeopithecus (?) sylvaticus, Pilgrim, 1927, Pal. Indica, N.S., XIV, pp. 9–10, fig. 3 of plate.
   Type.—G.S.I. No. D 199, a left mandibular ramus.
   Paratypes.—None.
   Horizon.—Middle Siwaliks, Nagri zone.
   Locality.—Hari Talyangar, Bilaspur State, Simla Hills.
   Diagnosis.—Molars characterized by a typical *Dryopithecus* pattern, but would seem to
be distinct because of their breadth.   Small fovea posterior.

Sivapithecus Pilgrim, 1927

Generic type, *Sivapithecus himalayensis* Pilgrim

Sivapithecus himalayensis Pilgrim

Sivapithecus himalayensis, Pilgrim, 1927, Pal. Indica, N.S., XIV, pp. 2–5, figs. 2, 2a, 2b of
      plate.
   Type.—G.S.I. No. D 197, left mandibular ramus and symphysis.
   Paratypes.—None.
   Horizon.—Middle Siwaliks.
   Locality.—Hari Talyangar, northwestern Bilaspur State.
   Diagnosis.—Distinguished from the typical *Dryopithecus* by the shortened symphysis.
P_4 and molars much as in the above mentioned genus, having wrinkled enamel on the
coronal surface.

Sivapithecus orientalis Pilgrim

Sivapithecus orientalis, Pilgrim, 1927, Pal. Indica, N.S., XIV, pp. 5–8, figs. 1, 1a of plate.
   Type.—G.S.I. No. D 196, right half of a palate.
   Paratypes.—None.
   Horizon.—Middle Siwaliks, Nagri zone.
   Locality.—Near Hari Talyangar, northwestern Bilaspur State.
   Diagnosis.—Upper molars very much like those of *Dryopithecus punjabicus.*   Canine
very large and pointed, extending well below the occlusal line.   Palate high.

Sivapithecus middlemissi Pilgrim

Sivapithecus middlemissi, Pilgrim, 1927, Pal. Indica, N.S., XIV, pp. 8–9, figs. 7, 7a of plate.
   Type.—G.S.I. No. D 198, posterior part of a left mandibular ramus.
   Paratypes.—None.
   Horizon.—Lower Siwaliks, Chinji zone.
   Locality.—From Rammagar, Jammu, Kashmir State Territory.
   Diagnosis.—Similar to other species of *Sivapithecus.*   Distinguished from *Dryopithecus*
by the relatively shorter molars, denoting a more advanced evolutionary stage.

Sivapithecus indicus Pilgrim

Sivapithecus indicus, Pilgrim 1910, Rec. Geol. Surv. India, XL, p. 63.
    Additional References.—
        Pilgrim, G. E., 1910B, p. 198; 1915A, pp. 34–54, figs. 7–9, Pl. II, figs. 1–3.
        Gregory, W. K., and Hellman, Milo, 1926.   Various pages.
        Pilgrim, G. E., 1927B, p. 9, fig. 5 of plate.
        Matthew, W. D., 1929, p. 453.
    Type.—G.S.I. No. D 176, a right M_3.
    Paratypes.—None.   [G.S.I. No. D 177, stated by Pilgrim (1915) to be the type, is a referred specimen.]
    Horizon.—Lower Siwaliks, Chinji zone.
    Locality.—In the neighborhood of Chinji village, Punjab.
    Diagnosis.—Mandible heavy and symphysis characterized by its shortness.   Molars similar to those of *Dryopithecus*, but relatively short antero-posteriorly, which denotes an advanced condition of evolutionary development.   Premolars relatively wide, canine large.

Incertae Sedis

Adaetontherium Lewis, 1934

Generic type, *Adaetontherium incognitum* Lewis

Adaetontherium incognitum Lewis

Adaetontherium incognitum, Lewis, 1934, Amer. Jour. Sci., XXVII, pp. 174–175, Pl. II, fig. 3.
    Type.—Y.P.M. No. 13808, a crown of a single tooth.
    Horizon.—Lower Siwaliks, Chinji zone.
    Locality.—Two and one half miles west southwest of Chinji village, Punjab.
    Diagnosis.—Doubtfully a primate.   Crown of tooth characterized by high relief and a complex system of folds.   Perhaps an unerupted aberrant third molar.

RODENTIA

Rodents are very rare among the fossils discovered in the Siwalik beds.   This is probably due to the fact that they are small, and easily destroyed in rapidly accumulating sediments such as those which constitute the Siwalik Series.   There are at present but four genera and six species of rodents (exclusive of the Lagomorpha) known from the Siwaliks, and of these one new genus and two new species are represented in the American Museum collection.   These new forms have been described elsewhere (Colbert, 1933F) and reference to the original descriptions should be made for a detailed account of the specimens.

The recently returned Yale North India Expedition has brought back a most excellent collection of Siwalik rodents, several of which are undoubtedly new.   These will be described in the near future by Mr. G. E. Lewis, who made the collection.

MYOMORPHA
MYOIDEA
SPALACIDAE
Rhizomys Gray, 1831

Generic type, *Rhizomys sinensis* Gray

Rhizomys sivalensis Lydekker

Rhizomys sivalensis, Lydekker, 1878, Rec. Geol. Surv. India, XI, p. 101.

Additional References.—

Lydekker, R., 1879, pp. 41, 52, fig. 3 of plate opposite p. 50; 1880B, p. 33; 1883C, pp. 84, 92; 1884D, pp. 106–108, figs. 1–3, p. 126; 1885B, p. 13; 1885C, pp. 233–234, fig. 31.

Major, C. J. F., 1897B, p. 720, fig. 9B.

Pilgrim, G. E., 1910B, p. 199; 1913B, p. 283.

Matthew, W. D., 1929, pp. 448, 558.

Type.—(Lectotype).—G.S.I. No. D. 97, Left mandibular ramus.

Cotypes.—G.S.I. No. D 97a, D 97b, two detached rami.

Referred Specimen.—G.S.I. No. D 98, a left calcaneum.

Horizon.—Middle Siwaliks.

Locality.—Jabi, Punjab.

Diagnosis.—Jaw slender and teeth relatively small.   Teeth narrower than in the modern species.

Rhizomys punjabiensis Colbert

Rhizomys punjabiensis, Colbert, 1933, Amer. Mus. Novitates, No. 633, pp. 1–3, fig. 1.

Type.—Amer. Mus. No. 19762, a right mandibular ramus containing M_{2-3}.

Paratypes.—None.

Horizon.—Lower Siwaliks, near the base.

Locality.—Found near the Sutlej River, 23 miles north and west of Bilaspur, Bilaspur State, Punjab.

Diagnosis.—A very small species of *Rhizomys*, about one half as large as *Rhizomys sivalensis*, and about equal in size to *Rhizomys sinensis*.   Molar teeth characterized by an external and two or three internal folds.

A full description is given in the type reference, cited above.

This is a very small species, showing the typical *Rhizomys* molar pattern.   In its structure and affinities, *Rhizomys punjabiensis* is quite close to the larger *Rhizomys sivalensis* of the Middle Siwalik beds.   The molar teeth of the Lower Siwalik form are moderately hypsodont.

The type figure and the measurements are reproduced below.

MEASUREMENTS

Rhizomys punjabiensis Colbert

Amer. Mus. No. 19762.

Length of RM_2	2.7 mm.
Width of RM_2	2.4
Length of RM_3	3.1
Width of RM_3	2.6
Length of M_2–M_3	5.8
Depth of ramus below M_3	6.0

A.M.19762

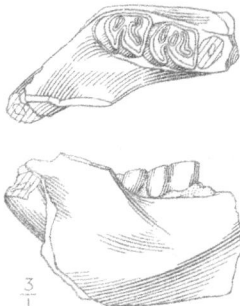

FIG. 31. *Rhizomys punjabiensis* Colbert. Type, Amer. Mus. No. 19762, right mandibular ramus with second and third molars. Crown view above, lateral view below. Three times natural size. From Colbert, 1933.

Rhizomys sp.

Specimens under Consideration.—B.M. No. 15925, a right mandibular ramus. B.M. No. 15926, a right mandibular ramus. B.M. No. 15927, a left mandibular ramus. B.M. No. 15927a, two molar teeth.

References.—

Lydekker, R., 1884D, p. 107; 1885C, pp. 233–234.

Pilgrim, G. E., 1910B, p. 199.

Matthew, W. D., 1929, pp. 444, 558.

The above listed specimens from the Siwalik Hills presumably came from Upper Siwalik strata. Probably, since they are considerably larger and later in age than the type of *Rhizomys sivalensis,* they represent a distinct species. Reference should be made to Dr. Matthew's paper of 1929, in which he indicates a possible comparison of this Upper Siwalik form to *Rhizomys troglodytes,* from the Pleistocene of China.

MURIDAE

Nesokia Gray, 1842

Generic type, *Mus hardwicki* Gray

Nesokia cf. hardwicki (Gray)

Specimen under Consideration.—B.M. No. 16529a, fragment of a left ramus of a mandible, with the incisor and the molars present. From the Upper Siwaliks.

References.—

Falconer, H., and Cautley, P.T., 1835, p. 706.

Lydekker, R., 1884D, p. 126; 1885C, p. 226; 1886F, p. xi.

Pilgrim, G. E., 1910B, p. 199.

Matthew, W. D., 1929, pp. 444, 558.

HYSTRICOMORPHA

HYSTRICOIDEA

HYSTRICIDAE

Hystrix Linnaeus, 1758

Generic type, *Hystrix cristata* Linnaeus

Hystrix sivalensis Lydekker

Hystrix sivalensis, Lydekker, 1878, Rec. Geol. Surv. India, XI, p. 98.

Additional References.—

Lydekker, R., 1880B, p. 33; 1883C, pp. 84, 91; 1884D, pp. 109–111, figs. 4–5, p. 126; 1885B, p. 13; 1885C, p. 248, fig. 32.

Pilgrim, G. E., 1910B, p. 200; 1913B, p. 283.

Matthew, W. D., 1929, pp. 444, 559.

Type.—G.S.I. No. D 96, a right mandibular ramus.

Horizon.—Middle Siwaliks.

Locality.—From near Hasnot, Salt Range, Jhelum District, Punjab.

Specimens in the American Museum.—A.M. No. 19909, a left M_1. From the Middle Siwaliks, $2\frac{1}{2}$ miles north of Hasnot.

Diagnosis.—Structure of cheek teeth complex, as in *Hystrix cristata*. Distinguished by its large size and by the separate roots on P_4.

Dr. Matthew gives a competent discussion of this species in his 1929 paper. He shows that it is closely comparable to *Hystrix primigenia* from Pikermi, but is less brachyodont than this latter species.

In the specimen at hand (left M_1) the tooth is distinguished by its separate roots.

A.M.19909

FIG. 32. *Hystrix sivalensis* Lydekker. Amer. Mus. No. 19909, left M_1. Crown view above, external lateral view below. Natural size.

MEASUREMENTS

Hystrix sivalensis, Amer. Mus. No. 19909

M_1	Length, 9.0 mm.	Width, 7.5 mm.	Height, 24.0 mm.

Hystrix cf. leucurus Sykes

Specimen under Consideration.—B.M. No. 15923, the skull and jaw of a young individual. From the Upper Siwaliks.

References.—

> Lydekker, R., 1884D, pp. 110–111, fig. 5. (Referred to *H. sivalensis.*) 1885C, pp. 248–249, fig. 33. (Referred to *H. sivalensis.*)
>
> Pilgrim, G. E., 1910B, p. 200. (*Hystrix* sp.)
>
> Matthew, W. D., pp. 444, 559–560, fig. 55.

This specimen was referred by Lydekker to *Hystrix sivalensis.* Dr. Matthew, after carefully considering the specimen, states that it probably belongs to a distinct species, and he refers it to *H. leucurus.*

The specimen was seemingly found in Upper Siwalik beds, while the type of *H. sivalensis* is from the Middle Siwaliks. Therefore the Upper Siwalik form probably represents a distinct species, as Dr. Matthew suggested.

Sivacanthion Colbert 1933

Generic type, *Sivacanthion complicatus* Colbert

Sivacanthion complicatus Colbert

Sivacanthion complicatus, Colbert, 1933, Amer. Mus. Novitates, No. 633, pp. 3–5, fig. 2.

Type.—A.M. No. 19626, a right and a left mandibular ramus each containing P_4, M_{1-2}.

Paratypes.—None.

Horizon.—Lower Siwaliks, at the level of Chinji Rest House (near the base of the Chinji zone).

Locality.—Four miles northeast of Chinji Rest House, Punjab.

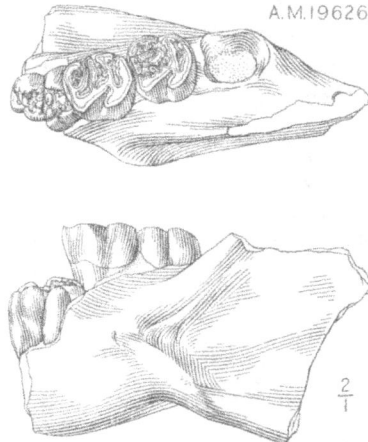

Fig. 33. *Sivacanthion complicatus* Colbert. Type, Amer. Mus. No. 19626, left mandibular ramus with fourth premolar and first two molars. Crown view above, lateral view below. Twice natural size. From Colbert, 1933.

Diagnosis.—An hystricomorph of medium size, considerably smaller than the modern species of *Hystrix* or *Acanthion.* Dental formula 1-0-1-3. Angle of mandibular ramus

very strong, as in other Hystricidae.   Hystricomorph pattern of the molar enamel complicated by secondary folding.

A distinct genus of the Hystricidae, typified by its brachyodonty and by the secondary
complication of the molar crowns.   This is a rather aberrant side branch in the evolution
of the hystricomorph rodents, most closely related to *Acanthion*.   A full account will
be found in the type reference, cited above.   The type figure and the measurements are
reproduced below.

MEASUREMENTS

Sivacanthion complicatus Colbert

Amer. Mus. No. 19626.

P_4	Length		6.1 mm.
	Width		5.7
	Height		6.3
M_1	Length		5.3
	Width		5.1
M_2	Length		5.2
	Width		5.4
Length of molar series (M_3 from alveolus)			16.8
Depth of ramus below M_1			13.8

LAGOMORPHA

LEPORIDAE

Caprolagus Blyth, 1845

Generic type, *Lepus hispidus* Pearson

Caprolagus sivalensis Major

Caprolagus sivalensis, Major, 1899, Trans. Linn. Soc., London, VII, Pl. 37, fig. 18.
   Additional References.—
      Pilgrim, G. E., 1910B, p. 200.
      Matthew, W. D., 1929, pp. 444, 560.
   Type.—B.M. No. 16529, a fragmentary mandible.
   Horizon and Locality.—Probably from the Upper Siwalik beds.
   "The only lagomorph remains consist of a fragment of jaw, probably from Upper Siwalik
beds, attributed by Major to *Caprolagus*."—Matthew, W. D., 1929, p. 560.

CARNIVORA

The carnivores are among the most interesting of the Siwalik mammals, for although
their remains are often of a fragmentary nature, they represent a great variety of genera and
species, which clearly demonstrate the development of heritage and habitus characters
through a period of geologic time extending from the upper Miocene into the early portion of
the Pleistocene.

Lydekker, in 1884, published a large monograph of the Siwalik carnivores known at that time. Since then, several authors have worked on this group, notably Bose, Pilgrim and Matthew. Matthew's paper of 1929 presents a complete survey of the Siwalik carnivores then known; consequently it forms the foundation of this study. Pilgrim's most recent monograph of 1932 is an exhaustive treatment of the subject.

In view of the recent work of Matthew and Pilgrim on the Siwalik carnivores, it will not be necessary to consider in detail all of the known species. The species represented by material in the American Museum collection will receive particular attention in the following pages. In discussing their relationships the views of Matthew will be followed, with particular reference to Pilgrim's work.

A type revision of the Siwalik carnivores will be presented below, for the purpose of clearing, wherever needed, questions of nomenclature and priority. The diagnoses of the species, accompanying this revision, have been taken largely from Pilgrim's monograph of 1932.

CREODONTA

OXYAENOIDEA [PSEUDOCREODI]

HYAENODONTIDAE

Dissopsalis Pilgrim, 1910

Generic type, *Dissopsalis carnifex* Pilgrim

Dissopsalis carnifex Pilgrim

Dissopsalis carnifex, Pilgrim, 1910, Rec. Geol. Surv. India, XL, p. 64.

Additional References.—

Pilgrim, G. E., 1910B, p. 198; 1913B, p. 349; 1914B, pp. 265–279, Pl. XXIX, figs. 1–4, 6.

Matthew, W. D., 1929, pp. 453, 459.

Pilgrim, G. E., 1932, pp. 11–12.

Type.—(Lectotype).—G.S.I. No. D 143, a right maxilla, with P^3–M^2.

Cotypes.—G.S.I. No. D 144–D 148, various maxillary and mandibular fragments.

Horizon.—Lower Siwaliks, Chinji zone.

Locality.—From the vicinity of Chinji Rest House, Salt Range, Attock District, Punjab.

Specimens in the American Museum.—Amer. Mus. No. 19401, a skull, the palate and cheek dentition complete; rostrum anterior to P^1 lacking; occipital portion shattered. Lower Siwaliks, 1600 feet above the level of Chinji Rest House, one mile north of Chinji Rest House.

Amer. Mus. No. 19402, portion of a right maxilla, with P^3–M^3; also left P^{2-3}. Lower Siwaliks, near base, east of Chinji Rest House.

Amer. Mus. No. 19403, mandibular fragment with left M_{2-3}. Lower Siwaliks, near base of Chinji zone, four miles west of Chinji Rest House.

Amer. Mus. No. 19339, left M^1. Lower Siwaliks, 600 feet above level of Chinji Rest House, one mile north of Chinji Rest House.

Amer. Mus. No. 19348, right P⁴.   Lower Siwaliks, about 1600 feet above the level of Chinji Rest House, one mile north of Chinji Rest House.

Amer. Mus. No. 19349, left P⁴, right M².   Lower Siwaliks, 1600 feet above the level of Chinji Rest House, one mile north of Chinji Rest House.

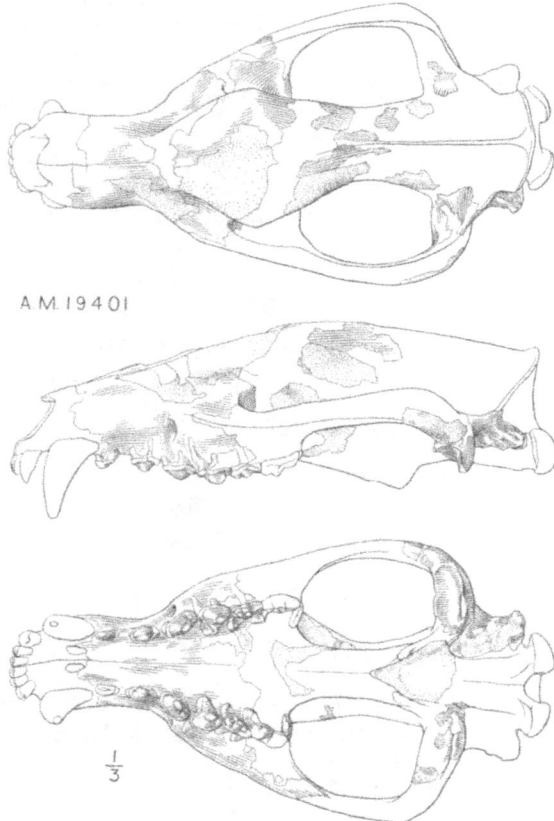

A.M. 19401

FIG. 34.   *Dissopsalis carnifex* Pilgrim.   Amer. Mus. No. 19401, skull.   Dorsal view above, lateral view in the middle, ventral view below.   One third natural size.   From Colbert, 1933.

The dotted areas represent matrix, which indicates the shape of the bone formerly covering these surfaces.

Diagnosis.—

1. A fairly large hyaenodont, comparable in size to *Hyaenodon cruentus*.

2. Dentition, I ?, C, P 4/4, M 3/3; carnassial shear on M², M₃.

3. Premolars robust, with well developed cingula.   P⁴ with a very large internal protocone.

4. Molars trenchant.   M¹ and M² with large protocone, appressed paracone and metacone, and with a metastyle shear.   M³ very small.   Lower first and second molars

with well developed trigonid and a basined talonid. In M$_3$ the trigonid has become trenchant, while the talonid is reduced to a small tubercle.

5. Skull heavy, palate wide. Frontals expanded above orbits, sagittal crest high.

6. Palate wide, pterygoids produced posteriorly, almost reaching the glenoid articulation. Bullae presumably cartilaginous.

The first mention of the Creodonta as occurring in Siwalik beds was in 1884, when Lydekker described *Hyaenodon indicus* from a supposed fourth lower premolar. In 1910, and later, in 1914, Pilgrim [16] expressed the opinion that the tooth described by Lydekker was really that of a suid, *Hyotherium*.

Consequently the first undoubted record of a creodont in the Siwalik Series is to be found in Pilgrim's note of 1910, in which he established the genus *Dissopsalis* containing two species, namely *D. carnifex* and *D. ruber*.

Pilgrim's knowledge of the genus *Dissopsalis* was based on rather fragmentary material, consisting for the most part of scattered teeth. Fortunately the American Museum collection contains a broken, but nevertheless a fairly complete skull, which has recently been described in detail and figured by the present author. (Colbert, E. H., 1933A.) The reader is referred to the above cited publication for a complete account of the skull of *Dissopsalis*. The measurements of the skull, and the figures appearing in the above mentioned paper, are reproduced below.

As was pointed out in the detailed description of the *Dissopsalis* skull in the American Museum collection, this genus represents the last known survivor of creodont stock. Although *Dissopsalis* is a late survivor in geologic time of the creodont line, it nevertheless retains many primitive characters typical of the early hyaenodonts; it does not show the advances in structure that characterize the later Oliogocene forms. It has already been shown [17] that *Dissopsalis* is derivable from *Sinopa* and that it is closely related to *Cynohyaenodon* and *Quercytherium*.

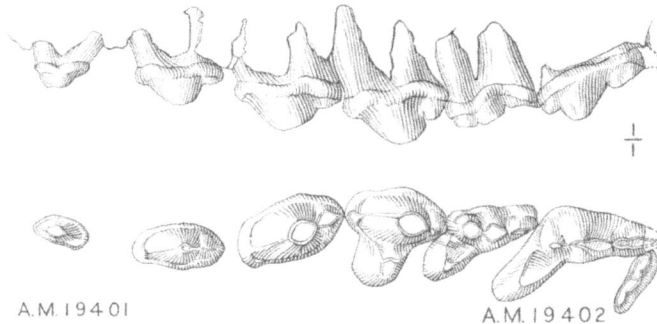

A.M. 19401   A.M. 19402

FIG. 35. *Dissopsalis carnifex* Pilgrim. Amer. Mus. No. 19401, right upper premolar and molar teeth. M^2 and M^3 drawn from Amer. Mus. No. 19402. Lateral view above, crown view below. Natural size. From Colbert, 1933.

The close resemblances between *Dissopsalis* and *Sinopa* are expressed below in a tabular form.

[16] Pilgrim, G. E., 1910A, p. 190; 1914B, pp. 265–266.
[17] Colbert, E. H., 1933A, pp. 7–8.

	Sinopa	*Dissopsalis*
Dental formula	I C P4/4 M3/3	I? C P4/4 M3/3

Premolars....First three upper premolars longer than wide, with central cone and encircling cingulum. — As in *Sinopa*.

P⁴ with enlarged protocone. — As in *Sinopa*.
P₄ with enlarged, backwardly pointing cusp and a trenchant heel. — As in *Sinopa*.

Molars.......M¹ and M² with large protocones and distinct para- and metacones. — As in *Sinopa*.
Metastyle shear short. — Metastyle shear elongated
Carnassial shear of M² at a high angle to midline of palate. — Carnassial shear of M² at a very low angle to midline of palate.
M³ large. — M³ greatly reduced.
M₁ and M₂ with high trigonid and basined talonid. — As in *Sinopa*.
M₃ with high trigonid and basined talonid. — M₃ with trigonid greatly enlarged, and talonid reduced to a small cusp.

Thus it may be seen that the salient primitive hyaenodont features have been passed from *Sinopa* to *Dissopsalis* as heritage characters, and in the latter genus have been but slightly modified during the period of time between the Eocene and the Pliocene.   And in this rather long stretch of geologic time, *Dissopsalis* has added very few habitus characters.

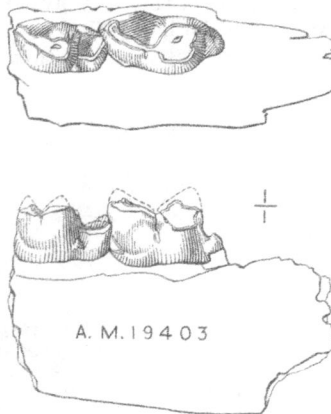

FIG. 36.   *Dissopsalis carnifex* Pilgrim.   Amer. Mus. No. 19403, left lower second and third molars.   Crown view above, lateral view below.   Natural size.   From Colbert, 1933.

Cynohyaenodon of the Phosphorites of Quercy, is closely comparable to *Dissopsalis*. It is a very small creodont, less than half the size of the Indian genus, but in the configuration of the skull and the development of the dentition the two forms are strikingly similar. In both genera the muzzle is constricted at the second premolar, the palate is wide between the molars, the frontals are wide and flat, and the infraorbital foramen is above the third premolar. Of course some of these are characters common to all of the hyaenodonts, but when the dentitions of the two are compared one to the other, they are seen to be very similar. In fact, many of the diagnostic characters given for the teeth of *Dissopsalis* apply equally well to those of *Cynohyaenodon*. The main features whereby the teeth of *Cynohyaenodon* differ from those of *Dissopsalis* are as follows:

1. The anterior premolars are not quite so wide in comparison to their length.
2. The heel of P^4 is relatively smaller.
3. The shearing blades of M^1 and M^2 are more transverse.
4. The paracone and metacone of M^1 and M^2 are more closely appressed.
5. M^3 is relatively larger.
6. P_4 is relatively smaller, with a less developed heel.
7. The talonids of M_1 and M_2 are relatively larger.
8. The talonid of M_3 retains some of its basined character.

These are differences that might be expected in forms as widely separated as the upper Eocene and the lower Pliocene. In all of the points outlined above with the exception of the fourth, *Cynohyaenodon* is more primitive than *Dissopsalis*. The close appression of the paracone and metacone in *Cynohyaenodon* is especially significant. Matthew [18] has shown that evolution in the hyaenodonts is marked by the ultimate fusion of the paracone and metacone into a single cusp, a condition well illustrated by *Hyaenodon* and *Pterodon*. These two cusps are relatively further apart in *Dissopsalis* than in any other of the hyaenodonts except *Sinopa*, a fact that points to the distinctness of the Siwalik genus from other genera in the family. Thus it seems necessary to realize that *Dissopsalis*, while seemingly derivable in most of its characters from *Cynohyaenodon*, must actually, because of its more primitive paracone and metacone, be recognized as an independent branch from the common ancestor of the European and Indian genera.

TABLE, TO SHOW RATIOS BETWEEN THE LENGTH OF THE PA—ME AND THE TOTAL LENGTH OF M^1.

(Tooth measured along the inner shearing blade, from protocone to metastyle.)

$$\text{Ratio } \frac{\text{Length of Pa}-\text{Me} \times 100}{\text{Length of } M^1}$$

Sinopa	50	*Tritemnodon*	39
Dissopsalis	44	*Cynohyaenodon*	33

Quercytherium represents a separate development from *Cynohyaenodon*, in which the premolars have become enlarged.

Considering now *Metapterodon*, the Miocene (?) creodont from Africa, it is at once seen to be quite distinct from *Dissopsalis*, and rather a form related to the *Pterodon* line. *Metapterodon* does offer an analogous comparison to *Dissopsalis* for like the latter, it is a persistent, primitive type, characterized by a well developed protocone and a slight metastyle shear.

[18] Matthew, W. D., 1909C, p. 464, fig. 72.

Evidently, in the course of hyaenodont evolution certain primitive forms persisted on into the later Tertiary, while the culmination of the development in this family was attained at the close of the Oligocene.

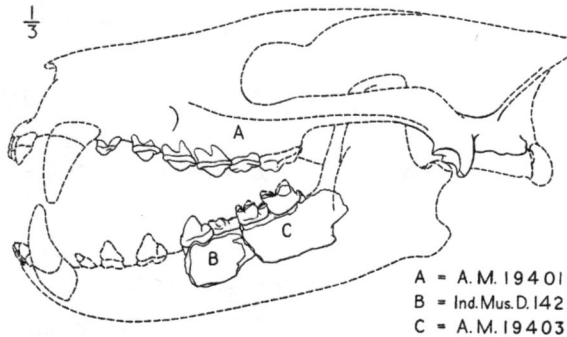

A = A.M. 19401
B = Ind. Mus. D. 142
C = A.M. 19403

FIG. 37.   *Dissopsalis carnifex* Pilgrim.   Restoration of skull and mandible, drawn from Amer. Mus. Nos. 19401, 19403, and Geol. Surv. India No. D 142.   Lateral view, one third natural size.   From Colbert, 1933.

The actual bone and tooth surfaces in place are represented by solid lines.   The restored portions, drawn either from crushed bone or matrix, or from hypothetical inferences gained by a study of related genera, are indicated by dotted lines.   This figure should be compared with figure 33, to see which portions of the skull were restored from bone and matrix, and which from hypothetical considerations.

TABLE OF MEASUREMENTS

A.M. 19401

P^1.....Antero-posterior diameter......................	9.0 mm.
Transverse diameter..........................	4.5
P^2.....Antero-posterior diameter......................	14.5
Transverse diameter..........................	8.0
P^3.....Antero-posterior diameter......................	18.5
Transverse diameter..........................	10.5
P^4.....Antero-posterior diameter......................	17.0
Transverse diameter..........................	17.5
M^1....Antero-posterior diameter......................	21.0
Transverse diameter..........................	14.0

A.M. 19402

M^2....Antero-posterior diameter......................	23.0
Transverse diameter..........................	13e
M^3....Antero-posterior diameter......................	4.0
Transverse diameter..........................	10.0

A.M. 19403

M_2....Antero-posterior diameter......................	20e
Transverse diameter..........................	8.5
M_3....Antero-posterior diameter......................	20.5
Transverse diameter..........................	9.5

e, Estimated measurement.

Pliocene

Dissopsalis

Miocene

Metapterodon

Hyaenodon

Oligocene

Quercytherium

Pterodon

Metasinopa

Galethylax

Apterodon

Cynohyaenodon

Tritemnodon

Eocene

Sinopa

Prorhyzaena

Proviverra

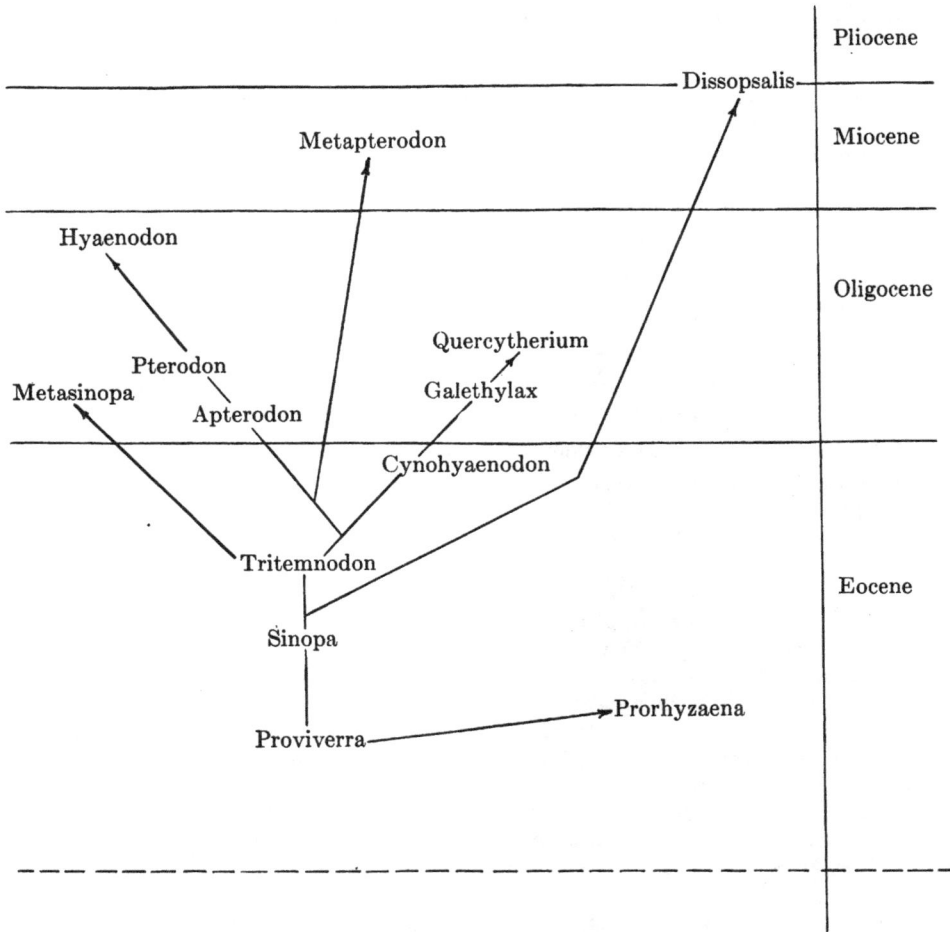

CHART, SHOWING THE EVOLUTION AND PHYLOGENY OF THE HYAENODONTIDAE. Modified after Matthew and
Pilgrim.

Dissopsalis ruber Pilgrim

Dissopsalis ruber, Pilgrim, 1910, Rec. Geol. Surv. India, XL, p. 54.

    Additional References.—

        Pilgrim, G. E., 1910B, p. 198; 1914B, pp. 268–275, Pl. XXIX, figs. 5, 7.

        Matthew, W. D., 1929, p. 453.

        Pilgrim, G. E., 1932A, pp. 12–13.

Type.—(Lectotype).—G.S.I. No. D 147, fragment of a maxilla with left M^{2-3}.

Cotypes.—G.S.I. No. D 148, left P^{2-4}; G.S.I. No. D 174, a fourth upper premolar.

Horizon.—Lower Siwaliks, Chinji zone.

Locality.—The precise locality of the type is unknown, but it came from the vicinity of Chinji in the Salt Range, Punjab.

Diagnosis.—"This species occurs in the same beds as the preceding one [*D. carnifex*] and is inferior to it in size, but differs in no very essential points of structure."—Pilgrim, G. E., 1910A, p. 54.

This species is regarded as a synonym of *Dissopsalis carnifex.* (See Matthew, W. D., 1929, p. 453.)

FISSIPEDIA

CANOIDEA [ARCTOIDEA]

CANIDAE

AMPHICYONINAE

Amphicyon Lartet, 1836

Generic type, *Amphicyon major* Blainville

Amphicyon palaeindicus Lydekker

Amphicyon palaeindicus, Lydekker, 1876, Pal. Indica (X), I, p. 84, Pl. VII, fig. 5.

Additional References.—

  Bose, P. N., 1880, p. 120.

  Lydekker, R., 1880B, p. 30; 1883C, pp. 82, 89; 1884A, p. 248, Pl. XXXII, fig. 8;
    1884D, p. 125; 1885B, pp. 8–9; 1885C, p. 138.

  Pilgrim, G. E., 1910B, p. 199.

  Matthew, W. D., 1929, pp. 453, 482–3.

  Pilgrim, G. E., 1932, pp. 15–16, Pl. I, figs. 9, 12.

Type.—G.S.I. No. D 26, a right M^2.

Paratype.—G.S.I. No. D 23, a lower jaw fragment. (From Nurpur).

Horizon.—Upper portion of the Lower Siwaliks, Chinji zone and lower portion of the Middle Siwaliks, Nagri zone.

Locality.—Kushalgarh, for type. Chinji and vicinity, Punjab for referred specimens.

Specimens in the American Museum.—Amer. Mus. No. 19404, a right M_1 from the Lower Siwaliks, about 100 feet above the level of Chinji Rest House, and four miles west of that place.

Amer. Mus. No. 19352, a right M_2 from the lower portion of the Middle Siwaliks, 1000 feet below the bone bed at Bhandar. One mile south of Nathot.

Amer. Mus. No. 19406, a left P^4 from the lower Siwaliks, Chinji zone, near the level of Chinji Rest House. At Chinji Rest House.

Diagnosis.—(See Pilgrim, G. E., 1932, p. 15.) A large species of the genus, with a reduced protocone and metacone. M_1 compressed, with paraconid, protoconid and hypoconid in a straight line. Hypoconid trenchant, entoconid low, narrow and not bifid. M_2 narrow; M_3 narrow, elongate with low cusps.

A complete discussion of the Amphicyonines is to be found in Matthew's "Third Contribution to the Snake Creek Fauna," [19] to which the reader is referred.

---

[19] Matthew, W. D., 1924A, pp. 104–128.

The lower carnassial, Amer. Mus. No. 19404, is a typical tooth of large size.

Amer. Mus. No. 19352 is probably more nearly like Pilgrim's *Amphicyon pithecophilus*, if this species is really distinct from the species now under consideration. The trigonid is high with an anterior rim circling the protoconid to the metaconid. A ridge descends transversely across from these cusps. The hypoconid is well developed and a rim circles inwardly from it enclosing a talonid basin. The back of the talonid is pointed, more so than in certain typical second molars of Amphicyonines. This feature is, however, one of individual variation, as may be seen by the examination of a series of teeth.

In this connection a series of undescribed teeth from the Tung Gur formation of Mongolia were studied.

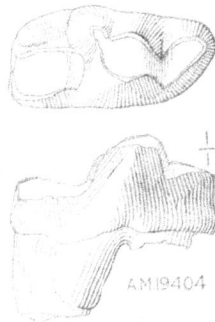

FIG. 38. *Amphicyon palaeindicus* Lydekker. Amer. Mus. No. 19404, right M₁. Crown view above, lateral view below. Natural size.

The tooth described above is of special interest in that it comes from the lower portion of the Middle Siwaliks, namely from the Nagri zone. It illustrates the persistence of a species from Chinji through Nagri times.

MEASUREMENTS

Amer. Mus. No. 19404 M₁
Length 34.0 mm.   Width 17.0 mm.
Amer. Mus. No. 19352 M₂
Length 21.0 mm.   Width 14.0 mm.

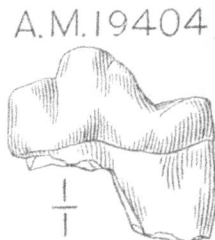

FIG. 39. *Amphicyon palaeindicus* Lydekker. Amer. Mus. No. 19404, right M₁. Internal view. Natural size.

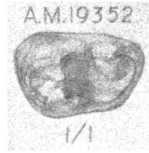

FIG. 40. *Amphicyon palaeindicus* Lydekker. Amer. Mus. No. 19352, right M₂. Crown view. Natural size.

The upper carnassial, Amer. Mus. No. 19406, is here referred to *A. palaeindicus* mainly on the basis of its size. This tooth was heretofore unknown in the Siwalik *Amphicyon*, and consequently there is no way of comparing it topographically with the described species. It is distinguished by the well developed internal cingulum and the small anterointernal tubercle. The measurements are given below. A figure illustrates the specimen.

MEASUREMENTS

Amer. Mus. No. 19406 P⁴
Length 27.0 mm.   Width 14.5 mm.   Greatest height of crown 18.0 mm.

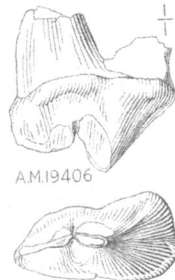

FIG. 41. *Amphicyon palaeindicus* Lydekker. Amer. Mus. No. 19406, left P⁴. Lateral view above, crown view below.
Natural size.

Amphicyon sindiensis Pilgrim

Amphicyon sindiensis, Pilgrim, 1932, Pal. Indica, N.S., XVIII, pp. 23–24.

Type.—G.S.I. No. D 25, a right mandibular ramus with M₂.

Paratypes.—None.

Horizon.—Basal beds of the Lower Manchars, equivalent to the Kamlial Zone of the Lower Siwaliks.

Locality.—Sind.

Diagnosis.—(After Pilgrim, G. E., 1932, p. 23.) "An *Amphicyon* with M₂ considerably smaller than in *A. pithecophilus*, elongated but narrower behind than in front; M₃ double-rooted, elongated."

This species is probably synonymous with *Amphicyon palaeindicus*, to which species the type specimen was originally referred by Lydekker.

Amphicyon pithecophilus Pilgrim

Amphicyon pithecophilus, Pilgrim, 1932, Pal. Indica, N.S., XVIII, pp. 16–20, Pl. II, figs. 1–4.

Several of the specimens referred by Pilgrim to this species were treated by former authors under the designation of *A. palaeindicus*.

See: Lydekker, R., 1876, p. 84, Pl. VII, figs. 8, 12; 1884A, p. 250, Pl. XXXII, fig. 5.

Matthew, W. D., 1929, pp. 482–483, figs. 19–20.

Type.—G.S.I. No. D 129, an M^2.

Paratypes.—G.S.I. No. D 155, an M^1, previously referred to *A. palaeindicus*. G.S.I. No. D 23, a right mandibular ramus from Nurpur. Designated by Matthew as the paratype of *A. palaeindicus*.

Horizon.—Lower Siwaliks, Chinji beds.

Locality.—From the region around Chinji, Salt Range, Attock District, Punjab.

Diagnosis.—(See Pilgrim, G. E., 1932, p. 17.) Comparable in size to *A. palaeindicus*. Upper molars with crenulated wings to protocone; displaying certain detailed differences from *A. giganteus*, *A. major* and *A. shahbazi*. M_1 with metaconid much lower than protoconid and placed behind it; M_3 elongate, with low cusps and talonid almost as long as trigonid.

This species is here considered as synonymous with *Amphicyon palaeindicus*.

Arctamphicyon Pilgrim, 1932

Generic type, *Arctamphicyon lydekkeri* Pilgrim

Arctamphicyon lydekkeri (Pilgrim)

Amphicyon lydekkeri, Pilgrim, 1910, Rec. Geol. Surv. India, XL, p. 64.

Arctamphicyon lydekkeri, Pilgrim, 1932, Pal. Indica., N.S., XVIII, pp. 24–26, Pl. II, figs. 7, 8.

Additional References.—

Pilgrim, G. E., 1910B, p. 199; 1913B, pp. 281, 292.

Matthew, W. D., 1929, pp. 481–2, fig. 18.

Type.—G.S.I. No. D 133, an M^2.

Paratypes.—None.

Horizon.—Middle Siwaliks, Dhok Pathan zone.

Locality.—Near the village of Padri, Salt Range, Punjab.

Diagnosis.—(See Pilgrim, G. E., 1932, p. 24.) Upper molars transversely extended. Crown and cusps lower than in the typical *Amphicyon*. Paracone stronger than metacone, inner cingulum crenulated and broad, external cingulum strong. M^2 with reduced metacone and reduced internal cingulum.

"This species is founded on M_1 from the Middle Siwaliks of Padhri, which differs from M_1 of *A. palaeindicus* Lyd. from Kushalgarh by its greater size and squareness."—Pilgrim, G. E., 1910A, p. 64.

(The specimen is actually an M^2. See Matthew, W. D., 1929, pp. 481–2.)

A very large member of the Amphicyoninae. Dr. Matthew indicated in 1929 that this species was probably generically distinct from the typical *Amphicyon*. Pilgrim's reference of *A. lydekkeri* to a separate genus would seem well founded.

Vishnucyon Pilgrim, 1932

Generic type, *Vishnucyon chinjiensis* Pilgrim

Vishnucyon chinjiensis Pilgrim

Vishnucyon chinjiensis, Pilgrim, 1932, Pal. Indica, N.S., XVIII, pp. 26–29, Pls. I, fig. 4, II, fig. 9.

Type.—G.S.I. No. D 123, a left maxilla with M^{1-2}.

Paratype.—G.S.I. No. D 127, a left maxilla with P^4.

Horizon.—Lower Siwaliks, Chinji zone.

Locality.—Near Chinji, Salt Range, Punjab.

Specimens in the American Museum.—Amer. Mus. No. 19343, a fragmentary upper first molar. From the Lower Siwaliks, about the level of Chinji Rest House. Four miles west of Chinji Rest House.

Amer. Mus. No. 19346, a right P^4. Lower Siwaliks, about level of Chinji Rest House. Four miles northeast of Chinji Rest House.

Amer. Mus. No. 19633, a left P^4. Lower Siwaliks, about the level of Chinji Rest House, near Rattu Kund, six miles west of Chinji Rest House.

Diagnosis.—(See Pilgrim, G. E., 1932, p. 26.) A small genus of the Amphicyoninae. Transverse and antero-posterior diameters of M^1 approximately equal. Internal cingulum of M^1 prominent. Protocone and metacone low, paracone high, hypocone absent. M^2 subrectangular and broad. Internal cingulum strong, but external cingulum absent. Outer cusps reduced in height. P^4 with a very weak protocone, and with a high paracone and metacone.

Several unassociated teeth in the American Museum collection would seem to belong to this species, rather than to the larger forms of Amphicyonines. The molar fragment is typically Amphicyon-like, but it is distinguished by its high paracone, a feature peculiar to this genus. The P^4 is characterized by the well developed shearing blade, the reduced anterior cusp and the cingulum.

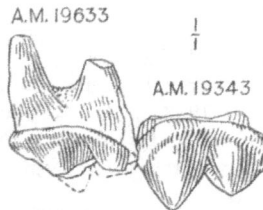

Fig. 42. *Vishnucyon chinjiensis* Pilgrim. Amer. Mus. Nos. 19633, 19343, left P^4, M^1, respectively. Lateral view. Natural size.

MEASUREMENTS

Amer. Mus. No. 19343 M^1
    Antero-posterior diameter 21.0 mm.   Height of paracone 16.0 mm.
Amer. Mus. No. 19633, P^4
    Antero-posterior diameter 21.5 mm.   Transverse diameter 11.5 mm.

Pilgrim in his recent memoir on the Siwalik Carnivora, demonstrates the aberrant character of the teeth assigned to this genus. He shows that the teeth have certain features that might indicate affinities to *Lycaon*, but on the totality of characters he would consider the form under discussion to be more closely related to *Temnocyon*.

Comparisons with *Temnocyon* show many similarities, such as the reduced "protocone" of the fourth upper premolar. The American Museum specimens of *Vishnucyon* are, however, of such fragmentary nature as to be of little use in solving the relationships of the genus.

CANINAE

Canis Linnaeus, 1758

Generic type, *Canis familiaris* Linnaeus

Canis cautleyi Bose

Canis cautleyi, Bose, 1880, Quar. Jour. Geol. Soc., London, XXXVI, p. 135, Pl. VI, figs. 7–9.
   Additional References.—
      Referred by Falconer to *Enhydriodon sivalensis*. See Falconer, Hugh, 1868A, p. 337.
      Lydekker, R., 1880B, p. 30; 1883C, pp. 83, 90; 1884A, p. 259, fig. 10, Pl. XXXII, figs. 3–6; 1884E, pp. 71–72, fig. 2; 1884D, p. 124; 1885B, p. 8; 1885C, pp. 128–129, fig. 18.
      Pilgrim, G. E., 1910B, p. 199; 1913B, p. 324.
      Matthew, W. D., 1929, pp. 443, 486–487.
      Pilgrim, G. E., 1932, pp. 31–32.
   Type.—(Lectotype).—B.M. No. 40181, a left mandibular ramus with M_{1-2}.
   *Cotype.—*B.M. No. 40182, a fragmentary left mandibular ramus with M_{1-3}.
   *Horizon.—*Probably Upper Siwaliks.
   *Locality.—*From the Siwalik Hills.
   Diagnosis.—(See Pilgrim, G. E., 1932, p. 31.) About the size of *Canis pallipes* of India. M^1 with large transverse diameter, M^2 with lesser transverse diameter. Antero-posterior diameter of P^4 equal to that of the first two molars combined. M_1 with antero-posterior diameter relatively less than in modern species of *Canis*. M_2 without paraconid or entoconid. M_3 small.

Sivacyon Pilgrim, 1932

Generic type, *Canis curvipalatus* Bose

Sivacyon curvipalatus (Bose)

Canis curvipalatus, Bose, 1880, Quar. Jour. Geol. Soc., London, XXXVI, pp. 134–136.
Vulpes curvipalata, Pilgrim, 1910B, Rec. Geol. Surv. India, XL, p. 199.
Nothocyon curvipalatus, Pohle, 1928, "Die Raubtiere von Oldoway," Wiss. Ergeb. d. Oldoway-Exped., p. 50.
Sivacyon curvipalatus, Pilgrim, 1932, Pal. Indica, N.S., XVIII, pp. 33–36, Pl. I, fig. 6.
   Additional References.—
      Lydekker, R., 1880B, p. 30; 1883C, pp. 82, 90; 1884A, p. 253, Pl. XXXII, figs. 1, 7; 1884D, p. 125; 1885C, p. 135.

Pilgrim, G. E., 1913B, p. 324.

Matthew, W. D., 1929, pp. 485–486, figs. 22, 23.

Originally referred by Baker and Durand to *Canis vulpes*. See Baker and Durand, 1836, p. 581, figs. 9, 10. Republished, Falconer, Hugh, 1868A, p. 341.

Type.—B.M. No. 37149, an associated skull and mandible.

Paratype.—None.

Horizon.—Upper Siwaliks.

Locality.—From the Siwalik Hills, between the Markanda Pass and Pinjor.

Diagnosis.—(See Pilgrim, G. E., 1932, pp. 32–33.) Skull low and lightly built, with a short muzzle. Interorbital region flat and occiput broad. Orbits small. Postorbital processes thin. Palate extending behind molars. Upper molars broad, M^2 only slightly reduced as compared to M^1. M^3 absent. P^4 large, but anterior premolars small. Canine slender. M_1 with protoconid and metaconid of about the same height, and with long talonid. M_2 without paraconid, M_3 small. Lower premolars elongate.

PROCYONIDAE

Sivanasua Pilgrim, 1931

Generic type, *Ailuravus viverroides* Schlosser

Sivanasua palaeindica Pilgrim

Sivanasua palaeindica, Pilgrim, 1932, Pal. Indica, N.S., XVIII, pp. 56–58, Pl. II, figs. 10–12.

Type.—G.S.I. No. D 224, a portion of a right mandibular ramus with M_1 and with P_4, M_2 erupting.

Paratype.—"An isolated M_2 (right side) . . . now in the British Museum."—Pilgrim, G. E., 1932, p. 56.

Horizon.—Lower Siwaliks, Chinji zone.

Locality.—Near Chinji, Salt Range, Attock District, Punjab.

Diagnosis.—(See Pilgrim, G. E., 1932, p. 56.) Trigonid of lower molars relatively short. Curved ridge connecting protoconid and paraconid, the latter cusp being much lower than the former. Hypoconulid prominent, entoconid and hypoconid almost as high as the trigonid cusps.

Sivanasua himalayensis Pilgrim

Sivanasua himalayensis, Pilgrim, 1932, Pal. Indica, N.S., XVIII, p. 59, Pl. II, fig. 13.

Type.—G.S.I. No. D 237, a fragmentary right mandibular ramus with ? M_1.

Paratype.—None.

Horizon.—Lower portion of the Middle Siwaliks, Nagri zone.

Locality.—Near Hari Talyangar, Bilaspur State, Simla Hills.

Diagnosis.—"A *Sivanasua* of slightly larger size than *S. palaeindica*; with relatively broader, lower molars, without a hypoconulid."—Pilgrim, G. E., 1932, p. 59.

This latter species may very possibly be synonymous with the former. Both species are founded on very fragmentary material, which makes their standing somewhat problematical.

URSIDAE

HEMICYONINAE

Agriotherium Wagner, 1837

Generic type, *Ursus sivalensis* Falconer and Cautley

Agriotherium sivalensis (Falconer and Cautley)

Ursus sivalensis, Falconer and Cautley, 1836, Asiatic Researches, XIX, pp. 193–200.

Agriotherium sivalensis, Wagner, 1837, Gelehrte Anziegen Bayer. Akad. Wiss., Munich, V, p. 335.

Amphiarctos sivalensis, Blainville, 1841, Osteographie, Fasc. IX, p. 96.

Sivalarctos sivalensis, Blainville, 1841, Osteographie, Fasc. IX, pp. 113, 114.

Hyaenarctos sivalensis, Owen, 1845, Odontography, Pt. 3, pp. 504, 505, Pl. CXXXI.

Additional References.—

>Falconer, Hugh, 1868A, p. 321, Pl. XXVI, figs. 1–4.
>
>Lydekker, R., 1880B, p. 30; 1883C, pp. 83, 91; 1884A, pp. 220–225, Pls. XXIX, fig. 1, XXX, fig. 5; 1884D, p. 125; 1885B, p. 10; 1885C, p. 150.
>
>Pilgrim, G. E., 1910B, p. 199; 1913B, p. 325; 1914A, p. 232.
>
>Matthew, W. D., 1929, pp. 443, 475–476, fig. 14.
>
>Pilgrim, G. E., 1932, pp. 38–40.

*Type.—*B.M. No. 39721, a skull.

*Paratypes.—*None.

*Horizon.—*Upper Siwaliks, probably Pinjor zone.

*Locality.—*From the Siwalik Hills.

Diagnosis.—(From Lydekker.) A very large species, larger than any of the living species of *Ursus*. Cranial profile straight. Frontals broad, orbits large, palate vaulted. Inner cingulum of molars strong.

Agriotherium palaeindicum (Lydekker)

Hyaenarctos palaeindicus, Lydekker, 1878, Rec. Geol. Surv. India, XI, p. 103.

Lydekkerion palaeindicus, Frick, 1926, Bull. Amer. Mus. Nat. Hist., LVI, p. 79.

Agriotherium palaeindicum, Pilgrim, 1932, Pal. Indica, N.S., XVIII, pp. 40–42.

Additional References.—

>Lydekker, R., 1880B, p. 30; 1883C, pp. 83, 91; 1884A, pp. 232–236, Pl. XXX, figs. 1, 3; 1884D, p. 125; 1885B, p. 9; 1885C, pp. 154–155.
>
>Pilgrim, G. E., 1910B, p. 199; 1913B, p. 281; 1914A, p. 228.
>
>Matthew, W. D., 1929, pp. 448, 476–478.

*Type.—*G.S.I. No. D 16, a maxilla with right P^4–M^2.

*Paratypes.—*None.

*Horizon.—*Middle Siwaliks, Dhok Pathan zone.

*Locality.—*The locality for the type is unknown. A referred specimen came from Hasnot.

Diagnosis.—(Pilgrim, G. E., 1932, pp. 40–41.) Cusps of upper molars very prominent. M^1 with rounded internal cones. M^2 with reduced metacone.

The reader is referred to Matthew, W. D., 1929, p. 474 (footnote), for a concise statement of the question regarding the priority of *Agriotherium* over *Hyaenarctos*.

Indarctos Pilgrim, 1913

Generic type, *Indarctos salmontanus* Pilgrim

Indarctos salmontanus Pilgrim

Indarctos salmontanus, Pilgrim, 1913, Rec. Geol. Surv. India, XLIII, pp. 281, 290.
  Additional References.—
      Pilgrim, G. E., 1914A, p. 225, Pl. XX, figs. 1–3.
      Matthew, W. D., 1929, p. 479.
      Pilgrim, G. E., 1932, pp. 43–44, Pl. III, fig. 9.
  Type.—G.S.I. No. D 158, a left maxilla with M^2.
  Paratypes.—None.
  Horizon.—Middle Siwaliks, Dhok Pathan zone.
  Locality.—Hasnot, Salt Range, Jhelum District, Punjab.
  Diagnosis.—A large species of the genus. Talon of M^2 larger than in *Indarctos punjabiensis*. Hypocone of M^2 strong. M_2 with a short talonid.
  Regarded by Dr. Matthew as synonymous with *Indarctos punjabiensis*. See the reference cited above.

Indarctos punjabiensis (Lydekker)

Hyaenarctos punjabiensis, Lydekker, 1884, Pal. Indica (X) II, pp. 226–232, Pl. XXX,
      figs. 2, 3, XXI, fig. 1.
Indarctos (H.) punjabiensis, Frick, 1926, Bull. Amer. Mus. Nat. Hist., LVI, p. 76.
  Additional References.—
      Lydekker, R., 1884D, p. 125; 1885B, p. 9; 1885C, p. 153.
      Pilgrim, G. E., 1910B, p. 199; 1913B, p. 280; 1914A, p. 227.
      Matthew, W. D., 1929, pp. 448, 479, fig. 16.
      Pilgrim, G. E., 1932, pp. 45–46.
  Type.—(Lectotype).—G.S.I. No. D 6, a right and left P^4–M^1.
  Cotypes.—G.S.I. No. D 12, a left maxilla with M^{1-3}.   G.S.I. No. D 8, a nearly complete mandible.
  Horizon.—Middle Siwaliks, Dhok Pathan zone.
  Locality.—From Hasnot, Salt Range, Jhelum District, Punjab.   The mandible came from Jabi in the Salt Range.
  Specimens in the American Museum.—Amer. Mus. No. 19340, a left M^1, from the upper portion of the Middle Siwaliks, Dhok Pathan zone, two miles south of Hasnot.
  Amer. Mus. No. 19350, a right M^2, from the upper portion of the Middle Siwaliks, Dhok Pathan zone, one mile northeast of Hasnot.
  Diagnosis.—(Pilgrim, G. E., 1932, p. 45.) Upper molars quadrate. Talon of M^2 less developed than in other species. P_3 with two roots.

  Two teeth from two individuals are the only remains of this species in the American Museum collection.   One of these teeth, a first upper molar, is definitely that of *Indarctos*, as shown by the square outline of the tooth.   This tooth is unworn and the enamel is crenulated.   The other tooth, a second molar, is that of a smaller individual.   It is well worn, and it shows the large heel, typical of this genus.   The dimensions of the teeth are given on page 92.

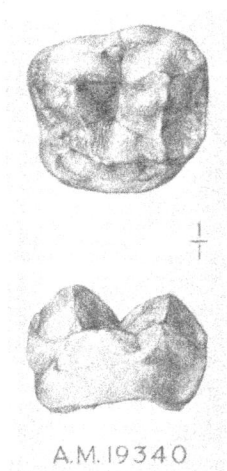

FIG. 43. *Indarctos punjabiensis* (Lydekker). Amer. Mus. No. 19340, left M[1]. Crown view above, internal view below. Natural size.

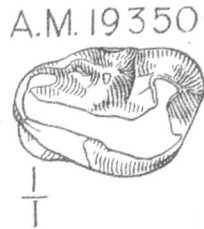

FIG. 44. *Indarctos punjabiensis* (Lydekker). Amer. Mus. No. 19350, right M[2]. Crown view, natural size.

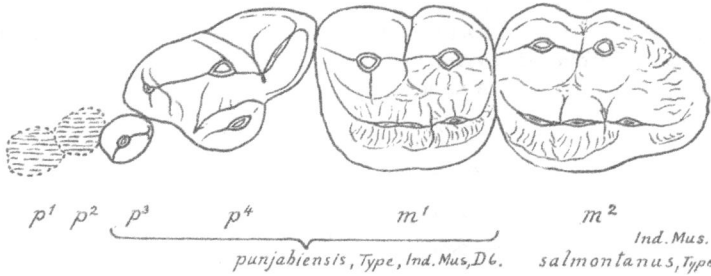

FIG. 45. *Indarctos punjabiensis* (Lydekker). Type, Geol. Surv. India, No. D 6, left P[2]–M[2]. Natural size. From Matthew, 1929.

MEASUREMENTS

Amer. Mus. No. 19340 M¹ . . . . . Antero-posterior diameter . . . . . 29.0 mm.
Transverse diameter . . . . 28.0 mm.
Amer. Mus. No. 19350 M² . . . . . Antero-posterior diameter . . . . . 28.0 mm.
Transverse diameter . . . . 23.0 mm.

One feature of special interest is the extension of the antero-internal border of the tooth, which is more pronounced in this specimen than in the type. A similar development would seem to be found in the tooth of *Hyaenarctos* sp. from the Red Crag of England, figured by Lydekker.

Matthew [20] has shown that the type of *Indarctos salmontanus* Pilgrim may be considered as synonymous with the type of *Hyaenarctos punjabiensis* Lydekker. In view, however, of the differences in the dentition of this form from *Hyaenarctos*, the generic name *Indarctos* is retained for *I. punjabiensis*.

The genus *Indarctos* was remarkably wide spread in its distribution, for it is to be found in India, China and North America.

THE MIGRATION OF INDARCTOS FROM ASIA TO NORTH AMERICA

It would seem that the genus *Indarctos* migrated from Asia to North America during Pliocene times. In the Middle Siwaliks the genus is represented by *Indarctos punjabiensis*, a species closely related to *Indarctos lagrelli*, found in the Lower Pliocene of China. Merriam, Stock and Moody have described a species from the west coast of North America, namely *Indarctos oregonensis*, a form also of Lower Pliocene age. Thus it would seem as if the genus became established first in eastern Asia in Pontian times, from whence it migrated during post-Pontian times to the west and to the east, arriving in India during Middle Siwalik times, and appearing in North America in the Rattlesnake, which is approximately correlative with the Middle Siwaliks.

Matthew has shown that *Indarctos lagrelli* of the North China Pontian, is somewhat more primitive than the Siwalik species. He therefore concludes that this is further proof of the post-Pontian age of the Dhok Pathan zone of India.

"As this Chinese fauna appears to be correlated rather closely with the Pontian, this would suggest a somewhat post-Pontian age for the Dhok Pathan zone of the Siwaliks, and equally for the Rattlesnake beds of Oregon, both more or less equidistant from the supposed palaearctic center of dispersal of the Ursidae." [21]

The genus *Arctotherium* of the Pleistocene of South America may be derived directly from *Indarctos*. Thus we have here a remarkable example of the intercontinental migration of a genus (*Indarctos*) from Asia to America, and the subsequent migration of its descendant genus (*Arctotherium*) into South America.

URSINAE

Melursus Meyer, 1793

Generic type, *Bradypus ursinus* Shaw and Nodder

Melursus theobaldi (Lydekker)

Ursus theobaldi, Lydekker, 1884, Pal. Indica (X), II, pp. 211–216, Pl. XXVIII, figs. 1, 2.
Melursus theobaldi, Pilgrim, 1910, Rec. Geol. Surv. India, XL, p. 199.

[20] Matthew, W. D., 1929, p. 479.
[21] Matthew, W.D., 1929, p. 479.

Additional References.—

Lydekker, R., 1884D, p. 125; 1885B, p. 10.

Pilgrim, G. E., 1913B, p. 324.

Matthew, W. D., 1929, pp. 443, 472–473.

Pilgrim, G. E., 1932, pp. 51–52.

Type.—G.S.I. No. D 17, a skull.

Paratypes.—None.

Horizon.—Probably Upper Siwaliks.

Locality.—From the region of Kangra.

Diagnosis.—A large species of *Melursus.* Auditory bulla lower than sphenoidal surface. Glenoid broad, palate concave and narrow. M² longer than M¹. P⁴ elongated.

The reader is referred to Dr. Matthew's remarks concerning this species.

PHYLOGENY OF THE URSIDAE
(PROVISIONAL)

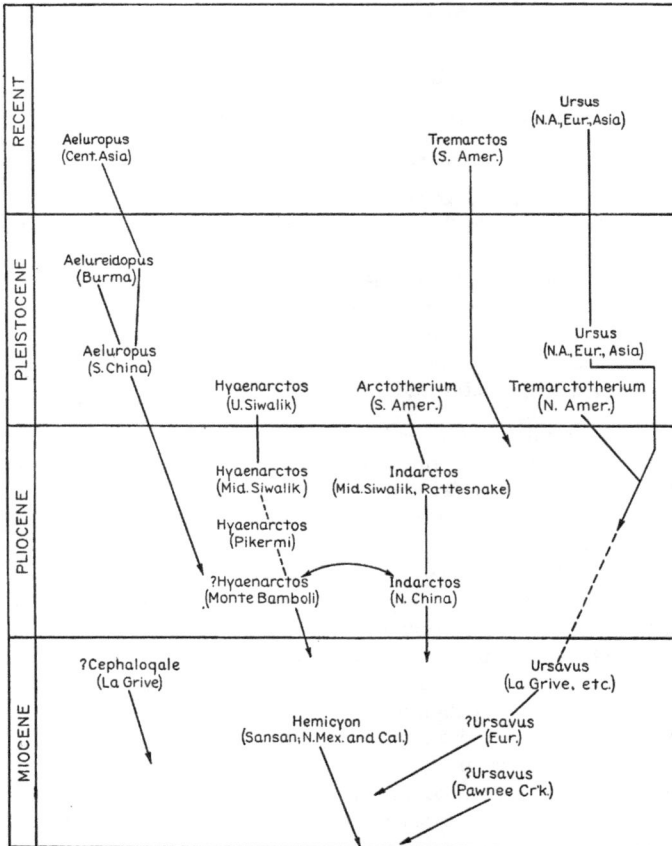

Fig. 46. Phylogeny of the Ursidae. From Matthew, 1929.

MUSTELIDAE

MUSTELINAE

Sinictis Zdansky, 1924

Generic type, *Sinictis dolichognathus* Zdansky

Sinictis lydekkeri Pilgrim

Sinictis lydekkeri, Pilgrim, 1932, Pal. Indica, N.S., XVIII, p. 62.
  Type.—B.M. No. 15914, a fragmentary left mandibular ramus.
  Paratypes.—None.
  Horizon.—Uncertain; probably Pinjor zone, Upper Siwaliks.
  Locality.—Siwalik Hills.
  Diagnosis.—(Pilgrim, G. E., 1932, p. 62.) Approximately of the same size as *Sinictis dolichognathus*. M_1 with a strong metaconid and a long talonid. Hypoconid high, immediately behind protoconid.

Martes Frisch, 1775

Generic type, *Martes foina* Erxleben

Martes lydekkeri (Colbert)

Mustela lydekkeri, Colbert, 1933, Amer. Mus. Novitates, No. 605, pp. 1–3.
  Type.—Amer. Mus. No. 19407, a fragmentary mandible with right and left M_1.
  Paratypes.—None.
  Horizon.—Lower Siwaliks, Chinji zone; 200 feet above the level of Chinji Rest House.
  Locality.—Three miles west of Chinji Rest House, Salt Range, Attock District, northern Punjab.
  Diagnosis.—Equal to *Martes martes* in size. Lower carnassial with a high trigonid and a low, basined talonid. Metaconid distinct and well developed. A small M_2 present.

This species was originally described as *Mustela lydekkeri*, having been compared with *"Mustela" palaeosinensis* Zdansky, rather than with the modern *Mustela*. However, since the Siwalik specimen has a well developed metaconid, as is the case in *M. palaeosinensis* as well, which structure is lacking in the modern *Mustela*, it can not be correctly regarded as of the latter genus.

In the type description of *Mustela lydekkeri*, the jaw in the British Museum, described by Lydekker as *Mustela sp.*, was referred to this species under discussion. The author had overlooked the recent description by Pilgrim, in which he assigned to the British Museum specimen (B.M. No. 15914) a definite identification, namely *Sinictis lydekkeri*.

A careful reexamination of the American Museum specimen and a comparison of it with Lydekker's description and figure, and with Pilgrim's description, leads the author to conclude that the reference of both specimens, that is the one in the American Museum and the one in the British Museum, to the same species is erroneous. Moreover, comparisons of the American Museum specimen with Zdansky's figures of the type of *Sinictis dolichognathus* leads me to the conclusion that *M. lydekkeri* is not of the genus *Sinictis*.

Not only is it considerably smaller than the typical *Sinictis*, but also it is characterized by a distinctly basined talonid, whereas *Sinictis* is defined as having a trenchant talonid.

"*Sinictis* seems to be distinguished from *Martes* mainly by the talonid of M_1 being trenchant instead of basin shaped, and from *Mustela* by the presence of a metaconid on M_1, and by the more pointed muzzle." [22]

"The coexistence of a metaconid with a submedian trenchant hypoconid are characters which this species [i.e., *Sinictis lydekkeri* Pilgrim] shares with *Sinictis dolichognathus*." [23]

Comparing Amer. Mus. No. 19407 further with other mustelines, it is seen to be almost identical in size and structure with the modern *Martes martes*. The chief distinction of the Siwalik form is in the fact that its metaconid is relatively larger, as would be expected in a geologically older and more primitive species.

Therefore it seems best for the present to express the relationships of the American Museum specimen by placing it in the genus *Martes*, and to regard it as a separate species, *Martes lydekkeri* (Colbert), which is quite distinct from *Sinictis lydekkeri* Pilgrim.

The measurements and figure of the type are reproduced below.

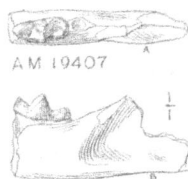

FIG. 47. *Martes lydekkeri* (Colbert). Type, Amer. Mus. No. 19407, left ramus with M_1. Crown view above, lateral view below. Natural size. From Colbert, 1933.

MEASUREMENTS

Amer. Mus. No. 19407.

	Right	Left
Length of M_1	9.9 mm.	9.4 mm.
Greatest width of M_1	4.4	4.3
Height of protoconid	4.8	4.5
Height of metaconid	3.2	3.4
Length of talonid [24]	4.5	4.0
Depth of ramus below pr^d	9.8	9.9
Ratio, height of pr^d to length of M_1	48	48
Ratio, length of talonid to length of M_1	45	42

[22] Pilgrim, G. E., and Hopwood, A. T., 1931, p. 38.

[23] Pilgrim, G. E., 1932, p. 62.

[24] Measured from the point of the metaconid to the back of the tooth.

MELLIVORINAE

Mellivora Storr, 1780

Generic type, *Mellivora ratel* Sparrmann

Mellivora sivalensis (Falconer)

Ursitaxus sivalensis, Falconer, 1868, Pal. Mem., Vol. I, p. 553, description of unpublished
  plate Q, fig. 4, of Fauna Antiqua Sivalensis.

Mellivora sivalensis, Lydekker, 1884, Pal. Indica, (X), II, pp. 180–183, fig. 1, Pl. XXVI,
  figs. 1–4.

  Additional References.—
    Lydekker, R., 1884D, p. 125; 1884E, p. 71, fig. 1; 1885B, p. 11; 1885C, pp. 188–189,
      fig. 27.
    Pilgrim, G. E., 1910B, p. 199; 1913B, p. 324.
    Matthew, W. D., 1929, pp. 443, 466–467, fig. 4.
    Pilgrim, G. E., 1932, pp. 63–65.

  Type.—Sci. and Art Museum, Dublin, No. 45, a skull.
  Paratypes.—None.
  Horizon.—Upper Siwaliks, Pinjor zone.
  Locality.—From the Siwalik Hills.
  Diagnosis.—(Pilgrim, G. E., 1932, pp. 63–64.)   Skull distinguished by flat and de-
pressed parietal region, inflated bulla, prominent mastoid and stout paroccipital.

Mellivora punjabiensis Lydekker

Mellivora punjabiensis, Lydekker, 1884, Pal. Indica (X), II, pp. 183–185, Pl. XXVII, fig. 6.
Promellivora punjabiensis, Pilgrim, 1932, Pal. Indica, N.S. XVIII, pp. 65–66.

  Additional References.—
    Lydekker, R., 1884D, p. 125; 1885B, p. 11.
    Pilgrim, G. E., 1910B, p. 199; 1913B, p. 282.
    Matthew, W. D., 1929, pp. 440, 467.

  Type.—G.S.I. No. D 20, a fragmentary right mandibular ramus.
  Paratypes.—None.
  Horizon.—Middle Siwaliks, Dhok Pathan zone.
  Locality.—From Hasnot, Jhelum District, Salt Range, Punjab.
  Diagnosis.—(See Pilgrim, G. E., 1932, p. 65.)   Mandible short, with a long symphysis.
M_1 relatively long and compressed. P_4 relatively short.   Anterior premolar small.   Canine
large and rather procumbent.

In speaking of this species, Dr. Matthew said: "*Mellivora punjabiensis* is known only
from a fragment of lower jaw showing P_{3-4} much worn, and somewhat battered roots of front
teeth and carnassial.   It may be, and very likely is, a distinct species from *sivalensis*, but
the type does not prove it; . . . "—Matthew, W. D., 1929, p. 467.

In view of the fragmentary condition of the type, Pilgrim's erection of a new genus.
based on this species seems very risky.   The more conservative attitude would be to regard
the form under discussion as a species of Mellivora.

Eomellivora Zdansky, 1924

Generic type, *Eomellivora wimani* Zdansky

Eomellivora necrophila Pilgrim

Eomellivora necrophila, Pilgrim, 1932, Pal. Indica, N.S., XVIII, pp. 67–71, Pl. III, figs. 4–7.

Type.—G.S.I. No. D 243, an isolated M_1.

Paratypes.—G.S.I. No. D 254, a left mandibular ramus with the roots of the teeth. G.S.I. No. D 255a, an M_1. G.S.I. No. D 255b, a lower premolar. G.S.I. No. D 256a, an upper premolar. G.S.I. No. D 256b, a lower premolar.

Horizon.—Lower Siwaliks, Chinji zone.

Locality.—Found near the village of Bhilomar, west of Chinji, Salt Range, Attock District, Punjab.

Diagnosis.—Smaller than *Eomellivora wimani*. Symphysis deep, masseteric fossa also deep. M_1 with high protoconid and relatively short talonid. M_2 relatively large. P_1 single rooted. Canine relatively large.

Eomellivora tenebrarum Pilgrim

Eomellivora tenebrarum, Pilgrim, 1932, Pal. Indica, N.S., XVIII, pp. 71–72.

The type of this species was referred by Lydekker to *Mellivorodon palaeindica*. See Lydekker, R., 1884A, p. 185, Pl. XXVII, fig. 8.

Type.—G.S.I. No. D 22, posterior portion of a left mandibular ramus.

Paratypes.—None.

Horizon.—Middle Siwaliks, Dhok Pathan zone.

Locality.—From Niki in the Salt Range.

Diagnosis.—(See Pilgrim, G. E., 1932, p. 71.) About the same size as *Eomellivora wimani*. Ramus shallow. M_1 relatively longer and narrower than in the above species; hypoconid lower and compressed.

This species is based on a very fragmentary specimen, so its status is at best, quite doubtful.

Lutrinae

Lutra Erxleben, 1777

Generic type, *Mustela lutra* Linnaeus

Lutra palaeindica Falconer

Lutra palaeindica, Falconer, 1868, Pal. Mem., I, p. 552, Pl. XXVII, figs. 6–8.

Additional References.—

Bose, P. N., 1880, p. 133.

Lydekker, R., 1880B, p. 31; 1883C, p. 83; 1884A, pp. 190–193, Pl. XXVII, figs. 1–2, 2a; 1884D, p. 125; 1885B, p. 11; 1885C, p. 191.

Pilgrim, G. E., 1910B, p. 199; 1913B, p. 325.

Pohle, H., 1919, p. 41, Pl. II, fig. 5.

Matthew, W. D., 1929, pp. 443, 469–470, figs. 6–7.

Pilgrim, G. E., 1932, pp. 73–74.

Type.—B.M. No. 37151, a skull.

Paratype.—B.M. No. 37152, a left mandibular ramus.

Horizon.—Upper Siwaliks.

Locality.—From the Siwalik Hills.

Diagnosis.—Skull very similar to that of the recent form, *Lutra sumatrana.*

Sivalictis Pilgrim, 1932

Generic type, *Sivalictis natans* Pilgrim

Sivalictis natans Pilgrim

Sivalictis natans, Pilgrim, 1932, Pal. Indica, N.S. XVIII, pp. 77–79, Pl. I, fig. 1.

Type.—G.S.I. No. D 227, an isolated tooth.

Paratypes.—None.

Horizon.—Lower Siwaliks, Chinji zone.

Locality.—From the vicinity of Chinji, Salt Range, Attock District, Punjab.

Diagnosis.—(See Pilgrim, G. E., 1932, p. 77.)   M[1] seemingly that of a mustelid, with three roots fused, low cusps, strong paracone, and protocone separated from outer cusps. Parastyle strong, internal cingulum and a slight anterior cingulum present.

This genus and species is based on a single tooth of obscure form and characters.

Enhydriodon Falconer, 1868

Generic type, *Enhydriodon sivalensis* Falconer

Enhydriodon sivalensis Falconer

Enhydriodon sivalensis, Falconer, 1868, Pal. Mem., Vol. I, pp. 331–338, Pl. XXVII, figs. 1–5.

Lutra sivalensis, Lydekker, 1884, Pal. Indica (X), II, pp. 195–201, fig. 3, Pl. XXVII, fig. 5, Pl. XLV, figs. 3, 3a.

Additional References.—

> Lydekker, R., 1880B, p. 31; 1883C, pp. 83, 90; 1884D, p. 125; 1885B, p. 12; 1885C, pp. 192–193, fig. 29.
>
> Pilgrim, G. E., 1910B, p. 199; 1913B, p. 325.
>
> Matthew, W. D., 1929, pp. 443, 471–472.
>
> Pilgrim, G. E., 1932, pp. 83–85.

Type.—(Lectotype).   B.M. No. 37153, a fairly complete skull.

Cotypes.—B.M. No. 37154, 37155, two skull fragments.

Horizon.—Upper Siwaliks.

Locality.—The exact locality for the above specimens is unknown, but they came from the Siwalik Hills.

Diagnosis.—(See Pilgrim, G. E., 1932, p. 83; Matthew, W. D., 1929, p. 471.)   A very large form, being one of the largest of the Mustelidae.   Protocone of M[1] often twinned. P[4] broad.   Cheek teeth relatively short.

The name *Enhydriodon ferox* was applied by Falconer to the above specimens in his description of the unpublished plates of the Fauna Antiqua Sivalensis.   Both names appeared first in 1868, in Falconer's Palaeontological Memoirs, Volume I, but since the use of *E. sivalensis* (p. 331) precedes that of *E. ferox* (p. 552), the former is established as the valid name.

Enhydriodon falconeri Pilgrim

Enhydriodon falconeri, Pilgrim, 1931, Cat. Pontian Carnivora of Europe (Brit. Mus. Cat.), p. 56, Pl. II, figs. 3, 5.

Additional References.—

Pilgrim, G. E., 1932, pp. 85–88, Pl. II, fig. 15.

The type of this species was referred to previously by Falconer and by Matthew.

Falconer, H., 1868A, p. 336.

Matthew, W. D., 1929, p. 472, fig. 11 (*Enhydriodon* sp.).

Type.—B.M. No. M 4847, an isolated P^4.

Paratypes.—G.S.I. No. D 161, an isolated M_1.

Horizon.—Probably Upper Siwaliks for the type. The paratype came from Hasnot, and may be of Middle Siwalik, Dhok Pathan age.

Locality.—Siwalik Hills for the type. Hasnot, Salt Range for the paratype.

Diagnosis.—(See Pilgrim, G. E., 1932, p. 85.) Smaller than *Enhydriodon sivalensis*. P^4 relatively narrow, protocone anterior in position; paracone and metacone less bunodont, and parastyle stronger than in *E. sivalensis*. (See Matthew, W. D., 1929, p. 472.)

This species is based on extremely scanty material. It does however, probably represent a distinct species of Middle Siwalik age.

Sivaonyx Pilgrim, 1931

Generic type, *Lutra bathygnathus* Lydekker

Sivaonyx bathygnathus (Lydekker)

Lutra bathygnathus, Lydekker, 1884, Pal. Indica (X), II, pp. 193–194, Pl. XXVII, figs. 3, 4.

Potamotherium bathygnathus, Pohle, 1919, Archiv. für Naturgesch. (A) IX, p. 26.

Sivaonyx bathygnathus, Pilgrim, G. E., 1931, Cat. Pontian Carnivora of Europe (Brit. Mus. Cat.), p. 74, Pl. II, figs. 4, 4a.

Additional References.—

Lydekker, R., 1884D, p. 125; 1885B, p. 11.

Pilgrim, G. E., 1910B, p. 199; 1913B, p. 281.

Matthew, W. D., 1929, pp. 448, 470, fig. 8.

Pilgrim, G. E., 1932, pp. 88–93, Pl. II, figs. 14, 16, Pl. IV, fig. 1.

Type.—G.S.I. No. D 33, a left mandibular ramus.

Paratype.—G.S.I. No. D 34, a fragmentary left mandibular ramus.

Horizon.—Middle Siwaliks, Dhok Pathan zone.

Locality. From Hasnot, Salt Range, Jhelum District, Punjab.

Specimen in the American Museum.—Amer. Mus. No. 19509, a fragment of a right mandibular ramus, with M_1 and the alveoli of P_4 and M_2. From the lower portion of the Upper Siwaliks, Tatrot zone, three miles south of Hasnot.

Diagnosis.—(See Pilgrim, G. E., 1932, p. 88.) A large lutrine. P^4 approximately quadrate, with a trenchant outer blade. Inner cusps lower than outer cusps and strong internal cingulum present. Mandibular ramus deeper than length of M_1. Canine large. P_4 broad posteriorly and M_1 very broad. Metaconid as high as paraconid, talonid broader than trigonid, equal to it in length, basin shaped and surrounded by a ring of cusps of which the entoconid is as well developed as the hypoconid. M_2 small.

The American Museum specimen compares closely, both as to size and configuration, with Lydekker's type figure.   The trigonid of M_1 is relatively low, occupying just half the length of the tooth, while the talonid is large and deeply basined.   Numerous small swellings or cusps succeed each other around the rim of the talonid, and a well developed cingulum runs from the inner side of the paraconid, around the front and along the external side of the tooth.   The alveolus shows that M_2 had two roots.

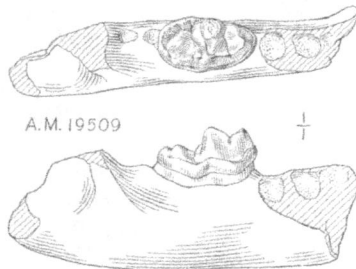

FIG. 48. *Sivaonyx bathygnathus* (Lydekker). Amer. Mus. No. 19509, right ramus with M_1. Crown view above, lateral view below. Natural size.

MEASUREMENTS

Amer. Mus. No. 19509.
Length of M_1 . . . . 16.5 mm.   Width of M_1 . . . . . .   9.5 mm.
Depth of mandible below anterior border of M_1 . . . . 17.0

The double rooted M_2 would seem to be unique, for in the other specimens of *Sivaonyx*, and in the other kinds of otters, the second molar is a small, single rooted tooth.   This jaw would indicate that M_2 was comparatively long.   Of course a double rooted M_2 might be expected in a primitive form, or perhaps it was a primitive persistent character in this Upper Pliocene animal.   Another suggestion is that M_2 was variable, and this specimen represents a throw back within the species to a more primitive condition.

Certainly the carnassial militates against the jaw in the American Museum being of any other species, and probably this tooth would have precedence over a single or double rooted M_2 in determining the question of specific relations.

Dr. Matthew, on the basis of his examination of the type specimen of *S. bathygnathus*, was inclined to be rather skeptical of the exact relationships of this species, for according to him "until the character of M_2 is known, it [i.e., the species under discussion] is too uncertain for generic reference." [25]

Pilgrim (1932) states that specimens from Hasnot, which he describes in his memoir on the Indian Carnivora, reveal additional information of the affinities of the species.   Accordingly he places it in a separate genus, *Sivaonyx*, of which the species in question is the single representative.   Perhaps in the creation of this new genus Dr. Pilgrim has ventured a bit too far, but even so it seems expedient, at the present time, to use the name *Sivaonyx*, mainly to indicate that this form is distinct from *Lutra*, and to specify the possibilities of a relationship to the cape otter, *Aonyx*.

[25] Matthew, W. D., 1929, p. 470.

Sivaonyx bathygnathus may be compared with the modern representatives of the genus *Lutra*. The lower carnassial of the Indian species differs from that of the modern forms mainly in its greater relative breadth, and in the configuration of its talonid. In the modern *Lutra* the outer edge of the talonid is high, forming a trenchant blade of a sort, while the inner edge is quite low. In *S. bathygnathus* the outer and inner rims of the talonid are of the same height, both being relatively lower than in the modern forms, and they enclose a well defined basin. As to the trigonid and the cingulum, the modern *Lutra* is much like the Indian species under discussion.

Pohle (1919) referred *S. bathygnathus* to the genus *Potamotherium*, mainly on the basis of a lower first premolar. Matthew pointed out the fallacy of this argument, and he suggested that "*Lutra*" *bathygnathus* might be close to *Brachypsalis*. A comparison of these two latter forms will show however, that they resemble each other only in a general way.

Brachypsalis approximates a more primitive condition in the lower carnassial than does *S. bathygnathus*, for in the former the trigonid is relatively long and high, and the talonid is short.

RATIO, LENGTH OF TRIGONID TO LENGTH OF M₁

Brachypsalis modicus, Amer. Mus. No. 17209, type.............. .69
Sivaonyx bathygnathus, Amer. Mus.No. 19509................ .57

Lydekker compared *S. bathygnathus* to the Cape otter, chiefly on the basis of the relative great depth of the mandible in both forms, and on the shape of the ascending ramus. Matthew (1929, p. 470) set forth arguments against the relation of the Siwalik form to the Cape otter.

A close comparison of the first lower molars in the American Museum specimen of *Sivaonyx bathygnathus* and *Aonyx capensis* shows that these two forms have lower carnassials quite similar to each other. The resemblances and differences between these two forms have been pointed out by Pilgrim (1932, pp. 89–93), the latter being found in the relatively longer trigonid of M₁ in the modern form, and in the internal shifting of the paraconid and in the absence of a strong cingulum and a postero-external accessory cusp.

Pilgrim would compare *S. bathygnathus* with *Lutra aonychoides* from North China, and with *Enhydriodon* of India. He postulates a close relationship to the latter species, suggesting that *Sivaonyx* may be an ancestral form.

The mandible found by Mr. Brown comes from a higher horizon than is typical for the species, it being recorded as from the lower portion of the Upper Siwaliks. Evidently *S. bathygnathus* persisted from the Dhok Pathan into the Tatrot zone.

Vishnuonyx Pilgrim, 1932

Generic type, *Vishnuonyx chinjiensis* Pilgrim

Vishnuonyx chinjiensis Pilgrim

Vishnuonyx chinjiensis, Pilgrim, 1932, Pal. Indica, N.S. XVIII, pp. 94–96, Pl. II, fig. 17, Pl. IV, fig. 6.

Type.—G.S.I. No. D 223, a left maxilla with P⁴.

Paratype.—G.S.I. No. D 245, a fragmentary right mandibular ramus.

Horizon.—Lower Siwaliks, Chinji zone.

Locality.—Near Chinji, Salt Range, Punjab.

Diagnosis.—(See Pilgrim, G. E., 1932, p. 94.)   A small lutrine having a triangular P^4. Protocone high, metacone elongated, parastyle weak, protocone and hypocone lower than outer cusps. M^1 small. Mandible with a deep ramus. P_4 elongate and not so broad posteriorly as in *Sivaonyx*. M_1 with talonid broader and shorter than trigonid, surrounded by a crenulated rim. M_2 elongated.

This genus and species are founded on such very fragmentary material that they can be accepted only as of doubtful standing.

FELOIDEA [AELUROIDEA]

VIVERRIDAE

VIVERRINAE

Viverra Linnaeus, 1758

Generic type, *Viverra zibetha* Linnaeus

Viverra bakerii Bose

Viverra bakerii, Bose, 1880, Quar. Jour. Geol. Soc. London, XXXVI, pp. 131–132.

Additional References.—

The type of this species was referred to previously by Falconer as *Canis* sp.

Falconer, H., 1868A, p. 553.

Lydekker, R., 1880B, p. 30; 1883C, pp. 82, 93; 1884A, pp. 268–271, Pl. XXXIII, figs. 1, 2; 1884D, p. 124; 1885B, p. 7; 1885C, p. 99.

Pilgrim, G. E., 1910B, p. 199; 1913B, p. 325.

Matthew, W. D., 1929, pp. 443, 487–488, fig. 24.

Pilgrim, G. E., 1932, pp. 98–99.

Type.—B.M. No. 40183, a skull.

Paratypes.—None.

Horizon.—Upper Siwaliks.

Locality.—From the Siwalik Hills.

Diagnosis.—(See Pilgrim, G. E., 1932, p. 98.)   A very large species of *Viverra*.   Skull similar to that of *V. zibetha*.   Antero-posterior diameter of P^4 exceeds that of the united M^1 and M^2. M^1 suboblong. P^3 large.

Viverra (?) chinjiensis Pilgrim

Viverra (?) *chinjiensis*, Pilgrim, 1932, Pal. Indica, N.S., XVIII, pp. 99–100, Pl. IV, figs. 9, 10.

Type.—G.S.I. No. D 214, a right mandibular ramus with almost complete dentition, and the associated M^1.

Paratypes.—None.

Horizon.—Lower Siwaliks, Chinji zone.

Locality.—Kukar Dhok, near Chinji, Salt Range, Punjab.

Diagnosis.—(See Pilgrim, G. E., 1932, p. 99.)   A small species of *Viverra* with a large M^1 and M_2.   Cusps on encircling rim of talonid less prominent than in other species.

Vishnuictis Pilgrim, 1932

Generic type, *Vishnuictis salmontanus* Pilgrim

Vishnuictis salmontanus Pilgrim

Vishnuictis salmontanus, Pilgrim, 1932, Pal. Indica N.S., XVIII, pp. 101–106, Pl. IV, figs. 7, 8.

Type.—G.S.I. No. D 160, a skull and mandible.

Paratypes.—None.

Horizon.—Middle Siwaliks, Dhok Pathan zone.

Locality.—From near Hasnot, Salt Range, Punjab.

Diagnosis.—(See Pilgrim, G. E., 1932, p. 101.) Medium sized viverrid. Skull elongated, rather high and narrow, with a slender muzzle. Teeth not laterally compressed, and with blunt cusps. Upper molars relatively large; premolars simple. Mandible stout and shallow. M_1 with relatively long trignoid; M_2 oblong.

Vishnuictis durandi (Lydekker)

Viverra durandi, Lydekker, 1884, Pal. Indica (X) II, pp. 271–274, Pl. XXXIII, fig. 3.

Vishnuictis durandi, Pilgrim, 1932, Pal. Indica, N.S., XVIII, pp. 106–108.

Additional References.—

> Lydekker, R., 1884D, p. 124; 1885B, p. 8; 1885C, pp. 99–100.
> Pilgrim, G. E., 1910B, p. 199; 1913B, p. 325.
> Matthew, W. D., 1929, pp. 443, 488, fig. 25.

Type.—B.M. No. M 1338, a skull.

Paratype.—B.M. No. 37150, a partial skull.

Horizon.—Upper Siwaliks.

Locality.—From the Siwalik Hills.

Diagnosis.—A very large species, with a comparatively heavy skull and a short muzzle.

"The molars differ very considerably from *V. bakeri*, the carnassial being relatively large, the carnassial angle much sharper, m_2 relatively reduced. On this, as on *bakeri*, there is a very weak parastyle on p^4, but less prominent here although the wear has opened a large worn space on it." [26]

VIVERRID (?), GENUS AND SPECIES INDETERMINATE.

Amer. Mus. No. 19526, a cranium, badly crushed. From the Middle Siwaliks, south of Nathot.

This very fragmentary specimen is the only cranium of a true carnivore in the American Museum Siwalik collection. The specimen has been broken transversely in several places, and it has suffered longitudinal compression, causing a certain amount of telescoping of the elements.

It would seem to be the skull of a viverrid, as shown by the long, narrow form of the brain case. Two supra-orbital ridges run back from the fronto-parietal suture. The lambdoidal crest is developed in the viverrid fashion, much as in *Arctictis*, and there is a supraoccipital crista joining the lambdoidal crest, at which junction there is a swelling or

---

[26] Matthew, W. D., 1929, p. 488.

tuberosity, a viverrid character. There is no sagittal crest. Slight remains of the alveolus of the last molar on the right side are to be seen.

Amer. Mus. No. 19342, a molar. From the Lower Siwaliks, four miles northeast of Chinji Rest House.

HYAENIDAE

Ictitherium Wagner, 1848

Generic type, *Ictitherium viverrinum* Wagner

Ictitherium sivalense Lydekker

Ictitherium sivalense, Lydekker, 1877, Rec. Geol. Surv. India, X, p. 32.
Lepthyaena sivalense, Lydekker, 1884, Pal. Indica (X), II, pp. 312–313, Pl. XLV, figs. 8, 9.
Palhyaena sivalense, Matthew, 1929, Bull. Amer. Mus. Nat. Hist., LVI, pp. 448, 493.

> *Additional References.—*
>
> > Lydekker, R., 1880B, p. 30; 1883C, pp. 82, 91; 1884D, p. 124; 1885B, p. 7; 1885C, p. 93.
> >
> > Pilgrim, G. E., 1910B, p. 199; 1913B, p. 282; 1931, p. 84; 1932, pp. 114–119, Pl. V, figs. 3, 4, 7, Pl. VI, fig. 5, Pl. IX, fig. 4.
>
> *Type.—*(Lectotype).—G.S.I. No. D 38a, a left mandibular ramus.
>
> *Paratype.—*G.S.I. No. D 38b, a right mandibular ramus.
>
> *Horizon.—*Middle Siwaliks, Dhok Pathan zone.
>
> *Locality.—*From near Hasnot, Salt Range, Jhelum District, Punjab.
>
> *Specimen in the American Museum.—*Amer. Mus. No. 19737, a fragment of a mandible with the right P_4, M_1, from the Middle Siwaliks, one mile west of Hasnot.
>
> *Diagnosis.—*(See Pilgrim, G. E., 1932, p. 114.) About equal to *Ictitherium wongi* in size. M_1 relatively long, with a moderately high metaconid. M_2 relatively small. Upper cheek teeth extended transversely.

The American Museum specimen shows the last lower premolar to be a large robust tooth, with a high protoconid, keeled anteriorly and posteriorly. There is a low anterior

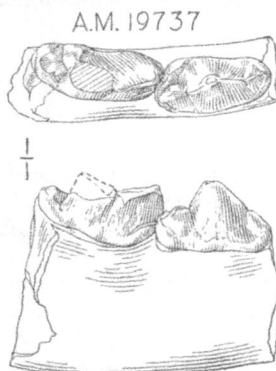

FIG. 49. *Ictitherium sivalense* Lydekker. Amer. Mus. No. 19737, right ramus with P_4, M_1. Crown view above, lateral view below. Natural size.

cusp, with a slight cingulum on either side of it, and in addition there is a posterior trenchant cusp, placed somewhat externally. This latter cusp is keeled, and its axis is in line with that of the protoconid, which in turn is somewhat oblique to the central axis of the tooth. An inner cusp springs from the cingulum, and it extends back as a cingular ridge.

The first lower molar is slightly longer than the preceding tooth, but it is not so wide. It consists of an anterior shearing paraconid, and behind that a shearing protoconid, the top portion of this latter element being broken away in the specimen at hand. Interior to the protoconid blade and opposite its posterior edge, is the smaller metaconid, a pointed cusp reaching but slightly higher than the talonid. The talonid is basined, and the edges of the basin are formed by the hypoconid and the entoconid, these cusps having a connecting ridge between them. There is a small cingulum external to the hypoconid. A well developed anterior cingulum encircles the front of the paraconid, having its greatest development on the external side of the tooth. The enamel is somewhat rugose on the external side of the carnassial shearing blade. The measurements of the specimen are given below.

MEASUREMENTS

Amer. Mus. No. 19737, mandibular ramus.

P_4....Antero-posterior diameter... 19.5 mm.	Transverse diameter... 9.5 mm.
M_1...Antero-posterior diameter... 21.5	Transverse diameter... 9.0
Depth of mandible below M_1...... 23.5	
Height of paraconid blade of M_1.... 8.5	
Height of P_4 above alveolus....... 12.0	

Ictitherium sivalense is certainly closely comparable to *I. wongi* from the Pliocene of China. Except for its larger size, the rugosity of the enamel and the stronger development of the cingula in the Siwalik species, the two forms are very close to each other. Perhaps if more of the American Museum specimen were preserved, greater differences would be manifested, but on the basis of the last lower premolar and the first lower molar the following distinctions can be drawn between the Siwalik and the Chinese species.

Ictitherium wongi differs from *Ictitherium sivalense* in that:

1. The two teeth in question are higher as compared to their length.
2. P_4 is somewhat smaller as compared to M_1.
3. The carnassial shearing blade is somewhat shorter, as compared to the total length of M_1.
4. The cingula are less developed.
5. The enamel is less rugose.

These are in the main, characters due to a more primitive stage of evolutionary development in the Chinese form.

When *Ictitherium sivalense* is compared to *Hyaena striata*, the resemblances are seen to be very close. *Hyaena striata* differs mainly by virtue of its having become more robust; hence the teeth are wider and higher in proportion to their length than is the case in *Ictitherium*. The differences that distinguish *Hyaena striata* from *I. sivalense* are as follows:

1. The teeth are heavier and more robust, and are proportionately higher.
2. The talonid of M_1 is shallower.
3. Enamel rugosities are less developed.

Similarities may be listed as follows:

1. P_4 is enlarged so that it is only slightly shorter than M_1.
2. The cingula are well developed.
3. The carnassial shear is proportionately long.

A comparison of the American Museum specimen of *I. sivalense* with a lower jaw of *I. hipparionum* from Samos, shows that these two species are very much like each other, point for point. The only differences are as follows:

1. The anterior cusp on P_4 of *I. hipparionum* is relatively larger than in *I. sivalense*.
2. In *I. sivalense* the cingulum is developed exteriorly to the anterior and posterior cusps of P_4, a feature not present in *I. hipparionum*.
3. In *I. sivalense* the paraconid blade is slightly straighter (that is, parallel to the axis of the ramus) and there is a slightly greater development of the anterior cingulum than is the case in *I. hipparionum*.

The resemblances and differences of *I. sivalense* to closely related forms, as set forth in the above paragraphs, are best shown in tabular form, as expressed by certain diagnostic ratios. The tables are presented below.

MEASUREMENTS

	P_4			M_1		
	Length	Width	Height	Length	Width	Height
Ictitherium wongi A.M., 26368..........	14.5 mm.	7 0 mm.	10.0 mm.	18.0 mm.	7.5 mm.	8.5 mm.
I. sivalense A.M., 19737.................	19.5	9.5	12.0	21.5	9.0	8.5
I. hipparionum A.M., 20554.............	16.0	8.0	11.5	20.0	9.0	13.0
Hyaena striata A.M., 1544 (D.C.A.)......	20.5	12.5	16.5	21.5	11.0	12.0

	I. wongi	*I. sivalense*	*I. hipparionum*	*H. striata*
1. Ratio, length of P_4 to length of M_1.....	.81	.91	.80	.95
2. Ratio, width of P_4 to length of P_4......	.48	.48	.50	.61
3. Ratio, width of M_1 to length of M_1.....	.41	.42	.45	.52
4. Ratio, height of P_4 to length of P_4......	.68	.61	.72	.81
5. Ratio, height of M_1 to length of M_1......	.47	.39	.65	.56
6. Ratio, length of carnassial shear to length of M_1.....................	.66	.72	.67	.74

MEASUREMENTS

Length of Carnassial Shear
    Ictitherium wongi..... 12.0 mm.   *I. sivalense*..... 15.5 mm.   *I. hipparionum*..... 13.5 mm.
    Hyaena striata....... 16.0 mm.

The actuality of the genus *Palhyaena* has been debated by numerous authors. This name was established by Gervais (1858) who applied it to the form that he had previously described as *Hyaena hipparionum*. The Indian species was first described by Lydekker as *Ictitherium*, but was subsequently placed in a new genus *Lepthyaena* by this author, who considered that it bore more resemblances to the hyaenids than it did to the viverrids.

Later, in 1903, Schlosser used the term *Palhyaena*, giving as his reason for differentiating it from *Ictitherium*, the greater reduction of M_1 and of both upper molars.

Gaudry, in 1862, showed how *Ictitherium hipparionum*, as he designated the form from Greece, is identical with the previously described *Hyaena hipparionum* of Gervais. In 1924, Zdansky, describing the fossil carnivores from North China, pointed out that *Ictitherium* and *Palhyaena* are in fact synonymous terms, a view supported by Pilgrim in his recent memoir on the Indian carnivora. Zdansky's final conclusions are as follows: "Es ist daher kaum länger die Trennung von *Ictitherium* und *Palhyaena* aufrechtzuerhalten." [27]

Ictitherium may be regarded as a transitional form, bridging the gap between the viverrids and the hyaenids. All in all, however, *Ictitherium* is a true hyaenid, a fact well demonstrated by the structure of the bulla and the distribution of the basicranial foramina. It demonstrates the close relationships existing between these two phyla of carnivores, and in addition it shows that the hyaenas branched off at a relatively late period from the ancestral viverrid stock, namely at some time during the Miocene.

Dr. Matthew, in 1901, suggested the possibility of the Hyaenidae originating from Palaeonictid stock. This view was followed by Schlosser in 1903, who further stipulated that the origin of the group might be in North America. Schlosser placed *Ictitherium* and *Palhyaena* as end stages of branches from the viverrid line, assigning their resemblances to *Hyaena* to convergence. In view of our now extended knowledge of these forms it seems more likely that *Ictitherium* represents a true transition between the hyaenids and the viverrids, and that the origin of the Hyaenidae was in Asia. This view has been summed up by Abel as follows:

"Die Hyänen sind ein Seitenzweig des Stammes der Viverriden und zwar dürfte die Gattung Ictitherium oder eine nahe verwandte Gattung das Bindeglied bilden. Echte Hyänen erscheinen zuerst in Unterpliozän Europas und Asiens und sind niemals nach Amerika ausgewandert, sondern stets auf Eurasien und Afrika beschränkt geblieben; in der Eiszeit Europas lebten drei Hyänenarten: Hyaena crocuta, H. spelaea, H. striata; die letzgenannte war die häufigste." [28]

Ictitherium indicum (Pilgrim)

Palhyaena indica, Pilgrim, 1910, Rec. Geol. Surv. India, XL, pp. 64–65.
Ictitherium indicum, Pilgrim, 1932, Pal. Indica, N.S. XVIII, pp. 119–122.

> *Additional References.—*
>> Pilgrim, G. E., 1910B, p. 199; 1913B, pp. 282, 289.
>> Matthew, W. D., 1929, pp. 448, 493.

> *Type.—*A maxilla, presumably in the collection of the Geological Survey of India.
> *Paratype.—*G.S.I. No. D 53, a mandible.
> *Horizon.—*Middle Siwaliks, Dhok Pathan zone.
> *Locality.—*From near Hasnot, Salt Range, Jhelum District, Punjab.
> *Diagnosis.—*(See Pilgrim, G. E., 1932, p. 120.) Larger than *I. sivalense*, and about

equal in size to *I. hipparionum*. M_1 relatively long, with a short talonid, and a low metaconid. M_2 very short. Premolar series short.

[27] Zdansky, O., 1924, p. 91.
[28] Abel, O., 1919, p. 741.

The type of this species is a maxilla, presumably in the Geological Survey of India collection, and *not* the mandible, G.S.I. No. D 53, which Pilgrim designates as the type in his 1932 memoir. In the original description (1910), Pilgrim definitely states that: "This species is established on a maxilla found at Asnot, which is somewhat inferior in size to *Palhyaena hipparionum* Gerv., and has rather broader molars." [29]

Hyaenictis Gaudry, 1861

Generic type, *Hyaenictis graeca* Gaudry

Hyaenictis bosei Matthew

Hyaenictis bosei, Matthew, 1929, Bull. Amer. Mus. Nat. Hist., LVI, p. 493, fig. 28.
  Additional References.—
      Pilgrim, G. E., 1932, pp. 122–125.
      The type of this species was originally described as *Hyaena sivalensis*. See:
      Falconer, H., 1868A, Pl. XXV, figs. 1–4. (*Felis cristata*, in errore.)
      Bose, P. N., 1880, p. 128.
      Lydekker, R., 1884A, p. 303, Pl. XXXIV.
Type.—B.M. No. 37133, a nearly complete skull.
Paratypes.—None.
Horizon.—Upper Siwaliks.
Locality.—From the Siwalik Hills.
  Diagnosis.—(See Matthew, W. D., 1929, p. 493.) "This skull is of very definitely primitive type, decidedly more so than *H. macrostoma*, comparable with *H. choeretis* or *Hyaenictis graeca* of Pikermi. These species are apparently nearly related to the striped Hyaena, *H. striata*.

"The whole aspect of the dentition is rather primitive, suggesting *Palhyaena hipparionum*. Referred specimens show M_2 sometimes present, sometimes absent." [30]

Lycyaena Henel, 1862

Generic type, *Hyaena chaeretis* Gaudry

Lycyaena macrostoma (Lydekker)

Hyaena macrostoma, Lydekker, 1884, Pal. Indica (X), II, pp. 298–303, Pls. XXXVI, fig. 2, XXXVII, XXXVIII, fig. 4, XXXIX, fig. 6.
Lycyaena macrostoma, Trouessart, 1887, Catalogus Mammalium, p. 320.
  Additional References.—
      Lydekker, R., 1884D, p. 124; 1885B, p. 7; 1885C, pp. 91–92.
      Pilgrim, G. E., 1910B, p. 199; 1913B, p. 282.
      Matthew, W. D., 1929, pp. 492–493.
      Pilgrim, G. E., 1932, pp. 125–129.
Type.—G.S.I. No. D 44, a skull.
Paratype.—G.S.I. No. D 52, a fragmentary left mandibular ramus.

[29] Pilgrim, G. E., 1910A, pp. 64–65.
[30] Matthew, W. D., 1929, p. 493.

Horizon.—Middle Siwaliks, Dhok Pathan zone.

Locality.—Jabi, Salt Range. Also Hasnot.

Diagnosis.—(See Pilgrim, G. E., 1932, p. 125.) Skull narrow and elongate with a large orbit. Sagittal crest low. Nasal opening relatively large and teeth relatively heavy. P_1 persistent whereas it is lost in the more advanced species. P^3 and P_4 relatively large.

Lycyaena macrostoma vinayaki Pilgrim

Lycyaena macrostoma vinayaki, Pilgrim, 1932, Pal. Indica, N.S. XVIII, pp. 129–130, Pl. VII, figs. 6–9, Pl. IX, fig. 3.

Cotypes.—G.S.I. No. D 137 and D 139, four associated fragments consisting of right P^3, left P^2, right P_4, left P_3. Also D 260, a fragment of a right maxilla.

Horizon.—Middle Siwaliks, Dhok Pathan zone.

Locality.—Hasnot, Salt Range, Jhelum District, Punjab.

Diagnosis.—(See Pilgrim, G. E., 1932, p. 129.) Larger than *L. macrostoma*. Certain minor differences in the dentition.

This subspecies is based on very fragmentary material. It would seem logical to group the above listed specimens with the preceding species, *L. macrostoma*.

Lycyaena (?) chinjiensis Pilgrim

Lycyaena (?) chinjiensis, Pilgrim, 1932, Pal. Indica, N.S. XVIII, pp. 133–134, Pl. VI, fig. 6.

Type.—G.S.I. No. D 233, a right mandibular ramus.

Paratypes.—None.

Horizon.—Lower Siwaliks, Chinji zone.

Locality.—Chinji, Salt Range, Attock District, Punjab.

Diagnosis.—(See Pilgrim, G. E., 1932, p. 133.) "A *Lycyaena* of about the same size as *L. proava;* with a more slender M_1; a much weaker metaconid; M_2 absent."

This species is probably synonymous with *P. proava*.

Progenetta Deperet, 1892

Generic type, *Mustela incerta* Lartet

Progenetta proava (Pilgrim)

Palhyaena proava, Pilgrim, 1910, Rec. Geol. Surv. India, XL, p. 65.

Progenetta proava, Pilgrim, 1913, Rec. Geol. Surv. India, XLIII, p. 282.

Lycyaena (?) proava, Pilgrim, 1932, Pal. Indica, N.S. XVIII, pp. 130–133, Pl. V, figs. 1, 6.

Additional References.—

Pilgrim, G. E., 1910B, p. 199.

Matthew, W. D., 1929, pp. 453, 488–489.

Type.—G.S.I. No. D 126, a left mandibular ramus with P_4–M_1.

Paratypes.—None.

Horizon.—Lower Siwaliks, Chinji zone.

Locality.—From near Chinji, Salt Range, Attock District, Punjab.

Diagnosis.—Distinguished mainly on the basis of its small size.

"This is a very much smaller species than the preceding [*i.e.*, *Palhyaena indica*] with relatively narrower teeth. It comes from the Lower Siwaliks of Chinji." [31]

Crocuta Kaup, 1828

Generic type, *Hyaena crocuta* Erxleben

Crocuta sivalensis (Falconer)

Hyaena sivalensis, Falconer, 1868, Pal. Mem., I, p. 548.
Crocuta sivalensis, Pilgrim, 1932, Pal. Indica, N.S. XVIII, pp. 134–137.

Additional References.—

The type specimen was mentioned by Baker and Durand in 1835, but was not at that time named. See:

Baker, W. E., and Durand, H. M., 1835B, p. 569, Pl. XLVI, figs. 22, 23.

Owen, R., 1870, p. 422, Pl. XXVIII, figs. 5–7 (*Hyaena sinensis*).

Bose, P. N., 1880, p. 128–130.

Lydekker, R., 1880B, p. 30; 1883C, pp. 82, 91; 1884A, p. 281 (*Hyaena felina*); 1884D, p. 124; 1885B, p. 6; 1885C, pp. 88–91. (Credits Bose as being the author of this species.)

Pilgrim, G. E., 1910B, p. 199; 1913B, p. 325.

Matthew, W. D., 1929, pp. 444, 489–491, figs. 26, 27.

*Type.—*Sci. and Art Museum, Dublin, No. 42, a skull and mandible.

*Paratypes.—*None.

*Horizon.—*Upper Siwaliks.

*Locality.—*Siwalik Hills, between the Markanda Pass and Pinjor.

*Specimens in the American Museum.—*Amer. Mus. No. 19824, a fragment of a mandible, with left P_3-M_1. From the Upper Siwaliks, upper clays below the conglomerates, two miles north of Siswan.

Amer. Mus. No. 19886, a fragment of a mandible, with left M_3. Upper Siwaliks, upper clays below conglomerates, near Siswan.

*Diagnosis.—*A large *Crocuta* with elongate and narrow skull. Teeth relatively low crowned, P_1 absent.

Dr. Matthew, in his 1929 paper gives a full discussion of the nomenclature of the various species of Siwalik hyaenids. This paper should be consulted.

He shows that there is a central type, *Hyaena sivalensis* in the Upper Siwalik beds. The so-called *Hyaena felina* is a synonym of *H. sivalensis*, while *H. colvini* is probably a small variety of the same species. This line is specialized towards the *Crocuta crocuta* type. *H. bosei* Matthew, is a more primitive form in the Upper Siwaliks, and it is related to the *Hyaena striata* line. *Hyaena macrostoma* of the Middle Siwaliks is comparable with *Hyaena eximia*.

The premolars of Amer. Mus. No. 19824, here referred to *H. sivalensis*, are robust and P_4 has a large posterior cusp. No trace of an anterior cingulum is found on the first molar. In fact, all of the teeth are without cingula.

[31] Pilgrim, G. E., 1910A, p. 65.

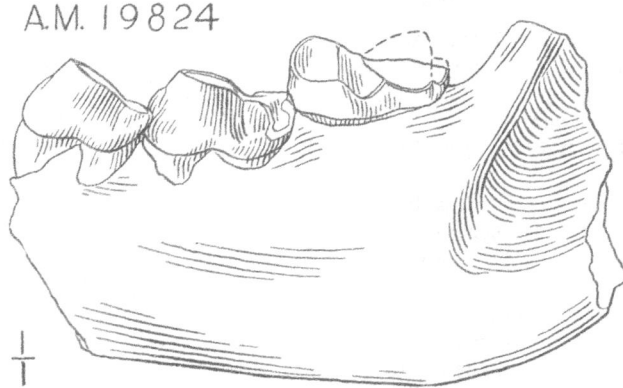

FIG. 50. *Crocuta sivalensis* (Falconer). Amer. Mus. No. 19824, left ramus with P$_3$–M$_1$. Lateral view. Natural size.

MEASUREMENTS

Amer. Mus. No. 19824.

P$_3$.....Antero-posterior diameter.. 21.0 mm.	Transverse diameter.. 15.0 mm.
P$_4$.....Antero-posterior diameter.. 23.0	Transverse diameter.. 14.5
M$_1$....Antero-posterior diameter.. 27.0	Transverse diameter.. 12.5
Depth of mandible below M$_1$...... 50.0	

The mandibular fragment, Amer. Mus. No. 19886, is closely comparable to No. 19824, described above. The lower carnassial of these specimens shows a possible derivation from the same tooth as developed in *C. carnifex*, a fact set forth by Pilgrim.

Crocuta felina (Bose)

Hyaena felina, Bose, 1880, Quar. Jour. Geol. Soc., London, XXXVI, pp. 130–131, Pl. VI, fig. 6.

Crocuta felina, Pilgrim, 1932, Pal. Indica, N.S., XVIII, pp. 137–139.

   Additional References.—

   Lydekker, R., 1880B, p. 30; 1883C, p. 91; 1884A, pp. 278–290, Pls. XXXVIII, fig. 1, XXXIX, fig. 1; 1884D, p. 124; 1884E, p. 72, fig. 3; 1885B, pp. 5–6; 1885C, pp. 80–84, fig. 7.

   Pilgrim, G. E., 1910B, p. 199; 1913B, p. 324.

   Matthew, W. D., 1929, pp. 444, 491.

Type.—B.M. No. 15902, a skull, the anterior portion complete.

Paratypes.—None.

Horizon.—Probably Upper Siwaliks.

Locality.—From the Siwalik Hills.

Diagnosis.—(See Pilgrim, G. E., 1932, p. 137.) Of large size; skull broad; M^1 small and two rooted; P^4 relatively long; P^1 absent; M$_1$ relatively larger than in *C. sivalensis*.

   Probably a synonym of *C. sivalensis*. Full discussion of the status of this species will be found in Matthew 1929, and Pilgrim 1932, cited above.

Crocuta colvini (Lydekker)

Hyaena colvini, Lydekker, 1884, Pal. Indica (X), II, pp. 290–298, Pl. XXXV, figs. 1, 2, 4, 5, Pl. XXXVa, Pl. XXXVI, fig. 1, Pl. XXXVIII, fig. 3, Pl. XXXIX, fig. 4.
Crocuta colvini, Pilgrim, 1932, Pal. Indica, N.S., XVIII, pp. 139–141.

Additional References.—

Lydekker, R., 1884D, p. 124; 1884E, pp. 72–73; 1885B, p. 6; 1885C, pp. 84–87, fig. 8.

Pilgrim, G. E., 1910B, p. 199; 1913B, p. 324.

Matthew, W. D., 1929, pp. 444, 491–492.

Type.—G.S.I. No. D 47, a skull lacking the occipital region.

Paratypes.—G.S.I. No. D 45, skull.   B.M. Nos. 37139 and 37140, two left maxillae. G.S.I. No. D 51, mandibular ramus.   B.M. Nos. 16526, 16578, mandibular rami.   B.M. No. 15413, a right upper carnassial.   Dublin Museum No. 41, part of a maxilla, stated by Lydekker to be a part of the type.   Dublin Museum No. B 9, a left ramus.

Horizon.—Upper Siwaliks.

Locality.—From the Siwalik Hills.

Diagnosis.—(See Pilgrim, G. E., 1932, p. 139.)   Of moderate size; skull slender; M^1 large, with three roots; mandible slender; M_1 relatively longer than in *C. sivalensis*.

Probably a variety of *C. sivalensis*.   See Matthew, W. D., 1929, cited above.

Crocuta carnifex (Pilgrim)

Hyaena carnifex, Pilgrim, 1913, Rec. Geol. Surv. India, XLIII, p. 312.
Crocuta carnifex, Pilgrim, 1932, Pal. Indica, N.S., XVIII, pp. 141–146, Pl. VII, figs. 1–5, 12; Pl. VIII, fig. 2.

Type.—G.S.I. No. D 172, a right mandibular ramus with the milk dentition exposed, and the permanent dentition in alveoli.

Paratypes.—G.S.I. Nos. D 173, a right maxilla; D 171, a left mandibular ramus; D 169, a left ramus; D 170, a left ramus; D 168, a right P^4; D 164, a right ramus of a juvenile specimen.

Horizon.—Lower Siwaliks, Chinji zone.

Locality.—From near Chinji, Salt Range, Attock District, Punjab.

Specimen in the American Museum.—Amer. Mus. No. 19405, a fragment of a left mandibular ramus, with P_4, M_1. From the Lower Siwaliks, Chinji zone, about the level of Chinji Rest House.   Four miles northeast of Chinji Rest House.

Diagnosis.—(See Pilgrim, G. E., 1932, p. 141.)   Of moderate size.   The first premolar above and below absent.   Canines heavy.   Upper carnassial with small, low protocone. Mandible robust anteriorly. M_1 without a metaconid. P_4 with strong anterior and posterior cusps.

The type specimen of this species was first supposed by Pilgrim to be an hyaenodont, as it was erroneously associated with some maxillary fragments of *Dissopsalis*.   Later Pilgrim recognized the true affinities of the specimen.

The American Museum specimen is quite similar to the various corresponding lower teeth figured by Pilgrim in his memoir of the Siwalik carnivores.   The salient characters, as Pilgrim shows, are the relatively small M_1, the reduced, single cusped talonid of this tooth, and the transverse posterior border and the tapering anterior portion of the fourth premolar.

There is little to add to Pilgrim's description, except to point out the shape of that portion of the mandible below the last two cheek teeth. The ramus is rather deep just behind M_1 and the lower border curves upward. The masseteric fossa extends forward to the posterior border of M_1.

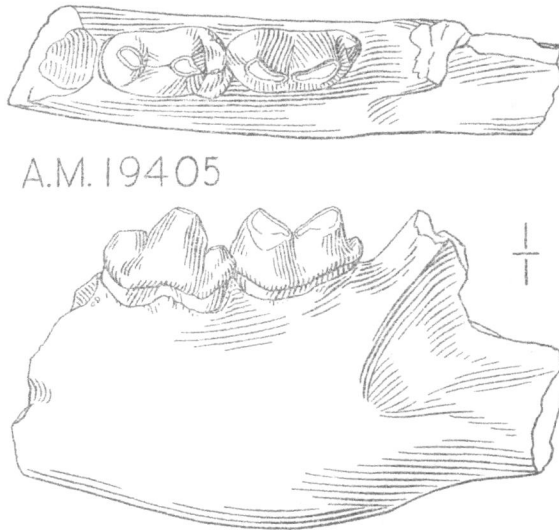

A.M. 19405

Fig. 51. *Crocuta carnifex* (Pilgrim).   Amer. Mus. No. 19405, left mandibular ramus with P_4, M_1.   Crown view above, lateral view below.   Natural size.

MEASUREMENTS

Amer. Mus. No. 19405.

P_4....Antero-posterior diameter.. 21.0 mm.   Transverse diameter.. 11.5 mm.
M_1...Antero-posterior diameter.. 22.0   Transverse diameter.. 11.0
Depth of ramus below anterior border of M_1.. 40.0 mm.

The American Museum specimen of *Crocuta carnifex* comes from the base of the Chinji zone, while the specimens described by Pilgrim were found at the top of the Chinji beds. Thus the American Museum specimen extends the range of this species downward.

Dr. Pilgrim has pointed out the close relationship between *Crocuta carnifex* and *Crocuta eximia*. The Chinji form would seem to be very close to *Hyaena variabilis* of North China. Evidently the *Crocuta* branch of the hyaenas developed at a relatively early date, probably from an *Ictitherium*-like ancestor, which later persisted on parallel to the advanced forms.

```
            Hyaena striata                    Crocuta crocuta
                 |                                  |
Ictitherium     Hyaena macrostoma                Crocuta eximia
     |               |                              |
     |               |                           Crocuta carnifex
     |               |_____     |
     |_____|_____|
                  Ictitherium
```

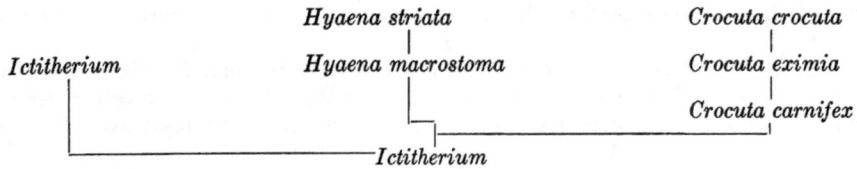

Crocuta gigantea (Schlosser)

Hyaena gigantea, Schlosser, 1903, Abhandl. Bayer. Akad. Wiss., Munich, XXII, p. 35,
    Pl. II, figs. 1–3, 6–8.
Additional References.—
    Pilgrim, G. E., 1913B, pp. 282, 289.
    Matthew, W. D., 1929, p. 448.

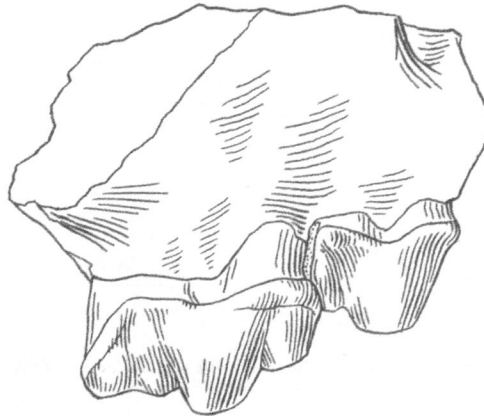

A.M. 19888

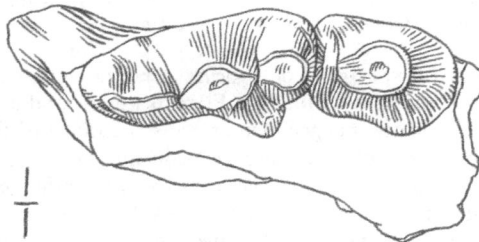

FIG. 52. *Crocuta gigantea* (Schlosser). Amer. Mus. No. 19888, fragment of maxilla with right P²⁻⁴. Lateral view
above, crown view below. Natural size.

Type.—Upper and lower teeth. See Schlosser, op. cit.

Horizon and Locality.—Pliocene of North China. Middle Siwaliks, Punjab, India.

Specimens in the American Museum.—Amer. Mus. No. 19888, a maxilla, with right P^{3-4}. From the upper portion of the Middle Siwaliks, or the lower portion of the Upper Siwaliks, three and one half miles north of Hasnot, Punjab.

Diagnosis.—Very large, and similar to *H. variabilis*, with robust metacone and a reduced protocone in the upper molars.

The American Museum specimen, here referred to this species, is distinguished by a robust P^3, having a large central cone, a small posterior cusp and a sloping anterior shelf. The carnassial is long, with a large robust paracone and metacone, and the protocone is reduced. The metastyle shear is somewhat less than half the length of the tooth.

MEASUREMENTS

Amer. Mus. No. 19888.

P^3....Antero-posterior diameter.. 26.0 mm.	Transverse diameter.. 18.0 mm.
P^4....Antero-posterior diameter.. 41.0	Transverse diameter.. 21.0

As pointed out above, the carnassial of this specimen is marked by a great reduction of the inner cusp. This character is typical of a group of Hyaenas of eastern Europe and Asia, namely *H. eximia, gigantea, variabilis,* and *mordax.* Lacking additional material it might be best to consider that all of these several species are closely related—perhaps some of them are local variations of one species. Zdansky has shown how variable are the molars of the North China form *H. variabilis,* and in the light of his studies it would seem possible that the distinctions between the various species named above may in some cases be more or less arbitrary.

Crocuta gigantea latro Pilgrim

Crocuta gigantea latro, Pilgrim, 1932, Pal. Indica, N.S., XVIII, pp. 146–149, Pl. VII, figs. 11, 13; Pl. VIII, figs. 1, 3.

Type.—G.S.I. No. D 206, a right maxilla.

Paratypes.—G.S.I. Nos. D 208, a left P^3; D 209, a fragmentary right mandibular ramus; D 162, a fragmentary left ramus; D 231, a fragmentary right ramus of a juvenile specimen.

Horizon.—Middle Siwaliks, running through the extent of the beds.

Locality.—Hasnot, Nila, Kadiraur and Hari Talyangar, northern India.

Diagnosis.—(See Pilgrim, G. E., 1932, p. 146.) A large form. Protocone of upper carnassial low, parastyle shorter than protocone. Mandible robust, with deep masseteric fossa. M_1 approximately of the same size as P_4.

Crocuta mordax Pilgrim

Crocuta mordax, Pilgrim, 1932, Pal. Indica, N.S., XVIII, pp. 150–153, Pl. VI, figs. 1, 3, 4; Pl. VII, fig. 10.

Type.—G.S.I. Nos. D 204, D 205, an associated mandible and left maxilla of a juvenile specimen.

Paratypes.—G.S.I. No. D 163, a fragmentary right ramus with P_4.

Horizon.—Middle Siwaliks, Dhok Pathan zone.

Locality.—From Dhok Pathan, a referred specimen from Hasnot.

Diagnosis.—(See Pilgrim, G. E., 1932, p. 150.) A very large species, differing from *C. sivalensis* in minor details of the dentition.

This species is very probably synonymous with *C. gigantea* or *C. gigantea latro.*

Hyaenids, Species Indeterminate

Amer. Mus. No. 19338, a fragment of a maxilla, with right P^2. From the Middle Siwaliks, Dhok Pathan zone, four miles east of Dhok Pathan.

Amer. Mus. No. 19344, fragments of right and left DM_4. Lower Siwaliks, Chinji zone, 200 feet above the level of Chinji Rest House, four miles west of Chinji Rest House, Punjab.

Amer. Mus. No. 19347, fragment of a left P^3. From the Lower Siwaliks, Chinji zone, at the level of Chinji Rest House, four miles northeast of Chinji Rest House, Punjab.

Amer. Mus. No. 19940, a right carnassial from the maxilla. Lower portion of the Middle Siwaliks, 1000 feet below the Bhandar bone bed, four and one half miles west of Hasnot, Punjab.

FELIDAE

Proailurinae (ex Pilgrim)

Mellivorodon Lydekker, 1884

Generic type, *Mellivorodon palaeindicus* Lydekker

Mellivorodon palaeindicus Lydekker

Mellivorodon palaeindicus, Lydekker, 1884, Pal. Indica (X), II, pp. 185–186, Pl. XXVII, figs. 7–8.

Additional References.—

Lydekker, R., 1884D, p. 125; 1885B, p. 11.

Pilgrim, G. E., 1910B, p. 199; 1913B, p. 282.

Matthew, W. D., 1929, pp. 466, 467–469, fig. 5.

Pilgrim, G. E., 1932, pp. 156–157.

Type.—(Lectotype).—G.S.I. No. D 21, an incomplete left mandibular ramus.

Cotype.—G.S.I. No. D 22, a fragment of a left mandibular ramus.

Horizon.—Middle Siwaliks, Dhok Pathan zone.

Locality.—Near Hasnot and Niki, Salt Range, Punjab.

Diagnosis.—(See Matthew, W. D., 1929, pp. 467–468.) Of large size. Two large, subequal premolars, compressed and elongated. Lower carnassial narrow and long, as in the cats. Lower border of mandible straight, with a slight angulation at the symphysis. Mental foramina arranged as in the Felidae.

As Dr. Matthew has pointed out, this genus and species have very little value, because of the extremely fragmentary material on which they are based. According to Matthew,

"No additional and more characteristic specimens having been referred to the species or genus, it appears that both should be suppressed."—Op. cit., p. 469.

Vinayakia Pilgrim, 1932

Generic type, *Vinayakia nocturna* Pilgrim

Vinayakia nocturna Pilgrim

Vinayakia nocturna, Pilgrim, 1932, Pal. Indica, N.S., XVIII, pp. 158–162, Pl. IV, figs. 2, 4.
   Type.—G.S.I. No. D 221, a right mandibular ramus.
   Paratype.—G.S.I. No. D 218, a fragmentary maxilla.
   Horizon.—Middle Siwaliks.
   Locality.—From Kaderpur, Attock District, Salt Range, and from Bahitta near Hasnot, Salt Range, Jhelum District, Punjab.
   Diagnosis.—(See Pilgrim, G. E., 1932, pp. 157–158.) Extremely large proailurine, with an anteriorly placed orbit. Lower border of mandible slightly concave, and symphysis deep. Lower premolars unreduced and spaced. P^4 compressed, with high, internal protocone. P^2 and P^3 expanded internally.

Vinayakia sarcophaga Pilgrim

Vinayakia sarcophaga, Pilgrim, 1932, Pal. Indica, N.S., XVIII, pp. 162–164, Pl. IV, fig. 5.
   Type.—G.S.I. No. D 217, associated left and right maxillae.
   Paratypes.—None.
   Horizon.—Lower Siwaliks, Chinji zone.
   Locality.—Kotal Kund, Jhelum district, Salt Range, Punjab.
   Diagnosis.—(See Pilgrim, G. E., 1932, pp. 162–164.) Anterior premolars less reduced than in the preceding species. Internal expansion of P^2, P^3 more prominent.
   The genus *Vinayakia* and its two species are based on very fragmentary material, thereby rendering the genus of doubtful value. According to Pilgrim, *Vinayakia* is probably directly ancestral to *Mellivorodon*.

INCERTAE SEDIS

Aeluropsis Lydekker, 1884

Generic type, *Aeluropsis annectans* Lydekker

Aeluropsis annectans Lydekker

Aeluropsis annectans, Lydekker, 1884, Pal. Indica (X), II, pp. 316–317, Pl. XXXIII, figs. 4, 4a.
   Additional References.—
      Lydekker, R., 1884D, p. 124; 1885B, p. 5.
      Pilgrim, G. E., 1910B, p. 199; 1913B, p. 283.
      Matthew, W. D., 1929, pp. 448, 496–497, fig. 29.
      Pilgrim, G. E., 1932, pp. 165–166.
   Type.—G.S.I. No. D 41, a right mandibular ramus.
   Paratypes.—None.

Horizon.—Probably Middle Siwaliks.

Locality.—From Hasnot, Salt Range, Jhelum District, Punjab.

Diagnosis.—(See Pilgrim, G. E., 1932, p. 165). Felidae with a long jaw, and a deep ramus. P_4 longer than M_1, with well developed anterior and posterior accessory cusps. M_1 short, with a relatively long talonid. M_2 present.

Both Matthew and Pilgrim have discussed this genus and species thoroughly. They show that *Aeluropsis* is a primitive true felid, comparable in development to *Aelurictis*.

Hyainailouros Biedermann, 1863

Hyaenailurus Rütimeyer, 1867

Hyaenaelurus Stehlin, 1907

Generic type, *Hyaenaelurus sulzeri* Biedermann

Hyaenaelurus lahirii Pilgrim

Hyainailouros lahirii, Pilgrim, 1932, Pal. Indica, N.S., XVIII, pp. 169–171, Pl. III, fig. 12.

Type.—G.S.I. No. D 236, a right mandibular ramus with P_4, M_1.

Paratypes.—None.

Horizon.—Kamlial zone of the Lower Siwaliks.

Locality.—One mile northwest of the village of Phamra Khalsa, Salt Range, Attock District.

Diagnosis.—(See Pilgrim, G.E., 1932, p. 169.) A very large species of the genus. P_4 and M_1 very broad in comparison to their lengths. Notch between paraconid and protoconid of M_1 relatively low, metaconid absent or rudimentary. P_4 with anterior and posterior accessory cusps placed further to the inside of the main cusp than in other species of the genus.

This extremely peculiar and aberrant genus is here placed in the Felidae with a full realization that it may more properly belong in the Hyaenidae or in a separate family of its own—as advocated by Helbing. The reader is referred to Dr. Pilgrim's monograph for a detailed description of the Siwalik form.

Machaerodontinae

Megantereon Croizet and Jobert, 1828

Generic type, *Felis megantereon* Croizet and Jobert

Megantereon falconeri (Pomel)

Meganthereodon falconeri, Pomel, 1853, Catalogue Méthodique et descriptif des vertébrés fossiles, etc., Paris, p. 56.

Machaerodus falconeri, Gaudry, 1862, Anim. Foss. et Geol. de l'Attique, p. 113.

Drepanodon sivalensis, Falconer, 1868, Pal. Mem., Vol. I, p. 550, Pl. XXV, figs. 5, 6.

Machaerodus sivalensis, Bose, P. N., 1880, Quar. Jour. Geol. Soc., London, XXXVI, pp. 122–125, Pl. VI, fig. 5.

Meganthereon falconeri, Matthew, 1929, Bull. Amer. Mus. Nat. Hist., LVI, pp. 444, 503–505, fig. 30.

Drepanodon cautleyi, Kretzoi, 1929, Proc. Congr. Internat. Zool., Budapest, 1927, p. 1331.

Megantereon falconeri, Pilgrim, 1932, Pal. Indica, N.S., XVIII, pp. 175–178.

Additional References.—

Lydekker, R., 1880B, p. 30; 1883C, pp. 82, 91; 1884A, pp. 334–341, Pl. XLIV, figs. 1, 2, 4, 5, 6; 1884D, p. 123; 1885B, p. 2; 1885C, pp. 44–46.

Pilgrim, G. E., 1910B, p. 199; 1913B, p. 324.

Type.—(Lectotype).—B.M. No. 16557, a left ramus with P_3–M_1.

Cotypes.—B.M. Nos. 16350, a right maxilla; 16554, a right mandibular ramus; 39370, portion of a maxilla.

Horizon.—Upper Siwaliks.

Locality.—From the Siwalik Hills.

Diagnosis.—(See Pilgrim, G. E., 1932, p. 175.) A fairly large species with crenulated upper canines. P^4 with greatly reduced protocone and with rudiment of an accessory cusp, anterior to the parastyle. P_4 with strong posterior cingulum.

The reader is referred to Matthew, W. D., 1929, pp. 503–505, for an illuminating discussion of this species.

Megantereon palaeindicus (Bose)

Machaerodus palaeindicus, Bose, 1880, Quar. Jour. Geol. Soc., London, XXXVI, pp. 125–126, Pl. VI, figs. 1–4.

Meganthereon palaeindicus, Matthew, 1929, Bull. Amer. Mus. Nat. Hist., LVI, pp. 444, 505–506.

Megantereon palaeindicus, Pilgrim, 1932, Pal. Indica, N.S., XVIII, pp. 178–179.

Additional References.—

Lydekker, R., 1880B, p. 30; 1883C, p. 91; 1884A, pp. 341–345, Pl. XLV, fig. 3; 1884D, p. 123; 1885B, p. 3; 1885C, pp. 46–47.

Pilgrim, G. E., 1910B, p. 199; 1913B, p. 325.

Type.—(Lectotype).—B.M. No. 48436, a left mandibular ramus.

Cotypes.—B.M. Nos. 48437, fragment of a right mandibular ramus; 39728, occipital portion of a skull; 39729, a damaged skull.

Horizon.—Upper Siwaliks.

Locality.—From the Siwalik Hills.

Diagnosis.—(See Pilgrim, G. E., 1932, p. 178.) Larger than *M. falconeri,* with a more robust mandible. P_4 stout, with a large posterior accessory cusp.

Megantereon praecox Pilgrim

Megantereon praecox, Pilgrim, 1932, Pal. Indica, N.S., XVIII, pp. 179–180, Pl. IX, fig. 2.

Type.—G.S.I. No. D 232, a left maxilla.

Paratypes.—None.

Horizon.—Middle Siwaliks, Nagri zone.

Locality.—From Hari Talyangar, Bilaspur State, Simla Hills.

Diagnosis.—(See Pilgrim, G. E., 1932, p. 179.) Slightly smaller than *M. falconeri.* P^4 with a more prominent protocone and a stronger anterior accessory cusp. M_1 larger.

Megantereon sp.

Amer. Mus. No. 19935, a left upper carnassial with the protocone broken away. From the Middle Siwaliks, near the top, two miles northwest of Hasnot.

This tooth is long and slender, and plainly that of a cat. The protocone is broken, but from the appearance of a fragment of its base, still remaining, it would seem to have been rather small. For this reason, as well as because of the general shape of the tooth, the specimen is referred to the genus *Megantereon*.

MEASUREMENTS

Amer. Mus. No. 19935, left P^4.
    Antero-posterior diameter.....36.0 mm.   Transverse diameter..... 13.0 mm.

Sansanosmilus Kretzoi, 1929

Generic type, *Felis palmidens* Blainville

Sansanosmilus serratus Pilgrim

Sansanosmilus serratus, Pilgrim, 1932, Pal. Indica, N.S., XVIII, pp. 181–183, Pl. VIII, fig. 9.
    Type.—G.S.I. No. D 165, a fragmentary mandibular ramus.
    Paratypes.—None.
    Horizon.—Lower Siwaliks, Chinji zone.
    Locality.—From near Chinji, Salt Range, Attock District, Punjab.
    Diagnosis.—(See Pilgrim, G. E., 1932, p. 181.) A small machaerodont. High crowned teeth, with crenulated and compressed cusps. M_1 with vestigial metaconid.

Sansanosmilus rhomboidalis Pilgrim

Sansanosmilus rhomboidalis, Pilgrim, 1932, Pal. Indica, N.S., XVIII, pp. 183–196, Pl. VIII, figs. 5, 10, 11.
    Type.—G.S.I. No. D 154, a fragmentary left maxilla.
    Paratypes.—G.S.I. Nos. D 152, a left DM^3; D 153, a left P^4.
    Horizon.—Lower Siwaliks, Chinji zone.
    Locality.—From near Chinji, Salt Range, Attock District, Punjab.
    Diagnosis.—(See Pilgrim, G. E., 1932, p. 183.) Upper canine very short and slender, with crenulated edges. P^4 slender.

Paramachaerodus Pilgrim, 1913

Generic type, *Machairodus orientalis* Kittl

Paramachaerodus pilgrimi Kretzoi

Paramachaerodus pilgrimi, Kretzoi, 1929, Proc. Internat. Congr. Zool., Budapest, 1927, p. 1300.
    Additional Reference.—
        Pilgrim, G. E., 1932, pp. 188–189, Pl. VIII, fig. 6.

Type.—G.S.I. No. D 140, a left mandibular ramus.
Paratypes.—None.
Horizon.—Middle Siwaliks, Dhok Pathan zone.
Locality.—From near Hasnot and Kotal Kund in the Salt Range, Jhelum District, Punjab.
.*Diagnosis.*—(See Pilgrim, G. E., 1932, p. 186.) A medium sized machaerodont. Post canine diastemata short. Mandible relatively heavy.

Dr. Matthew was opposed to the erection of a new species for the above mentioned type, which specimen had been referred by Pilgrim (1915) to *Paramachaerodus schlosseri.* Matthew referred this specimen (G.S.I. No. D 140) to *Machaerodus sivalensis.* The reader is referred to Matthew, W. D., 1929, p. 506.

Paramachaerodus indicus (Kretzoi)

Pontosmilus indicus, Kretzoi, 1929, Proc. Internat. Congr. Zool., Budapest, 1927, p. 1300.
Paramachaerodus indicus, Pilgrim, 1932, Pal. Indica, N.S., XVIII, pp. 189–190, Pl. VIII, fig. 7.
Type.—G.S.I. No. D 141, a left mandibular ramus.
Paratypes.—None.
Horizon.—Middle Siwaliks, Dhok Pathan zone.
Locality.—Bahitta, south of Hasnot, Salt Range, Jhelum District, Punjab.
Diagnosis.—(See Pilgrim, G. E., 1932, p. 189.) About equal in size to *P. pilgrimi,* but with more slender ramus. M₁ relatively short. Canine crenulated on both keels.

As in the case of *P. pilgrimi,* Dr. Matthew opposed the erection of a new species for the above named type. He referred it to *Pseudaelurus.* The reader is referred to Matthew, W. D., 1929, p. 506.

Propontosmilus Kretzoi, 1929

Generic type, *Pseudaelurus sivalensis* Lydekker

Propontosmilus sivalensis (Lydekker)

Pseudaelurus sivalensis, Lydekker, 1877, Rec. Geol. Surv. India, X, p. 83.
Aelurogale sivalensis, Lydekker, 1884, Pal. Indica (X) II, pp. 317–319, Pl. XLIV, fig. 7.
Aelurictis sivalensis, Pilgrim, 1910, Rec. Geol. Surv. India, XL, p. 199.
Sivaelurus sivalensis, Pilgrim, 1913, Rec. Geol. Surv. India, XLIII, pp. 282, 291.
Paramachaerodus sivalensis, Pilgrim, 1915, Rec. Geol. Surv. India, XLV, p. 142.
Machaerodus sivalensis, Matthew, 1929, Bull. Amer. Mus. Nat. Hist., LVI, p. 506.
Propontosmilus sivalensis, Kretzoi, 1929, Proc. Internat. Congr. Zool., Budapest, 1927, p. 1298.
Additional References.—
    Lydekker, R., 1884D, p. 124; 1885B, p. 5.
    Pilgrim, G. E., 1932, pp. 191–192, Pl. VIII, fig. 8.
Type.—G.S.I. No. D 95, a right mandibular ramus with M₁.
Paratypes.—None.
Horizon.—Middle Siwaliks, Dhok Pathan zone.

Locality.—From an unknown locality in the Salt Range, Punjab.

Diagnosis.—(See Pilgrim, G. E., 1932, pp. 190–191.)   Of small size, with stout but shallow mandible. M_1 relatively robust and long.

Full discussions of this species will be found in Matthew (1929) and Pilgrim (1932), cited above.

"*P. sivalensis* is too imperfectly known to be certain of its affinities, but probably is near to the D 140 jaw [type of *Paramachaerodus pilgrimi* Kretzoi], differing chiefly in somewhat more shallow jaw, less vertical symphyseal ridge, presence of minute p_2."— Matthew, W. D., 1929, p. 506.

Sivasmilus Kretzoi, 1929

Generic type, *Sivasmilus copei* Kretzoi

Sivasmilus copei Kretzoi

Sivasmilus copei, Kretzoi, 1929, Proc. Internat. Congr. Zool., Budapest, 1927, p. 1297.

Additional Reference.—

Pilgrim, G. E., 1932, p. 193.

Type.—G.S.I. No. D 151, a left mandibular ramus.

Paratypes.—None.

Horizon.—Lower Siwalik, Chinji zone.

Locality.—From near Chinji, Salt Range, Attock District, Punjab.

Diagnosis.—(See Pilgrim, G. E., 1932, p. 192.)   A small machaerodont with relatively slender mandible.   Chin forming more obtuse angle with lower border of ramus than is the case with other genera.   Canine large, with oval cross section. P_2 rudimentary, and P_3 and P_4 reduced.

FELINAE

Felis Linnaeus, 1758

Generic type, *Felis catus* Linnaeus

Felis subhimalayana Bronn

Felis subhimalayana, Bronn, 1848, Index Palaeontologicus, p. 492.

Additional References.—

Lydekker, R., 1884A, pp. 330–331; 1884D, p. 124; 1884E, p. 73.

Pilgrim, G. E., 1910B, p. 199; 1913B, p. 324.

Matthew, W. D., 1929, pp. 444, 495.

Pilgrim, G. E., 1932, pp. 196–197.

Type.—(Lectotype).—Sci. and Art Mus., Dublin, No. 47, a skull.

Cotypes.—Sci. and Art Museum, Dublin, No. 48, a mandible.

Horizon.—Probably Upper Siwaliks.

Locality.—From the Siwalik Hills.

A small species, of which little is known.

Panthera Frish, 1775

Generic type, *Felis panthera* Pallas

Panthera cristata (Falconer and Cautley)

Felis cristata, Falconer and Cautley, 1836, Asiatic Researches, XIX, p. 135, Pl. XXI, figs. 1, 2.

Felis palaeotigris, Falconer, 1868, Pal. Mem., Vol. I, p. 315, also, p. xxi.

Felis grandicristata, Bose, 1880, Quar. Jour. Geol. Soc., London, XXXVI, pp. 127–128.

Uncia cristata, Cope, 1880, Amer. Nat., XIV, p. 853.

Felis cristata, Lydekker, 1884, Pal. Indica (X), II, pp. 320–326, Pl. XL, figs. 1, 2; Pls. XLI, XLII.

Panthera cristata, Pilgrim, 1932, Pal. Indica, N.S., XVIII, pp. 198–199.

Additional References.—

Lydekker, R., 1880B, p. 30; 1883C, pp. 82, 90; 1884D, p. 124; 1885B, p. 3; 1885C, p. 58.

Pilgrim, G. E., 1910B, p. 199; 1913B, p. 324.

Matthew, W. D., 1929, pp. 444, 494.

Kretzoi, N., 1929B, p. 14.

Type.—A skull in the Museum of the Royal College of Surgeons, London. (British Museum cast No. 28913.)

Paratypes.—None.

Horizon.—Upper Siwaliks.

Locality.—From the Siwalik Hills.

Diagnosis.—(See Pilgrim, G. E., 1932, p. 198.) Slightly smaller than *F. tigris*. Facial portion of skull considerably shorter than cranial portion. Skull closely comparable to that of *F. tigris*.

Sivafelis Pilgrim, 1932

Generic type, *Sivafelis potens* Pilgrim

Sivafelis potens Pilgrim

Sivafelis potens, Pilgrim, 1932, Pal. Indica, N.S., XVIII, pp. 200–202, Pl. III, figs. 2, 10.

Type.—G.S.I. No. D 222, a right mandibular ramus.

Paratypes.—B.M. No. 16537a, a left mandibular ramus. G.S.I. No. D 265, a left maxilla.

Horizon.—Upper Siwaliks, Pinjor zone.

Locality.—From near Moginand in the Siwalik Hills. The maxilla came from Jharakki, in the Salt Range.

Diagnosis.—(See Pilgrim, G. E., 1932, p. 200.) Of moderate size, with a very deep, stout mandible. Canine oval in cross section, and relatively small. Metaconid of M_1 absent or rudimentary.

Sivafelis brachygnathus (Lydekker)

Felis (?Cynaelurus) brachygnathus, Lydekker, 1884, Pal. Indica (X), II, pp. 326–328, Pl. XLIII, figs. 1, 2.

Cynaelurus brachygnathus, Pilgrim, 1910, Rec. Geol. Surv. India, XL, p. 199.

Felis brachygnatha, Matthew, 1929, Bull. Amer. Mus. Nat. Hist., LVI, pp. 444, 494.

Acinonyx brachygnathus, Kretzoi, 1929, Proc. Internat. Congr. Zool., Budapest, 1927, p. 1330.

Abacinonyx brachygnathus, Kretzoi, 1929, M. Kir. Foldt. Int. Haxiny, Budapest, V, 24, p. 11.

Sivafelis brachygnathus, Pilgrim, 1932, Pal. Indica, N.S., XVIII, pp. 202–203.

> *Additional References.—*
>> Lydekker, R., 1884D, p. 124; 1885B, p. 3; 1885C, pp. 58–59.
>> Pilgrim, G. E., 1913B, p. 325.
>
> *Type.—*(Lectotype).—B.M. No. 16573, a right mandibular ramus.
>
> *Cotype.—*B.M. No. 16537, a right mandibular ramus.
>
> *Horizon.—*Upper Siwaliks, Pinjor zone.
>
> *Locality.—*From the Siwalik Hills.
>
> *Diagnosis.—*(See Pilgrim, G. E., 1932, p. 202.)   Larger than *S. potens*, with a less

robust mandible.

An illuminating discussion will be found in Matthew (1929), cited above.   He shows that *Felis brachygnatha* is very close, if not identical to *Felis arvernensis* Croizet and Jobert, and to *Cynaelurus pleistocaenicus* Zdansky.

Sivaelurus Pilgrim, 1913

Generic type, *Pseudaelurus chinjiensis* Pilgrim

Sivaelurus chinjiensis (Pilgrim)

Pseudaelurus chinjiensis, Pilgrim, 1910, Rec. Geol. Surv. India, XL, p. 65.

Sivaelurus chinjiensis, Pilgrim, 1913, Rec. Geol. Surv. India, XLIII, pp. 282, 291, 314.

Aeluropsis chinjiensis, Matthew, 1929, Bull. Amer. Mus. Nat. Hist., LVI, pp. 453, 498–499.

Sivaelurus chinjiensis, Pilgrim, 1932, Pal. Indica, N.S., XVIII, pp. 204–206.

> *Additional References.—*
>> Pilgrim, G. E., 1910B, p. 199; 1915B, p. 145, Pl. VI, figs. 1, 1a.
>
> *Type.—*G.S.I. No. D 150, a right maxilla.
>
> *Paratypes.—*None.
>
> *Horizon.—*Lower Siwaliks, Chinji zone.
>
> *Locality.—*From near Chinji, Salt Range, Attock District, Punjab.
>
> *Diagnosis.—*(See Matthew, W. D., 1929, p. 498.)   Upper canine oval, probably not

greatly elongate.   Upper and lower second premolars present.   Upper and lower third premolars unreduced.   Upper carnassial with a distinct inner cusp, small parastyle and rudimentary fourth cusp. M^1 narrow, elongate, set transversely.   Infraorbital foramen small.

Dr. Matthew, cited above, gives a comprehensive discussion of this species.   He shows that this species is more primitive than the Siwalik sabre tooth cats, and is probably nearer to the true cats.

Vishnufelis Pilgrim, 1932

Generic type, *Vishnufelis laticeps* Pilgrim

Vishnufelis laticeps Pilgrim

Vishnufelis laticeps, Pilgrim, 1932, Pal. Indica, N.S., XVIII, pp. 206–209, Pl. IX, fig. 1.

Type.—G.S.I. No. D 266, associated portions of a skull.

Paratypes.—None.

Horizon.—Lower Siwaliks, Chinji zone.

Locality.—Two and three fourths miles east of Paridarwaza in the Jhelum District, Salt Range.

Diagnosis.—(See Pilgrim, G. E., 1932, p. 206.) Small, skull low and face elongated. Very broad at zygomatic arches. Nasals short and narrow. Infraorbital foramen of moderate size. Brain case relatively narrow. Mastoid prominent. Upper canine rather procumbent, lower canine small. P^2 rudimentary. P^4 with prominent protocone.

Felis (?) sp.

Amer. Mus. No. 19337, a fragment of a mandible from the Middle Siwaliks, three miles east of Dhok Pathan.

The crowns of the teeth are missing on this specimen, making an exact identification impossible.

Felis (?) sp.

Amer. Mus. No. 19344, two lower milk carnassials (?), from the Lower Siwaliks, Chinji zone, from a level about 200 feet above Chinji Rest House. Four miles west of Chinji Rest House.

There are two lower carnassials, a right and a left, of a small felid. From their small size and slender build, they may be inferred to be milk teeth. They may be referable to Pilgrim's *Sivaelurus* or to his *Vishnufelis* but because of the want of better and more definite material they can not be more exactly determined.

MEASUREMENTS

Amer. Mus. No. 19344. Left DM_1.
Length...... 13.0 mm.   Width...... 6.0 mm.   Height...... 11.0 mm.

FELID

Amer. Mus. No. 19835, a fragment of a lower jaw, with P_2 of the right side. From the base of the Upper Siwaliks, or the top of the Middle Siwaliks, three miles south of Hasnot.

The premolar tooth is small. It consists of a larger central cusp and a smaller posterior one.

This specimen would seem to be pathologic. The bone of the mandible in front of the tooth is swollen by exostosis, while the canine alveolus is filled with cancellous tissue.

MEASUREMENTS

Amer. Mus. No. 19835.
Length of right P_2.................. 15.0 mm.   Width.......... 8.5 mm.
Depth of mandible below P_2.......... 34.0

CARNIVORA, INCERTAE SEDIS

Amer. Mus. No. 19341, an upper third molar. Lower Siwaliks, Chinji zone, about 100 feet above the level of Chinji Rest House. Four miles west of Chinji Rest House.

Amer. Mus. No. 19345, a tooth. Lower Siwaliks, five miles east of Chinji Rest House.

TUBULIDENTATA

Tubulidentates in the Siwalik beds were unknown until the time of Mr. Brown's work in the Punjab and the Salt Range area. There he discovered some scattered remains of *Orycteropus* from the lower portion of the Middle Siwaliks. These have recently been described (Colbert, E. H., 1933B) as representing two species.

The chief interest of *Orycteropus* in the Siwaliks is that it extends the range of the order Tubulidentata much to the east of its previous known distribution. This in turn suggests the probability of the order having extended into eastern Asia during the earlier portion of the Tertiary period, and from thence across into North America in the Eocene, where it very likely had its origin. The following quotation is taken from the paper cited above.

"If Jepsen's newly described form [i.e., *Tubulodon taylori* Jepsen, Wind River, Eocene, Wyoming] is really a tubulidentate, . . . it would seem that this peculiar order of mammals had its origin in North America, during the late Mesozoic or the Eocene. Consequently it migrated from thence, westward through Asia to Africa. . . . This is an example of a westward migration of an order, somewhat the opposite of the commonly postulated radial migration from central Asia, applied to so many groups of mammals."[32]

ORYCTEROPODIDAE

Orycteropus Geoffroy, 1795

Generic type, *Myrmecophaga capensis* Gmelin

Orycteropus browni Colbert

Orycteropus browni, Colbert, 1933, Amer. Mus. Novitates, No. 604, pp. 1–6, figs. 1–4.

Type.—A.M. No. 29840, a left maxilla with M^{2-3}.

Paratypes.—None.

Horizon.—Middle Siwaliks, near base. Nagri zone.

Locality.—One half mile south of Nathot, Salt Range, Jhelum District, Punjab.

Diagnosis.—A small species, about three fifths the size of *Orycteropus gaudryi*. Third molar characterized by the extreme reduction of the posterior column. In a microscopic cross section the tubules of the molar are seen to be closely appressed and irregularly polygonal. Tubules quite variable in size. In a longitudinal section the tubules are seen to branch to a certain extent which indicates the retention of a primitive character.

A detailed account of this species is given in the type description, to which the reader is referred. The type figures and the measurements are reproduced below.

[32] Colbert, E. H., 1933B, p. 8.

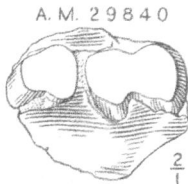

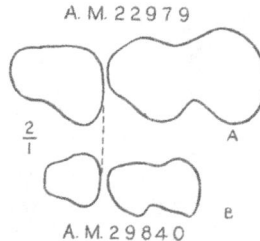

FIG. 53.

FIG. 54.

FIG. 53. *Orycteropus browni* Colbert. Type, Amer. Mus. No. 29840, maxilla with left M²⁻³. Crown view. Twice natural size. From Colbert, 1933.

FIG. 54. Outlines of the crowns of the left upper second and third molars in (A) *Orycteropus gaudryi* Forsyth Major, and in (B) *Orycteropus browni* Colbert. Twice natural size. From Colbert, 1933.

MEASUREMENTS

Orycteropus browni.   A.M. No. 29840.

M²....Length.... 7.7 mm.   Width.... 5.3 mm.   Height.... 10.0 mm.
M³.... "   .... 4.7   "   .... 4.7   " 8.0
Ratio, M³/M² × 100 = 61
Radial diameter of tubules (average)................................... .10–.50 mm.
Tangential diameter of tubules (average)............................... .10–.25
Width of external cement band... .10

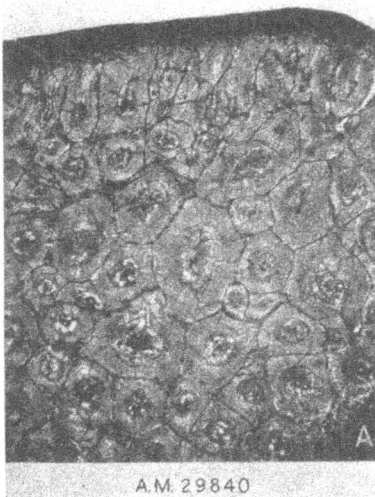

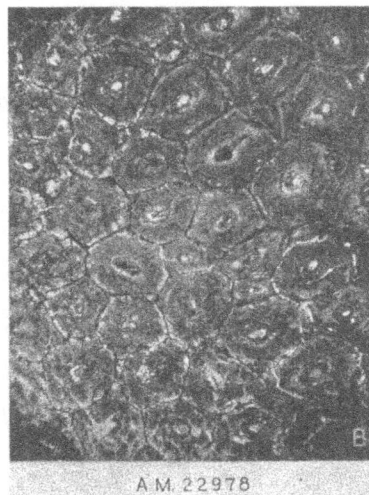

FIG. 55. Photomicrographs of cross sections of the left upper second molar in (A) *Orycteropus browni* Colbert, and in (B) *Orycteropus gaudryi* Forsyth Major. In A, a section of the thin band of cementum, which surrounds the tooth, is shown at the top, and below are irregular tubules characteristic of *O. browni*. Section B shows the rather regular tubules characteristic of *O. gaudryi*. Both sections about forty times natural size. From Colbert, 1933.

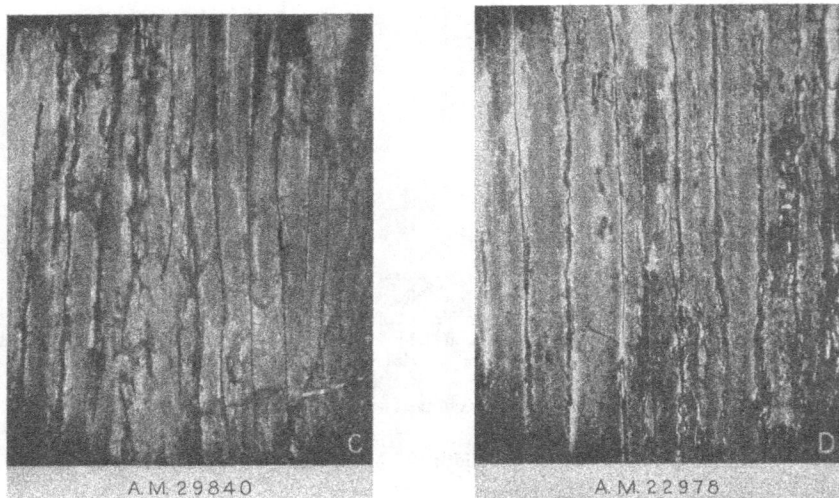

Fig. 56. Photomicrographs of longitudinal sections of the right upper second molar in (C) *Orycteropus browni* Colbert, and in (D) *Orycteropus gaudryi* Forsyth Major. Section C shows the branching and irregular tubules characteristic of *O. browni*. Section D shows the parallel tubules of *O. gaudryi*. Both sections about forty times natural size. From Colbert, 1933.

Orycteropus pilgrimi Colbert

Orycteropus pilgrimi, Colbert, 1933, Amer. Mus. Novitates, No. 604, pp. 6–7, figs. 5–7.

Type.—A.M. No. 29997, a right M_2.

Paratypes.—None.

Horizon.—Lower portion of the Middle Siwaliks, Nagri zone.

Locality.—Four and a half miles west of Hasnot, Salt Range, Jhelum District, Punjab.

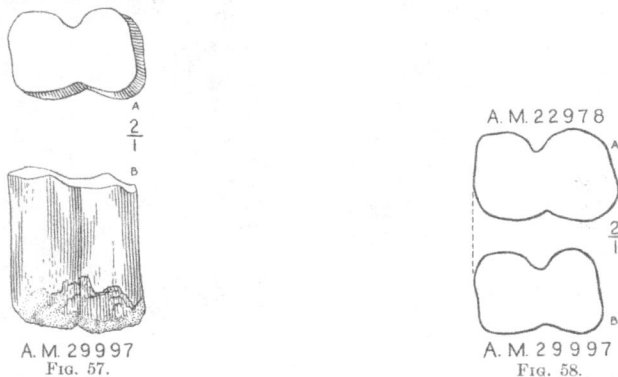

Fig. 57.

Fig. 58.

Fig. 57. *Orycteropus pilgrimi* Colbert. Type, Amer. Mus. No. 29997. A, crown view of right M_2, anterior edge of tooth facing the left. B, side view of lingual surface of the same, showing on the broken basal portion the parallel tubules. Twice natural size. From Colbert, 1933.

Fig. 58. Outlines of crown of the right second lower molar in (A) *Orycteropus gaudryi* Forsyth Major, and in (B) *Orycteropus pilgrimi* Colbert. Twice natural size. From Colbert, 1933.

Diagnosis.—Comparable in size to *Orycteropus gaudryi.* In a microscopic cross section the tubules are seen to be rather regularly hexagonal, as is characteristic of the later species of *Orycteropus.* There is no great variation in the size of the tubules, and they are strictly parallel longitudinally. The tubules are slightly elongated radially, as seen in cross section, near the periphery of the tooth.

A detailed description of this species is given in the type description, to which the reader is referred. The type figures and the measurements are reproduced below.

A.M. 29997

Fig. 59. Section of the occlusal surface of the right second lower molar of *Orycteropus pilgrimi* Colbert, showing the rather regular, hexagonal tubules. Twenty times natural size. From Colbert, 1933.

MEASUREMENTS

Orycteropus pilgrimi   A.M. No. 29997

M_2....Length.... 10.0 mm.   Width.... 7.5 mm.   Height.... 12.8 mm.

Average diameter of tubules, .25–.40 mm.

PROBOSCIDEA

A complete systematic revision of the Siwalik Proboscidea is being presented by Professor Osborn in his "Proboscidea Monograph." Consequently it is not thought advisable to attempt a discussion of the order here. In the Siwalik faunal lists tabulated on some preceding pages of this present work, a list of the Siwalik Proboscidea, as Professor Osborn has revised and classified them, will be found.

PERISSODACTYLA

EQUOIDEA

EQUIDAE

EQUINAE

Hipparion Christol, 1832

Generic type, *Hipparion prostylum* Gervais

Hipparion antelopinum (Falconer and Cautley)

Hippotherium antelopinum, Falconer and Cautley, 1849, Fauna Antiqua Sivalensis, plates LXXXII–LXXXV.

Hipparion antelopinum, Lydekker, 1885, Cat. Siw. Vert. Ind. Mus., pp. 57–58.

Hippodactylus antelopinum, Pilgrim, 1910, Rec. Geol. Surv. India, XL, p. 201.

Hipparion antelopinum, Matthew, 1929, Bull. Amer. Mus. Nat. Hist., LVI, pp. 448, 451, 526–528.

    Additional References.—

        Falconer, H., 1868A, pp. 186–189, 527–532.

        Lydekker, R., 1880B, p. 31; 1882A, pp. 75–80, Pls. XI, XII, XIII; 1883C, pp. 83, 91; 1886A, pp. 59–64.

        von Meyer, H., 1865, p. 17.

        Gaudry, A., 1873, p. 40.

        Cope, E. D., 1888, p. 449.

        Pilgrim, G. E., 1910B, p. 201.

        Siwalik equines of this genus had been originally considered as *H. gracile* by Owen in 1846. See Owen, R., 1846, p. 395.

    Type.—(Lectotype).—B.M. No. M 2647, a right maxilla with P²–M³.

    Cotypes.—B.M. Nos. 16170, a portion of a cranium; M 2652, a mandible; M 2653, a mandible; M 2648, fragment of an upper molar; various limb bones, figured by Falconer and Cautley, 1849, Pls. LXXXIII–LXXXV.

    Horizon.—Middle Siwaliks, typically from the Dhok Pathan zone.

    Locality.—From the Siwalik Hills and from the Salt Range, Punjab.

    Specimens in the American Museum:—Amer. Mus. No. 19292. Fragment of a jaw. Middle Siwaliks, 4 miles east of Dhok Pathan.

19478. Miscellaneous teeth. Middle Siwaliks, 3 miles east of Dhok Pathan.

19492. Palate, with complete right and left cheek dentitions. Middle Siwaliks, near Dhok Pathan.

19513. Miscellaneous teeth. Upper portion of Middle Siwaliks, 1 mile northeast of Hasnot.

19523. Miscellaneous teeth. Upper portion of the Middle Siwaliks, 500 feet below the Bhandar bone bed. Two miles west of Hasnot.

19536. Miscellaneous teeth. Upper portion of the Middle Siwaliks, 1½ miles northeast of Hasnot.

19548. Various teeth and a phalanx. Upper portion of the Middle Siwaliks, near Hasnot.

19550. Miscellaneous teeth. Upper portion of the Middle Siwaliks, just above the Bhandar bone bed. 1½ miles northeast of Hasnot.

19661. Milk teeth and a mandibular fragment. Middle Siwaliks, ½ mile south of Dhok Pathan.

19668. Palatal fragment with right MM²⁻⁴. Middle Siwaliks, ½ mile south of Dhok Pathan.

19670. Miscellaneous teeth and vertebrae. Middle Siwaliks, ½ mile south of Dhok Pathan.

19676. Right maxillary with P⁴–M³. Middle Siwaliks, ½ mile south of Dhok Pathan.

19704. Miscellaneous teeth. Middle Siwaliks, 1 mile west of Dhok Pathan.

19717. Miscellaneous teeth. Middle Siwaliks, 4 miles east of Dhok Pathan.

19723. Crushed maxillaries, with cheek teeth on both sides. Middle Siwaliks, 2000 feet above the base, 1½ miles northeast of Hasnot.

19727. Fragments of a lower jaw. Middle Siwaliks, 1 mile west of Hasnot.

19735. Right P⁴–M². Middle Siwaliks, ½ mile southwest of Hasnot.

19752. Section of a palate with P³–M². Upper portion of Middle Siwaliks, 1½ miles north of Hasnot.

19761. A complete skull. Upper portion of Middle Siwaliks, ½ mile southwest of Dhok Pathan.

19766. Right P². Middle Siwaliks, near Karsai, northwest of Bilaspur, Simla Hills States.

19767. Molar. Middle Siwaliks, 12 miles west of Bilaspur.

19836. Molar. Middle Siwaliks, ½ mile south of Dhok Pathan.

19843. Miscellaneous teeth. Middle Siwaliks, 3 miles east of Dhok Pathan.

19855. Mandible, with right and left P₃–M₃. Upper portion of the Middle Siwaliks, 4 miles west of Dhok Pathan.

19907. Miscellaneous teeth. Upper portion of the Middle Siwaliks, 2½ miles north of Hasnot.

19945. Lower right molar. Upper portion of the Middle Siwaliks, near Tatrot.

19959. Two upper molars. Upper portion of the Middle Siwaliks, 1½ miles northeast of Hasnot.

29807. Left M¹. Upper portion of the Middle Siwaliks, 1 mile northeast of Hasnot.

29808. Fragment of a right ramus, with M₁₋₂. Upper portion of the Middle Siwaliks, 2 miles northeast of Hasnot.

29809. Fragment of a left ramus, with M₂₋₃. Middle Siwaliks, 1 mile northeast of Hasnot.

The following specimens are attributed to *Hipparion antelopinum*.

Amer. Mus. No. 19465. Ten dorso-lumbar vertebrae and several ribs. Upper portion of the Middle Siwaliks, ½ mile southwest of Dhok Pathan. These specimens are assigned to this species because of their small size.

19667. Metatarsal. Middle Siwaliks, ½ mile south of Dhok Pathan. This is the only element of the postcranial skeleton in the collection, which is definitely identified as belonging to *H. antelopinum*.

19902. Miscellaneous teeth and skeletal fragments. Middle Siwaliks, about 1000 feet below the level of the Bhandar bone bed, four and one half miles west of Hasnot.

Diagnosis.—A large species of *Hipparion*, characterized by the oval protocone and the crenulated fossette borders of the upper molars, by the well developed preorbital fossa of the skull and by the slender medial metapodials of the feet.

Hipparion punjabiense Lydekker

Hipparion punjabiense, Lydekker, 1886, Cat. Foss. Mam., Brit. Mus., III, p. 60.

Additional References.—

Pilgrim, G. E., 1910A, p. 66; 1910B, p. 201; 1913B, p. 284.

Matthew, W. D., 1929, pp. 526–528.

Cotypes.—G.S.I. No. C 139, a left maxilla with P³–M²; G.S.I. No. C 138, a left maxilla with a milk dentition and M¹. B.M. No. M 2646, a maxilla with the milk dentition and M¹.

Horizon.—Middle Siwaliks.

Locality.—From Niki in the Punjab and from the Siwalik Hills.

Diagnosis.—"A specimen (No. M 2646), noticed below, indicates that these specimens may be specifically distinct, this being confirmed by the Punjab teeth being found in association with the distal articular surfaces of small-sized lateral metapodials, and with first phalangeals of a stouter type than those referred to the present form. If this inference should be correct, the Punjab form may be named *H. punjabiense.*" [34]

Hipparion perimense Pilgrim

Hipparion perimense, Pilgrim, 1910, Rec. Geol. Surv. India, XL, p. 66.

   Additional References.—

      Pilgrim, G. E., 1910B, p. 201; 1913B, p. 321.

      Matthew, W. D., 1929, pp. 527–528.

   Type.—A skull described and figured by Lydekker in Pal. Indica (X), III, pp. 11, 14, Pl. III, figs. 1, 2.

   Paratypes.—None.

   Horizon.—Middle Siwaliks.

   Locality.—Perim Island.

   Diagnosis.—"This specific name is intended to receive the Perim Island skull described by Lydekker. Very perfect skulls of *H. punjabiense* Lyd. and *H. theobaldi* Lyd. obtained from the Middle Siwaliks of Dhok Pathan have shown that these two species are quite distinct. *H. punjabiense* has a much larger higher skull and the facial cavity is deeper and farther from the teeth." [35]

Hipparion chisholmi (Pilgrim)

Hippodactylus chisholmi, Pilgrim, 1910, Rec. Geol. Surv. India, XL, p. 67.
Hipparion chisholmi, Pilgrim, 1913, Rec. Geol. Surv. India, XLIII, p. 284.

   Additional References.—

      Pilgrim, G. E., 1910B, p. 201.

      Matthew, W. D., 1929, pp. 448, 526–527.

   Type.—Not specified.

   Horizon.—Middle Siwaliks.

   Locality.—From near Dhok Pathan, Punjab.

   Diagnosis.—"A skull of this species was found in the Middle Siwaliks at Dhok Pathan. It is smaller than either of the other species of *Hipparion*. Pm_1 is present. The enamel folds of the fossettes are less complex. The protocone is larger and united with the protoconule in a very advanced state of wear. There is a large facial cavity, somewhat nearer the orbit than in the other Indian species. It was probably monodactyl. It differs from *H. antelopinus*, so far as this species is known, by the larger M_3 and the much squarer-crowned teeth." [36]

   Dr. Matthew has shown in his paper of 1929 (pp. 526–528) that the above three species are synonyms of *Hipparion antelopinum* (Falconer and Cautley).

[34] Lydekker, R., 1886, Cat. Foss. Mam., Brit. Mus., III, p. 60.
[35] Pilgrim, G. E.—1910A, p. 66.
[36] Pilgrim, G. E., 1910A, p. 67.

Hipparion theobaldi (Lydekker)

Sivalhippus theobaldi, Lydekker, 1877, Rec. Geol. Surv. India, X, p. 31.

Hippotherium theobaldi, Lydekker, 1882, Pal. Indica (X), II, pp. 81–87, Pl. II, fig. 3, 4; XII, figs. 2, 4; XIII, figs. 1–3.

Hipparion theobaldi, Lydekker, 1885, Cat. Siw. Vert. Indian Mus., pp. 58–60.

Additional References.—

> Lydekker, R., 1877B, p. 81; 1880B, p. 31; 1883C, pp. 83, 91; 1884D, p. 132; 1886A, pp. 64–65.
>
> Pilgrim, G. E., 1910B, p. 201; 1913B, pp. 284, 296.
>
> Matthew, W. D., 1929, pp. 448, 450, 524–526.

Type.—G.S.I. No. C 153, a left maxilla containing the milk molars.

Horizon.—Middle Siwaliks, typically from the Dhok Pathan zone. This species extends down to the base of the Chinji beds.

Locality.—The type is from Keypar in the Punjab. The species is widespread throughout the Middle Siwalik exposures.

> *Specimens in the American Museum.* Amer. Mus. No. 19466. A skull, complete except for the portion anterior to P^2. Middle Siwaliks, 2 miles east of Dhok Pathan.

19481. Miscellaneous teeth. Middle Siwaliks, 2 miles east of Dhok Pathan.

19491. Miscellaneous teeth, and skeletal fragments. Middle Siwaliks, 3 miles east of Dhok Pathan.

19500. Right P^2. Upper portion of the Middle Siwaliks, 200 feet above the Bhandar bone bed, 6 miles east of Hasnot.

19501. Right upper molar. Middle Siwaliks, locality uncertain.

19505. Right lower molar. Upper portion of the Middle Siwaliks, 1 mile northeast of Hasnot.

19518. Fragment of ramus with left MM_{3-4}. Upper portion of the Middle Siwaliks, near Padri, ½ mile north of Hasnot.

19656. Right maxilla with MM^{2-3}. Middle Siwaliks, ½ mile south of Dhok Pathan.

19657. Miscellaneous teeth and a jaw fragment. Middle Siwaliks, ½ mile south of Dhok Pathan.

19659. Right ramus with M_{2-3}. Middle Siwaliks, ½ mile south of Dhok Pathan.

19674. Various teeth and vertebrae. Middle Siwaliks, ½ mile south of Dhok Pathan.

19693. Miscellaneous teeth. Upper portion of Middle Siwaliks, 4 miles east of Dhok Pathan.

19695. Miscellaneous teeth. Middle Siwaliks, 3 miles east of Dhok Pathan.

19700. Miscellaneous teeth. Middle Siwaliks, 4 miles west of Dhok Pathan.

19706. Miscellaneous teeth and jaw fragments. Middle Siwaliks 4 miles east of Dhok Pathan.

19708. Miscellaneous teeth. Middle Siwaliks, 3 miles east of Dhok Pathan.

19711. Miscellaneous teeth. Middle Siwaliks, 1 mile west of Dhok Pathan.

19719. Miscellaneous teeth and skeletal fragments. Middle Siwaliks, near Dhok Pathan.

19740. Miscellaneous teeth, and fragmentary right ramus. Middle Siwaliks, near top of series, 2 miles north of Hasnot.

19749. Miscellaneous teeth. Middle Siwaliks, 200 feet below the Bhandar bone bed, 1 mile east of Hasnot.

19759. Right upper molar. Middle Siwaliks, about 200 feet above the Bhandar bone bed, 1½ miles north of Hasnot.

19847. Miscellaneous teeth and jaw fragments. Upper portion of the Middle Siwaliks. Locality uncertain.

19853. Two incisors, and LP$_4$, RM^3. Upper portion of the Middle Siwaliks, 4 miles west of Dhok Pathan.

19857. A skull and an associated mandible, both badly broken. Upper portion of Middle Siwaliks, 200 feet above Hari Temple. Hari Talyangar, northwest of Bilaspur.

19895. Right M^3. Middle Siwaliks, 100 feet above the Bhandar bone bed, near Bhandar.

19899. Left P^2, fragmentary ramus. Upper portion of the Middle Siwaliks, 1½ miles north of Hasnot.

19905. Upper and lower molar. Middle Siwaliks, 1 mile east of Hasnot.

19924. Two lower molars. Upper portion of the Middle Siwaliks, 2 miles west of Hasnot.

19925. Miscellaneous teeth. Upper portion of the Middle Siwaliks, 2 miles south of Hasnot.

29802. Palatal and mandibular fragments. Middle Siwaliks, ½ mile south of Dhok Pathan.

29806. Various teeth and jaw fragments. Middle Siwaliks, 4 miles west of Dhok Pathan.

The following specimens are from the lower Siwaliks.

Amer. Mus. No. 19555. Right upper molar. Lower Siwaliks, about 200 feet above Chinji Rest House, 2 miles west of Chinji Rest House.

19573. Various teeth. Horizon uncertain, either from the top of the Lower Siwaliks, or the base of the Middle Siwaliks, 3 miles northwest of Chinji Rest House.

19584. Miscellaneous teeth. Lower Siwaliks, about the level of Chinji Rest House, 5 miles east of Chinji Rest House.

19590. Left upper molar. Lower Siwaliks, about 600 feet above the level of Chinji Rest House, 1 mile north of Chinji Rest House.

The following specimens, comprising various portions of the postcranial skeleton, are attributed to *Hipparion theobaldi*, on the basis of size and the development of the metapodials.

Amer. Mus. No. 19466. Astragalus, phalanges, two lunars and a cuboid. Middle Siwaliks, 2 miles east of Dhok Pathan.

19671. Right metatarsals, II, III, IV. Middle Siwaliks, ½ mile south of Dhok Pathan.

19683. Left femur, tibia and pes. Upper portion of the Middle Siwaliks, ½ mile southwest of Dhok Pathan.

19685. Right humerus, ulna-radius, carpus and manus. Upper portion of the Middle Siwaliks, ½ mile southwest of Dhok Pathan.

19757. Right femur. Lower portion of the Middle Siwaliks, near Hasnot.

19769. Various phalanges. Upper portion of the Middle Siwaliks, 300 feet above the temple, 4 miles east of Hari, in turn about twenty miles northwest of Bilaspur.

19842. Left femur. Middle Siwaliks, ½ mile south of Dhok Pathan.

19856. Right femur of juvenile. Middle Siwaliks, ½ mile south of Dhok Pathan.

19969. Right scapula, humerus, ulna-radius, carpus and manus, right femur, left femur, cervical series, four dorsals. Upper portion of Middle Siwaliks, ½ mile southwest of Dhok Pathan.

29803. Miscellaneous foot bones. Upper portion of the Middle Siwaliks, ½ mile southwest of Dhok Pathan.

29810. Left metacarpals II, III, IV, and sesamoids. Middle Siwaliks, ½ mile south of Dhok Pathan.

29811. Right metatarsals III, IV. Middle Siwaliks, ½ mile south of Dhok Pathan.

29819. Distal end of left humerus, left radius-ulna, carpus and manus. Upper portion of Middle Siwaliks, ½ mile southwest of Dhok Pathan.

29822. Right femur, tibia, tarsus and pes. Upper portion of the Middle Siwaliks, ½ mile south of Dhok Pathan.

29823. Right tibia, tarsus and pes. Upper portion of the Middle Siwaliks, ½ mile south of Dhok Pathan.

29824. Left tibia, tarsus and pes. Upper portion of the Middle Siwaliks, ½ mile southwest of Dhok Pathan.

29825. Right humerus, ulna-radius, carpus and manus. Upper portion of the Middle Siwaliks, ½ mile southwest of Dhok Pathan.

29826. Metatarsal and navicular. Upper portion of the Middle Siwaliks, ½ mile southwest of Dhok Pathan.

29827. Ribs and vertebrae. Upper portion of the Middle Siwaliks, ½ mile southwest of Dhok Pathan.

29828. Distal end of tibia. Upper portion of the Middle Siwaliks, ½ mile southwest of Dhok Pathan.

29829. Astragalus and vertebra. Upper portion of the Middle Siwaliks, ½ mile southwest of Dhok Pathan.

29830. Phalanges. Upper portion of the Middle Siwaliks, ½ mile southwest of Dhok Pathan.

29831. Distal end of left femur, tibia, tarsus and pes. Upper portion of Middle Siwaliks, ½ mile southwest of Dhok Pathan.

Diagnosis.—A very large, heavy type of *Hipparion*. Similar in skull and tooth structure to *Hipparion antelopinum*, but considerably larger. Median metapodials heavy.

"None of the characters adduced by Lydekker to separate *theobaldi* from *antelopinum* appear to be valid specific distinctions. Nevertheless, the *type* of *theobaldi* is too large to represent the milk dentition of the *type* of *antelopinum*, and comparison of various permanent dentitions from the Siwaliks supports Lydekker's view that there are a larger and a smaller form, the former decidedly more robust and with heavier limb bones and larger lateral digits." [33]

[33] Matthew, W. D., 1929, p. 525.

RESEMBLANCES AND DIFFERENCES IN *Hipparion antelopinum* AND *Hipparion theobaldi*

Hipparion antelopinum is typically a medium sized species of slender build, as shown especially by the metapodials. *Hipparion theobaldi* is typically a large, robust species with heavy metapodials. But the differences between these two species are mainly differences in size, and in any large series of teeth an almost perfect gradation may be found between them. Of course this intergradation between the two species introduces many difficulties into the study of the genus, as represented in the Siwalik deposits, difficulties that may be solved only by establishing arbitrary lines of distinction.

Lydekker separated *Hipparion theobaldi* from *Hipparion antelopinum* on the basis of the following distinctions.

1. Greater size of the molars, and their more oblong and less square shape in *Hipparion theobaldi*.

2. Protocone compressed, as compared to a round-oval protocone in *Hipparion antelopinum*.

3. Hypocone extending back to the posterior border of the molar crown in *Hipparion theobaldi*.

4. Hypocone united to posterior crescent in DP^2 of *Hipparion theobaldi*.

5. Enamel borders of fossettes relatively simple in *Hipparion theobaldi;* complicated in *Hipparion antelopinum*.

Matthew has shown (1929, p. 525) that the supposed differentiations listed above, are entirely dependent upon age and the degree of occlusion, and consequently they become invalid as specific distinctions. To quote from the above citation:

"None of the characters adduced by Lydekker to separated *theobaldi* from *antelopinum* appear to be valid distinctions. Nevertheless the *type* of *theobaldi* is too large to represent the milk dentition of the *type* of *antelopinum*, and comparison of various permanent dentitions from the Siwaliks supports Lydekker's view that there are a larger and a smaller form, the former decidedly more robust and with heavier limb bones and larger lateral digits." [37]

A careful study of the material in the American Museum collection supports Dr. Matthew's conclusion, namely that there are two forms of *Hipparion* in the Siwaliks, a larger and a smaller species, and that they can be separated only on the basis of size differences.

As stated above, a continuous gradation exists in the teeth, from the smallest specimens, typical of *H. antelopinum*, to the largest ones, typical of *H. theobaldi*. The increase in the size of the molars is not marked by any changes in pattern, the smaller teeth being quite similar to the larger ones.

On the other hand, a sharp distinction exists in the foot bones of the two species. In *Hipparion antelopinum* the metapodials are long and slender, as in *H. whitneyi* or *H. gracile*, while in *Hipparion theobaldi* the medial and lateral metapodials are very robust. Nor does there seem to be any gradation between these two sets of conditions in the American Museum specimens. Perhaps a *very* large series of foot bones from the Siwaliks might show gradations here, as in the teeth, but such a series has as yet to be collected.

It might be well to say at this point that although the lateral digits of *Hipparion theobaldi* are very robust, they are not enlarged out of proportion to the rest of the foot. Rather, they are heavy because the entire foot is heavy, and conversely, in *Hipparion ante-*

---

[37] Matthew, W. D., 1929, p. 525.

lopinum (and in the many species similar to it) the lateral digits are very slender because the whole foot is slender.

It is not possible to make further comparisons of the skeletons of the two Siwalik species of *Hipparion* because of the lack of material.

Matthew stated in 1929 (p. 525) that "most of the Indian material belongs to *theobaldi* . . . ." Studies on the American Museum collection would seem to bear him out. In the light of our present knowledge, we may picture *Hipparion theobaldi* as the dominant equine in the Lower and Middle Siwaliks, and *Hipparion antelopinum* as a rarer form.

With the foregoing evidence in mind, the following conclusions may be reached.

1. There are two species of *Hipparion* in the Siwalik series, a larger one, *Hipparion theobaldi*, and a smaller one, *Hipparion antelopinum*.

2. The teeth of these two species differ only as regards size. There are no appreciable differences in the molar patterns.

3. There is a constant gradation in the teeth between the two species.

4. The foot bones of these two species are markedly different, being robust in *H. theobaldi* and very slender in *H. antelopinum*.

5. Until associated teeth and feet of both species are found, it will be impossible to draw definite lines of distinction between them on the basis of teeth alone. Consequently it becomes necessary to establish an arbitrary line in dividing a series of teeth between these two species. In general, molar teeth that measure more than 24 millimeters in an anteroposterior direction and 25 millimeters transversely, may be assigned to the species, *Hipparion theobaldi*.

On the following pages are presented two graphs of the teeth of *Hipparion* from the Siwaliks. These graphs show the gradation in size from the smaller species, *H. antelopinum* into the larger form, *H. theobaldi*.

The first graph is based on measurements of molar lengths and widths, the former being plotted along the horizontal axis and the latter along the vertical axis. The second graph is presented to show lengths and widths of the protocone, here again the dimensions being represented along the two axes as in the preceding graph. Only first and second upper molars were used in the compilation of these graphs, since these two teeth are almost always of uniform size. In the case of isolated teeth (which constitute the bulk of the collection) the molars were identified as to their position in the maxilla as best as could be done by comparisons with complete series.

The measurements of the teeth and of their single element the protocone, as plotted on these two graphs, are contained in the two keys on page 138. These keys include also the museum numbers of the specimens represented on the graphs by consecutive numbers, affording an opportunity to identify the specimens as to horizon and locality.

The Skull and Skeleton of *Hipparion antelopinum* and *Hipparion theobaldi*

Falconer never published descriptions of *Hipparion antelopinum*. Lydekker in 1882, gave a detailed account of this species, as known from the teeth and the foot bones, and at the same time he described the teeth and foot bones of his new species, *Hipparion theobaldi*. He did not, however, describe the skull of any of the Siwalik *Hipparion*, nor did he present a systematic description of the skeleton. In view of these deficiencies in the osteological

KEY TO GRAPH

Hipparion theobaldi and *Hipparion antelopinum*.   M^1–M^2

Key No.	Number	Length	Width	Key No.	Number	Length	Width
1	29807	20.5 mm.	21.0 mm.	19	19693	24.0 mm.	25.0 mm.
2	19836	22.0	21.0	20	19711	24.0	25.0
3	19717	21.5	21.5	21	19695	24.0	25.0
4	19492	20.0	22.0	22	19843	24.5	24.0
5	19704	20.5	22.5	23	19657	25.0	26.0
6	19857	22.0	22.0	24	19759	25.0	27.0
7	19843	22.0	22.5	25	19706	25.5	26.5
8	19752	23.0	22.5	26	19466	26.0	26.0
9	19723	23.0	23.0	27	19501	25.5	27.5
10	19550	22.0	23.5	28	19740	26.5	28.0
11	19661	22.0	24.0	29	19491	27.5	27.0
12	19907	23.0	23.5	30	19700	28.5	27.0
13	19857	20.0	25.0	31	19719	27.5	29.0
14	19761	22.0	24.5	32	19711	28.0	29.0
15	19676	23.0	24.0	33	19749	28.0	30.0
16	19735	23.0	25.0	34	19555	26.5	28.0
17	19478	24.0	24.0	35	19706	24.5	25.5
18	19708	23.5	25.0				

KEY TO GRAPH

Hipparion theobaldi and *Hipparion antelopinum*.   Protocone of M^1–M^2

Key No.	Number	Length	Width	Key No.	Number	Length	Width
1	29807	5.4 mm.	3.2 mm.	19	19693	7.3 mm.	4.0 mm.
2	19836	6.0	3.2	20	19711	6.5	4.0
3	19717	6.5	4.0	21	19695	6.5	3.8
4	19492	5.8	3.7	22	19843	6.0	3.8
5	19704	6.5	3.7	23	19657	7.7	4.8
6	19857	—	—	24	19759	10.0	5.5
7	19843	7.0	4.0	25	19706	7.0	3.5
8	19752	6.6	3.7	26	19466	9.2	4.2
9	19723	6.2	4.0	27	19501	7.7	4.5
10	19550	7.8	4.3	28	19740	8.0	4.5
11	19661	5.5	4.2	29	19491	—	—
12	19907	6.8	3.5	30	19700	6.8	4.0
13	19857	9.0	5.8	31	19719	8.0	4.5
14	19761	8.2	4.0	32	19711	—	—
15	19676	6.0	3.9	33	19749	9.5	4.5
16	19735	7.0	4.0	34	19555	9.0	4.8
17	19478	7.1	4.0	35	19706	7.5	4.8
18	19708	—	—				

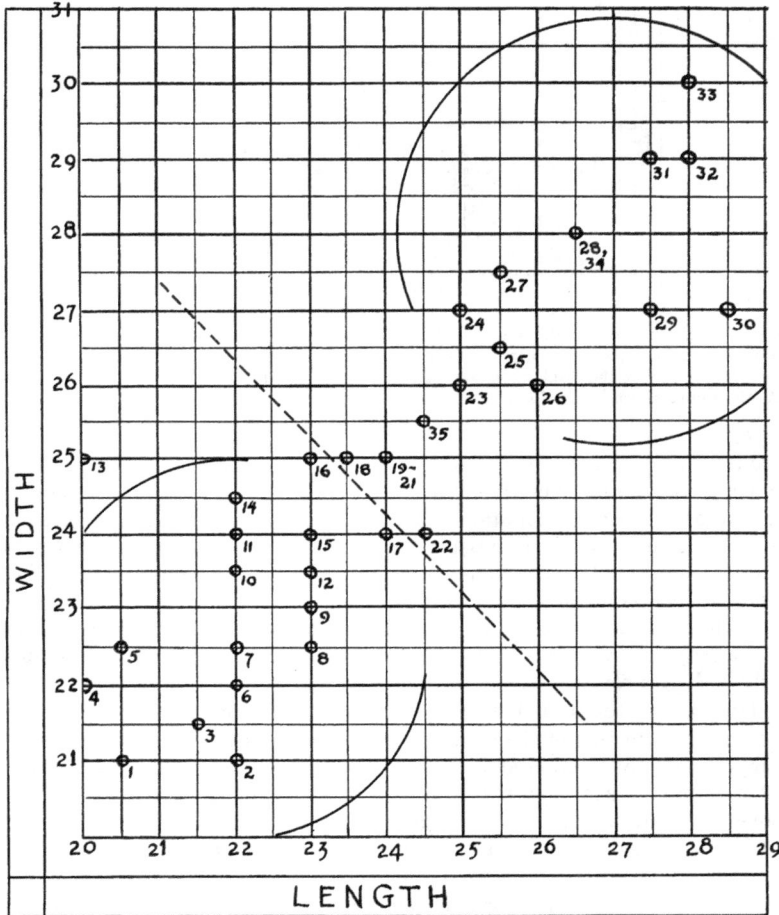

Fig. 60. Variation in size of the first two upper molars in *Hipparion antelopinum* and *Hipparion theobaldi* molar length in millimeters along horizontal axis, molar width along vertical axis. Typical *H. antelopinum* molars are enclosed within the lower left hand circle; typical *H. theobaldi* molars are enclosed within the upper right hand circle. The arbitrary division between the two species is indicated by the broken line.

studies of *Hipparion* from the Siwalik beds, it may be well to offer at this place, short anatomical descriptions of the genus, as characterized by the two species named above, on the basis of the excellent material collected by Mr. Brown for the American Museum.

The main purpose of the following descriptions will be to elucidate our knowledge of the skull and mandible of the Siwalik *Hipparion*. The account of the postcranial skeleton will be concerned with proportional differences, rather than with topographic anatomy, since the detailed anatomical characters of the Indian species are not distinctive enough from those of other Eurasiatic species of *Hipparion* to warrant a long description.

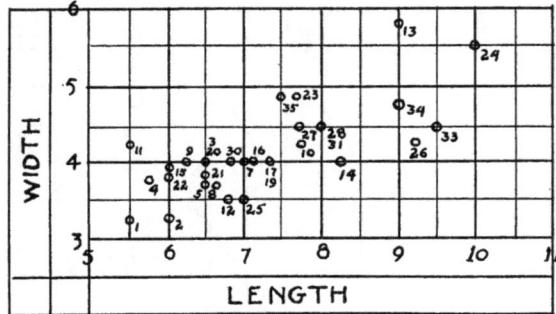

FIG. 61. Variation in size of the protocone of the first two upper molars in *Hipparion antelopinum* and *Hipparion theobaldi*. Length of protocone in millimeters along horizontal axis, width along vertical axis. The numbers correspond to the numbers on the graph in Fig. 60. The protocone measurements fall into two groups, generally correlative with the groups shown in Fig. 60.

The Skull

There is an exceptionally fine skull in the collection, Amer. Mus. No. 19761, which is virtually complete, and but slightly deformed by crushing. The cheek teeth on both sides are all present and in perfect condition. The left canine is present, as well as a stub of the right one. All of the incisors are missing but the alveoli are preserved, filled with matrix. The tips of the nasals are broken off, as are the paroccipital processes, and a small portion of the lambdoidal crest is missing.

This skull is of intermediate size, being about fifteen per cent smaller than Amer. Mus. No. 19466, a large and typical specimen of *H. theobaldi*, and is closely comparable in form to the skull of *H. gracile*.

In a lateral view the orbit is seen to be placed well back, having its anterior border above the posterior portion of the third molar. There is a small but deep lacrymal fossa, situated high up on the maxilla above the last premolar and the first molar, and directly in front of this fossa, on a level with its lower border, is the infraorbital foramen, this opening being above the third premolar. The zygomatic arch is typical of *Hipparion* in that it is short and heavy, though proportionately longer than in *Equus*.

The brain case is expanded, as would be expected of an equine from the upper portion of the Tertiary, and the sagittal crest is well developed. The incisor foramina are long, stretching from the posterior border of the third incisor to the mid-portion of the canine. The posterior nares reach far forward in the palate to a point opposite the middle portion of the first molar.

The basicranium needs no special description, as it is quite typical of the advanced Equinae. Suffice it to say that there has been a process of compression and conjunction of the basicranial foramina, causing the exits of the optic foramen, the foramen lacerum anterius, the foramen rotundum and the alisphenoid canal to occur very close to each other. Moreover, there is a fusion of the foramen ovale, foramen lacerum medius and the foramen lacerum posterius to form a large open space around the inner edge of the bulla, as is the case in *Equus*. The bulla is of good size. The paroccipitals though broken off, were probably long.

The Mandible

There is not much to say about the lower jaw, as typified by Amer. Mus. No. 29806. It is typically equine, having a deep horizontal ramus for the accommodation of the increasingly lengthening cheek teeth.

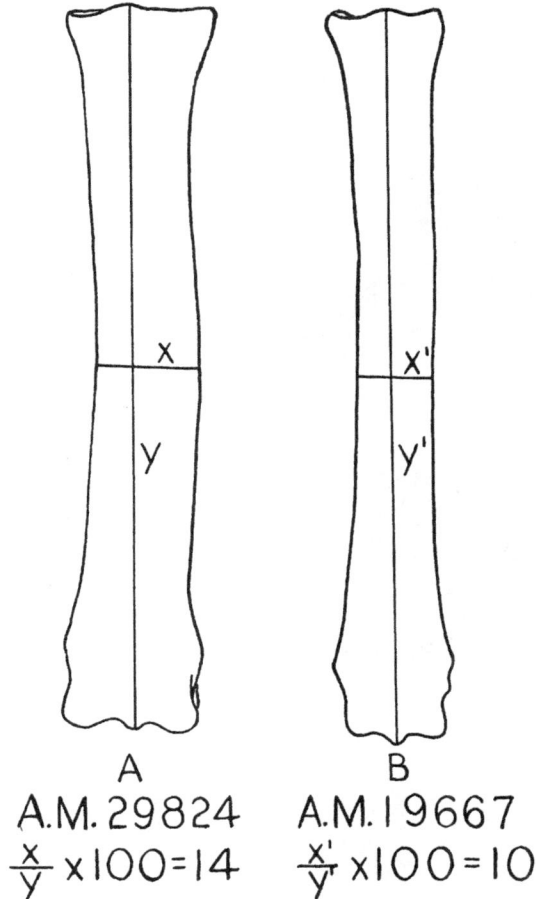

Fɪɢ. 62. Comparison of the median metacarpal in (A) *Hipparion theobaldi* and (B) *Hipparion antelopinum*. Length of metacarpals reduced to unity.

The Dentition

Lydekker (1882A, pp. 81–86) has described the cheek teeth of this species in considerable detail, and his descriptions have been recently supplemented by the lucid remarks of Dr. Matthew (1929, pp. 524–526).

The upper canine is well developed and separated by a considerable diastema from the second premolar.   There is a marked decrease in size in the cheek teeth from front to back, a character of *Hipparion* that is well shown by the Siwalik species.   The cheek teeth are large, as would be expected in this large *Hipparion*, and fairly hypsodont.   An accompanying figure (Fig. 66) illustrates the degree of hypsodonty in this Indian *Hipparion*.

Fig. 63. *Hipparion antelopinum* (Falconer and Cautley). Amer. Mus. No. 19761, skull. Top view above, lateral view in middle, ventral view below. One third natural size.

FIG. 64. *Hipparion theobaldi* (Lydekker). Amer. Mus. No. 19466, skull. Top view above, lateral view in middle, ventral view below. One third natural size.

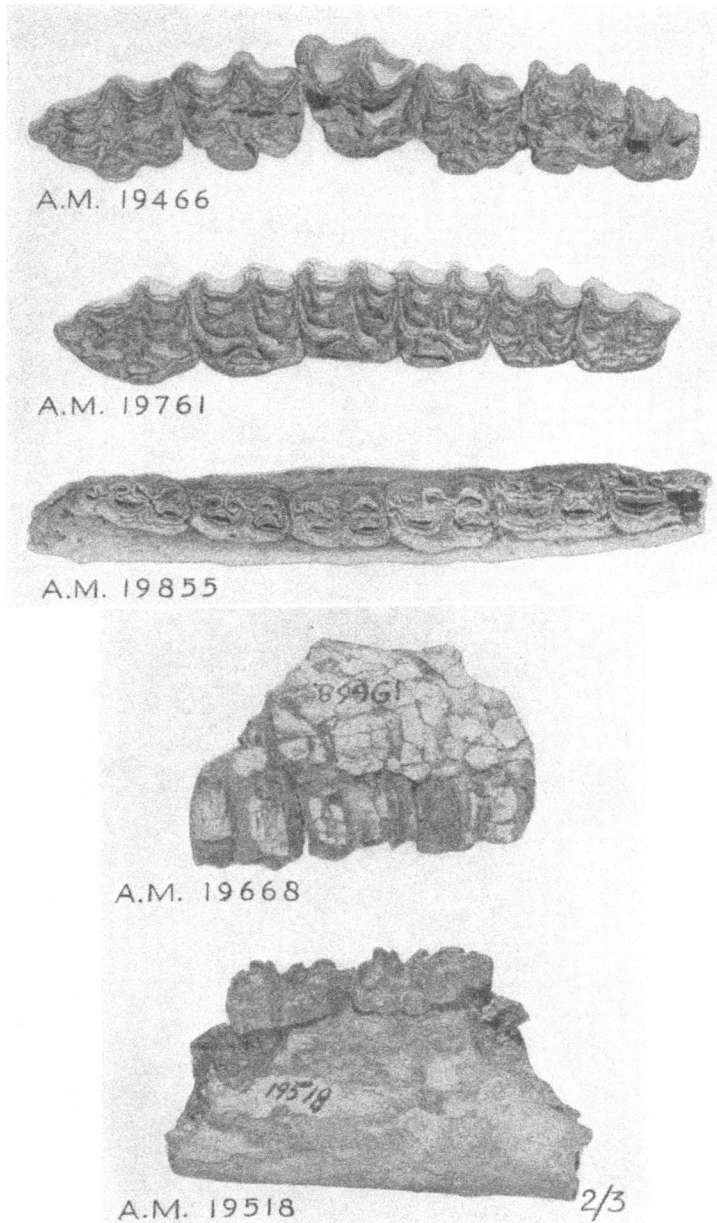

FIG. 65. Teeth of *Hipparion antelopinum* and *Hipparion theobaldi*.

Reading from the top down:

    Hipparion theobaldi (Lydekker). Amer. Mus. No. 19466, left P^2 – M^3. Crown view.

    Hipparion antelopinum (Falconer and Cautley). Amer. Mus. No. 19761, left P^2 – M^3. Crown view.

    Hipparion antelopinum (Falconer and Cautley). Amer. Mus. No. 19855, right P$_2$ – M$_3$. Crown view.

    Hipparion antelopinum (Falconer and Cautley). Amer. Mus. No. 19668, right MM$^{2-4}$. Lateral view.

    Hipparion theobaldi (Lydekker). Amer. Mus. No. 19518, left MM$_{2-4}$. Lateral view.

All figures two thirds natural size.

The protocone is elongated and somewhat flattened, and the external styles are pronounced in their development. The enamel borders of the fossettes are complicated to a degree closely comparable to the condition found in *Hipparion gracile*.

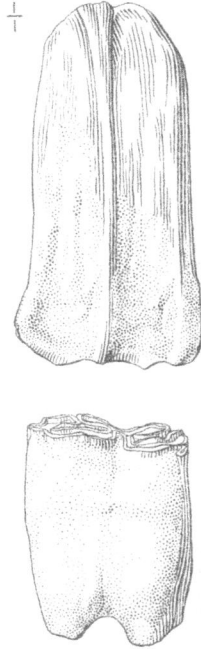

FIG. 66. Upper and lower molars of *Hipparion theobaldi* (Lydekker). Amer. Mus. No. 19584. Lateral view to show hypsodonty. Natural size.

In the lower cheek teeth the metaconid and metastylid are rounded, and separated from each other by a deep internal vertical groove.

Additional Remarks

A second skull, that of a young individual with the fourth premolar just erupting, Amer. Mus. No. 19466, is somewhat larger than the skull described above, and it may be assigned to *Hipparion theobaldi*. In all of its features it is essentially similar to the specimen already described.

Several portions of skulls referable to *Hipparion antelopinum* are in the collection, but separate descriptions of them are not deemed necessary. Measurements of various specimens of the two species of *Hipparion* from the Siwaliks are given in the following tables.

Hipparion antelopinum

Measurements of A.M. 19761.   Skull

Basilar length (back of occipital condyle to alveolus of I^1)....456.0 mm.
Width at glenoids....................................179.0
Width across zygoma.................................185.5
Length P^2-M^3....................................155.5
      P^2-P^4.................................... 89.0
      M^1-M^3.................................... 69.0
      P^2-C diastema............................ 54.0
Antero-posterior diameter canine...................... 13.0
Transverse diameter canine........................... 11.0
Height of canine (above alveolus)..................... 21.0
Antero-posterior diameter of P^2..................... 35.0
Transverse diameter of P^2.......................... 26.5
Antero-posterior diameter of P^3..................... 27.0
Transverse diameter of P^3.......................... 28.0
Antero-posterior diameter of P^4..................... 25.0
Transverse diameter of P^4.......................... 27.0
Antero-posterior diameter of M^1..................... 22.0
Transverse diameter of M^1.......................... 26.5
Antero-posterior diameter of M^2..................... 22.0
Transverse diameter of M^2.......................... 24.5
Antero-posterior diameter of M^3..................... 23.0
Transverse diameter of M^3.......................... 23.0
Transverse diameter occipital condyle................. 68.0
Width of palate at M^1.............................. 60.0
Height of orbit..................................... 32.5
Length of preorbital fossa........................... 46.0
Height of preorbital fossa........................... 26.0
Preorbital length of skull...........................280.0
Postorbital length of skull..........................196.0
Width of cranium................................... 95.0
Height of maxilla above M^1..........................100 [38]

Hipparion antelopinum

Measurements of A.M. 19723

Length P^2-M^3.......................................150.0 mm.
      P^2-P^4.................................... 82.5
      M^1-M^3.................................... 67.5

Antero-posterior diameter R. P^2..................... 33.0
Transverse diameter R. P^2.......................... 23.0
Antero-posterior diameter R. P^3..................... 25.0
Transverse diameter R. P^3.......................... 25.0
Antero-posterior diameter R. P^4..................... 23.5
Transverse diameter R. P^4.......................... 26.0
Antero-posterior diameter R. M^1..................... 23.0
Transverse diameter R. M^1.......................... 23.0
Antero-posterior diameter R. M^2..................... 22.5
Transverse diameter R. M^2.......................... 22.5
Antero-posterior R. M^3............................. 22.0
Transverse diameter R. M^3.......................... 21.0

[38] Approximate.

Hipparion antelopinum

Measurements of A.M. 19752

Antero-posterior diameter R. P³.......................... 27.0 mm.
Transverse diameter R. P³............................... 24.5
Antero-posterior diameter R. P⁴.......................... 24.0
Transverse diameter R. P⁴............................... 24.0
Antero-posterior diameter R. M¹.......................... 23.5
Transverse diameter R. M¹............................... 23.0
Antero-posterior diameter R. M².......................... 23.0
Transverse diameter R. M²............................... 22.0

Hipparion antelopinum

Measurements of A.M. 19676

Length molar series.................................... 64.5 mm.

Antero-posterior diameter R. P⁴.......................... 26.0
Transverse diameter R. P⁴............................... 27.0
Antero-posterior diameter R. M¹.......................... 23.0
Transverse diameter R. M¹............................... 24.0
Antero-posterior diameter R. M².......................... 21.0
Transverse diameter R. M²............................... 22.0
Antero-posterior diameter R. M³.......................... 20.0
Transverse diameter R. M³............................... 19.0

Hipparion antelopinum

Measurements of A.M. 19492

Length P²-M³..134.0 mm.
    P²-P⁴... 73.0
    M¹-M³... 61.0

Antero-posterior diameter R. P².......................... 29.0
Transverse diameter R. P²............................... 22.0
Antero-posterior diameter R. P³.......................... 22.5
Transverse diameter R. P³............................... 25.0
Antero-posterior diameter R. P⁴.......................... 21.0
Transverse diameter R. P⁴............................... 24.5
Antero-posterior diameter R. M¹.......................... 20.0
Transverse diameter R. M¹............................... 22.0
Antero-posterior diameter R. M².......................... 20.0
Transverse diameter R. M²............................... 21.5
Antero-posterior diameter R. M³.......................... 20.5
Transverse diameter R. M³............................... 19.5

Hipparion antelopinum

Measurements of A.M. 19668. (Milk Dentition)

Antero-posterior diameter R. DM².......................... 35 mm. (est.)
Transverse diameter R. DM²............................... 22.5
Antero-posterior diameter R. DM³.......................... 27.0
Transverse diameter R. DM³............................... 23.5
Antero-posterior diameter R. DM⁴.......................... 27.0
Transverse diameter R. DM⁴............................... 23.0

Hipparion antelopinum

Measurements of A.M. 19855

Length P_2-M_3...164.0 mm.

P_2-P_4....................................... 85.0

M_1-M_3....................................... 79.0

Antero-posterior diameter R. P_3......................... 27.0

Transverse diameter R. P_3.............................. 15.5

Antero-posterior diameter R. P_4......................... 26.0

Transverse diameter R. P_4.............................. 15.5

Antero-posterior diameter R. M_1........................ 24.5

Transverse diameter R. M_1............................. 15.0

Antero-posterior diameter R. M_2........................ 25.0

Transverse diameter R. M_2............................. 14.0

Antero-posterior diameter R. M_3........................ 28.0

Transverse diameter R. M_3............................. 12.5

Hipparion theobaldi

Measurements of A.M. 29806

Length molar series................................. 83.5 mm.

Antero-posterior diameter L. M_1....................... 25.5

Transverse diameter L. M_1............................ 15.5

Antero-posterior diameter L. M_2....................... 25.0

Transverse diameter L. M_2............................ 15.0

Antero-posterior diameter L. M_3....................... 30.0

Transverse diameter L. M_3............................ 13.0

Hipparion theobaldi

Measurements of A.M. 19740.   (Milk Dentition)

Antero-posterior diameter R. DM_3...................... 33.0 mm.

Transverse diameter R. DM_3........................... 17.0

Antero-posterior diameter R. DM_4...................... 35.0

Transverse diameter R. DM_4........................... 15.0

Depth of ramus below R. DM_4......................... 54.0

Hipparion theobaldi

Measurements of A.M. 19466.   Skull

Postorbital length...................................193.0 mm.

Width of condyles................................... 81.0

Width at glenoids...................................206.0

Width of palate at M^1.............................. 77.0

Height of orbit...................................... 50.0

Length of preorbital fossa............................ 55.0

Height of preorbital fossa............................ 40.0

Width of cranium....................................100.0

Height of maxilla above R. M^1......................115.0

Length R. P^2-M^3................................173.0

R. P^2-P^4.................................... 97.0

R. M^1-M^3.................................... 76.0

Antero-posterior diameter R. P². . . . . 38.5 mm.
Transverse diameter R. P². 26.0
Antero-posterior diameter R. P³. 32.0
Transverse diameter R. P³. 29.0
Antero-posterior diameter R. P⁴. 30.0
Transverse diameter R. P⁴. 29.0
Antero-posterior diameter R. M¹. 26.0
Transverse diameter R. M¹. 26.0
Antero-posterior diameter R. M². 26.5
Transverse diameter R. M². 26.0
Antero-posterior diameter R. M³. 24.0
Transverse diameter R. M³. 22.0

Hipparion theobaldi

Measurements of A.M. 19857

Length P²-M³. . . . . .149.0 mm.
P²-P⁴. 82.0
M¹-M³. 68.0

Antero-posterior diameter R. P². 32.0
Transverse diameter R. P². 26.5
Antero-posterior diameter R. P³. 25.0
Transverse diameter R. P³. 27.5
Antero-posterior diameter R. P⁴. 24.0
Transverse diameter R. P⁴. 27.5
Antero-posterior diameter R. M¹. 21.0
Transverse diameter R. M¹. 25.0
Antero-posterior diameter R. M². 21.0
Transverse diameter R. M². 25.0
Antero-posterior diameter R. M³. 25.0
Transverse diameter R. M³. 23.0

The Post Cranial Skeleton

The skeleton of the genus *Hipparion* is known from numerous descriptions, so little need be given here in the way of detailed descriptions. A few outstanding facts regarding the skeleton of *Hipparion theobaldi* are recorded below.

The skeleton represents an animal of large size for the genus *Hipparion*. In fact, *H. theobaldi* was comparable in size to the smaller species of *Equus*.

The vertebrae are quite like those of other upper Tertiary Equinae.

The ulna is complete.

The scapula and humerus do not show any points of special interest.

The femur is of ordinary form.

The fibula extends about halfway down the tibia.

The lateral digits are heavy, but this is due to the fact that the entire foot is heavy.

One of the most reliable methods of comparison, when working with various genera and species of mammals, is that whereby the ratios of homologous parts are considered. A certain skeletal element may seem to show considerable variations in two separate species when the comparisons are made on the basis of that element alone, but when it is considered in its relation to some other element of functional affinities the supposed differences often turn out to be more apparent than real.

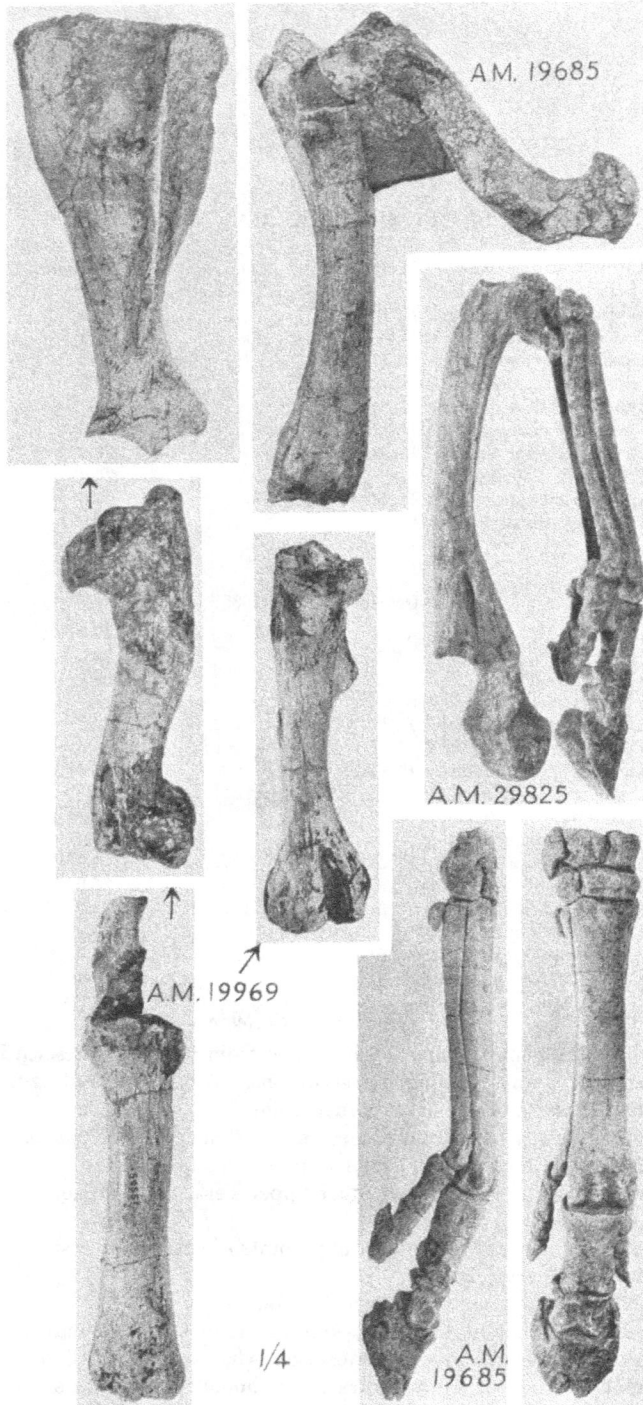

Fig. 67. *Hipparion theobaldi* (Lydekker). On the left, Amer. Mus. No. 19969, scapula, humerus, radius-ulna. Lateral and anterior views. On the right, Amer. Mus. No. 29825, radius-ulna and fore foot. Lateral views. Amer. Mus. No. 19685, humerus, radius-ulna. Lateral view. Lower right, Amer. Mus. No. 19685, fore foot. Lateral and anterior views. All figures one fourth natural size.

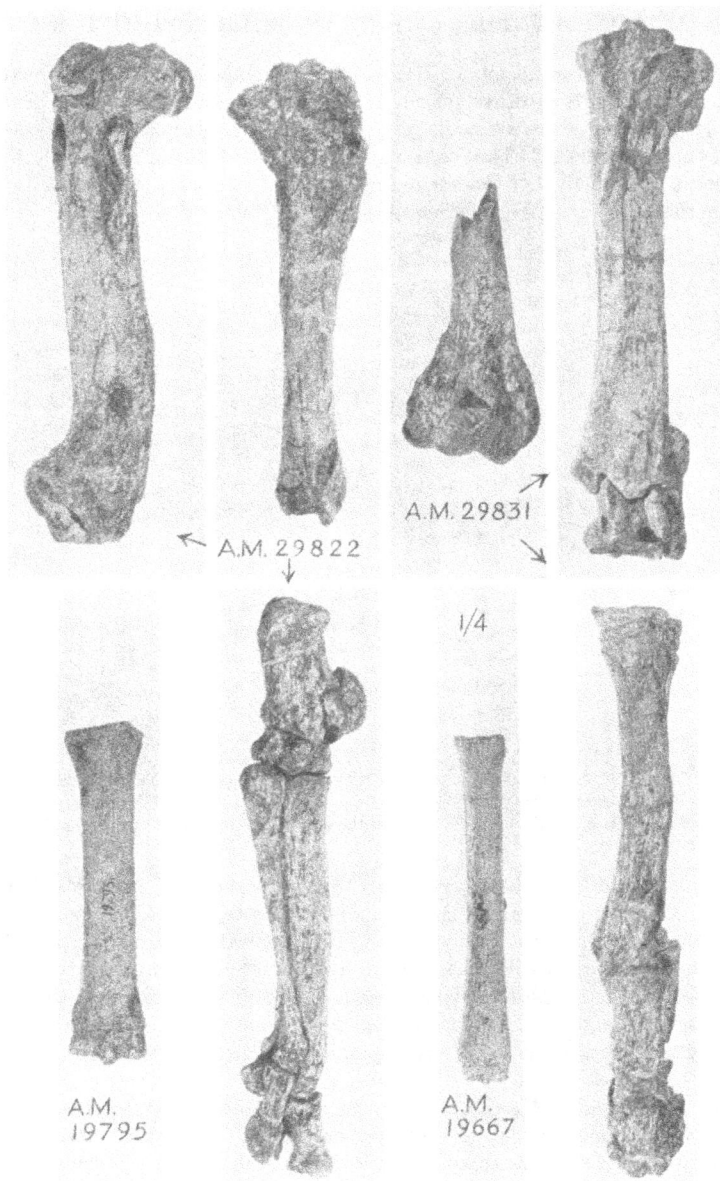

Fig. 68. *Hipparion theobaldi* (Lydekker). Amer. Mus. No. 29822, femur, tibia and hind foot. Lateral views. Amer. Mus. No. 29831, femur, tibia and hind foot. Anterior views.

Hipparion antelopinum (Falconer and Cautley). Amer. Mus. No. 19667, metatarsal. Anterior view.

Equus sivalensis Falconer and Cautley. Amer. Mus. No. 19795, metacarpal. Anterior view.

All figures one fourth natural size.

For instance, the lateral digits of *Hipparion theobaldi* seem to be very heavy as compared to the lateral digits of other species of *Hipparion*, say *H. whitneyi*. But when the lateral digits are compared to the medial digit it is seen that they are proportionately about the same in the two species. That is to say, the lateral digits of *Hipparion theobaldi* are heavy by virtue of the fact that the entire foot and limb is heavy, but proportionately they are no larger than in other more slender species of the genus.

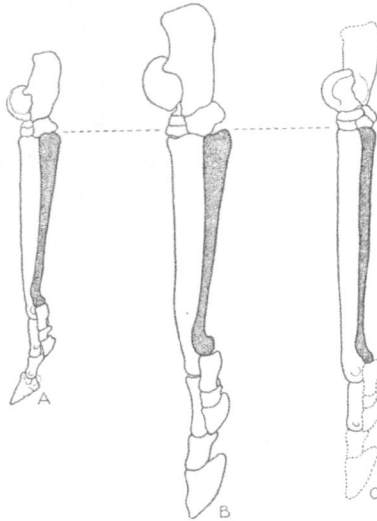

Fig. 69.   Hind feet of (A) *Merychippus sejunctus* (Cope); (B) *Hipparion theobaldi* (Lydekker), and (C) *Hipparion whitneyi* Gidley.   A and C after Osborn, 1918.   B from Amer. Mus. No. 29824.
Comparative figures to show differences in size, and the degree of development of the lateral metatarsals.   One fourth natural size.

To continue the argument, we may turn to a comparison of *H. theobaldi* with *Merychippus*. Dr. Matthew maintained that in many of the Old World Hipparions the lateral digits are secondarily enlarged over the condition typical of *Merychippus*. This is undoubtedly true as a general rule, but measurements and comparative ratios show that in *Hipparion theobaldi* the supposed enlargement is not so much confined to the lateral digits as it is a general increase in the entire foot and in the skeleton. The Siwalik form is an extremely large *Hipparion*, as large as a small wild *Equus* and with correspondingly heavy bones.

Thus we may imagine *Merychippus* giving rise to several phylogenetic lines of *Hipparion*. In North America the several species tended to remain small and slenderly proportioned, while in India one representative of the genus became unusually large. Now it is a seemingly valid principle in vertebrate evolution, that an increase in size in a phylogenetic line is accompanied by a proportionately great increase in the width and heaviness of the bones. Consequently, since *Hipparion theobaldi* is a very large species, we would expect it to be characterized by heavy limb and foot bones. This is in accordance with Matthew's contention that the Old World *Hipparion* is progressive, but rather aberrantly specialized.[39]

[39] Matthew, W. D., 1929, p. 529.

The following tables give the data concerning the feet of *Hipparion theobaldi*.

MEASUREMENTS OF *Hipparion theobaldi* LIMBS
(Articular Lengths in mm.)

Number	Humerus	Radius	Forefoot	Femur	Tibia	Hindfoot
19685	240 mm.	283 mm.	415 mm.	....	....	....
29822	....	....	....	354 mm.	313 mm.	400 mm.
29825	244	267	398	....	....	....
29824	....	....	....	....	335	457
29819	....	278	388	....	....	....
29831	....	....	....	....	327	423
29823	....	....	....	....	297	428
19969	260	273	....	357	....	....
19842	....	....	....	346	....	....
19683	....	....	....	....	325	....

MEASUREMENTS OF *Hipparion* PHALANGES
(Transverse diameters in mm.)

	Number	Proximal			Medial			Ungual		
		Med.	Lat.	Index	Med.	Lat.	Index	Med.	Lat.	Index
Manus.....	29825	33.0 mm.	12.5 mm.	38	38.0 mm.	12.0 mm.	32	63.0 mm.	14.0 mm.	22
Manus.....	29819	32.0	11.0	34	38.5	14.0	36	62.0	13.0	21
Manus.....	19685	32.5	10.0	31	39.0	14.5	37	68.0	15.0	22
Averages.				34			35			21.5
Pes........	29823	33.5 mm.	11.0 mm.	33	33.0 mm.	14.0 mm.	42	49.0 mm.	12.5 mm.	25
Pes........	29824	34.0	11.5	34	36.0	13.0	36	60.0	11.5	19
Pes........	29822	33.0	14.0	42	....	14.0	......	....	13.0	......
Pes........	29831	33.5	13.0	39	36.0	14.0	39	55.0	13.0	24
Averages.				37			39			22.5

MEASUREMENTS OF *Hipparion* FEET
Measured in mm.

Number	Art. height of carpus	Art. height of tarsus	Length of Mc III	Length of Mt III	Length of Mc II	Length of Mt II	Length of Mc IV	Length of Mt IV	Ext. length of med. phal. Manus	Ext. length of med. phal. Pes	Ext. length of int. lat. phal. Manus	Ext. length of int. lat. phal. Pes
A.M. 29825	43 mm.	....	209 mm.	....	193 mm.	....	187 mm.	....	146 mm.	....	79 mm.	......
19685	47	....	215	....	198	....	190	....	158	....	82	......
29819	43	....	204	....	190	....	184	....	141	....	78	......
29823	....	75 mm.	....	230 mm.	....	210 mm.	....	212 mm.	....	130 mm.	......	65 mm.
29824	....	75	....	242	....	220	....	222	....	145	......	82
29331	....	75*	....	235*	....	215*	....	218	....	140	......	70
29822	....	70	....	230	....	209	....	212	....	....	......	67
19683	....	....	....	232	....	....	....	208	....	130	......	70
29810	....	....	192	....	174	....	170	....	....	....	......	......
29671	....	....	215	....	....	....	180	....	....	....	......	......
29811	....	....	234	....	....	....	....	....	....	....	......	......
.....	....	....	245	....	....	....	....	....	....	....	......	......
19667	....	....	248	....	....	....	....	....	....	....	......	......

* Approximate

It was mentioned above that an increase in size in a phylogenetic line is accompanied by a proportionately great increase in the width and the heaviness of the bones of the skeleton. This rule applies especially to the limb bones, which necessarily form the supports for the body.

Numerous studies by Dr. W. K. Gregory over a period of years have led him to believe that the increase in width of a skeletal element, in an evolving animal often succeeds the increase in length. That is, the bone or structure first becomes elongated, which accomplishes the advantageous size increase, for which the animal has been striving, and then it becomes wide, as an adaptation to the support of an increasingly greater weight. This principle has been outlined by Gregory, with reference to certain of the Titanotheres, as follows:

"The changing proportions of the feet in the same phylum [i.e., *Rhadinorhinus* and the *Brontotheriinae*] were hypothetically as follows: the narrow foot of *Eotitanops* gave rise, chiefly by an increase in size, but also by some degree of broadening, to the narrow foot of *Rhadinorhinus;* the narrow foot of *Rhadinorhinus* may have broadened out, especially the magnum and astragalus, into the broad foot of *Brontotherium*." [40]

This same principle may be seen at work in the Equidae, not only in a general way throughout the entire family, but also within limited bounds, such as in the genus *Hipparion.* This genus developed from *Merychippus*, first by an increase in shoulder height, and therefore by an elongation of the limb bones, as is seen in the small and more primitive American species, and subsequently by an increase in weight, as typified by the Siwalik species. This weight increase, resulting in *Hipparion theobaldi* attaining the size of a small *Equus*, caused a consequent increase in the width of all of the limb bones and foot bones, as illustrated by the accompanying tables and figures.

Limb ratios and the Mechanics of Locomotion in *Hipparion*

Various authors, notably Matthew, Osborn and Gregory, have shown that running speed in the hoofed animals is dependent to a considerable degree on the proportions of the several limb elements to each other. That is, if the lower leg is long and the upper leg short, the animal will be speedy, because of the concentration of muscular power near the body and the development of the lower leg into a stilt, capable of long, rapid strides. This line of development has been followed by the horses.

A comparison of several Upper Tertiary equids shows that generally speaking the limb proportions are very similar among them. When detailed comparisons are made, slight differences are seen to exist. If *Merychippus* is taken as a central type, ancestral to the later Tertiary Equinae, it may be seen that in the small slender forms there is a slight tendency towards increase in the lower leg lengths and a correlative decrease in upper leg lengths, while in the large, heavy forms the trend was in the direction of slightly shorter lower leg lengths and longer upper leg lengths.

Whether these changes and differences were of any evolutionary significance is a debatable question. Certainly they were not of sufficient magnitude to enable us to say that some forms were swifter runners than other forms—perhaps they were features of hereditary development without any particular advantages or disadvatages in the history of the animals. The comparisons discussed above are shown graphically in the accompanying table.

[40] Osborn, H. F., 1929B p. 832.

COMPARISON OF LIMB PROPORTIONS IN THE EQUINAE

	Fore Limb			Hind Limb		
	Percentage to Total Limb Length			Percentage to Total Limb Length		
	Humerus	Radius	Manus	Femur	Tibia	Pes
Merychippus isonesus quintus...........	28	28	44	29	30	41
Hipparion whitneyi.....................	25	32	43	27	32	41
Hipparion gracile......................	28	31	41	29	30	41
Hipparion theobaldi....................	27	29	44	33	29	38
Plesippus simplicidens.................	27	31	42	31	28	41
Equus caballus........................	30	30	40	31	29	40

THE APPEARANCE OF *Hipparion* IN INDIA

The discovery by Mr. Brown of *Hipparion* teeth at Chinji Rest House has proved beyond all doubt that the genus was present in India during the earlier portions of Siwalik times. This stratigraphic occurrence of *Hipparion* has been recently substantiated through discoveries made by Mr. G. E. Lewis, palaeontologist of the Yale North India Expedition.[41]

The presence of the genus *Hipparion* in the lower portion of the Chinji beds at once brings up two questions of great importance.

1. How did *Hipparion* get to India?
2. When did *Hipparion* arrive there?

The first question is one concerning phylogenetic relationships and migrations. The second has its most important bearing on the problem of correlation. These questions will be considered in order.

The results of forty years of exploration and research by various institutions in North America, have proven beyond much doubt that the evolutionary history of the Equidae went through all of its major important phases in North America. Moreover, careful studies of the upper Tertiary genera of America have shown that *Hipparion* is pretty certainly derived directly from *Merychippus*. There is a gradual and perfect gradation in the teeth, skulls and skeletons from the advanced species of *Merychippus*, typical of the upper Miocene into the most primitive species of *Hipparion*, typical of the lower Pliocene of North America. This transition is in fact so well graded, that the question of a dividing line between the two genera assumes an academic aspect.

(The development of the North American Equinae has been so well traced by various authors that it need not be elaborated on here. For the details bearing on *Merychippus* and on the primitive North American species of *Hipparion* derived from the former genus, the reader is referred especially to the gollowing works.

Osborn, H. F., 1918, "Equidae of the Oligocene, Miocene and Pliocene of North America," *Mem. Amer. Mus. Nat. Hist.*, N.S., II, Pt. I.

Matthew, W. D., 1924, "Third Contribution to the Snake Creek Fauna," *Bull. Amer. Mus. Nat. Hist.*, L, Art II, pp. 153–175.

Matthew, W. D., 1926, "The Evolution of the Horse: A Record and its Interpretation," *Quar. Review of Biology*, I, No. 2, pp. 139–185.)

[41] Private communication.

Therefore all of our evidence points to the conclusion that *Hipparion* must have arisen in North America as a direct development from *Merychippus*, subsequently migrating to Asia and Europe. This was the view taken by Matthew, who stated it in the following way.

"I conclude therefore that the Equinae are surely of American evolution and dispersal and appeared in the Old World as immigrant types." [42]

This migration might have taken place either by a trans-Bering land bridge to Asia, or by a trans-Icelandic bridge to Europe. The point can not be definitely settled, in the light of our present knowledge (although all evidences seem to point to the former interpretation), nor does it bear directly on the problem at hand. The fact to keep in mind is that *Hipparion* did migrate from North America, the place of its origin, to Asia and Europe, where it spread over great areas and developed to a large size.

Now there comes the question of the time when this migration took place. *Hipparion* first appears in the Valentine formation of North America, which represents the basal portion of the Pliocene on that continent. The Valentine forms, *Hipparion gratum* particularly, are relatively primitive, showing an elongated oval protocone in the upper molars and simple to moderately complex foldings of the enamel on the fossette borders.

On the other hand, the *Hipparion* found at the base of the Chinji beds is a well advanced form, in all respects comparable to *Hipparion theobaldi* the typical Middle Siwalik species. It has a molar typified by an elongate oval protocone and complex enamel foldings on the fossette borders. All in all, the Chinji *Hipparion* is more closely comparable to the American species from the Republican River beds than it is to the earlier Valentine forms. This is what we might expect, because there would obviously be a certain time element involved during the migration of the genus from North America to Asia.

Hipparion occurs in Europe according to Borissiak, at Sebastopol in Sarmatian times. Therefore, if we accept Borissiak's correlation of the Sebastopol deposits, it seems logical to regard the Sarmatian as no older than the Valentine of North America, and probably a little younger. At least, the Sarmatian may be the equivalent of the upper portion of the Valentine. Considering this to be the case, the Pontian would be about equivalent to the Republican River of North America, a view that was advocated by Dr. Matthew in 1929. [43]

Now the base of the Chinji, since it contains an advanced *Hipparion* can be no older than the Sarmatian, and very probably it is equivalent to the upper portion of the Sarmatian or the lower part of the Pontian. But here we meet a difficulty, in that the Chinji fauna as a whole is typically more primitive than a Sarmatian or a Pontian fauna should be. However, the presence of *Hipparion* in the Chinji fauna would seem to be incontrevertible, so that it becomes necessary to regard this fauna as relict, stratigraphically of upper Sarmatian or lower Pontian age, but homotaxially of more primitive affinities. Thus it would seem probable that *Hipparion* developed in North America in the Valentine, and arrived in India in lower Chinji times, which consequently must be either of upper Sarmatian or of Pontian age.

Various European authorities have considered Hipparion as of Old World origin, a view quite the opposite of that outlined above. Dr. Pilgrim, considers the possibility of *Hipparion* being not only of Old World origin, but also of predating the genus in the New World.

[42] Matthew, W. D., 1929, p. 529.
[43] Matthew, W. D., 1929, p. 529.

His view is expressed in the following remarks, which were written in a private communication to the author. With his kind permission I take the liberty of reproducing them here.

"The occurrence of *Hipparion* in the Lower Chinji certainly makes it more difficult to correlate that stage with the Tortonian of Europe. But I cannot see how one can disregard the evidence of a whole fauna because one cannot find an explanation of the occurrence of a single form. Questions of migration, character of the fauna, i.e., whether plains or forest enter into the problem of correlation in a way which Matthew himself has rendered abundantly clear. It is a fact that *Hipparion* occurs in Europe in the Sarmatian, and one cannot deny that it is possible that *Hipparion* may have occurred in Central Asia and India even previously to this. I am inclined to think myself that the *Hipparion* of the Old World may be distinct from and have predated the *Hipparion* of America. This has more behind it than mere supposition. The lateral digits of the Old World *Hipparion* are certainly stronger than those of the American *Hipparion*. Where are we to find the links between the earlier American forms and this *Hipparion* with strong lateral digits? Apparently not in America. Whether the lateral digits have been strengthened secondarily, as Matthew thinks, or are a relic of an earlier condition, as I think, does not in the least matter. The point is that somewhere or other the ancestral form of the *Hipparion* of Sebastopol and Pikermi must have existed, and since the American deposits are so well known that its presence could hardly escape notice there, it seems more likely that it will one day be found in the Old World. It may be that the Chinji form might even turn out to be the required link, since we do not know the foot. I cannot even regard it as proved that such a type of *Hipparion* did not live in Tung Gur times. Its absence proves nothing definite, since the fauna is of a forest rather than a plains type, and as such is more likely to have contained *Anchitherium* than *Hipparion*. The rest of the fauna shows nothing, so far as I am aware, which would militate against a Tortonian age, the equivalent of La Grive St. Alban."

The Tung Gur formation, referred to by Dr. Pilgrim, is an horizon of upper Miocene age, probably correlative with the Sarmatian of Europe and the Pawnee Creek, Lower Snake Creek and Mascall of North America. (See Colbert, E. H., 1934A.)

A reply to Dr. Pilgrim's arguments is hereby presented in the following paragraphs.

In the first place, as pointed out above, there is a perfect gradation from *Merychippus* into *Hipparion* in North America, and the primitive *Hipparion* of North America is more primitive than the earliest Eurasiatic species. As to the lateral digits mentioned by Dr. Pilgrim, the pages preceding, 152–154, have shown that the enlargement in the Siwalik forms is due to an increase in the size of the entire skeleton.

As to the argument that the American deposits are so well known that an ancestor of the Pikermi *Hipparion* could "hardly escape notice there," this line of reasoning may be reversed with equal facility. Certainly the Eurasiatic deposits are pretty well known, and if an ancestor of *Hipparion* were to be found in them it probably would have turned up by this time. As a matter of fact, there is no equine in the European or Asiatic Tertiary that is ancestral to *Hipparion*, and since such atavistic forms are actually present in North America it seems only reasonable, at least on the basis of our present knowledge, to suggest that the origin of the genus was in the New World.

Nor can the Chinji *Hipparion* be a primitive link, as Dr. Pilgrim suggests, because it is fully as advanced in structure as the later Middle Siwalik *Hipparion*.

In the light of these considerations, there seems to be but one course to take, and that has been indicated above.

1. *Hipparion* is of North American origin.

2. It migrated to Eurasia.

3. It arrived in Eurasia subsequent to its appearance in the Valentine (basal Pliocene) of North America.

Fig. 70.   Upper and lower molar teeth of North American and Indian species of *Hipparion*.

Upper row, from left to right: *Hipparion theobaldi* (Lydekker), right upper molar, Amer. Mus. No. 19584; right lower molar, Amer. Mus. No. 19584; left upper molar, Amer. Mus. No. 19590.   (These teeth are from the Chinji zone of the Lower Siwaliks.)

Middle row, from left to right: *Hipparion antelopinum* (Falconer and Cautley), Amer. Mus. No. 19478, right upper molar; *Hipparion occidentale* Leidy, left upper molar; *Hipparion antelopinum* (Falconer and Cautley), Amer. Mus. No. 19478, left upper molar.   Amer. Mus. No. 19478 is from the Middle Siwaliks.

Bottom row, from left to right: *Hipparion mohavense* Merriam, right upper molar; *Hipparion gratum* Leidy, right lower molar; *Hipparion gratum* Leidy, left upper molar.

All teeth natural size.

This figure shows:

1. The primitive character of the earliest North American *Hipparion* (*H. gratum*).

2. The close resemblances between the earliest Siwalik *Hipparion* and the more advanced North American Hipparion.

3. The large size and the advanced character of the earliest Siwalik *Hipparion*.

Finally, it may be well to quote Dr. Matthew regarding this question.

"I conclude therefore that the Equinae are surely of American evolution and dispersal and appeared in the Old World as immigrant types. If this be so, by all homotaxial principles, they should be at least as advanced, and usually more advanced, in America at synchronous horizons. Nothing in the Valentine horizon of the American succession is as advanced as the *Hipparions* of the Old World; even in the Upper Chinji the *Hipparions* are more advanced than anything in the Valentine, and equivalent rather to the Republican River species (although I know of no American species that have the secondarily enlarged lateral metapodials).

"I think that Pilgrim may be mistaken in setting his correlations of Indian horizons so far back as he does. It would seem probable to me that India had, as it still has, the characters of a partly relict fauna, where older types survive than in the Holarctic world. On the other hand, the American succession has been judged younger than it is. If Santa Fé = Sansan, as Frick's work seems to indicate, then Republican River may = Pontian. I do not see how it can be any later, although it could be earlier on a general review of the fauna.

"But I do not see under the circumstances how any portion of the Siwalik fauna that carries *Hipparion* can be older than Pontian, unless we accept the highly improbable and quite unsupported theory of Pilgrim that *Hipparion* appeared earlier in India than in Europe—and, as matters now stand, earlier than it did in China." [44]

The reader is referred to a foregoing portion of this present work, "Correlation of the Siwaliks," pp. 21–27, for further remarks regarding the appearance of *Hipparion* in the Siwaliks.[45]

NOTES ON THE AGE OF THE SEBASTOPOL FAUNA

The Sebastopol fauna certainly seems, on the basis of its mammalian remains, to be Pontian (or possibly later) in age. However Borissiak has stated quite definitely that it comes from the upper portion of the Middle Sarmatian, and his conclusions are founded not only on the mammalian fossils, but also on the stratigraphic evidence and on the occurrence of marine and fresh water shells associated with the mammals. So, lacking further first hand evidence, it seems best to accept Borissiak's correlation.

The mammalian fauna is as follows:

Ictitherium tauricum Bor.
Aceratherium zernowi Bor.
Aceratherium zernowi asiaticum Bor.
Hipparion gracile sebastopolitanum Bor.
Achtiaria expectans Bor.
Tragocerus leskewitschi Bor.
Tragoreas sp.
Gazella sp.

Borissiak's description of the occurrence of the Sebastopol fauna is given below.

"Quant aux conditions de gisement de la faune décrit, elle a été trouveé dans une brèche ossifère qui représentait une petite intercalation en forme de lentille de deux sagènes de

[44] Matthew, W. D., 1929, p. 529.
[45] Some of the remarks appearing in the above argument are taken from Colbert, Edwin H., 1935, Amer. Mus. Novitates, No. 797.

diamètre sur 1 archive d'epaisseur formeé d'un calcaire dur, blanc, un peu jaunâtre, fine-
ment grenu, en partie oolithique, parsemé de rares coquilles marines, et d'eau douce et tout
rempli d'ossements.   Stratigraphiquement cette lentille appartient à la zone supérieure de
Sarmatien moyen." [46]

Equus Linnaeus, 1758

Generic type, *Equus caballus* Linnaeus

Equus sivalensis Falconer and Cautley

Equus sivalensis, Falconer and Cautley, 1849, Fauna Antiqua Sivalensis, Pls. LXXXI–
LXXXV.

> *Additional References.*—
>
> Falconer, Hugh, 1868A, pp. 186–188.
> Lydekker, R., 1880B, p. 31; 1882A, pp. 87–92, Pls. XIV, figs. 1, 2, XV, fig. 1;
>   1883C, pp. 83, 90; 1885B, p. 54; 1886A, pp. 66–69.
> Major, C. J. F., 1885, p. 2.
> Pilgrim, G. E., 1910B, p. 201; 1913B, p. 324.
> Matthew, W. D., 1929, pp. 444, 530.

Type.—(Lectotype.)   B.M. No. 16160, a cranium.

Cotypes.—B.M. Nos. 16227, posterior portion of a cranium; M 2666, left maxilla; M 2698,
premaxilla and symphysis; 22107, portion of right ramus; 22108, portion of mandible;
M 2667, premaxilla with incisors; 16171, left maxilla; various limb and foot bones, vertebrae and other skeletal parts figured in Fauna Antiqua Sivalensis, Pls. LXXXIII–LXXXV.

Horizon.—Upper Siwaliks.

Locality.—Siwalik Hills, Salt Range, Punjab.

Specimens in the American Museum.—Amer. Mus. No. 19795.   Left third metacarpal.
      From the Upper Siwaliks, seven miles west of Kalka.

19796.   Fragmentary ramus with left M_3.   Upper Siwaliks, below conglomerate,
      two miles south of Charnian.

19806.   Portions of the right and left maxillae, with contained right P^4–M^3, and left
      P^{2-3}, M^3.   Variegated beds, below the conglomerate, seven miles west of
      Kalka.

19827.   Left first upper molar.   Upper Siwaliks, below conglomerates, two and one
      half miles south of Chandigarh.

19884.   Fragment of ramus with left P_4–M_3.   Upper Siwaliks, below conglomerates,
      near Siswan.

Diagnosis.—"Dist. Characters, auct. Lydekker.—Protocone of premolars small, never
larger than in m^2.   This distinguishes from *E. caballus;* resembles *E. hemionus,* but larger
size and p^1 less reduced.   A distinct trace of a preorbital fossa ('larmial cavity' of Lydekker,
but it certainly is not the larmier of ruminants).   Muzzle shorter than in *E. caballus,* jaw
deeper, thereby approaching *hemionus.*   Limbs and feet also are relatively slender.

"The short muzzle and deep jaw are characteristics of early Pleistocene species, both
in America and the Old World, as compared with *E. caballus.*   They are in varying degree
approached by *E. prjevalskii,* the zebras, etc." [47]

---

[46] Borissiak, A., 1914, p. 105.
[47] Matthew, W. D., 1929, p. 530.

Nothing need here be added to the full descriptions of this species already published by Falconer and Cautley and by Lydekker. *Equus sivalensis* is a large species of horse, typified by its elongated protocone. Measurements of the American Museum specimens are given below.

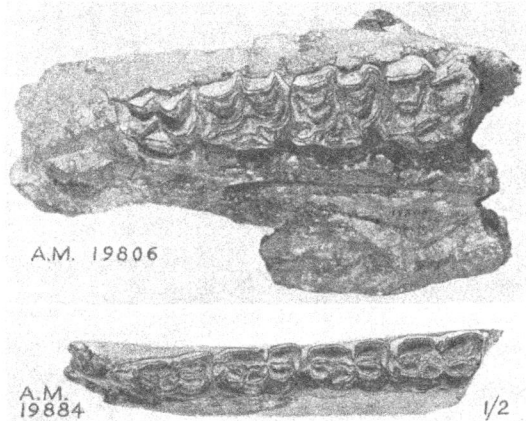

FIG. 71. *Equus sivalensis* Falconer and Cautley. Amer. Mus. No. 19806, palate with right P⁴-M³. Crown view. Amer. Mus. No. 19884, mandible with left P₄-M₃. Crown view. Both figures one half natural size.

Amer. Mus. No. 19806, maxilla.

P²....Antero-posterior diameter... 41.5 mm.	Transverse diameter... 29.0 mm.
P⁴....Antero-posterior diameter... 30.0	Transverse diameter... 33 0
M¹...Antero-posterior diameter... 30.5	Transverse diameter... 29.0
M²...Antero-posterior diameter... 31.5	Transverse diameter... 27.0

Amer. Mus. No. 19827 M¹.

Antero-posterior dia.. 30 mm. Transverse dia.. 29 mm. Height.. 84 mm.

Amer. Mus. No. 19884, mandible.

P₄....Antero-posterior diameter.... 32 mm.	Transverse diameter.... 20 mm.
M₁....Antero-posterior diameter.... 29	Transverse diameter.... 19
M₂....Antero-posterior diameter.... 29	Transverse diameter.... 18
M₃...Antero-posterior diameter.... 35	Transverse diameter.... 15

Depth of ramus below M₁........ 100

Amer. Mus. No. 19795, left metacarpal III.

Length............ 241 mm. Width at middle of shaft............ 36 mm.

Equus namadicus Falconer and Cautley

Equus namadicus, Falconer and Cautley, 1849, Fauna Antiqua Sivalensis, Pls. LXXXI, LXXXII.

Additional References.—

Lydekker, R., 1882A, pp. 92–96, Pls. XIV, figs. 3, 4, XV, fig. 2–4; 1883C, pp. 83, 90; 1885B, pp. 55–57; 1886A, pp. 71–73.

Pilgrim, G. E., 1910B, p. 201.

Matthew, W. D., 1929, pp. 444, 530.

Type.—(Lectotype.)   B.M. No. M 2683, a skull.

Cotypes.—B.M. Nos. M 2684, a left maxilla; M 2685, premaxilla; M 2686, premaxilla; M 2687, left ramus; M 2689, left ramus; M 2691 and M 2692, radii.

Horizon.—Upper Pleistocene, Narbada Valley.   This would be stratigraphically higher than the Upper Siwalik beds.   Also recorded by Lydekker from the Upper Siwaliks of the Siwalik Hills.

Locality.—Narbada Valley, India.   Siwalik Hills.

Diagnosis.—(See Matthew, W. D., 1929, p. 531.)   Protocone of premolars and molars much longer than in *E. sivalensis*.

This species is possibly synonymous with *E. sivalensis*.   In view of its younger geologic age, however, it may very possibly be a progressive *Equus*, more like the recent *E. caballus*.   (See Matthew, W. D., 1929, pp. 530, 531.)

Equus palaeonus Falconer and Cautley

Equus palaeonus, Falconer and Cautley, 1849, Fauna Antiqua Sivalensis, Pl. LXXXII, figs. 9, 10, 11.

Additional References.—

Falconer, Hugh, 1868A, p. 186.

Lydekker, R., 1882A, p. 92; 1883C, p. 90; 1885B, p. 55; 1886A, pp. 72–73.

Cotypes.—B.M. Nos. M 2685, premaxilla; M 2686, premaxilla; M 2689, left ramus.

Horizon and Locality.—From the Upper Pleistocene of the Narbada Valley.

Diagnosis.—See the diagnosis for *Equus namadicus*.

This species is synonymous with *E. namadicus*, and the specimens are listed under the latter form.   *E. namadicus* is, in turn, possibly synonymous with *E. sivalensis*.

CHALICOTHERIOIDEA

GENERAL CONSIDERATIONS [48]

The taxonomic history of the chalicotheres is long and involved, and can not be taken up here.   A full account of it may be found in the monograph by Holland and Peterson, and the reader should refer to that publication for the necessary details.[49]

As the result of their studies on the chalicotheres, these authors made the following classification:

Superfamily Chalicotherioidea
Family Chalicotheriidae
Subfamily Schizotheriinae
Genera *Schizotherium*
Pernatherium
Eomoropus
Phylotillon

[48] The remarks on the following pages concerning the classification and the evolution of the chalicotheres are adapted in part from a recent publication by the present author.   See Colbert, Edwin H., 1935, Amer. Mus. Novitates, No. 798.

[49] Holland, W. J., and Peterson, O. A., 1914, "Osteology of the Chalicotheroidea," *Mem. Carnegie Mus.*, III, No. 2, pp. 189–403.

Subfamily Moropodinae
Genera *Moropus*
    Nestoritherium
Subfamily Macrotheriinae
Genera *Macrotherium*
    Chalicotherium
    Circotherium

In 1929 Dr. W. D. Matthew classified this group along somewhat different lines.   His arrangement is presented below.[50]

Family Chalicotheriidae
Subfamily Eomoropinae
Genus *Eomoropus*
Subfamily Chalicotheriinae
A.   Brachyodont Series
Genera *?Olsenia*
    ?Pernatherium
    Schizotherium
    Macrotherium
    Chalicotherium
    Circotherium
B.   Hypsodont Series
Genera *Schizotherium* (tentative for certain species)
    Moropus
    Phylotillon
    Nestoritherium

Dr. Matthew pointed out a fact in connection with chalicotherine relationships which other authors had seemingly missed, namely that *Eomoropus* differs far more from the later Tertiary forms than they do from each other.   Consequently he divided the family into two subfamilies, one containing *Eomoropus* and the other containing all of the later Tertiary genera.   This second subfamily he again split into series A and series B, based on the brachyodonty and the hypsodonty of the teeth respectively.   In making this division of the Chalicotheriinae it may be possible that Matthew fell into one error by placing the genus *Schizotherium* in the same group with *Macrotherium* and *Chalicotherium* (series A of his classification).   It would seem that *Schizotherium* is more truly referred to the second series, along with *Moropus, Phylotillon* and *Nestoritherium*.   Evidence for this statement will be brought out below.

Of course Matthew had in mind *Schizotherium pilgrimi* especially, a form of lower Miocene age in which the molars are brachyodont and quadrate, as in *Macrotherium*. This species would certainly fall into Matthew's "series A" of the Chalicotheriinae.   A close examination of the figure of *S. pilgrimi* leads me to think, however, that this form may not belong to the above mentioned genus, but that it may be rather a primitive chalicotherine directly ancestral to *Macrotherium*.   The questionable relationship of *S. pilgrimi* was recognized by Forster Cooper and by Matthew.

[50] Matthew, W. D., 1929, pp. 516–522.

"*Schizotherium pilgrimi* certainly not the milk dentition of *Phylotillon*, and has every appearance of being permanent dentition of a brachyodont chalicothere of quite small size and very primitive construction of the teeth, the anterior transverse crest being more normally developed and protocone less isolated and less shifted in position than in any Miocene genus. I suspect that Cooper's identification indicates that *Schizotherium* belongs, some species at least, in the chalicotheriine series as defined below, as the earliest stage of its development." [51]

Dr. Matthew's statement, quoted above, bears directly on the question of the relationships of *Schizotherium turgaicum* Borissiak, an Oligocene form from eastern Asia. This species was originally described as belonging to the genus *Schizotherium*, but Koenigswald, in 1932, referred it to *Macrotherium*, especially on the basis of its quadrate, brachyodont upper molars.

A close scrutiny of the figures of *S. turgaicum* leads to the conclusion that it is not a *Macrotherium*, but rather a much more primitive genus. It is quite possible that this form is a separate genus, more primitive even than *Schizotherium*, for it shows certain characters that relate it to the Eocene chalicotheres *Eomoropus* and *Grangeria*. The quadrate upper molar may very well be a primitive chalicothere character, inherited from an Eocene ancestor such as *Eomoropus*. In the lower molars the metastylid is distinct, being rather separate from the anterior spur from the hypoconid—a primitive character found in most of the early Eocene perissodactyls. In the hindfoot the astragalus is primitive by reason of its narrowness and the relative depth of its trochlea. On the other hand the neck of the astragalus is reduced—an advanced character, and its lower articular surfaces has a facet for the navicular only. In this last feature, *S. turgaicum* shows a decided trend towards the development found typified in *Moropus* and *Nestoritherium*. The metapodials of *S. turgaicum* are long and in general they show a definite trend towards the long footed kind of development, as found in *Moropus* and related genera. The phalanges are rather primitive.

Thus we see that *Schizotherium turgaicum* is a primitive Oligocene chalicothere, obviously descended from the Eocene types and showing certain trends towards the typical *Schizotherium* of Europe and Asia. If *S. turgaicum* is of the genus *Schizotherium*, then we must postulate that the genus shows two broad stages of development, an earlier one retaining many holdovers of Eocene characters, and a later one in which the primitive characters are lost and the definitive *Schizotherium* features are established.

Turning now to the classification formulated by von Koenigswald in 1932, we see that the genus *Schizotherium* is grouped with *Moropus* and *Phylotillon*, which it seems to me is the correct view.[52]

The classification followed by von Koenigswald is here presented in a summary form. Only genera are listed.

Family Chalicotheriidae
Subfamily Eomoropinae
Genus *Eomoropus*
Subfamily Chalicotheriinae

[51] Matthew, W. D., 1929, p. 518.
[52] Koenigswald, H. G. R. von, 1932, p. 22.

Genera *Chalicotherium* (including *Macrotherium, Schizotherium turgaicum, Schizotherium pilgrimi*)
Nestoritherium (*Circotherium*)
Subfamily Schizotheriinae
Genera *Schizotherium*
Metaschizotherium
Colodus
Phylotillon
Moropus
Postschizotherium

With the foregoing information before us, we may now turn to a brief review of the chalicotheres, thereby attempting an evaluation of the various genera comprising this group of perissodactyls.

The earliest known chalicotheres are of middle to upper Eocene age, and are found in North America and Asia. As Dr. Matthew has pointed out, these animals are really more like the Eocene titanotheres and the other primitive Perissodactyla than they are like the later chalicotheres. These Eocene forms are characterized by unspecialized skulls and feet, and a primitive perissodactyl dentition with the canine well developed.

As the chalicotheres continue into Oligocene and later Tertiary times they are seen to split into two seemingly well defined groups. One group is typified by *Schizotherium* and *Moropus*, and is characterized by a rather elongate skull, elongated, hypsodont cheek teeth, a skeleton in which the fore and the hindlimbs are of subequal length, and feet having long metapodials.

The other group, as typified by *Macrotherium* and *Chalicotherium*, is characterized by a skull in which the facial portion has become quite short, quadrate, brachyodont cheek teeth, a skeleton in which the forelimbs are much longer than the hindlimbs, and feet having short metapodials and flattened phalanges.

This twofold division of the advanced chalicotheres would seem to be a natural one, for when it is tested in the light of our present available information, it would seem to hold true. Therefore, on this basis of the division of the chalicotheres into a primitive group and two advanced groups, the following classification is presented. It is really a modification of Matthew's classification of 1929.

Order Perissodactyla
Superfamily Chalicotherioidea
Family Chalicotheriidae
Subfamily Eomoropinae
Genera *Eomoropus* Upper Eocene; North America, China.
Grangeria Upper Eocene-Oligocene; China, Mongolia.
Subfamily Chalicotheriinae
Tribe Chalicotherini
Genera *Chalicotherium* Miocene-Pliocene; Europe.
Nestoritherium Pleistocene; India, China.
Macrotherium Miocene-Pliocene; Eurasia, North America.
Oreinotherium Oligocene; North America.

Tribe Schizotherini
>Genera *Schizotherium* Oligocene; Eurasia.
>>*Metaschizotherium* Upper Miocene; Europe.
>>*Moropus* Lower Miocene; North America.
>>*Phylotillon* Lower Miocene; Baluchistan.
>>*Ancylotherium* Pliocene; Europe.
>>*Postschizotherium* Pleistocene; China.

Incertae Sedis
>*Pernatherium* Oligocene; Europe.

The characters on which the larger groups in the above classification are based, are presented below.

Family CHALICOTHERIIDAE

Cheek teeth bunoselenodont; last upper premolars with two outer and one inner cusp, last lower premolars with double crescents; upper molars with W-shaped ectoloph, with protoloph connecting protocone and paracone, and metaloph connecting metacone and hypocone; lower molars doubly crescentic with a separate metastylid and the third molar without a talonid except in the primitive genera. Auditory bulla large; orbit open behind; strong postglenoid and paroccipital processes; foramen lacerum anterius and foramen rotundum enclosed in a common vestibule; alisphenoid canal present; mandible with a broad ascending ramus. Cervical vertebrae keeled. Pelvis elongated; femur with or without a third trochanter. Distal face of astragalus articulating with the navicular and cuboid, or with the navicular only. Manus either tetradactyl or tridactyl; pes tridactyl; distal ends of metapodials with convex articulating surfaces; terminal phalanges deeply bifid, except (?) in the primitive genera.

Subfamily EOMOROPINAE

Primitive, and of small size. Quadrate, brachyodont molars, with protoloph connecting protocone and metacone, and metaloph connecting paracone and hypocone; lower molars with a separate metastylid; third lower molar with a talonid. First upper premolar and both upper and lower canines present; lower canine more or less in series with the incisors; incisor formula variable. Manus tetradactyl, pes tridactyl; astragalus with or without a cuboid facet. Metapodials and phalanges not highly modified as in the later chalicotheres. Limbs subequal in length.

Subfamily CHALICOTHERIINAE

Advanced genera of large size. Premaxillaries often, and probably always edentulous. Canines and first upper premolar absent; upper molars quadrate to elongate; no third lobe on the last lower molar. Manus and pes highly modified; femur with third trochanter.

Tribe *Chalicotherini*

Advanced genera of medium to very large size. Quadrate, brachyodont upper molars, with ectoloph bent lingually beyond the median line of the tooth; molar indices usually above 90; metastylid reduced in lower molars. Manus and pes tridactyl, manus longer than pes; trapezium wanting; astragalus with a cuboid facet; articulating facets of proximal phalanges tending to be parallel with the long axis of the bone; claws short; limbs unequal, the forelegs being very long.

Tribe *Schizotherini*

Advanced genera of medium to very large size. Elongated, hypsodont molars with ectoloph tending to be vertical, thus making the paracone and metacone on or outside of the median line of the tooth; molar indices usually below 90; metastylid not reduced. Manus tetradactyl, pes tridactyl; trapezium present; astragalus articulating with navicular only; articulating facets of proximal phalanges inclined to median axis of bones; limbs subequal in length.

INDICES OF MOLAR TEETH IN THE CHALICOTHERIINAE

The indices listed in the table below were obtained by dividing the width of the tooth by its length and multiplying the quotient by 100. It will be noticed that the genera belonging to the Chalicotherini have a high index, indicating a short, quadrate molar, while the genera representative of the Schizotherini have a lower index, denoting an elongated molar.

INDICES OF M^3

	Index
Chalicotherini	
Macrotherium grande. From cast of type skull	110
Macrotherium brevirostris. A.M. 26518	100
Macrotherium salinum (M^2). A.M. 19467	97
Chalicotherium goldfussi. From Zittel	104
Nestoritherium sivalense. From Falconer and Cautley	100
Schizotherini	
Schizotherium modicum. A.M. 10411, cast	87
Moropus elatus. A.M. 14424	84
Phylotillon naricus. A.M. 9946, cast	85
Ancylotherium pentelici. A.M. 10564, cast	90

CHALICOTHERIIDAE

CHALICOTHERIINAE

CHALICOTHERINI

Macrotherium Lartet, 1837

Generic type, *Macrotherium sansaniense* Lartet

Macrotherium salinum Forster Cooper

Macrotherium salinum, F. Cooper, 1922, Ann. and Mag. of Nat. Hist., Ser. 9, X, pp. 542–544, 3 figs.

Additional References.—

Matthew, W. D., 1929, pp. 517, 519.

*Type.—*B.M. No. M 12239, a left third molar, superior.

*Paratypes.—*None.

*Horizon.—*Lower Siwaliks.

*Locality.—*Near Chinji, Salt Range, Attock District, Punjab.

Specimens in the American Museum.—Amer. Mus. No. 29834.   Left M². From the
lower portion of the Middle Siwaliks, 1000 feet below the Bhandar bone
beds.   One mile south of Nathot, Salt Range, Jhelum District, Punjab.

19437.   Left M₂ and a fragment of the ramus.   Lower Siwaliks, 1600 feet above the
level of Chinji Rest House.   Twelve miles east of Chinji Rest House, Salt
Range, Attock District, Punjab.

19577.   Right ramus of mandible of a young individual with DM₂₋₄, M₁.   Lower
Siwaliks, 1600 feet above level of Chinji Rest House.   One and one half
miles northwest of Chinji Rest House, Attock District, Punjab.

19647.   Right M² and fragments of left M³.   Lower Siwaliks, about the level of Chinji
Rest House.   Four miles east of Chinji Rest House, Attock District,
Punjab.

29816.   Left second metacarpal.   Lower Middle Siwaliks, 1000 feet below the level
of the Bhandar bone bed.   One mile south of Nathot, Salt Range, Jhelum
District, Punjab.

19436.   Various foot bones, among which are: a left calcaneum and astragalus, a pisi-
form, the distal end of a metacarpal, nine proximal, seven median and two
ungual phalanges.   Also two patellae.   Lower Siwaliks, 1600 feet above
the level of Chinji Rest House.   Twelve miles east of Chinji Rest House,
Salt Range, Attock District, Punjab.

Diagnosis.—A typical *Macrotherium*, closely comparable to *M. grande*, but smaller
than this latter species.   Upper molars quadrate and brachyodont, and protoconule more
distinct than in the other species of the genus.

The original description of *M. salinum* was based on a single third upper molar.   Consequently the material in the American Museum collection, listed above, affords a considerable amount of information not hitherto known about the dentition and the feet of
this species.

The second upper molar of *Macrotherium salinum* is more quadrate than is the third
molar, that is to say, the transverse diameter across the protocone-paracone is about equal
to the width across the metacone-hypocone, whereas in the last tooth the posterior portion
is somewhat reduced.   Otherwise the second molar is quite similar to the following tooth.
The protocone is rounded, and from its anterior surface a ridge curves forward to the protoconule.   A broad anterior cingulum runs from the paracone to the protocone, stopping
short at this point and not continuing around the protocone, as it does in the third molar.
The crests of the paracone and the metacone are placed far in towards the lingual side of
the tooth, and their external slopes, forming the ectoloph, are very low.   There is a narrow
vertical ridge on the outer slope of the paracone, terminating in the crest of this cusp, but
the slope of the metacone is smooth.   The hypocone is compressed and slightly higher than
the protocone.

Forster Cooper's description of the third molar in this species is quoted below.

"The crown surface shows a well-marked protocone, from which a sharply defined
ridge runs in a wide curve to the protoconule.   The latter cusp is rather more sharply
defined from the paracone than is usually the case.   The cingulum is broad in front and
runs round the protocone, and ends in the valley between the protocone and hypocone.

It is not interrupted by the protocone as it is in *C. sinensis* and, to some extent, in *P. naricus*.

"The hypocone is compressed, bent rather forwards, and is higher than the protocone. The external wall of the paracone is much bent inwards." [53]

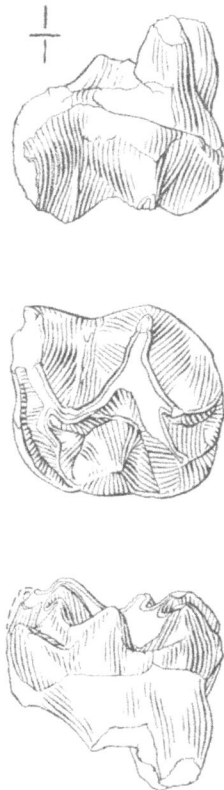

FIG. 72. *Macrotherium salinum* Forster Cooper. Amer. Mus. No. 29834, a left M². Outer view above, crown view in middle, inner view below. Natural size.

A lower jaw of a young individual (A.M. 19577), shows that the premolar-canine diastema was short. The second milk molar is a small tooth, the third one is long and narrow tending to be doubly crescentic, while the last milk molar simulates a true molar in its form. The first lower molar consists of two crescents, one behind the other. There is a rather broad posterior cingulum.

The second lower molar, as shown by A.M. 19437, is like the preceding tooth. In this species the metastylid is reduced to the point of being obliterated, whereas in most chalicotheres this element is quite distinct from the metaconid.

[53] Cooper, C. Forster, 1922, pp. 542–544.

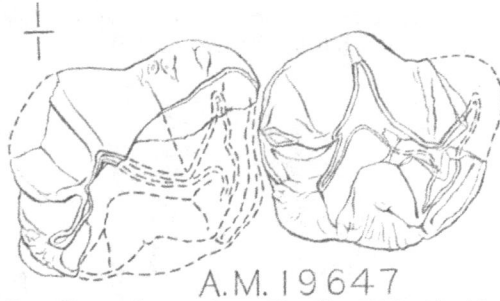

Fig. 73. *Macrotherium salinum* Forster Cooper. Amer. Mus. No. 19647, right M² and left M³ (fragmentary).
M³ reversed in figure. Crown view. Natural size.

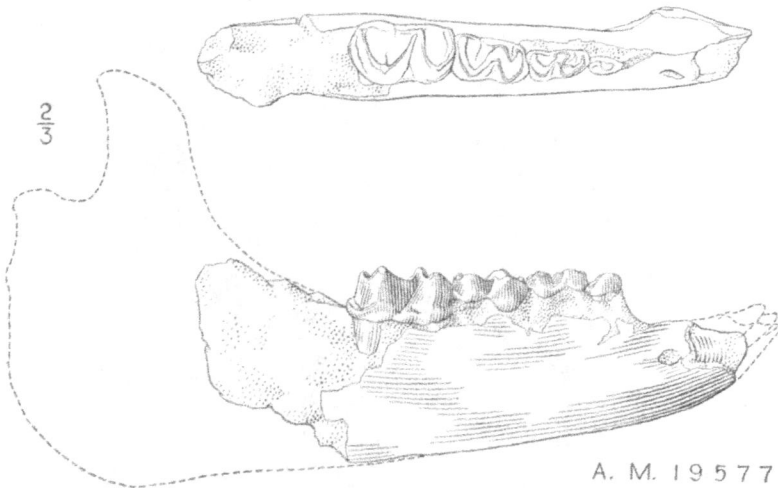

Fig. 74. *Macrotherium salinum* Forster Cooper. Amer. Mus. No. 19577, mandibular ramus with right DM₂₋₄, M₁.
Crown view above, lateral view below. Two thirds natural size.

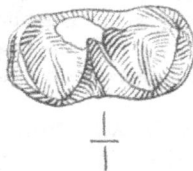

Fig. 75. *Macrotherium salinum* Forster Cooper. Amer. Mus. No. 19437, left M₂. Crown view. Natural size.

MEASUREMENTS

		Length	Width	Height
A.M. 29834	M²	34.0 mm.	32.5 mm.	22.0 mm.
A.M. 19647	M²	37.5	36.0	24.0
	M³	43.0	39e	
A.M. 19577	DM₂	9.0	5.0	
	DM₃	16.0	8.0	
	DM₄	19.0	12.5	
	M₁	27.0	17.0	14.0
	Depth of ramus at M₁			37
A.M. 19437	M₂	31.0	16.0	16.0

e—Estimated

The Siwalik material in the American Museum gives us only a partial knowledge of the foot structure in *Macrotherium salinum*, but such as it is it represents an advance over our previous information.

There are no carpal elements present in the material collected by Mr. Brown, but two tarsal bones, namely the calcaneum and the astragalus, afford us some knowledge of the ankle. The astragalus is extraordinarily low and wide, with a shallow trochlea. It shows articular facets for both the navicular and the cuboid bones, as is characteristic of the genus *Macrotherium*. The sustentacular facet is broad and convex and meets the cuboid facet along the lower portion of its border. The inner facet has a concave surface. The navicular facet is very large, and it is bounded along one edge by the sustentacular and the cuboid facets.

The calcaneum has a very peculiar shape, due to the great transverse width of its distal facet. The tuber is narrow and short as compared to the length of the peroneal and sustentacular parts. Together these two portions give the bone a V shape. There is a broad cuboid facet, which is transversely concave.

A single metapodial has been identified as the second metacarpal of the left manus. This element is typically chalicotherine, being relatively short and having both its distal and proximal ends greatly expanded. The proximal articular surface consists of two facets, one for the trapezoid and one for the magnum, and beneath the magnum facet is a large surface which articulates with the third metacarpal. The phalangeal articulation has the form of a hemisphere with a sharp keel along its back surface, thus allowing a great freedom of movement for the toe.

A number of proximal phalanges show typical chalicothere characters. Each is short and wide, and the metapodial articulation is practically parallel with the long axis of the bone. The distal articulation is deeply grooved, to receive the keeled articular surface of the median phalanx. The various median phalanges are similar to one another except as to size, as indeed is the case with the proximal phalanges. There is a large median phalanx which is quite short and high, which would indicate that it came from the inner digit of the manus. In the median phalanges the proximal articular surface is keeled and the distal one is cleft.

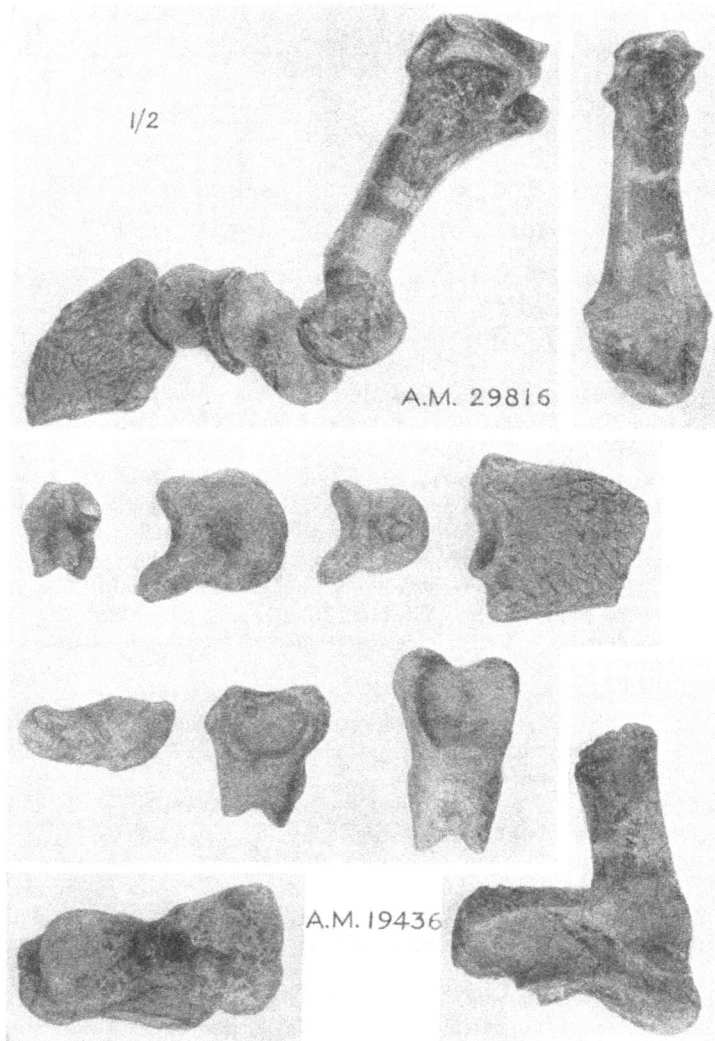

FIG. 76. *Macrotherium salinum* Forster Cooper. Amer. Mus. Nos. 29816 and 19436, left second metacarpal and various foot bones, respectively. Top: metacarpal and phalanges, lateral view; metacarpal, dorsal view. Middle: Median phalanges, dorsal and lateral views; ungual phalanx, lateral view; proximal phalanges, lateral and dorsal views. Bottom: left astragalus and calcaneum. All figures one half natural size.

A large claw, seemingly belonging to the second digit of the manus, is very high and narrow. It is deeply cleft, as is characteristic of the chalicotheres.

A patella associated with the above described foot bones is here provisionally assigned to *Macrotherium salinum*. It is of medium size and the lower end is produced into a point. The accompanying measurements and figures give further information about the podial elements described above.

Macrotherium salinum

MEASUREMENTS

A.M. 29816, left metacarpal II
Length.. 131 mm.
Transverse diameter, proximal end.................. 32
Antero-posterior diameter, proximal end............. 66
Transverse diameter, distal end..................... 52
Antero-posterior diameter, distal end................ 33
A.M. 19436, calcaneum
Greatest length.................................... 107
Width... 81
A.M. 19436, astragalus
Greatest antero-posterior diameter.................. 56
Width... 91
Height.. 50
A.M. 19436, digit II of manus (as identified)
Greatest length of proximal phalanx................. 70
Greatest length of median phalanx.................. 49
Height of median phalanx.......................... 36
Length of ungual phalanx.......................... 73e
Height of ungual phalanx.......................... 60

e—Estimated

Nestoritherium Kaup, 1859

Generic type, *Nestoritherium sivalense* (Falconer and Cautley)

Nestoritherium sivalense (Falconer and Cautley)

Anoplotherium sivalense, Falconer and Cautley, 1843, Proc. Geol. Soc., London, No. 98, Pl. II.

Anoplotherium posterogenium, Falconer and Cautley, 1835, Jour. Asiatic Soc. Bengal, IV, p. 706. (Nomen nudum.)

Anoplotherium sivalense, Falconer and Cautley, 1836, Trans. Geol. Soc., London (II), V, p. 502. (Nomen nudum.)

Chalicotherium sivalense, Falconer and Cautley, 1847, Fauna Antiqua Sivalensis, Pl. LXXX.

Circotherium sivalense, Holland and Peterson, 1914, Mem. Carnegie Mus., III, No. 2, pp. 211–212.

Additional References.—

Falconer, H., 1868A, pp. 191–197, 208–226, Pl. XVII.

Lydekker, R., 1880B, p. 31; 1883C, pp. 83, 90; 1884D, p. 132; 1885B, p. 73; 1886B, pp. 164–165.

Pilgrim, G. E., 1910B, p. 201.

Matthew, W. D., 1929, pp. 444, 516, 520.

Koenigswald, G. H. R. von, 1932, p. 22.

Type.—B.M. Nos. 15366, 15367, right and left maxillae of the same individual, with left P²–M³ and right P⁴–M³.

Paratypes.—None.

Referred Specimens.—Front of skull, and mandible, the originals of which are probably lost. B.M. cast No. M 2710. Also a mandible, B.M. No. 36734.

Horizon.—Upper Siwaliks.

Locality.—Siwalik Hills.

Diagnosis.—Somewhat smaller than the typical *Chalicotherium*. Brachyodont, with protoloph of molars absent. Anterior crest of premolars absent. Protoselenid of lower molars better developed than in the other genera.

Dr. Matthew has pointed out the generic distinctions between *Nestoritherium* (= *Circotherium*) and *Chalicotherium*, on page 518 of his "Critical Observations upon Siwalik Mammals." The reader is referred to this paper.

Circotherium Holland and Peterson, 1914, is a synonym of *Nestoritherium* and is therefore invalid. In Kaup's original description of the genus *Nestoritherium* he clearly indicated the Siwalik form, *Chalicotherium sivalense*, as the type species. The Pikermi chalicothere is *Ancylotherium pentelici*.

The reader is referred to the following reference: Colbert, Edwin H., 1935, "The Proper Use of the Generic Name *Nestoritherium*," Jour. Mammalogy, XVI, pp. 233–234.

Nestoritherium (?) sindiense (Lydekker)

Manis sindiense, Lydekker, 1876, Pal. Indica (X), I, p. 64, Pl. VIII, figs. 11–14.

Chalicotherium sindiense, Pilgrim, 1910, Rec. Geol. Surv. India, XL, p. 201.

Circotherium sindiense, Matthew, 1929, Bull. Amer. Mus. Nat. Hist., LVI, p. 520.

Additional References.—

Lydekker, R., 1880B, pp. 26, 33; 1885B, p. 109.

Holland, W. J., and Peterson, O. A., 1914, p. 218.

Type.—G.S.I. No. D 99, a median phalanx.

Paratypes.—None.

Horizon.—Manchar Beds (Siwaliks) of Sind. Exact level unknown.

Locality.—Sind.

This species, originally described as an edentate, was based on a single phalangeal bone. This bone might be referred to any of several genera of chalicotheres, but in size and shape, as well as probable horizon, it would seem likely to be referable to the genus *Nestoritherium*. The species must be considered as of little or no value.

The Origin and Evolution of the Chalicotheres

The chalicotheres would seem to have had their beginnings as small, unspecialized lophiodont forms in North America. *Eomoropus amarorum* from the upper Eocene of Wyoming represents our first record of a definite chalicothere. During the closing stages of the Eocene period these small chalicotheres migrated from North America to Asia, probably by way of a trans-Bering land bridge, and in the oriental continent they spread over a broad area, and enjoyed a period of untrammeled development, persisting through the uppermost stages of the Eocene and into the lowermost portions of the Oligocene. The primitive Asiatic chalicotheres, which may be regarded as directly derived from the North American *Eomoropus*, are represented by two genera, *Eomoropus* and *Grangeria*.

From these generalized Eocene chalicotheres two specialized phylogenetic branches arose and developed through the middle and the upper portions of the Tertiary. One of these groups, the Chalicotherini, typified by brachyodont quadrate upper molars and short feet, was mainly Eurasiatic in its distribution. It enjoyed a long period of development through the Miocene and Pliocene and into the Pleistocene, and it spread throughout Europe and Asia. It would seem, also, that in the Miocene certain members of this group (*Macrotherium*) crossed from Asia to North America.

The other group, the Schizotherini, typified by hypsodont, elongated molars, and long feet, inhabited both Eurasia and North America. In the Oligocene the members of this group are well defined in Eurasia, but whether they were living in North America or not, is at present a question difficult to answer. Chalicothere remains in the Oligocene of North America are extremely rare. In the Miocene, however, we find members of this group in the Old World (*Phylotillon*) and in the New World (*Moropus*). This line persisted through the lower Pliocene in Europe and Asia.

A question now arises—did the chalicotheres originate in North America, migrate to Asia, and then become extinct in America, only to reappear in a subsequent counter-migration from Asia? Our present knowledge of the distribution of the group would lend weight to this theory. For instance, *Eomoropus* appears in the Eocene of North America, and slightly later, in the topmost Eocene of Mongolia and China. Then through the Oligocene we have no definite records of chalicotheres in North America, although their remains are relatively common in Europe and Asia.[54] Finally, the two branches of advanced chalicotheres are found in North America, where they persist but for a short time. In Eurasia, however, these two branches of chalicotheres undergo a broad range of adaptive radiation, and persist until and through the early Pleistocene.

Perhaps the above records constitute the evidence for a series of migrations and counter migrations between the two hemispheres. Perhaps again, it shows that the Old World was the center of chalicotherine evolution, at least for the advanced types, while the New World offered a haven for certain immigrant and restricted forms.

Of course, the above speculations may be changed by subsequent discoveries. It must be admitted that the lack of chalicotheres in the Oligocene of North America may be circumstantial, dependent upon the chances of discovery, and consequently not a real index as to the presence or absence of the group in the New World. If some Oligocene genera of the Schizotherini were to be discovered in North America we would be justified in regarding *Moropus* as a truly autochtonous type. But until such a discovery is made, the possibility of *Moropus* being an immigrant from Asia remains strong. At any rate, the above information shows us that the chalicotheres are a group of North American origin, but primarily of Eurasiatic development.

[54] Cope has described "*Chalicotherium*" *bilobatum* from the Oligocene of Saskatchewan, but the reference of this species to the chalicotheres is very doubtful, because of the fragmentary nature of the type. The John Day chalicotheres come from the upper levels and are quite likely of lower Miocene age. Russell has recently (1934) placed *C. bilobatum* in a new genus, *Oreinotherium*, which he considers as of macrotherine relationships.

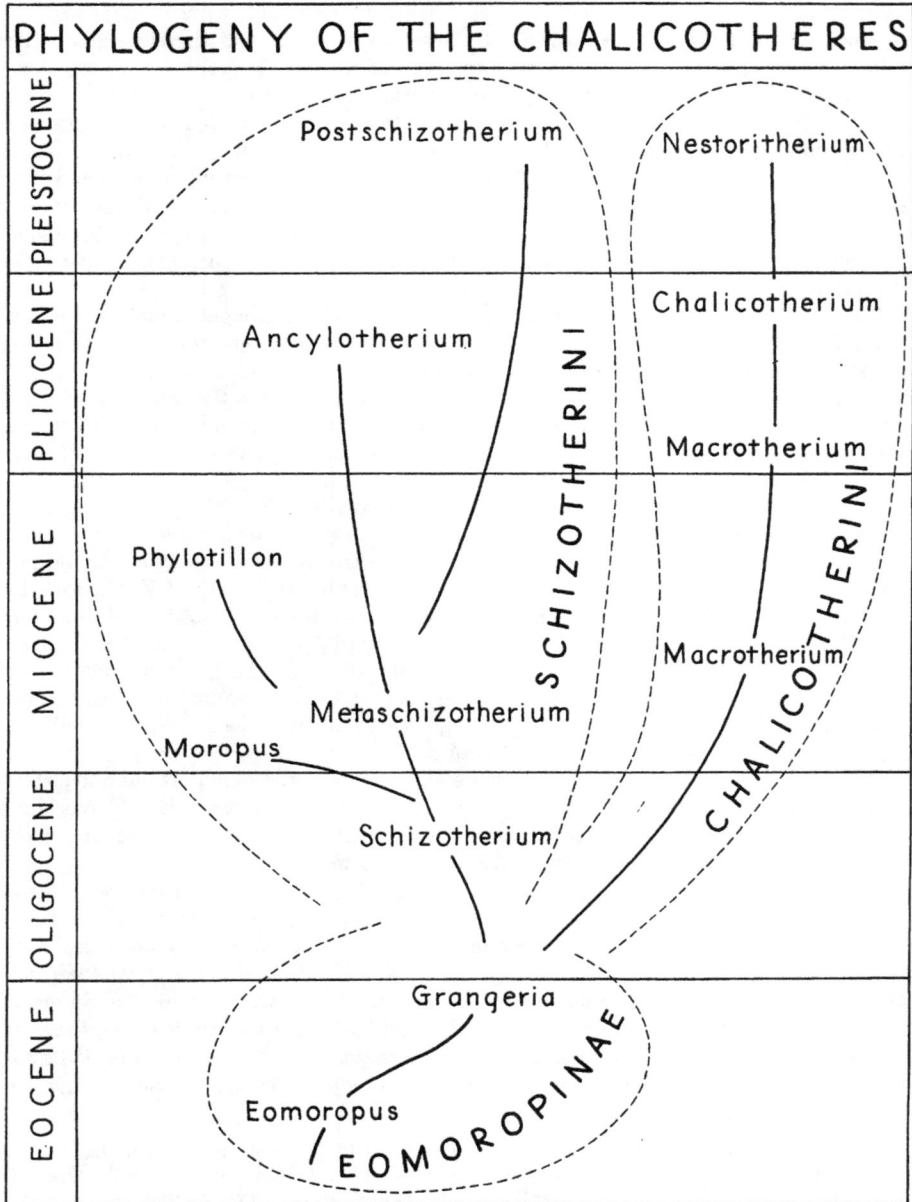

Fig. 77. Phylogeny of the Chalicotherioidea.

RHINOCEROTOIDEA

General Considerations

The rhinoceroses from the Siwalik deposits have long been in a state of confusion, and in view of the prevalent difficulties standing in the way of a correct interpretation of our knowledge of any of the fossil rhinoceroses, it is doubtful whether we are in a position as yet to arrive at a clear picture of the history of this group in the Indian region.

Quite a number of species of fossil rhinoceroses have been described from the Siwaliks during the course of time since the first work of Falconer and Cautley. These species are listed below as they were originally described.

> *Rhinoceros sivalensis* Falconer and Cautley, 1847.
> *Rhinoceros palaeindicus* Falconer and Cautley, 1847.
> *Rhinoceros platyrhinus* Falconer and Cautley, 1847.
> *Rhinoceros perimensis* Falconer and Cautley, 1847.
> *Rhinoceros iravadicus* Lydekker, 1876.
> *Rhinoceros planidens* Lydekker, 1880.
> *Aceratherium blanfordi* Lydekker, 1884.
> *Rhinoceros sivalensis intermedius* Lydekker, 1884.
> *Aceratherium lydekkeri* Pilgrim, 1910.
> *Teleoceras blanfordi mihi* Pilgrim, 1910 (nomen nudum).
> *Gaindatherium browni* Colbert, 1934.

Certain species from the Bugti beds are not included in this list, as they come from a series lower than the Siwaliks.

Pilgrim, in 1913, suggested that an acerathere close to *Aceratherium tetradactylum* of the Tortonian and Sarmatian of Europe is present in the Lower Siwaliks.

"In the Chinji beds a much smaller species is found, which is very nearly allied to *Aceratherium tetradactylum* of the Tortonian and Sarmatian of Europe, and which may be ancestral to both *A. perimense* and *A. lydekkeri*." [55]

Dr. Matthew, in 1929, suggested that the so-called *Aceratherium tetradactylum* in the Lower Siwalik beds is more properly referable to the genus *Chilotherium*. Dr. Pilgrim, however, did not describe any *actual* material as being representative of the genus and species in question. Consequently *Aceratherium tetradactylum* can not be placed on a list of *described* rhinoceroses from the Siwalik Series.

The list of Siwalik rhinoceroses, presented above, may be revised in the following manner.

> *Rhinoceros sivalensis* Falconer and Cautley . . . . Upper Siwaliks
>   Synonym: *Rhinoceros palaeindicus* Falconer and Cautley
> *Gaindatherium browni* Colbert . . . . Lower Siwaliks
> *Coelodonta platyrhinus* (Falconer and Cautley) . . . . Upper Siwaliks
> *Aceratherium perimense* (Falconer and Cautley) . . . . Middle Siwaliks
>   Synonyms: *Rhinoceros planidens* Lydekker
>   *Rhinoceros iravadicus* Lydekker
> *Aceratherium lydekkeri* Pilgrim . . . . Middle Siwaliks

[55] Pilgrim, G. E., 1913B, p. 297.

Chilotherium blanfordi (Lydekker)..........................Lower Siwaliks
    Synonym: *Teleoceras blanfordi mihi* Pilgrim
Chilotherium intermedium (Lydekker) ....Middle Siwaliks

From the above list it may be seen that the rhinoceroses that lived during Siwalik times were representative of several distinct phylogenetic groups.   Thus they show a gathering of various species that were evolving in different portions of Eurasia; a congregating of different forms in a single haven that afforded ample protection for their existence.   One phylogenetic line, that of the *Rhinoceros*, would seem to have undergone its later evolutionary development in India.   The others are, for the most part, immigrant forms, migrating in from outside regions.

RHINOCEROTIDAE

Rhinocerinae

Coelodonta Bronn, 1831

Generic type, *Coelodonta boiei* Bronn

Coelodonta platyrhinus (Falconer and Cautley)

Rhinoceros platyrhinus, Falconer and Cautley, 1847, Fauna Antiqua Sivalensis, Pl. LXXII, figs. 1–7, Pl. LXXV, figs. 9–12.
Dicerorhinus platyrhinus, Pilgrim, 1910, Rec. Geol. Surv. India, XL, p. 201.
Coelodonta platyrhinus, Matthew, 1929, Bull. Amer. Mus. Nat. Hist., LVI, pp. 444, 534–535.
    Additional References.—
        Falconer, H., 1868A, pp. 157–169, Pl. XIV, figs. 3, 4.
        Lydekker, R., 1876A, pp. 29–32, Pl. IV, fig. 4; 1880B, p. 31; 1883C, p. 83; 1884D, p. 132; 1884E, p. 82, Pl. III, fig. 2; 1885B, p. 65; 1886A, pp. 99–101.
        Pilgrim, G. E., 1913B, pp. 305–306, 324.
    Type.—(Lectotype.)   Brit. Mus. No. 33662, a battered skull.
    *Cotypes.—*Brit. Mus. Nos. M 2731, back portion of a skull, possibly associated with No. 33662; 39620, anterior portion of a mandible; 39640, right M³; 39641, right upper molar; 39642, symphysis and right ramus of a mandible; 39643, right maxilla.
    *Neotype.—*Brit. Mus. No. 36661, a nearly complete skull.
    *Horizon.—*Upper Siwaliks.
    *Locality.—*From the Siwalik Hills.
    *Specimens in the American Museum.—*Amer. Mus. No. 19777, fragment of a right maxilla with P^4, M^{1-3}.   From the Upper Siwaliks, one mile east of Mirzapur.
    19822, fragment of a left maxilla with MM^{2-4}.   From the Upper Siwaliks, three miles northeast of Siswan.
    19875, fragment of left maxilla with M^{1-2}.   Upper Siwaliks, three miles northeast of Mirzapur.
    *Diagnosis.—*A large rhinoceros, showing many characters that would seem to ally it with *Coelodonta antiquitatus.*   Skull without nasal septum.   Premaxillaries heavy.   Teeth hypsodont.   Evidently two horns were present.

The species under consideration is closely related to the modern *Dicerorhius sumatrensis* and to the Pleistocene form, *Coelodonta antiquitatus*. As to which of these two genera it bears the closest resemblance is a question difficult to decide, due to the fact that *Dicerorhinus* and *Coelodonta* are very close to each other. In this regard the reader is referred to Breuning, Stephan, 1924, p. 27.

Breuning explains that although the genera *Coelodonta* and *Ceratorhinus* (*Dicerorhinus*) are separate, there are numerous characters in the various species that bridge the differences between the two genera. He then goes on to point out that certain characters generally considered as typical of *Coelodonta* and not of *Dicerorhinus*, such as the ossified nasal septum, the reduction of the incisors and the backward development of the occiput, are in reality quite variable and consequently of little phylogenetic or taxonomic importance. For instance, although *Dicerorhinus sumatrensis* is typically without a nasal septum, such a structure may occur in this species. Therefore it may be possible that *Coelodonta* has been derived directly from *Dicerorhinus*, and with this consideration in mind the two genera may be grouped together in the subfamily Ceratorhinae of Osborn.

The Siwalik form under discussion shows many approaches to the modern African Rhinoceros, *Diceros bicornis*, a fact pointed out first by Lydekker.

The several taxonomic possibilities concerning the relationships of "*Rhinoceros*" *platyrhinus* were fully recognized by Dr. Matthew, and at various times he referred this species successively to the genera *Dicerorhinus*, *Coelodonta*, and *Diceros*.[56]

The evidence may be summed up as follows:

1. The presence of incisors (Matthew, W. D., 1929, p. 535) would place *platyrhinus* with the genus *Dicerorhinus*. Moreover, in general skull form it is very similar to the Sumatran species. As a third point of resemblance there might be mentioned the separation of the postglenoid and the posttympanic, leaving the external auditory meatus open below. This latter is a primitive feature, characteristic of *D. sumatrensis* (a primitive form) and retained in *D. platyrhinus*.

2. As to the pattern and hypsodonty of the cheek teeth, the Siwalik species would seem to be close to *Diceros*, *Ceratotherium* and *Coelodonta*. Moreover it resembles these genera also in the separated postglenoid and posttympanic.

"The teeth are rather closely related to *Coelodonta* and *Ceratotherium*, not to *Dicerorhinus*." [57]

In view of the above evidence, it would seem that a safe course is to assign provisionally the species under consideration to the genus *Coelodonta*, recognizing however that it probably represents a separate phylogenetic side branch retaining certain primitive characters. *Coelodonta* (?) *platyrhinus* may well deserve a new generic designation, but it is not thought advisable to make such a distinction at this time.

Some measurements and figures of the specimens in the American Museum collection are presented below.

[56] See Matthew, W. D., 1929, pp. 444, *Dicerorhinus;* 462, 534, *Coelodonta;* 1931, p. 7, *Diceros.*
[57] Matthew, W. D., 1929, p. 462.

MEASUREMENTS

Coelodonta platyrhinus

Amer. Mus. No. 19777.

	Length	Width
Molar series	175 mm.	
M^1	51	78 mm.
M^2	61	74
M^3	66	63

Amer. Mus. No. 19822.

MM^2	34	40
MM^3	49	49
MM^4	55	52

Rhinoceros Linnaeus, 1758

Generic type, *Rhinoceros unicornis* Linnaeus

Rhinoceros sivalensis Falconer and Cautley

Rhinoceros sivalensis, Falconer and Cautley, 1847, Fauna Antiqua Sivalensis, Pl. LXXIII, figs. 2, 3; Pl. LXXIV, figs. 5, 6; Pl. LXXV, figs. 5, 6.

Additional References.—

Falconer, H., 1868A, pp. 157–169, Pl. XIV, figs. 1, 2; pp. 514–516.

Lydekker, R., 1876A, pp. 26–29, Pl. V, figs. 2, 5; 1880B, p. 31; 1881A, pp. 28–42, Pls. V, VI, figs. 2, 3, Pl. VII, fig. 1, Pl. X, fig. 4; 1883C, p. 92; 1884D, p. 132; 1885B, pp. 61–64; 1886A, pp. 130–132.

Pilgrim, G. E., 1910B, p. 201; 1913B, p. 324.

Matthew, W. D., 1929, pp. 444, 531.

Type.—(Lectotype.)   Brit. Mus. No. 39626, part of a skull.

Cotypes.—Brit. Mus. Nos. 39625, a skull; 39646, a mandibular symphysis; 39647, part of a skull.

Horizon.—Upper Siwaliks.

Locality.—Siwalik Hills.

Specimen in the American Museum.—Amer. Mus. No. 19793, a right first upper molar. From the Upper Siwaliks, six miles east of Chandigarh.

Diagnosis.—A large species of the genus.   Molars with a parastyle buttress, distinct crochet which may unite with the protoloph to enclose a fossette, and without a crista.

The specimen in the American Museum collection need not be described.   Reference should be made to the descriptions of Falconer and Cautley, Lydekker and Matthew. Measurements and a figure of the American Museum specimen are presented here.

Rhinoceros sivalensis, Amer. Mus. No. 19793.
Right M^1 Length... 61 mm.   Width... 80 mm.

Rhinoceros palaeindicus Falconer and Cautley

Rhinoceros palaeindicus, Falconer and Cautley, 1847, Fauna Antiqua Sivalensis, Pl. LXXIII, fig. 1, Pl. LXXIV, figs. 1–4, Pl. LXXV, figs. 1–4.

Additional References.—

Falconer, H., 1868A, pp. 157, 514–516.

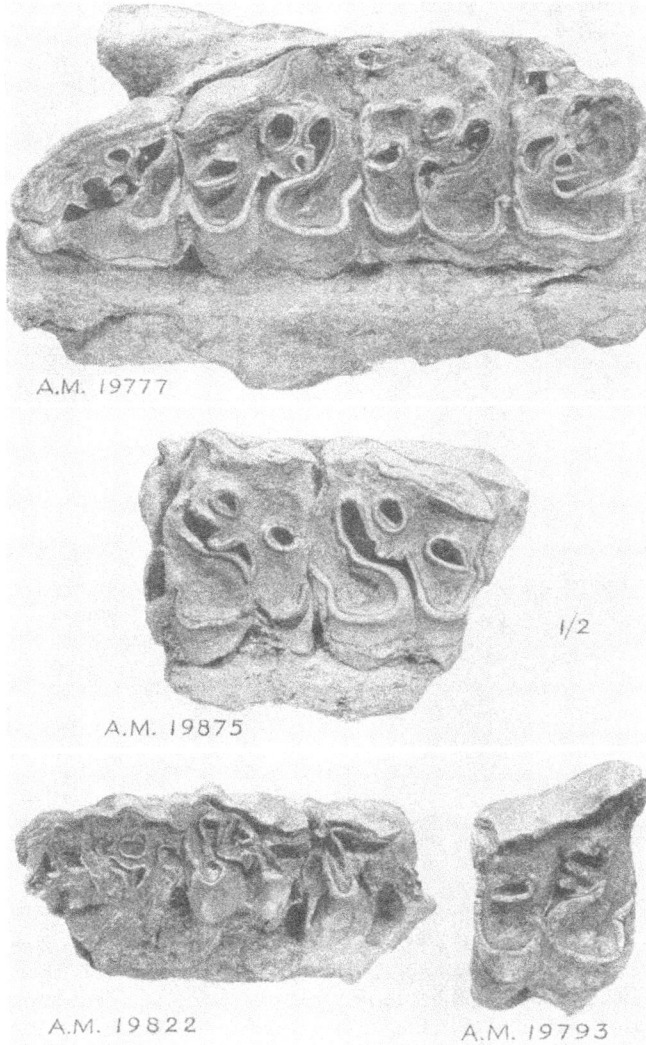

FIG. 78. *Coelodonta platyrhinus* (Falconer and Cautley). Amer. Mus. No. 19777, maxilla with right P⁴, M¹⁻³; Amer. Mus. No. 19875, left M¹⁻²; Amer. Mus. No. 19822, maxilla with left MM²⁻⁴. Crown views. *Rhinoceros sivalensis* Falconer and Cautley. Amer. Mus. No. 19793, right M¹. Crown view. All figures one half natural size.

Lydekker, R., 1876A, pp. 22–26, Pl. IV, figs. 3, 4; 1880B, p. 31; 1881A, pp. 42–48, Pl. VI, fig. 1, Pl. VII, figs. 2, 3, Pl. X, fig. 3; 1883C, p. 92; 1884D, p. 132; 1884E, pp. 82–83, Pl. III, fig. 1, 3; 1885B, p. 64; 1886A, pp. 132–135, fig. 15.

Pilgrim, G. E., 1910B, p. 201; 1913B, p. 324.

Matthew, W. D., 1929, pp. 444, 531–532.

Type.—(Lectotype.)   Brit. Mus. No. 16444, a skull.

Cotypes.—Brit. Mus. Nos. M 2727, a skull; 36740, a skull; 39644, back portion of a left mandibular ramus; 39645, portion of a right mandibular ramus; 39646, mandibular symphysis; 39740, a skull.   Also specimens figured in Pl. LXXV, figs. 1, 2, of the Fauna Antiqua Sivalensis.

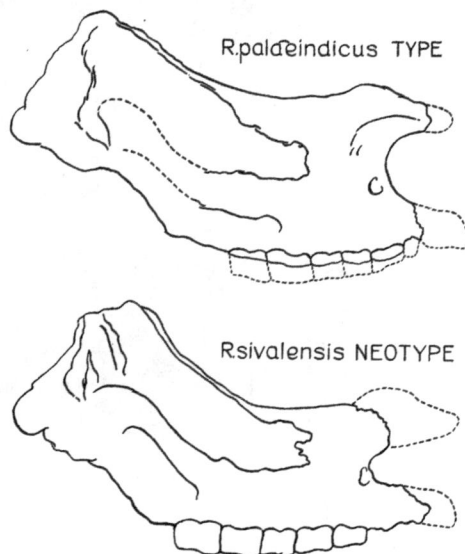

FIG. 79.   Comparison of skulls of the type of *Rhinoceros palaeindicus* Falconer and Cautley (above), and the neotype of *Rhinoceros sivalensis* Falconer and Cautley (below).   One eighth natural size.   From Matthew, 1929.

Horizon.—Upper Siwaliks.

Locality.—Siwalik Hills.

Diagnosis.—Like *R. sivalensis*, but wider across the frontals, with a slightly different cranial profile and with a flat ectoloph to molars, without parastyle buttress.

Matthew (1929, pp. 531–532) has shown that this species is probably synonymous with *R. sivalensis*.   He points out that the supposed differences in the dentition given above really do not exist, and that the skull differences may very well be within the bounds of individual variation.   Consequently the two species are here considered as one, its designation being *Rhinoceros sivalensis*.

Gaindatherium Colbert, 1934
Generic type, *Gaindatherium browni* Colbert
Gaindatherium browni Colbert

Gaindatherium browni, Colbert, 1934, Amer. Mus. Novitates, No. 749.

Type.—Amer. Mus. No. 19409, an almost complete skull. From the Lower Siwaliks, Chinji zone, near Chinji Rest House, Punjab.

Paratypes.—Amer. Mus. No. 29838, associated right and left upper and lower dentitions. Lower Siwaliks, Chinji zone, near Chinji Rest House, Punjab. Amer. Mus. No. 19471, a mandibular symphysis. From the lower portion of the Middle Siwaliks, Nagri zone, one mile south of Nathot, Punjab. Amer. Mus. No. 29793, an upper incisor tooth Lower Siwaliks, Chinji zone, about 500 feet above the level of Chinji Rest House, one and one half miles west of Chinji Rest House, Punjab.

Horizon.—Lower Siwaliks, Chinji zone. The species may extend up into the lower portion of the Middle Siwaliks, into the Nagri zone.

Locality.—Vicinity of Chinji Rest House, south of Chinji village, Salt Range, Attock District, Punjab.

Specimens in the American Museum.—The type and paratypes, listed above.

Amer. Mus. No. 29837. Various teeth from the lower dentition. Lower Siwaliks, near Chinji Rest House, Punjab.

19422. A right M². Lower Siwaliks, Chinji zone, four miles west of Chinji Rest House, Punjab.

29792. A left P⁴. Lower Siwaliks, Chinji zone, ten miles east of Chinji Rest House, Punjab.

Also miscellaneous teeth from the Lower Siwaliks, near Chinji Rest House.

Diagnosis.—An upper Tertiary rhinoceros of medium size, with a "saddle-shaped" skull having a single horn on the nasals, and with brachyodont, simple molar teeth. The orbit is located in an approximately central position above the first molar; the occiput is vertical; the postglenoid and posttympanic are fused, forming a closed tube for the external auditory meatus. There are two upper incisors, of which the lateral one is quite small; the upper molars are without an antecrochet or crista, and the crochet is but slightly developed.

This species has been described elsewhere (Colbert, E. H., 1934) in some detail. Consequently it need not be discussed at length here. The salient characters distinguishing *Gaindatherium* are listed below.

1. The skull is of medium size and of comparatively primitive structure. It is relatively long and narrow, with the orbit in an approximately central position. That is, the facial and the cranial portions of the skull are subequal in length.

2. There is a single nasal horn. The nasals are expanded laterally and vertically for the accommodation of this horn, thus giving the cranial profile a saddle shaped outline.

3. The occiput is vertical as is common among primitive rhinoceroses.

4. The postglenoid and posttympanic are fused, forming a closed tube for the external auditory meatus.

5. There are two upper incisors, of which the lateral one is small.

6. The cheek teeth are brachyodont and relatively simple, without antecrochet or crista, but with a crochet present in the last molar.

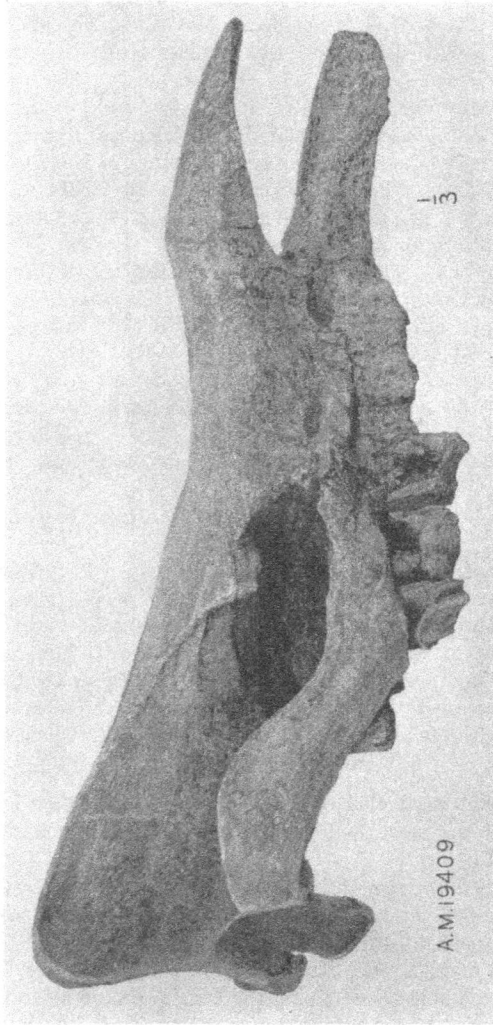

Fig. 80. *Gaindatherium browni* Colbert. Amer. Mus. No. 19409, type skull. Lateral view. One third natural size. From Colbert, 1934.

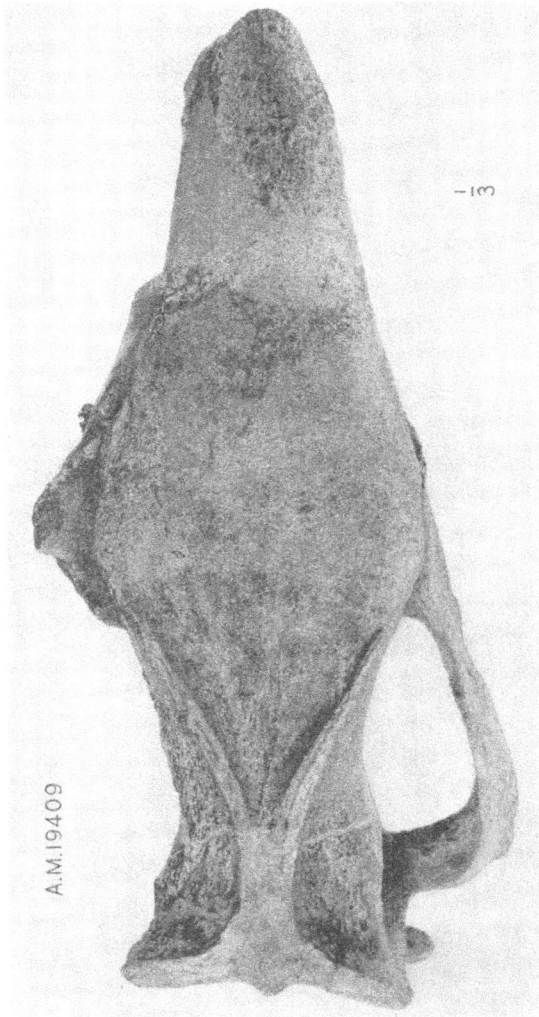

Fig. 81. *Gaindatherium browni* Colbert. Amer. Mus. No. 19409, type skull. Dorsal view. One third natural size. From Colbert, 1934.

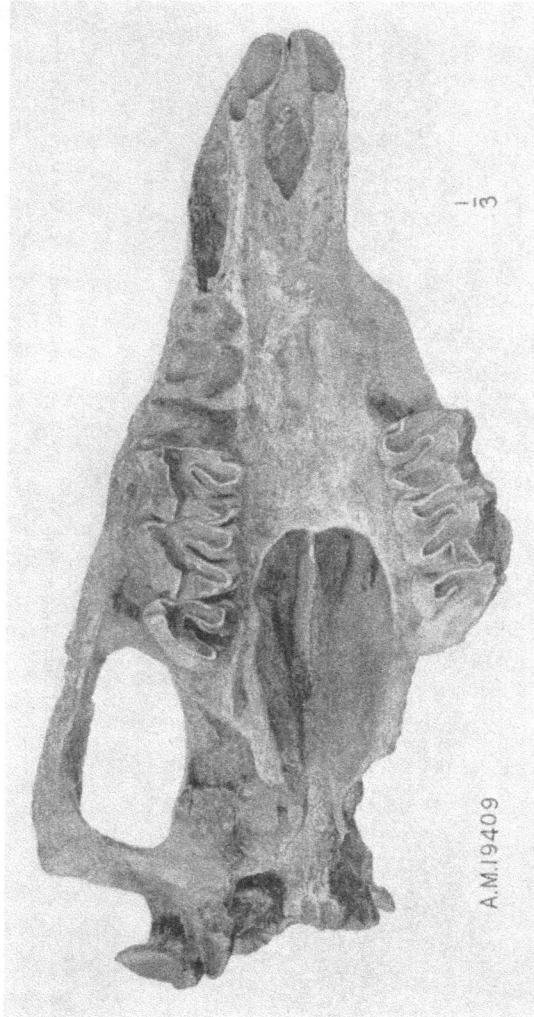

FIG. 82. *Gaindatherium browni* Colbert.   Amer. Mus. No. 19409, type skull.   Ventral view.   One third natural size.
From Colbert, 1934.

Measurements and figures are given below.

Gaindatherium browni, Amer. Mus. No. 19409, type

Skull.

Length, lamboidal crest to tip of nasals..............	496.0 mm.
Length, condyles to incisor alveolus (Estimated)......	520.0
Length, anterior border of orbit to incisor alveolus....	243.0
Length, anterior border of orbit to condyles.........	290.0
Width at glenoids.................................	298.0
Width of parietals, narrowest portion..............	93.0
Width of frontals, supraorbital....................	168.0
Width of palate at M^1...........................	68.0
M^1 length.....................................	40.0
width......................................	51.0
M^2 length....................................	42.0
width......................................	52.0
M^3 length....................................	37.0
width......................................	48.0

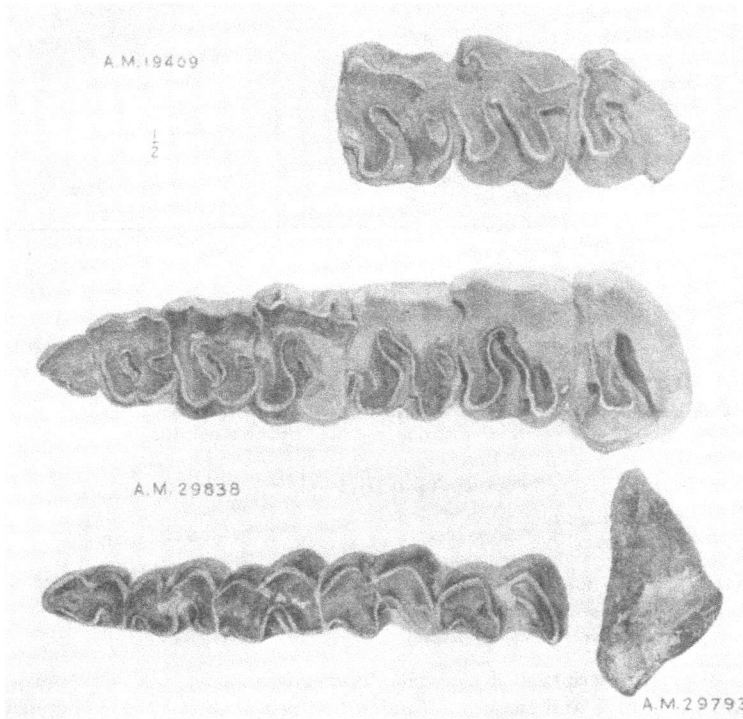

FIG. 83. *Gaindatherium browni* Colbert. Upper and lower dentitions. At top: type, Amer. Mus. No. 19409, left M^{1-3}, crown view. In middle: Amer. Mus. No. 29838, left P^1–M^3, crown view. At bottom: Amer. Mus. No. 29838, right P_2–M_2, crown view, and Amer. Mus. No. 29793, upper incisor, lateral view. All figures one half natural size. From Colbert, 1934.

Amer. Mus. No. 29838, paratype

	Length	Width
P^1	19.0 mm.	22.5 mm.
P^2	28.0	34.5
P^3	32.0	43.0
P^4	37.0	49.0
P_2	28.5	21.5
P_3	30.0	26.0
P_4	36.0	28.0
M_1	40.0	30.0
M_2	43.0	28.0

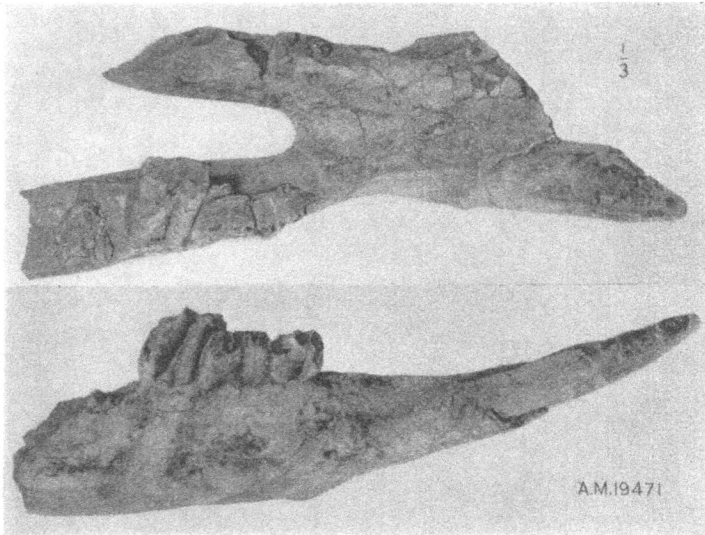

FIG. 84. *Gaindatherium browni* Colbert. Amer. Mus. No. 19471, symphysis of mandible. Superior view above, lateral view below. One third natural size. From Colbert, 1934.

Amer. Mus. No. 19471, paratype

Mandibular symphysis.
    Depth of symphysis at P_2............................ 66.0 mm.
    Width of symphysis at narrowest part............... 79.0
    Length of symphysis................................135.0
    Transverse diameter of incisor...................... 39.0
    Vertical diameter of incisor........................ 27.0

In the original description of *Gaindatherium* it was shown that this genus, although in many ways primitive and thereby retaining heritage characters in common with other primitive Rhinocerinae, possessed certain habitus characters that definitely point in the direction of *Rhinoceros*. Therefore it was suggested that *Gaindatherium* is a form more or less directly ancestral to the modern Asiatic rhinoceros and that it represents an intermediate link between the stem *Caenopus* type of true rhinocerine and the modern one horned rhinoceros. A list of heritage characters that demonstrate the derivation of *Gaindatherium* from a primitive *Caenopus*-like ancestor, and also a list of habitus characters that show the trend of the Siwalik genus towards *Rhinoceros*, are presented below.

A. HERITAGE CHARACTERS IN *Gaindatherium*

1. The light, slenderly built skull is an heritage character derived from an ancestor of relatively small size and slender proportions.

2. The centrally placed orbit is a character derived from a primitive ancestor. In the primitive perissodactyls the preorbital portion of the skull is approximately equal in length to the postorbital region. In advanced forms the orbit tends to lose its central position.

3. The slight sagittal crest is a primitive character, due to the fact that the brain case has not expanded to any great degree.

4. The vertical occiput is a primitive heritage character.

5. The presence of the second upper incisor is primitive.

6. The brachyodont, simple molars show heritage characters of an ancestor similar to *Caenopus*.

B. HABITUS CHARACTERS IN *Gaindatherium*

1. The "saddle shaped" skull is a definite advance towards *Rhinoceros*.

2. The presence of one nasal horn is an habitus character in the direction of *Rhinoceros*.

3. The union of the postglenoid and the posttympanic is again an habitus character that is also found in *Rhinoceros*.

4. The presence of a crochet on the last molar in *Gaindatherium* is a character that would seem to point towards *Rhinoceros*. In the latter genus the crochet and crista are well developed, but the antecrochet is not distinct. In *Gaindatherium* the crochet is present on the last molar, and the antecrochet is not distinct.

5. The relatively narrow, shallow symphysis and the straight lower incisor would seem to be characters indicative of a relationship with *Rhinoceros*.

In order to demonstrate more clearly the gross anatomical characters of the skull of *Gaindatherium* that define it as a form intermediate between the *Caenopus* type and *Rhinoceros*, the accompanying chart (Fig. 85) has been prepared.

As the basis for this chart a skull of *Caenopus*, here considered as approximating in a general way the stem form of the Rhinocerinae, was drawn to scale. It was then overlain by a system of quadrants or squares of arbitrary size (A of figure). In the arrangement of these squares certain bases were established, namely the occlusal line of the upper molars for the horizontal components and the anterior border of the orbit for the vertical components. The horizontal lines were lettered A, B, C, etc., above the base or zero line, and A', B', C', etc., below the base line. The vertical lines were numbered 1, 2, 3, etc., to the left of the base or zero line, and $1'$, $2'$, $3'$, etc., to the right of this datum. Then the skulls of *Gaindatherium* (B of figure), *Rhinoceros* (C of figure) and *Dicerorhinus* (D of figure) were drawn to the same scale as the first skull. The intersections of the various lines making up the various squares were plotted on the other skulls, retaining their special relationships to anatomical details as in the case of the *Caenopus* skull. Then when the numerous points

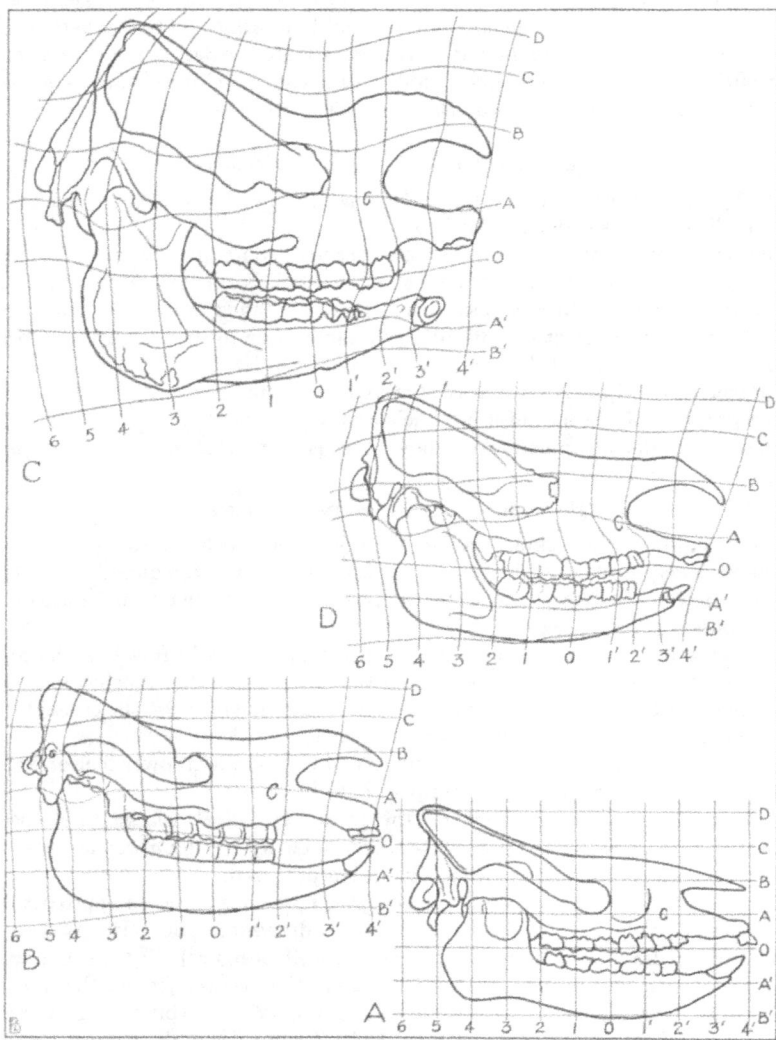

FIG. 85.   Cartesian coordinate chart to illustrate the manner in which the skull of *Rhinoceros* might have evolved through *Gaindatherium* from a primitive form such as *Caenopus*.

    A. *Caenopus occidentalis* Leidy = *Subhyracodon occidentalis* (Leidy).

    B. *Gaindatherium browni* Colbert.

    C. *Rhinoceros unicornis* Linnaeus.

    D. *Dicerorhinus sumatrensis* (Cuvier).

    The skull of *Dicerorhinus* is included in the chart for comparison with the skull of *Gaindatherium*.   Both of these genera, being relatively primitive, have skulls that are somewhat similar to each other, although they belong to two different branches of rhinocerotid evolution.

so located were connected, figures of various shapes were obtained that show the proportional change of each portion of the skull in its relations to other portions of the skull.

This is essentially the method used so widely by d'Arcy Thompson, but here it is applied in a more detailed manner than was done by that author. Thus it may be seen that the skull of *Gaindatherium* is not greatly changed from the primitive *Caenopus* skull. Nor is the skull of *Dicerorhinus* greatly different from that of *Caenopus* or *Gaindatherium*. These are the skulls of relatively primitive animals, so that they show similarities indicating the community of their origin.

It may be noticed, however, that the *Dicerorhinus* skull, although primitive in a general way, does show characters that set it apart from *Gaindatherium* and *Rhinoceros*. It is marked by the forward position of the cheek teeth and the lack of a strong forward inclination of the occiput.

The *Rhinoceros* skull is, on the other hand, an exaggerated accentuation of the *Gaindatherium* skull, characterized by its great depth.

In making the accompanying chart, the skull of *Gaindatherium* was restored as nearly as possible to its original form.

Aceratherium Kaup, 1832

Generic type, *Rhinoceros incisivus* Cuvier

Aceratherium perimense (Falconer and Cautley)

Rhinoceros (Acerotherium?) perimensis, Falconer and Cautley, 1847, Fauna Antiqua Sivalensis, Pl. LXXV, figs. 13–16; Pl. LXXVI, figs. 14–16.
Aceratherium perimense, Lydekker, 1876, Pal. Indica (X), I, pp. 51–55, Pl. VI, figs. 2, 5.
> *Additional References.*—
>> Falconer, H., 1868A, pp. 157, 171, 517–519.
>> Lydekker, R., 1880B, p. 31; 1880C, p. xiii, Pl. IV, figs. 7, 9; 1881A, pp. 9–28, Pls. I–IV; 1883C, p. 89; 1884D, p. 132; 1885B, pp. 66–68; 1886A, pp. 155–157, fig. 19.
>> Pilgrim, G. E., 1910B, p. 200; 1913B, p. 297.
>> Matthew, W. D., 1929, p. 507.

Cotypes.—The various specimens figured in Pls. LXXV, figs. 13–16, and LXXVI, figs. 14–17, of the Fauna Antiqua Sivalensis.

Matthew lists as the type of this species the two upper molars figured by Lydekker in the Palaeontologica Indica, Series X, Volume I, Pl. IV, figures 7 and 9. It is not quite clear why Matthew should have regarded these specimens as the type.

Aceratherium perimense was first published in the plates of the Fauna Antiqua Sivalensis, by Falconer and Cautley in 1847. No descriptions accompanied the figures, but nevertheless the figures alone constituted publication and thus the specimens so illustrated must be considered as the cotypes of the species.

Horizon.—Presumably from the Middle Siwaliks. Also from the Lower Siwaliks.

Locality.—Perim Island for the type specimens. The Punjab for referred specimens.

Specimens in the American Museum.—Amer. Mus. No. 19410. A right P^1, left P^3, right DM^3. Lower Siwaliks, 600 feet above the level of Chinji Rest House, 12 miles east of Chinji Rest House.

19418. A right M_{2-3}. Lower Siwaliks, locality uncertain.

19454. A right upper dentition and two mandibular rami. Lower Siwaliks, 1600 feet above the level of Chinji Rest House, one and one half miles northwest of Chinji Rest House.

19470. A skull, complete back of the premolars. Lower portion of the Middle Siwaliks, 1000 feet below the Bhandar bone bed, one mile south of Nathot.

19527. Teeth, vertebrae, foot bones, Middle Siwaliks, south of Nathot.

19528. Fragment of mandibular ramus with left P_1. Middle Siwaliks, one mile south of Kohala.

19531. A right P_3. Base of Middle Siwaliks, near Nathot.

19544. Miscellaneous teeth. At base of Middle Siwaliks, near Nathot.

19571. Second and third left upper deciduous molars. Lower Siwaliks, 600 feet above the level of Chinji Rest House, five miles west of Chinji Rest House.

19576. Right upper incisor. Lower Siwaliks, 1100 feet above the level of Chinji Rest House, one mile north of Chinji Rest House.

19585. Miscellaneous teeth. Lower Siwaliks, five miles east of Chinji Rest House.

19589. A right P_2 Lower Siwaliks, 1600 feet above the level of Chinji Rest House, one and one half miles west of Chinji Rest House.

19692. Palate with left P^4–M^2 and right M^2. Lower Siwaliks, six miles north of Dhok Pathan.

19743. A right P^4. Middle Siwaliks, two and one half miles northeast of Hasnot.

19921. A right P_2. Lower Siwaliks, near Rammagar.

19942. Miscellaneous teeth. Lower portion of the Middle Siwaliks four and one half miles west of Hasnot.

29790. A right DM^4. Lower Siwaliks, 1600 feet above the level of Chinji Rest House, twelve miles east of Chinji Rest House.

29794. Right M^3 and associated fragments. Lower Siwaliks, 500 feet above the level of Chinji Rest House, one and one half miles west of Chinji Rest House.

Diagnosis.—A rhinoceros of gigantic size with hypsodont teeth. Skull rather short and deep, with retracted nasals; zygomatic arch heavy; postglenoid separate from posttympanic. Upper incisor present and well developed. Molars with moderately developed crochet, weaker antecrochet and rudimentary crista. Protocone somewhat pinched off. Lower molars narrow and compressed. Mandibular symphysis narrow.

Aceratherium perimense is distinguished from the other Siwalik rhinoceroses by its great size, a fact pointed out by Lydekker in 1881, in his detailed description of the skull of the species. Number 19470 in the American Museum collection is quite indicative of the large proportions characteristic of this species. Perhaps it might be well to present at this juncture a short description of the skull just mentioned, to serve as a corollary to the original detailed description written by Lydekker in 1881.

The impression of gigantic size in *Aceratherium perimense* is due to the great mass of the skull, rather than to any preponderance over other large rhinoceroses in linear measurements. For instance, the length from the front of the orbit to the back of the occipital condyles is not much different in the specimen being considered than is the corresponding length in a modern *Rhinoceros unicornis*. But owing to the massiveness of the skull of *Aceratherium perimense* it gives the impression of being extraordinarily great in size.

The skull is short, as would be expected in a form belonging to the *Aceratherium* group, and the nasals are retracted.   The back of the narial notch is above the last premolar, and it is but slightly separated from the front border of the orbit, which latter is above the anterior portion of the first molar.   The position of the orbit above the first molar would seem to be a result of the shortening of the face in this species.   A comparison between various genera and species of rhinoceroses would seem to show that as the face tends to become short, the orbit tends to move forward in its position relative to the cheek teeth.   Conversely, when the face and skull elongate, the orbit tends to migrate backward in relation to the cheek teeth.   These evolutionary trends are to be seen in other perisosdactyls, notably the horses and the titanotheres.   In the horses the face becomes long and the orbit moves back to a position above the posterior molars, while in the titanotheres the face becomes short and the orbit moves forward to a position above the premolars.

The zygomatic arch is heavy, especially below the orbit, so that its lower edge in the forward portion extends below the alveolus of the molars.   The arches are not widely spread.

The glenoid is transverse and heavy, and from its inner side an extremely heavy postglenoid process extends down.

The top of the skull as seen in profile is bowed into a shallow saddle, or rather, the frontals are approximately flat, while the parietals rise sharply to the lambdoidal crest.   In the skull under consideration the portions anterior to the first molar are missing, but evidently the nasals, the maxillaries and the premaxillaries were extremely short.   The skull roof, consisting of the frontals and parietals, is quite narrow.   The parietal crests, running from the blunt postorbital processes to the lambdoidal crest, approach each other very closely but they do not meet.

The occiput is vertical and the condyles, large but not of unduly great size, project back beyond it.   The supraoccipital is indented for the attachment of powerful neck muscles, and since the lambdoidal crest follows the superior edge of this bone, it is strongly bowed forward, as seen from the top, forming a broad V.   The junction of the squamosal and the exoccipital is flared out to form a crest projecting down from the lambdoidal crest, and this continues ventrally in the plate-like posttympanic.   The paroccipital process is rather short and blunt.

As seen from behind the occiput is somewhat rectangular in shape, due to the great development of the lambdoidal crests, an adaptation for the attachment of strong neck muscles.

The postglenoid process is long and is quite separated from the posttympanic, thus leaving the external auditory meatus open below.   The basioccipital is strongly keeled, and at its anterior termination is expanded into a large rugose tuberosity for the attachment of the rectus capitis ventralis major muscles.   The pterygoids are extremely heavy, as might be expected in a large animal requiring heavy pterygoid muscles, and at their bases each is expanded and pierced by a large alisphenoid canal.   The palate is narrow, and the posterior nares extend far forward to the anterior border of the second molars.

The basicranial foramina follow the usual rhinocerotid plan; that is the foramen lacerum anterius and the foramen rotundum open within a single common vestibule, there is a large alisphenoid canal, and the posterior foramina including the foramen ovale and the foramen lacerum medius are confluent.

Only the molars are present in the skull being discussed. The first two molars are fully erupted and but slightly worn, and the third molar is in the process of eruption. These teeth are remarkable for their large size and their hypsodonty. The antero-posterior length is relatively great, and the ectoloph is flat. At the antero-external corner of the tooth there is a well defined parastyle groove or fold, running vertically for the height of the tooth along its anterior edge. This fold in the parastyle is a characteristic *Aceratherium* feature. There are two oblique cross crests, the protoloph and the metaloph, and both the crochet and the antecrochet are strongly developed. There is a well developed anterior cingulum which runs around the internal side of the protocone. As in the other advanced rhinoceroses, the third molar is of triangular form, due to the bending back of the ectoloph and its fusion with the metaloph.

A.M. 19470   1/3

FIG. 86. *Aceratherium perimense* (Falconer and Cautley). Amer. Mus. No. 19470, a skull. Lateral view. One third natural size.

A second specimen, Amer. Mus. No. 19692, shows the anterior premolars and the premaxillaries. It conclusively shows that in this form the upper incisor was present. The incisor premolar diastema was relatively shorter than in *Aceratherium incisivum*.

A third specimen, Amer. Mus. No. 19454, an associated palate and mandible, shows the characters of the premolars. The upper premolars are, with the exception of the first one, completely molariform. The premolar incisor diastema is short. In the mandible the last two premolars are molariform, the second one is compressed and the first is absent.

The diastema is relatively short. The mandibular symphysis is not expanded, and the lower incisors point forward and upward.

Measurements and figures of the several specimens considered in the above discussion, are given below.

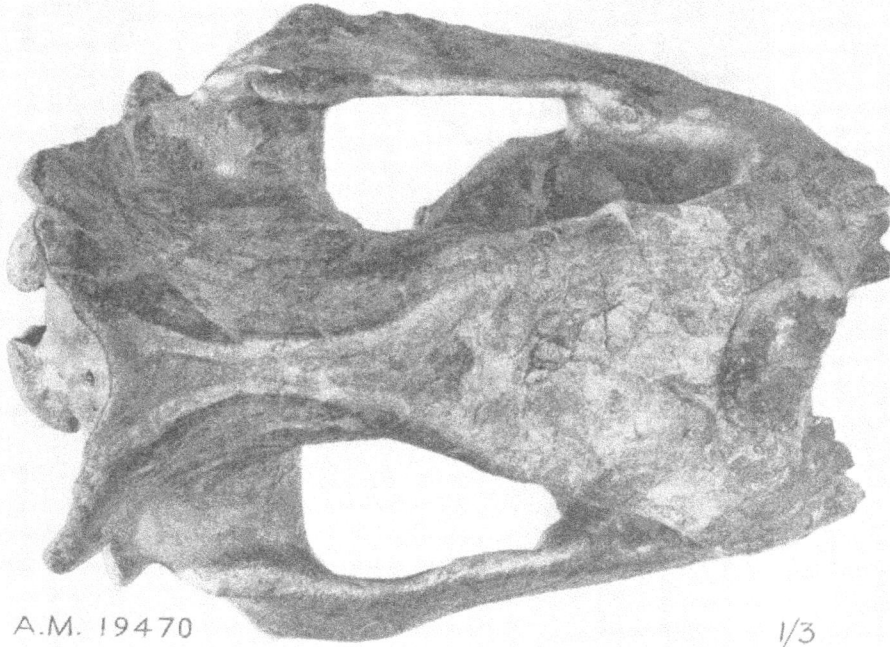

Fɪɢ. 87. *Aceratherium perimense* (Falconer and Cautley). Amer. Mus. No. 19470, skull. Dorsal view. One third natural size.

Mᴇᴀsᴜʀᴇᴍᴇɴᴛs ᴏғ *Aceratherium perimense*

A.M. 19470. Skull.

Length, lambdoidal crest to posterior boundary of narial notch	400 mm.
Length from anterior border of orbit to condyles	395
Width at glenoids	363
Width at parietals, narrowest portion	108
Width of frontals, at postorbital process	188
Width of palate at M^1	85
Height of skull at M^1	230
Diameter of orbit	83
Greatest depth of zygomatic arch	124
Greatest width of occiput	290
Height of occiput above foramen magnum	195
Transverse diameter of condyles	134

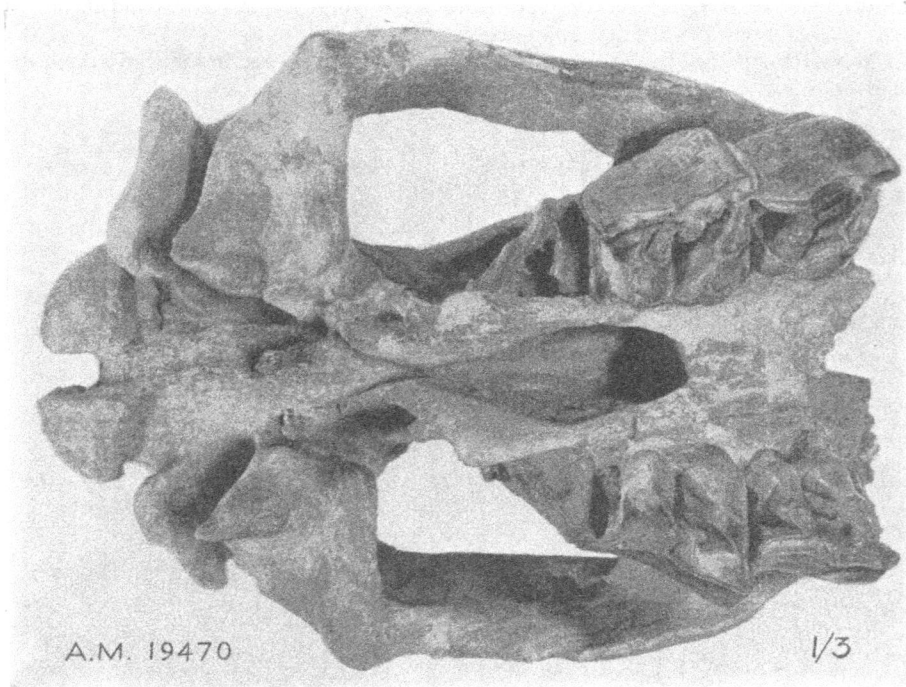

FIG. 88.   *Aceratherium perimense* (Falconer and Cautley).   Amer. Mus. No. 19470, skull.   Ventral view.   One third natural size.

Dentition

M¹....length...... 81 mm.	width...... 88 mm.	height...... 91 mm.
M²....length...... 87	width...... 94	height...... 95
M³.`....(in alveolus)		

COMPARATIVE MEASUREMENTS OF *Aceratherium perimense*

	A.M. 19470	After Lydekker
Height of occiput from base of foramen magnum to crest..................................	259 mm.	254 mm.
Greatest width of occiput....................	290	305
Width of frontals at postorbital process........	188	266
Interval between anterior angle of orbit and auditory fissure...........................	285	335
Vertical diameter of orbit....................	83	95
Breadth of base of nasals....................	85*	101
Width of palate at M¹......................	85	76
Width of palate at M³......................	84	112
Long diameter of occipital condyles..........	87	82

* Estimated.

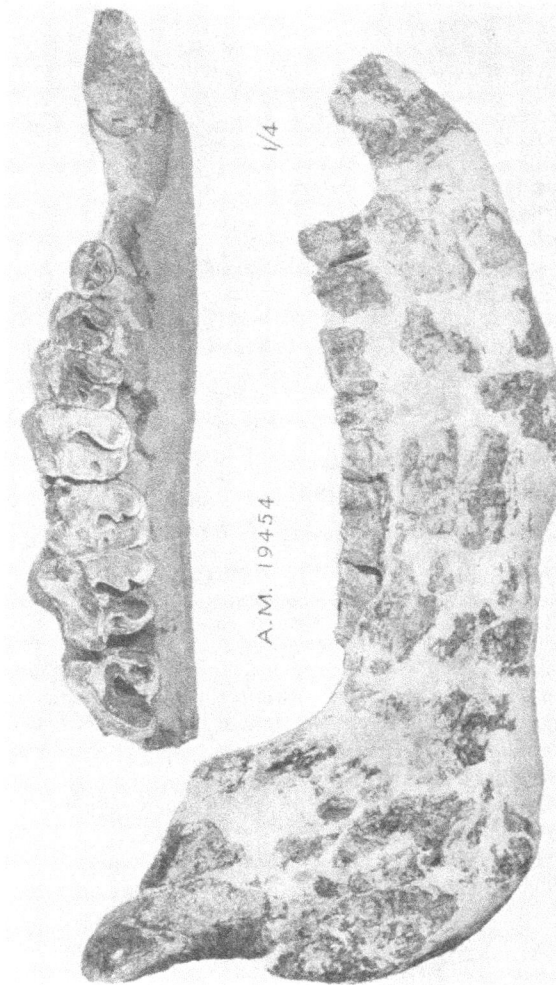

FIG. 89. *Aceratherium perimense* (Falconer and Cautley). Amer. Mus. No. 19454, right maxilla and mandible. Crown view of upper dentition above, lateral view of mandible below. One fourth natural size.

A.M. 19454.   *Aceratherium perimense*.   Maxilla.

	Length	Width
P¹	37 mm.	25 mm.
P²	38	46
P³	46	62
P⁴	51	74*
M¹	60	80*
M²	69	78
M³	63	63
Length premolar series	161	
Length molar series	183	
Length premolar-incisor diastema	90	
Incisor	71	31

* Estimated

A.M. 19454.   *Aceratherium perimense*.   Mandible.

	Length	Width
P₂	32 mm.	23 mm.
P₃	40	26
P₄	49	37
M₁	53	36
M₂	64	40
M₃	72	35
Length premolar series	123	
Length molar series	187	
Incisor	58	45
Length from symphysis to condyle	625	
Length premolar-incisor diastema	48	
Depth of ramus at M₁	106	
Height of condyle above lower border of ramus	305	

The question of the generic and specific identity of the American Museum skull, No. 19470, has indeed been a perplexing one.   At first sight it would seem to be considerably different from the large skull of *A. perimensis*, figured by Lydekker.   On the other hand, a more careful comparison would seem to indicate that the apparent differences between the two specimens may not be as great as first they appeared to be.   Supposing the two specimens to belong to one species, how should they be generically classified?

In speaking of the characters and affinities of the Middle Siwalik rhinoceroses Dr. Matthew pointed out the inappropriateness of some of the previous identifications applied to certain genera.   "The so-called Aceratheria from India were referred to *Aceratherium* by Lydekker on the quite arbitrary ground that they were hornless.   They appear to me to be gigantic species of *Chilotherium*, and whether or not they are placed within that genus (the skull differences are considerable) they have nothing to do with the true *Aceratherium*, but belong in the Oriental rhinoceros group." [58]

Certain anatomical characters of the American Museum material would seem to ally it, if not with the skull described by Lydekker, at least with the genus *Aceratherium*.   Such for instance is the presence of an upper incisor, which is definitely established by the American Museum specimens.   The retracted narial notch, the close approach to each

[58] Matthew, W. D., 1929, p. 451.

other of the parietal crests, the proportions of the upper teeth, especially as regards length and breadth, and the seeming lack of any expansion in the mandibular symphysis, are all characters that would seem to link this form with *Aceratherium* rather than with *Chilotherium*.

At this point a table is presented to show the comparative features of *Chilotherium*, *Aceratherium* and *Aceratherium perimense*. This table was prepared by the writer. Since its preparation, a very similar table has been published by Forster Cooper (1934) in his excellent paper on the rhinoceroses of Baluchistan, and some of the facts brought out by Forster Cooper appear on the present table. They were, however, deduced quite independently of Forster Cooper's work.

Chilotherium	*Aceratherium*	*Aceratherium perimense* (A.M. 19470, 19454)
1. Skull of moderately large size, brachycephalic.	1. Skull of moderately large size, dolichocephalic.	1. Skull extremely large, brachycephalic.
2. Horns absent.	2. Horns absent.	2. Horns absent.
3. Nasals straight and pointed.	3. Nasals straight and pointed.	3. Nasals straight and pointed.
4. Parietal cristae widely separated.	4. Parietal cristae close together.	4. Parietal cristae close together.
5. Narial notch moderately retracted.	5. Narial notch greatly retracted.	5. Narial notch greatly retracted.
6. Occiput vertical.	6. Occiput vertical.	6. Occiput vertical.
7. Lambdoidal crest transversely expanded.	7. Lambdoidal crest not expanded.	7. Lambdoidal crest expanded.
8. Zygomatic arch of average size.	8. Zygomatic arch slightly expanded.	8. Zygomatic arch greatly expanded.
9. Postglenoid and posttympanic separated, or in some cases touching.	9. Postglenoid and posttympanic separated, or in some cases touching.	9. Postglenoid and posttympanic separated.
10. Premaxillaries short and attenuated.	10. Premaxillaries long and heavy.	10. Premaxillaries of moderate length, and heavy.
11. Upper incisor absent.	11. Upper incisor present.	11. Upper incisor present.
12. Cheek teeth hypsodont.	12. Cheek teeth sub-hypsodont.	12. Cheek teeth hypsodont.
13. Parastyle fold indistinct or lacking	13. Parastyle fold stong.	13. Parastyle fold strong.
14. Protocone constricted.	14. Protocone constricted.	14. Protocone constricted.
15. Ectoloph greatly elongated.	15. Ectoloph not elongated.	15. Ectoloph moderately elongated.
16. Mandibular symphysis transversely expanded.	16. Mandibular symphysis not expanded.	16. Mandibular symphysis not expanded.
17. Lower incisors directed up and outwardly.	17. Lower incisors not directed outwardly.	17. Lower incisors not directed outwardly.

From the above table it may be seen that the definitive characters of the American Museum material referable to *Aceratherium perimense* would seem to ally this material with the genus *Aceratherium* rather than with *Chilotherium*.

The presence of an upper incisor in the American Museum material is especially significant. That an incisor was present is definitely shown by Amer. Mus. No. 19692, in which specimen the molar teeth are certainly like the molars of the large skull, Amer. Mus. No. 19470. An associated palate and mandible, Amer. Mus. No. 19454 also show the upper incisor as a well developed tooth. Although this latter specimen is somewhat smaller than Amer. Mus. No. 19470, there seems to be no valid reason for separating it specifically from the large skull.

All of the large Middle Siwalik and Lower Siwalik rhinoceros material in the American Museum collection is hereby referred to *Aceratherium perimense*. The fact is recognized, however, that this material does show certain important diagnostic characters, such as the hypsodont cheek teeth, the great development of the lambdoidal crest, etc., that set it apart from the typical *Aceratherium*. Consequently it is referred to the genus *Aceratherium* with a full realization that when more material makes the species under consideration better known, it may be transferred to a separate and a new genus. It may be, too, that more complete studies will prove that the Perim Island material, the true *Aceratherium perimense*, is separate from the gigantic rhinocerotid from the Punjab.

Aceratherium lydekkeri Pilgrim

Aceratherium lydekkeri, Pilgrim, 1910, Rec. Geol. Surv. India, XL, pp. 65–66.
    Additional References.—
        Pilgrim, G. E., 1910B, p. 200; 1913B, pp. 284, 297.
        Matthew, W. D., 1929, pp. 449, 507.
    Type.—Not definitely designated. "It is therefore necessary to refer the Dhariala skull to Falconer's species *A. perimense* and to establish a fresh specific name for the Middle Siwalik skull and teeth described by Lydekker." According to this it would be inferred that Pilgrim meant to include G.S.I. Nos. C 1, C 2, C 3, C 4, C 7, C 14, C 18, C 238 in the new species.
    Cotypes.—See remarks under type.
    Horizon.—Middle Siwaliks.
    Locality.—From the Punjab, particularly around Hasnot.
    Diagnosis.—Like *Aceratherium perimense* but smaller and more dolichocephalic.

This species is probably synonymous with *Aceratherium perimense*. The reader is referred to Matthew, W. D., 1929, p. 507.

Aceratherium planidens (Lydekker)

Rhinoceros planidens, Lydekker, 1876, Pal. Indica (X), I, pp. 41–43, Pl. IV, figs. 7, 9.
Aceratherium perimense, Lydekker, 1880, Pal. Indica (X), I, p. xiii.
    Type.—G.S.I. No. C 13, two imperfect upper molars.
    Horizon.—Middle Siwaliks.
    Locality.—Padri, Punjab.
    Diagnosis.—(After Lydekker, R., 1876A, p. 41.) Median valley of upper molars wide at entrance; crochet blunt and simple; antecrochet large; anterior cingulum well developed.
    Synonymous with *Aceratherium perimense*.

Aceratherium iravadicus (Lydekker)

Rhinoceros iravadicus, Lydekker, 1876, Pal. Indica (X), I, pp. 36–41, Pl. V, figs. 1–4.
Aceratherium perimense, Lydekker, 1881, Pal. Indica (X), II, p. 10.
    Cotypes.—G.S.I. Nos. C 74, a left M^2; C 73, portion of a skull; C 75, a right M^2; C 76, a fragmentary maxilla.

Horizon.—Irrawaddy beds, probably an equivalent of the Middle Siwaliks.

Locality.—Burma, Irrawaddy River.

Diagnosis.—(After Lydekker, R., 1876A, p. 36.) Entrance to median valley in upper molars blocked by a large tubercle; prominent parastyle groove; crochet simple; well developed antecrochet and anterior cingulum.

A synonym of *Aceratherium perimense*, as was recognized by Lydekker in 1881. "In consequence of the above re-determinations the species *R. planidens* and *R. iravadicus* must be removed from the list of Siwalik mammals." [59]

Chilotherium Ringström, 1924

Generic type, *Chilotherium anderssoni* Ringström

Chilotherium intermedium (Lydekker)

Rhinoceros sivalensis intermedius, Lydekker, 1884, Pal. Indica (X), III, p. 5, Pl. I, fig. 3.

Aceratherium gajense intermedium, Pilgrim, 1910, Rec. Geol. Surv. India, XL, p. 200.

Chilotherium intermedium, Matthew, 1929, Bull. Amer. Mus. Nat. Hist., LVI, p. 508, fig. 32.

Additional References.—

Lydekker, R., 1885B, p. 64.

Type.—G.S.I. No. C 34, a second right upper molar.

Paratypes.—None.

Horizon.—Lower Siwaliks, for the type. Lower and Middle Siwaliks for referred specimens.

Locality.—Sind, for the type. Punjab for referred specimens.

Specimens in the American Museum.—Amer. Mus. No. 19477. Anterior portion of a mandible with right and left P_{2-3}. From the Middle Siwaliks, near Dhok Pathan.

19483. A left P_3. Middle Siwaliks, two miles east of Dhok Pathan.

19563. Left M^3. Lower Siwaliks, 200 feet above the level of Chinji Rest House, one half mile north of Chinji Rest House.

19580. Portions of the right upper dentition. Lower Siwaliks, 100 feet above the level of Chinji Rest House, two miles west of Chinji Rest House.

19680. Right lower molar. Middle Siwaliks, four miles west of Dhok Pathan.

19681. Fragment of a left ramus with M_{1-2}. Middle Siwaliks, one mile west of Dhok Pathan.

19689-19690. A complete skull and mandible of a juvenile animal. Milk dentition. Middle Siwaliks, one half mile southwest of Dhok Pathan.

19722. Fragment of right ramus with broken cheek teeth. Middle Siwaliks, three miles east of Dhok Pathan.

19898. Right lower molar. Middle Siwaliks, two miles north of Hasnot.

29795. Left M^3. Middle Siwaliks, one mile west of Hasnot.

29797. Left M^3. Lower Siwaliks, 1600 feet above the level of Chinji Rest House, one mile west of Chinji Rest House.

29799. Various associated cheek teeth. Lower Siwaliks, near Chinji Rest House.

Diagnosis.—A *Chilotherium* of medium size, very close to *C. blanfordi*. Distinguished by its rather prominent parastyle fold, and the slight constriction of the protocone.

[59] Lydekker, R., 1881, Pal. Indica (X), II, p. 10.

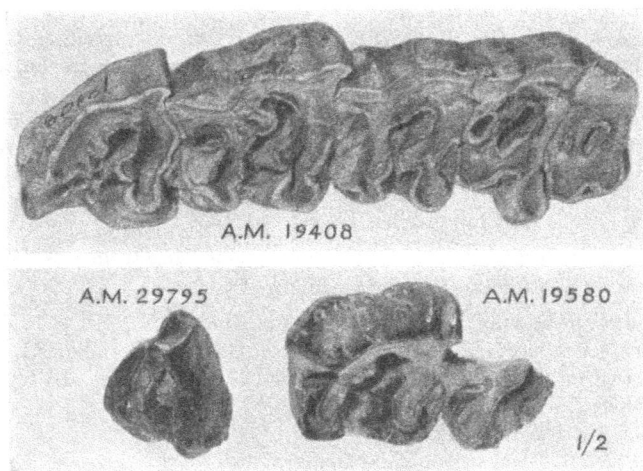

FIG. 90. *Chilotherium blanfordi* (Lydekker). Amer. Mus. No. 19408, right P²–M³. Crown view.
Chilotherium intermedium (Lydekker). Amer. Mus. No. 29795, left M³, crown view; Amer. Mus. No. 19580, right upper molars, crown view.
Figures one half natural size.

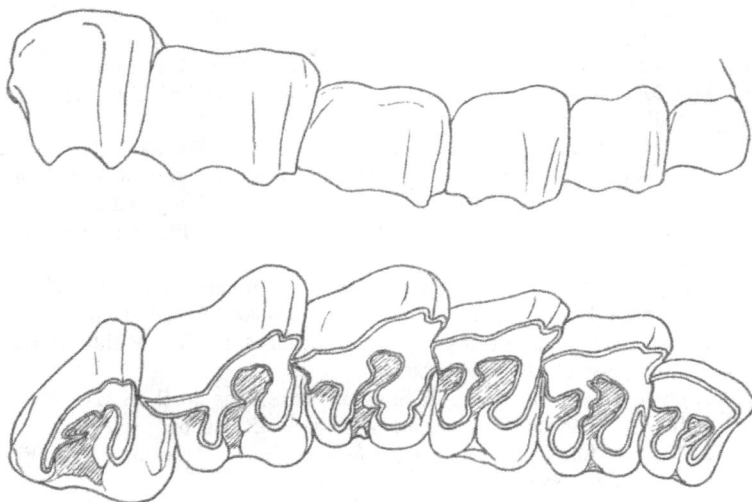

FIG. 91. *Chilotherium intermedium* (Lydekker). Geol. Surv. India, No. C 100, right upper dentition. External view above, crown view below. One half natural size. From Matthew, 1929.

Matthew first suggested that this species, assigned by Lydekker to the genus *Rhinoceros*, and by Pilgrim to the genus *Aceratherium*, should be properly classified in the genus *Chilotherium*. Matthew's opinion is followed in this present work. Although *Chilotherium intermedium* is typically of Lower Siwalik age, there are several specimens from the Middle Siwaliks in the American Museum collection that would seem to be referable to this species. The differences between these specimens and the typical *C. intermedium* do not seem to be enough to warrant their separation as a distinct form, so they are included within the species under discussion, and this species is thereby considered as ranging through the Chinji and the Middle Siwalik beds.

A JUVENILE SKULL AND MANDIBLE, REFERRED TO *Chilotherium intermedium*

The associated skull and mandible (Amer. Mus. Nos. 19690, 19689) would seem to be referable to *Chilotherium intermedium*. The skull is slightly crushed and it lacks the floor of the brain case. The mandible lacks only the tip of the coronoid processes and the borders of the angles. The full milk dentition is preserved above and below.

This skull bears out Ringström's interpretation of the dental formula for the milk dentition of *Chilotherium*, in that it shows DI 0/2, DC 0/0, DM 4/4. The skull is rather small, being about 300 millimeters in length, which is slightly smaller than a skull of *Chilotherium anderssoni* of comparable ontogenetic development, figured by Ringström.

The skull and mandible under consideration are quite similar to the skull and mandible of *Chilotherium anderssoni*; the nasals are straight and hornless and the premaxillaries are very much reduced, an indication of the complete reduction of the incisors in this genus. An interesting character of this skull is the division of the infraorbital foramen, so that there are two separate exits near the narial notch. This division of the infraorbital foramen is a *Chilotherium* character, as has been pointed out by Ringström in his original description of the genus. The postglenoid process is long and is separated from the posttympanic. The frontal region has begun to show a concavity between the orbits, a feature typical of the adult *Chilotherium*. The mandible shows the beginning of the increase in symphyseal width which is characteristic of the adult animal. The symphysis, however, is not broadened to the same degree as is the symphysis in the juvenile *C. anderssoni*, so it would seem likely that *C. intermedium* showed a lesser degree of specialization in this particular feature than did the species from North China.

The first upper milk molar of the specimen under consideration, a tooth just erupting, is small and of triangular outline. It consists essentially of an outer ectoloph and of a posterior transverse cross crest. The second upper milk molar is quadrate with a convex ectoloph and a fairly strong parastyle. There is a large anterior cingulum in this tooth, separated by a valley from the protoloph, and it continues around to the lingual side of the tooth to close the median valley. An antecrochet and a crochet are present.

The third and fourth upper milk molars are essentially similar to the permanent molars. They have a broad and flat ectoloph with a strong parastyle well developed, a somewhat oblique protoloph and metaloph, and strong crochet. There is no internal cingulum. It might be well to point out the fact that in the specimen at hand the protocone is not divided by a posterior vertical fissure, as in the juvenile specimen of *C. anderssoni*. This is as it should be, for we find that in the adult *C. intermedium* the protocone is not strongly divided as in the adult *C. anderssoni*.

The first lower milk incisor was seemingly very small and rudimentary, as may be judged from the alveolus, which is located medially to the erupting second milk incisor. A very small alveolus in front of DM_2 would indicate that a rudimentary DM_1 was present, but was soon lost. The remaining milk molars show the typical rhinocerotid pattern, each tooth being composed of two crescents. Measurements of the specimen are given in the accompanying tables.

FIG. 92. *Chilotherium intermedium* (Lydekker). Amer. Mus. No. 19690, juvenile skull. Dorsal view above, ventral view below. One third natural size.

The question arises as to whether the skull and mandible considered above really do belong to the species *C. intermedium*, or whether they represent a distinct new species of *Chilotherium*. As compared to the adult teeth of *C. intermedium* in the American Museum collection, the milk molars of the specimen under consideration seem rather small. The proportions between milk and permanent teeth in *C. anderssoni*, as taken from measurements given by Ringström, are as follows:

DM⁴....length.... 56 mm.   M²....length.... 65 mm.   Ratio.... 86

A comparison between juvenile and adult of *C. intermedium*, as demonstrated by the American Museum material, shows the following ratio.

DM⁴....length.... 40 mm.   M²....length.... 58 mm.   Ratio.... 69

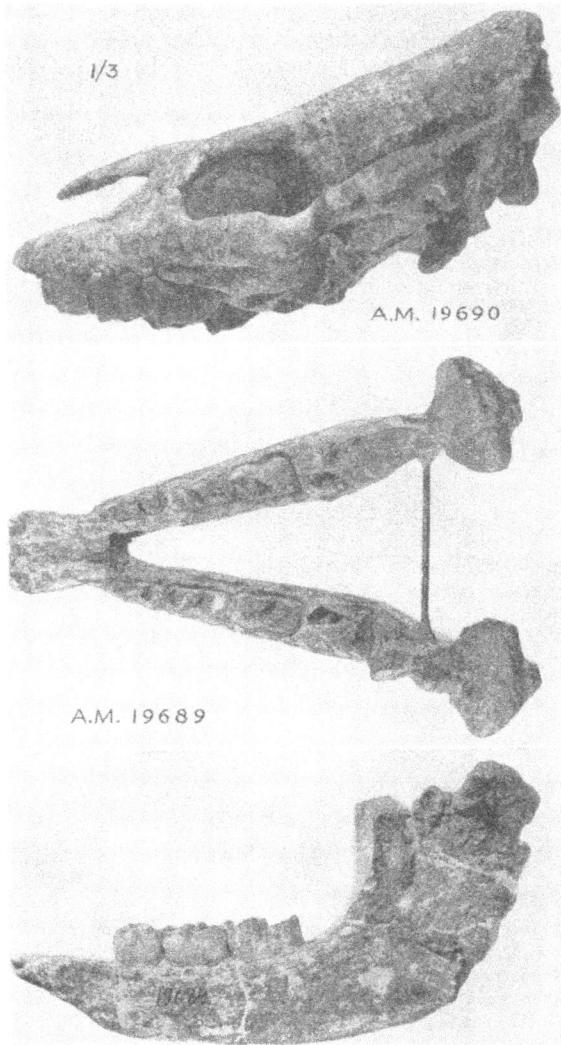

FIG. 93. *Chilotherium intermedium* (Lydekker). Amer. Mus. Nos. 19690, 19689, juvenile skull and mandible. Lateral view of skull at top, dorsal view of mandible in middle, lateral view of mandible at bottom. One third natural size.

Thus there is a greater discrepancy between the supposed juvenile of *C. intermedium* and the adult in the American Museum collection, than exists in *C. anderssoni*.   In view, however, of the inadequacy of material for measurements the above figures can not be taken too seriously.   Consequently it seems best for the time being to consider the juvenile skull and mandible, discussed above, as belonging to the species *C. intermedium*, rather than to a new species.

MEASUREMENTS (IN MM.) OF *Chilotherium*

	A.M. 19690 *C. intermedium*	A.M. 26340 *C. anderssoni*
Length of skull (premaxilla-lambdoidal crest)...	298 mm.	
Width across glenoids......................	165	
Width at frontals (postorbital process)........	106	113 mm.
DM^2....Length...........................	34	38
Width.............................	27	33
Height............................	25	28
DM^3....Length...........................	35	43
Width.............................	31	39
Height............................	28	34
DM^4....Length...........................	40	52
Width.............................	33	44
Height............................	31	43

	A.M. 19689 *C. intermedium*	A.M. 26341 *C. anderssoni*
Length of mandible (condyle-symphysis)......	273 mm.	285 mm.
Width across condyles......................	175	
Width of symphysis........................	39	48
DM_2....Length...........................	27	30*
Width.............................	14	16*
Height............................	18	
DM_3....Length...........................	31	38*
Width.............................	16	20*
Height............................	20	
DM_4....Length...........................	36	45*
Width............................:	18	22*
Height............................	23	

* Ringström, T., 1924, Pal. Sinica, Ser. C, I, Fas. 4, p. 37.

A.M. 19477.   *Chilotherium intermedium.*   Mandible.

Transverse diameter between outside surfaces of rami at
   P_3.. 105 mm.
Transverse diameter of symphysis at narrowest part.... 83
Transverse diameter of symphysis at incisors.......... 78
Depth of ramus at P_2............................... 56

A.M. 29798.   *Chilotherium blanfordi.*

   DM^4....Length..... 50 mm.   Width..... 45 mm.

Chilotherium blanfordi (Lydekker)

Aceratherium blanfordi, Lydekker, 1884, Pal. Indica (X), III, pp. 2–11, figs. 1–3, Pl. I, figs. 1, 2, 6, Pl. II, figs. 1, 2, 3.

Teleoceras blanfordi, Pilgrim, 1910, Rec. Geol. Surv. India, XL, p. 200.

Chilotherium blanfordi, Ringström, 1924, Pal. Sinica, Ser. C, I, Fas. 4, pp. 75–76.

Aceratherium blanfordi, Forster Cooper, 1934, Phil. Trans. Royal Soc., London, Ser. B, CCXXIII, pp. 589–594, fig. 9.

Additional References.—

Lydekker, R., 1885B, pp. 68–69; 1886A, pp. 154–155, fig. 18.

Pilgrim, G. E., 1912, pp. 30–32, Pl. VII, figs. 4–7; 1913B, p. 312.

Matthew, W. D., 1929, p. 508.

Type.—(Lectotype.)—G.S.I. No. C 268, a left maxilla with M^{1-3}.

Cotypes.—G.S.I. Nos. C 50, left maxilla; C 258, milk molar; C 260, right MM^{1-2}; C 262, molar; C 267, right ramus; C 271, right ramus; C 269 a left maxilla; C 270, left mandibular ramus. The last two specimens were figured as *A. blanfordi minus*.

Horizon.—Bugti beds for the type. Kamlial and Chinji zones, Lower Siwaliks, for the referred specimens. Also Middle Siwaliks.

Locality.—Dera Bugti and Gandoi in Baluchistan for the type and cotypes. Northern Punjab for referred specimens.

Specimens in the American Museum.—Amer. Mus. No. 19408. A portion of the palate and the basicranium, with right P^3–M^3. Lower Siwaliks, 100 feet below the level of Chinji Rest House, near the contact between the Chinji and the Kamlial formations, near Chinji Rest House.

19469. Right mandibular ramus with P_2–M_3. Lower portion of Middle Siwaliks, 1000 feet below the Bhandar bone bed, one mile south of Nathot.

19532. Fragments of teeth and a mandible. Lower portion of the Middle Siwaliks 1000 feet below the Bhandar bone bed, five miles west of Hasnot.

19538. Fragments of mandibular ramus and teeth. Middle Siwaliks, one mile south of Nathot.

19539. Right M^2, and miscellaneous fragments. Middle Siwaliks, one mile south of Nathot.

29788. A left DM^3. Lower Siwaliks, at the level of Chinji Rest House, two miles west of that place.

29789. A right P^2. Middle Siwaliks, one half mile north of Hasnot.

29791. A left P^3. Middle Siwaliks, one mile northeast of Hasnot.

29798. A left DM^4. Middle Siwaliks, four and one half miles west of Hasnot.

Diagnosis.—A rather large species of *Chilotherium*. Molars with flat ectoloph, and with rather sharply constricted protocone.

Ringström first placed this species in the genus *Chilotherium* and in this he was followed by Matthew, in 1929. Later, however, Matthew referred this species to *Rhinoceros*, and subsequently it was classified by Forster Cooper as *Aceratherium* and as "Rhinoceros" *sensu lato*.

It is thought most expedient for the present to consider it as belonging to the genus *Chilotherium*. This designation is provisional at best.

It will be noticed that a considerable number of teeth referred to this species in the American Museum collection range up into the Middle Siwaliks. Although this extends the range of the species beyond its hitherto known limits, there seems to be no other very logical solution as to the identification of the teeth in question. The discovery of additional skulls and skeletal material will aid materially in the interpretation of the confusion of Siwalik rhinoceros remains. Until these skulls can be found, the isolated teeth must needs be assigned in the best way possible to the genera and species that they would seem to represent. Naturally, on the basis of teeth alone, a great many errors will appear in the identification of the American Museum specimens, when at some time they can be studied in the light of supplementary and more detailed knowledge.

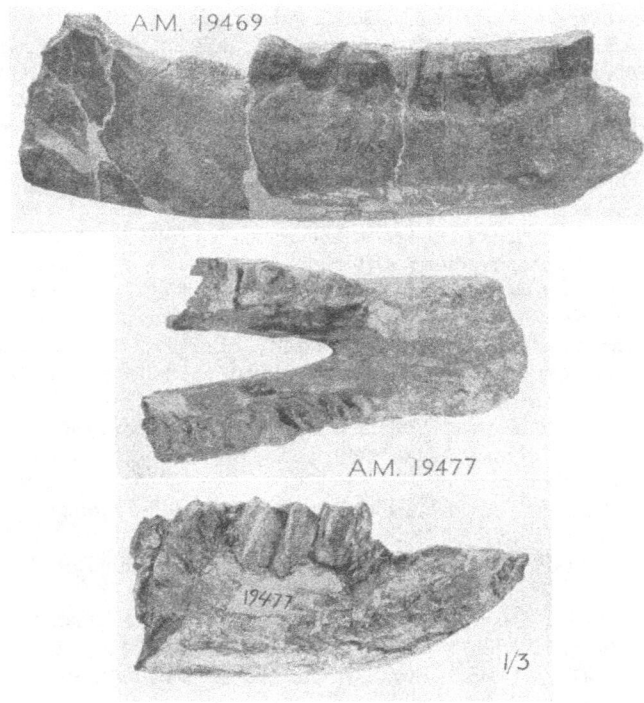

FIG. 94. *Chilotherium blanfordi* (Lydekker). Amer. Mus. No. 19469, mandibular ramus. Lateral view.
Chilotherium intermedium (Lydekker). Amer. Mus. No. 19477, symphysis of mandible. Dorsal view above in middle, lateral view below.
Figures one third natural size.

The two species supposedly representing the genus *Chilotherium* in the Siwalik deposits are very similar to each other, a fact recognized by Dr. Matthew in 1929.

"This [i.e., *Chilotherium intermedium*] is close to *C. blanfordi* Lydekker of the Bugti Hills, differs chiefly in more prominent antero-external pillar and protocone less constricted off. Doubtful if really separable." [60]

A review of the literature and of the material in the American Museum would seem to point to the following facts.

1. *C. blanfordi* and *C. intermedium* are probably distinct species.

2. Both species were geologically of long range, and persistent. The former appeared in the Bugti beds and lasted through the Kamlial, Chinji and into the Middle Siwaliks, while the latter appeared in the Chinji and was well developed in the Middle Siwaliks.

Several characters serve to separate *C. blanfordi* from *C. intermedium*.

1. *C. intermedium* is typically somewhat smaller than *C. blanfordi*.

2. In *C. intermedium* the antero-external pillar is prominent while in *C. blanfordi* the ectoloph is relatively flat.

3. In *C. intermedium* the protocone is much less constricted off from the protoloph than is the case in *C. blanfordi*.

4. In *C. intermedium* the metaloph of the third upper molar is longer than the protoloph, while in *C. blanfordi* the two crests are more nearly equal in length.

Forster Cooper (1934) has made the following helpful remarks about these two species.

"The animal here under discussion [*A. blanfordi*] has been placed in the genus *Chilotherium* by Ringström (1924) and by Matthew (1929), presumably on the evidence of the structure of the molars. This structure, however, the constricted protocone, the large crochet, the heavy cingulum, etc., is found in so many genera that it has little value as evidence of affinity. There is, moreover, no evidence that the lower jaw, to which a symphysis is here referred, fig. 10, A, had that exceptionally wide symphysis which is such a leading characteristic of *Chilotherium*. The structure of the feet and skeleton, from which much information might be obtained, is entirely unknown. Matthew (1929) states that '*Chilotherium*' *intermedium* (Lydekker) is close to '*C*.' *blanfordi* (Lydekker). The type of the first-named is no more than a second upper molar, but Matthew (1929, fig. 32, p. 508) refers to this species a specimen in the Indian Museum. According to him it is doubtful if the two species are really separable. They differ only in that the first-named has a more prominent antero-external pillar and a less constricted protocone. In the light of the specimens here described it can be seen that *C. intermedium* is much more advanced in the evolution of the premolars, in the lesser development of the cingulum and in the development of a much larger crochet on the molars, and appears therefore to be a separate species. All that can be said, therefore, about the generic position of *R. blanfordi* is that it appears to have been derived from some *Caenopus* stock and that, as far as the evidence goes, there is nothing to prevent it from being regarded as having some affinity with *Aphelops*." [61]

[60] Matthew, W. D., 1929, p. 508.
[61] Forster Cooper, C., 1934, p. 594.

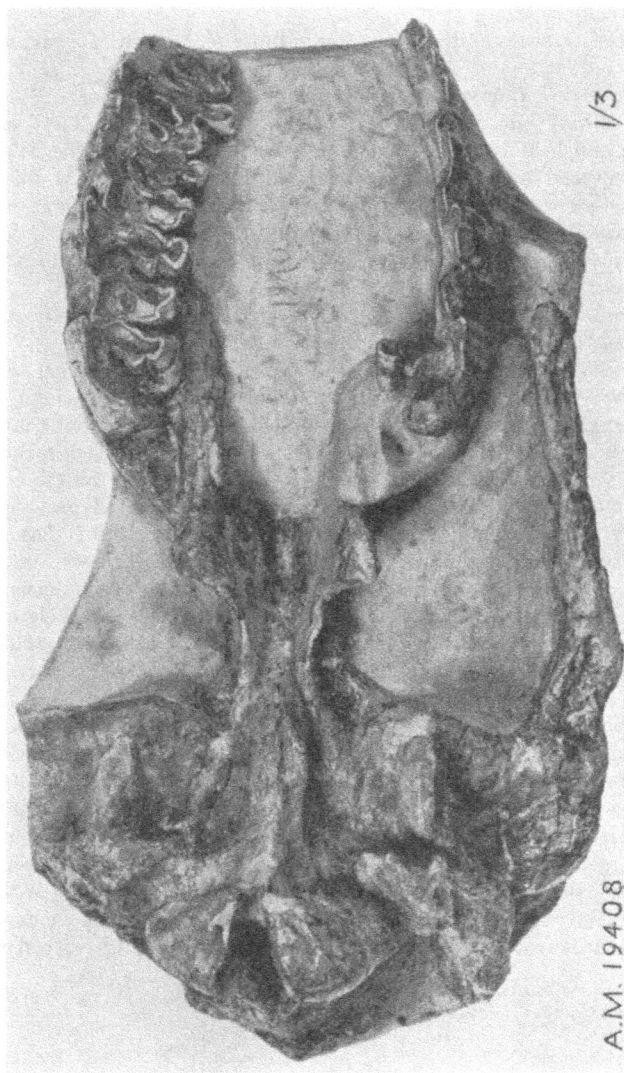

FIG. 95. *Chilotherium blanfordi* (Lydekker). Amer. Mus. No. 19408, lower portion of skull. Ventral view. One third natural size. (See Fig. 90 for detail of teeth.)

MEASUREMENTS

Chilotherium blanfordi

Amer. Mus. No. 19408.

	Length	Width
Right M¹	52 mm.	64 mm.
M²	64	66
M³	56	61
Transverse diameter, occipital condyles	122	
Width of skull at glenoids	352	

Amer. Mus. No. 19469, mandible.

Length of premolar series	83	
Length of molar series	143	
Depth of ramus at M₁	77	
Thickness of ramus at M₁	49	

Amer. Mus. No. 29798.

DM⁴....Length 50 mm.   Width....45 mm.

Limb and Foot bones Referable to Chilotherium.

Amer. Mus. No. 19435. Various foot bones from the Lower Siwaliks.

29818.   Phalanges and proximal end of a metacarpal. Middle Siwaliks, fifteen miles northwest of Bilaspur.

29832.   Right femur, tibia, tarsus, pes; left radius and ulna. Middle Siwaliks, one half mile southwest of Dhok Pathan.

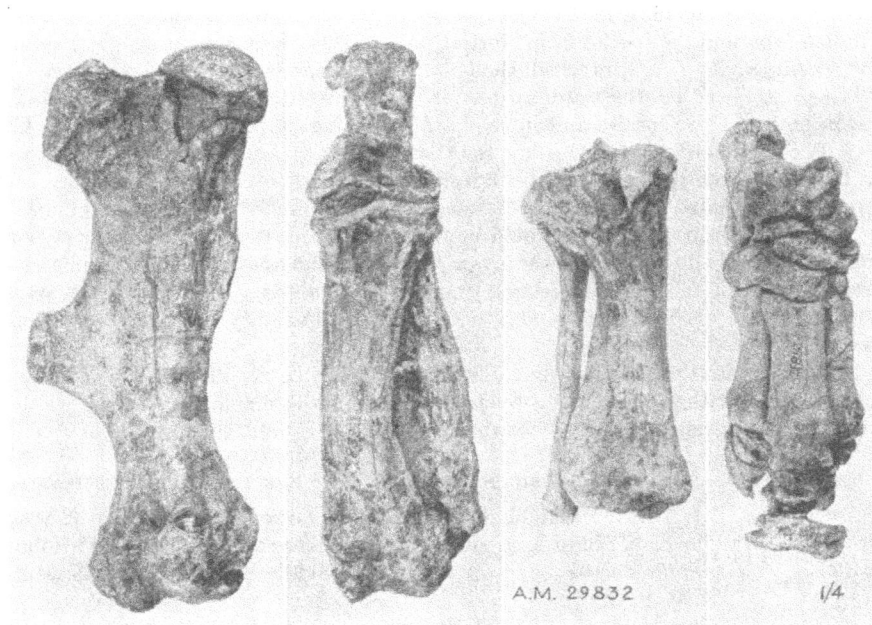

FIG. 96.   Associated limb and foot bones referable to *Chilotherium*. Amer. Mus. No. 29832. From left to right: right femur, left radius-ulna, right tibia-fibula, right pes. Anterior views. Figures one fourth natural size.

Of the skeletal elements in the American Museum collection, referable to *Chilotherium*, the associated hindlimb and foot show particularly well the characters of the genus. The limb bones and the pes are very short. The median metatarsal is much wider than the lateral ones, a character shown by Ringström to be especially characteristic of the genus under discussion.

MEASUREMENTS

Amer. Mus. No. 29832.

Radius, articular length	249 mm.
Ulna, greatest length	325
Femur, greatest length	360
Femur, breadth at mid-shaft, below trochanter	64
Tibia, greatest length	245
Tibia, breadth at mid-shaft	43
Calcaneum, length	65
Calcaneum, breadth	65
Astragalus, breadth	77
Metatarsal II, length	92
Metatarsal II, breadth at middle of bone	25
Metatarsal III, length	103
Metatarsal III, breadth in middle of bone	33
Metatarsal IV, length	90
Metatarsal IV, breadth in middle of bone	24

A comparison of the Siwalik *Chilotherium* hindlimb with that of the *Chilotherium* figured by Ringström (1924, Pl. IX, figs. 3, 4, 5) shows that a great similarity exists between the Indian form and the species from North China. This comparison is shown graphically in the accompanying figure, in which the femur in both species is reduced to unity.

It will be seen that the third trochanter is well down towards the distal end of the femur in both of these species, but in the Siwalik one it is lower than in the North China form. The portion of the femur below the trochanter is correspondingly short. It would seem that in the shortening of the limbs in this brachypodine rhinoceros there has been a differential shortening in the femur. That is, the upper end of the bone has retained to a large extent the proportions that would be found in a rhinoceros having limbs of normal length, whereas the lower end of the femur, the tibia and the foot have become greatly abbreviated. This offers an interesting example of the differential growth in one working unit, whereby an evolutionary trend is accomplished differently by separate portions of the same unit.

In both species the tibia and fibula are short, but in the Siwalik form the tibia is proportionately wider than in the Chinese species. The pes is similar in both species, but it would seem to be proportionately somewhat longer in the Siwalik than in the Chinese form.

RHINOCEROTIDAE IN THE AMERICAN MUSEUM, NOT SPECIFICALLY IDENTIFIABLE

Amer. Mus. No. 19427. Mandibular fragments. Lower Siwaliks, 1600 feet above the level of Chinji Rest House, twelve miles east of Chinji Rest House.

19473. *Chilotherium* carpals. Lower portion of Middle Siwaliks, one mile south of Nathot.

19532. Miscellaneous teeth. Middle Siwaliks, 1000 feet below the level of the Bhandar bone bed, five miles west of Hasnot.

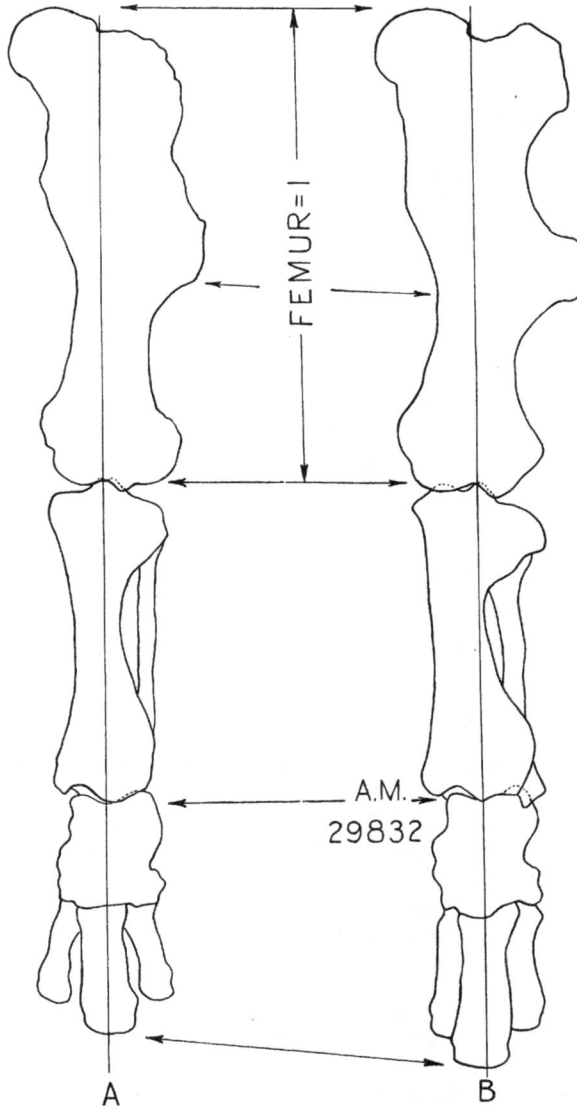

FIG. 97. Comparison of the hind limb in *Chilotherium* from North China (left) with *Chilotherium* from the Siwaliks, Amer. Mus. No. 29832 (right). Femur reduced to unity. North China form (*A*) taken from Ringström, 1924, Pl. IX, figs. 3, 4, 5.

19547. Carpal bones. Middle Siwaliks, near Hasnot.

19619. *Chilotherium*, mandibular ramus. Lower Siwaliks, 1600 feet above the level of Chinji Rest House, one mile west of Chinji Rest House.

19684. Fragmentary mandibular ramus. Middle Siwaliks, one half mile southwest of Dhok Pathan.

19696. Broken teeth. Middle Siwaliks, three miles east of Dhok Pathan.

19718. Miscellaneous teeth. Middle Siwaliks, one mile east of Dhok Pathan.

19903. Teeth. Middle Siwaliks, 1000 feet below the level of the Bhandar bone bed, four and one half miles west of Hasnot.

19941. Teeth. Middle Siwaliks, 1000 feet below the level of the Bhandar bone bed, four and one half miles west of Hasnot.

19290. Tooth. Middle Siwaliks, four miles east of Dhok Pathan.

29812. *Chilotherium*, symphysis of mandible. Lower Siwaliks, locality not known.

ARTIODACTYLA

BUNODONTA

SUOIDEA

TAYASSUIDAE (?)

Pecarichoerus Colbert, 1933

Generic type, *Pecarichoerus orientalis* Colbert

Pecarichoerus orientalis Colbert

Pecarichoerus orientalis, Colbert, 1933, Amer. Mus. Novitates, No. 635.

Type.—Amer. Mus. No. 29955, various isolated cheek teeth from a single individual.

Paratypes.—None.

Horizon.—Lower Siwaliks, Chinji zone, about 1600 feet above the level of Chinji Rest House.

Locality.—Three miles west of Chinji Rest House, Salt Range, Attock District, Punjab.

Specimens in the American Museum.—The type, listed above.

Diagnosis.—(Colbert, E. H., 1933G, p. 2.) "Molar teeth short, brachyodont and quadricuspid. Cusps conical and separated from each other. Median valley of the third molar occupied by sharp, oblique ridges, which run between the anterior and the posterior pairs of cusps. Enamel smooth."

This genus and species would seem to be indicative of a true peccary in India in Lower Siwalik times. The paper quoted above, gives a full description, as well as the evidence for considering *Pecarichoerus* a member of the Tayassuidae.

The type figures and the measurements are reproduced below.

MEASUREMENTS

Pecarichoerus orientalis, Amer. Mus. No. 29955

M^1....Length... 13.5 mm.   Width... 12.5 mm.   Index... 93
M^3....Length... 14.0   Width... 12.0   Index... 86

In the original description of *Pecarichoerus* the following remarks, by way of conclusion, were made.

"There now comes the question of the derivation of *Pecarichoerus* and the implications raised by its presence in India. The older theory as to the separate origin of the pigs and peccaries (mentioned above) has been more or less disproven by the recent work of Miss

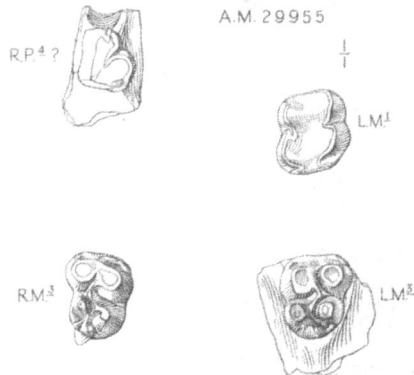

FIG. 98. *Pecarichoerus orientalis* Colbert. Amer. Mus. No. 29955, isolated upper cheek teeth, placed in the approximate positions they would occupy in the palate. Crown views above, lateral views below. Natural size. From Colbert, 1933.

Pearson. She has clearly demonstrated that the American genus, *Perchoerus*, and the European form, *Palaeochoerus*, are indeed closely related, and that they have descended from a not very distant common ancestor. According to Miss Pearson, *Palaeocheorus* is the European derivative of this common ancestral stock, and it gave rise to the true pigs, while *Perchoerus* is the American derivative which gave rise to the peccaries.

"At the same time, *Doliochoerus* is a European cousin of *Perchoerus*, and must be regarded as a persistent remnant staying on near the center of origin of the group. *Pecarichoerus* now takes a position similar to that of *Doliochoerus*, in that it represents a persistent remnant near the center of origin for the peccaries.

"We may imagine an Eurasiatic origin for the common ancestor of the pigs and the peccaries. Early in the history of the Tayassuidae, the group migrated to the Western Hemisphere (possibly by way of a trans-Bering land bridge), while a few hardy forms lingered on for a short time in the Old World. These holdovers we see in the genera *Doliochoerus* and *Pecarichoerus*, in Europe and Asia respectively."

Of course, there is an opposite consideration, namely that *Pecarichoerus* is a migrant from the New World to Asia. Certain reasons militate against this argument, the chief one of which is the fact that *Pecarichoerus* would seem to have its closest relationship in *Doliochoerus*, and this latter genus is quite probably of Old World origin.

Both *Doliochoerus* and *Pecarichoerus* are rather aberrant forms, distinct from the typical peccaries, but nevertheless definitely related to them. *Doliochoerus* is of Oligocene age, while *Pecarichoerus* is found in the lower Pliocene. *Perchoerus* the ancestor of the North American peccaries, did not arrive in the New World until the Oligocene. Therefore, if *Doliochoerus* entered Europe as a derivative of *Perchoerus*, a great deal of migration and counter migration within a relatively short geologic period of time must be supposed. Of course, these movements back and forth are not at all improbable in a group of mammals as active as the peccaries.

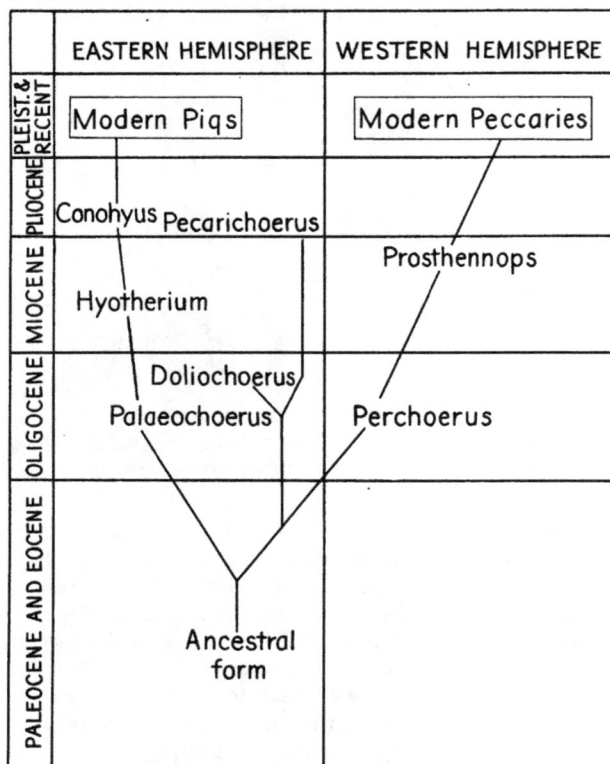

FIG. 99. Diagram to show in an abbreviated form the geologic and geographic relations of the pigs and peccaries. From Colbert, 1933.

On the other hand, it seems quite logical to assume that, in view of the rather aberrant form of *Doliochoerus*, it arose as a separate Old World branch from the primitive peccary ancestor, before the latter had migrated to the New World. *Pecarichoerus* would then represent a descendant of *Doliochoerus* and therefore would be of Eurasiatic origin.

SUIDAE

Palaeochoerus Pomel, 1847

Generic type, *Palaeochoerus typus* Pomel

Palaeochoerus perimensis (Lydekker)

Hyotherium perimense, Lydekker, 1887, Quar. Jour. Geol. Soc., London, XLIII, p. 19.
Palaeochoerus perimensis, Stehlin, 1899, Abhand. Schweiz. Pal. Gesell., Zurich, XXVI, p. 54.

Additional References.—

Pilgrim, G. E., 1910B, p. 202; 1926A, p. 47.

Type.—Brit. Mus. No. M 3051, a maxilla.

Paratypes.—None.

Horizon.—From the Middle Siwalik beds of Perim Island. Equivalent to the Nagri or Dhok Pathan beds of the Punjab.

Locality.—From Perim Island, Gulf of Cambay, India.

Specimens in the American Museum.—Amer. Mus. No. 19594. A right M³ and a left M³, not associated. From the Lower Siwalik beds, about 100 feet above the level of Chinji Rest House, one mile west of Chinji Rest House, Salt Range, Attock District, Punjab.

Amer. Mus. No. 19431. A left M₃. From the Lower Siwaliks, Chinji zone, about 100 feet above the level of Chinji Rest House, two miles west of Chinji Rest House, Salt Range, Attock District, Punjab.

Diagnosis.—Small and primitive; molars quadricuspid; last upper molar with small talon.

Three teeth, two upper third molars and a lower third molar, from the lower Siwalik beds near Chinji Rest House are referred to the genus *Palaeochoerus*. The upper teeth are placed in this genus because of the simplicity of their structure, and the extreme shortness of the talon. The lower tooth would seem to be referable to this genus by virtue of its small and simple form, and the narrowness of its transverse diameter. The teeth are referred to *Palaeochoerus perimensis*, although this species is typically of Middle Siwalik age. It seems better, however, to regard *Palaeochoerus perimensis* as a primitive persistent species, that might very well have lasted through several geologic horizons, rather than to create a new species on very fragmentary evidence, such as that afforded by these few teeth. The teeth in question compare quite closely to the type of *Palaeochoerus perimensis*.

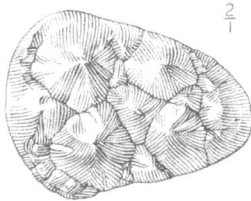

A.M. 19594

Fig. 100.

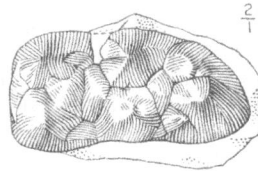

A.M. 19431

Fig. 101.

Fig. 100. *Palaeochoerus perimensis* (Lydekker). Amer. Mus. No. 19594, right upper third molar. Crown view. Twice natural size.

Fig. 101. *Palaeochoerus perimensis* (Lydekker). Amer. Mus. No. 19431, left lower third molar. Crown view. Twice natural size.

MEASUREMENTS

	Length	Width	Index
Amer. Mus. No. 19594 R. M³.................	19.5 mm.	14.5 mm.	74
Amer. Mus. No. 19594 L. M³.................	20.0	15.5	77
Amer. Mus. No. 19431 L. M₃.................	20.0	10.5	52

Palaeochoerus lahirii Pilgrim

Palaeochoerus lahirii, Pilgrim, 1926, Pal. Indica, N.S., VIII, No. 4, p. 47, Pl. XV, figs. 5, 6.

Type.—(Lectotype.)   G.S.I. No. B 718, a right M^3.

Cotype.—G.S.I. No. B 719, a left M_3.

Horizon.—From the Kamlial zone, base of the Lower Siwaliks.

Locality.—Sind.

Diagnosis.—Doubtfully of this genus.   The type specimens may be representative of the genus *Conohyus.*   Like *Palaeochoerus perimensis*, but with a cusp in the median valley of M^3, and with a slightly larger talon.

The few teeth presumably representative of the genus *Palaeochoerus* in the Siwaliks, indicate a curious persistence of this primitive suid, which is typically of Oligocene age, into the upper Miocene and the lower Pliocene.   This continuation of a primitive form into a relatively late geological period is in keeping with the trend of the entire Lower and Middle Siwalik faunas.   *Dissopsalis*, among the creodonts, is an example of the persistence of a genus of essentially Eocene affinities into the Pliocene.   Other examples are numerous. Thus *Palaeochoerus* adds evidence to that already accumulated, showing that the several Siwalik faunas are in reality relict associations, continuing on in geologic time.

Conohyus Pilgrim, 1926

Generic type, *Hyotherium simorrense* Lartet

Conohyus sindiense (Lydekker)

Hyotherium sindiense, Lydekker, 1884, Pal. Indica (X), III, p. 95, Pl. XII, fig. 6.

Conohyus sindiense, Pilgrim, 1926, Pal. Indica, N.S., VIII, pp. 12–13, Pl. II, figs. 1–6.

Additional References.—

      Lydekker, R., 1885B, pp. 43–44; 1887, pp. 19–23.

      Stehlin, H., 1899, p. 12, and various other pages in the text.

      Pilgrim, G. E., 1910B, p. 202; 1913B, pp. 284, 313; 1914B, p. 266.

      Matthew, W. D., 1929, pp. 453, 459.

Type.—G.S.I. No. B 102, fragment of a maxilla, with left M^{2-3}.

Paratypes.—G.S.I. No. B 96b, a right M^2; G.S.I. No. B 98, a fourth premolar, questionably referred.

Horizon.—Lower Siwaliks.   Also lower Middle Siwaliks.

Locality.—The type came from the Laki Hills, Sind.   Referred specimens are from the Punjab.

Specimens in the American Museum.—Amer. Mus. No. 19616.   A broken skull, with the cheek teeth well preserved, and the canine and incisor alveoli present. From the lower Siwaliks, about 1600 feet above the level of Chinji Rest House, and one and one half miles northwest of that place.

19739.   A right mandibular ramus, with M_{2-3} and roots of anterior teeth.   Lower Middle Siwaliks, about one thousand feet below the level of the bone bed at Bhandar, four and one half miles west of Hasnot.

19386.   Fragment of mandibular ramus, with right P_3–M_3.   Lower Siwaliks, near Chinji.

19387. Fragment of mandibular ramus, with right P_{3-4}. Lower Siwaliks, 100 feet above the level of Chinji Rest House, two miles west of Chinji Rest House.

19388. Left M_3. Lower Siwaliks, about level of Chinji Rest House, ten miles east of Chinji Rest House.

19389. Miscellaneous lower teeth. Lower Siwaliks, 100 feet above the level of Chinji Rest House, two miles west of Chinji Rest House.

19390. Jaw fragment, with right M_{1-3}. 100 feet above the level of Chinji Rest House, two miles west of Chinji Rest House.

19511. Fragment of mandibular ramus with left P_3–M_3. Lower portion of the Middle Siwaliks, south of Phadial.

19557. Left P_4 and left M_3. Lower Siwaliks, at level of Chinji Rest House, three miles west of Chinji Rest House.

19558. Left M^2 and an astragalus. Lower Siwaliks, about 200 feet above the level of Chinji Rest House, four miles west of Chinji Rest House.

19591. Jaw fragment with left M_{1-2}. About 600 feet above the level of Chinji Rest House, one mile north of Chinji Rest House.

19901. Jaw fragment with left M_3. Lower portion of Middle Siwaliks, about 1000 feet below the level of the bone bed at Bhandar, four and one half miles west of Hasnot.

19643 (doubtfully referred). Palatal fragments with left M^{1-3} and right M^3. About level of Chinji Rest House, five miles east of Chinji Rest House.

19393. Fragmentary mandibular ramus. Lower Siwaliks, 200 feet above the level of Chinji Rest House, one half mile north of Chinji Rest House.

19428, M^3. Lower Siwaliks, 1600 feet above the level of Chinji Rest House, thirteen miles east of Chinji Rest House.

Diagnosis.—*Conohyus sindiense* is an upper Tertiary suid of rather primitive form. It is closely related to *Hyotherium*, and may be considered as a rather direct derivative from *Palaeochoerus*.

The skull is distinguished by its essentially primitive form. It is moderately long and narrow, and the orbit is placed in a comparatively central position—that is, the pre-orbital portion of the skull is not elongated to the degree found in the more advanced Suidae. The brain case is much restricted, and consequently the sagittal and the lambdoi-dal crests are quite prominent, as in *Palaeochoerus*. The zygomatic arches are heavy. The basicranium shows the typical suilline characters, namely the glenoids are widely expanded, the bulla is elongated downward and is narrow, as in the modern suids, while the basicranial foramina are arranged much as in the recent genera. The pterygoids are heavy and the paroccipitals are presumably long. The mandible has a short but heavy symphysis.

The teeth are but slightly advanced over the condition found in *Palaeochoerus*. The molars are essentially quadricuspid and the third molars have but slight talons or talonids. The premolars are characterized by the enlargement of the third and fourth members of the series. The canines were presumably rather large, and turned outward above; they were smaller in the mandible. The incisors were well developed.

A detailed description of the complete skull and mandible of *Conohyus sindiense* has been published elsewhere (Colbert, 1933D) and the reader is referred to this publication. For descriptions of the teeth of this form, the reader is referred to Pilgrim's memoir on the Siwalik Suidae (1926). The figures and measurements of the American Museum specimen are reproduced below.

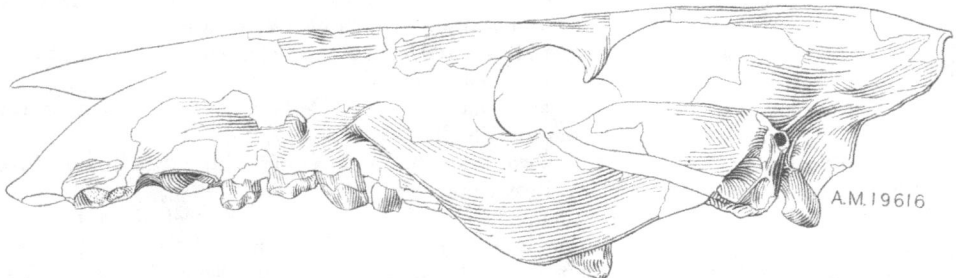

FIG. 102. *Conohyus sindiense* (Lydekker). Amer. Mus. No. 19616. Side view of skull, one half natural size. The specimen has been crushed, so that its true height would be somewhat greater than shown in the figure. (See Fig. 106 for restoration.) Restored portions unshaded. From Colbert, 1933.

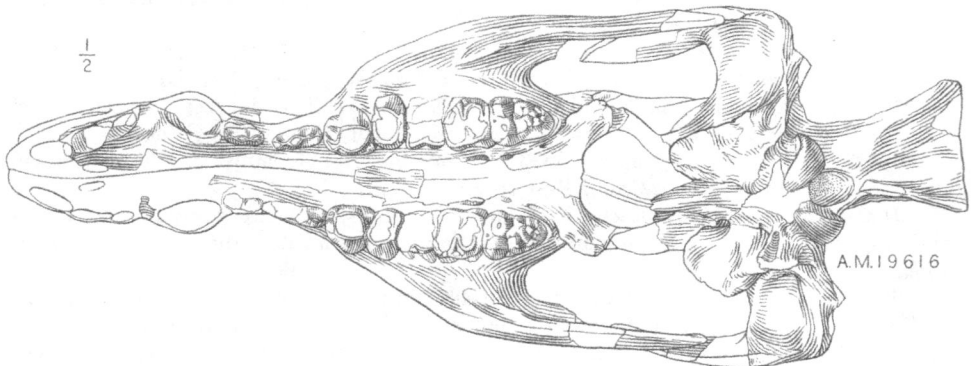

FIG. 103. *Conohyus sindiense* (Lydekker). Amer. Mus. No. 19616, palatal view of skull, one half natural size. From Colbert, 1933.

MEASUREMENTS OF *Conohyus sindiense*

A.M. 19616, skull. A.M. 19739, mandible.
Skull.
Length, occ. cond. to tip of premaxilla (approximate)... 277 mm.
Width, at glenoids................................. 123
Preorbital length.................................. 164
Postorbital length (to condyles)...................... 123
Ratio-Preorbital: Postorbital : : 133 : 100
Vertical diameter of orbit.......................... 37
Width across occ. condyles.......................... 40
Over all length (back of occiput-premaxilla).......... 317
Mandible.
Length, condyle to anterior border of first premolar.... 182 mm.
Depth of ramus below first molar.................... 45
Height of condyle above occlusal line................ 54

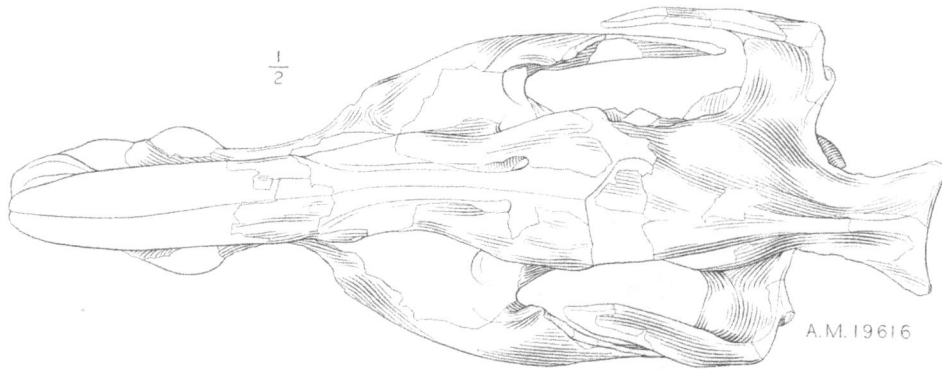

FIG. 104. *Conohyus sindiense* (Lydekker). Amer. Mus. No. 19616, top view of skull, one half natural size. From Colbert, 1933.

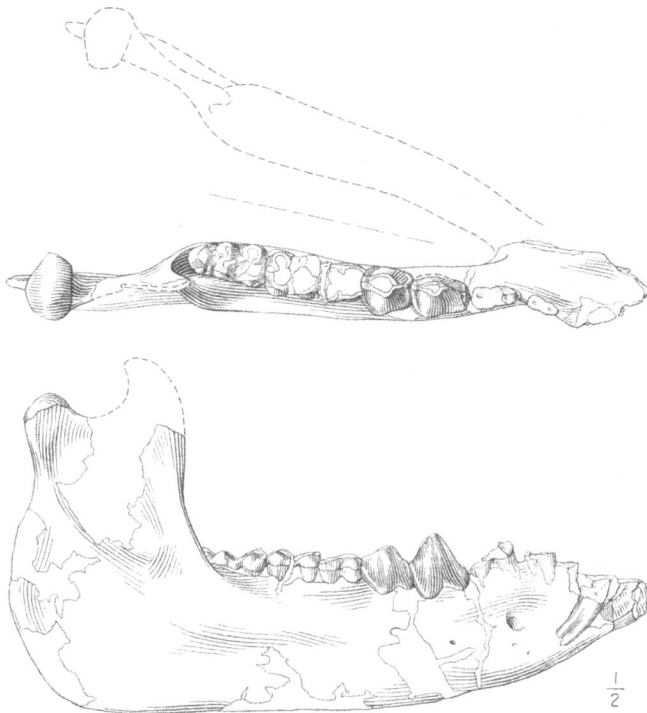

FIG. 105. *Conohyus sindiense* (Lydekker). Amer. Mus. No. 19739, mandible, with the third and fourth premolars and the first molar drawn in from Amer. Mus. No. 19386. Crown view above, and side view below. One half natural size. From Colbert, 1933.

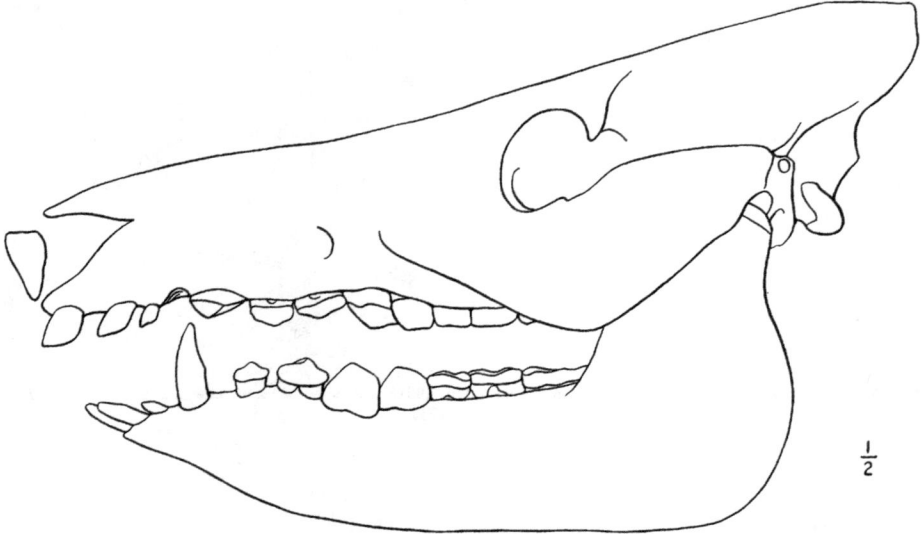

FIG. 106. *Conohyus sindiense* (Lydekker). Restoration of skull and mandible, based on Amer. Mus. Nos. 19616, 19739, 19386, and figures by Filhol, Pearson and others. The shape of the os rostri is hypothetical. One half natural size. From Colbert, 1933.

A.M. 19616, Upper Dentition.

	Antero-posterior	Transverse
I^2 estimated from alveolus	10.0 mm.	5.0 mm.
I^3 estimated from alveolus	7.0	4.0
C estimated from alveolus	17.0	13.0
P^1	13.0	6.0
P^2	17.0	7.0
P^3	16.0	16.3
P^4	11.5	17.5
M^1	12.8	16.5
M^2	18.0	18.0
M^3	20.0	17.0

Premolar length . . . . 60.0 mm.
Molar length 52.0
Ratio . . . . 115 : 110

A.M. 19739, Lower Dentition.

	Antero-posterior	Transverse
C	15.0 mm.	7.5 mm.
P$_1$	11.0	4.0
P$_2$	17.0	7.5
P$_3$ (From A.M. 19386)	18.0	14.0
P$_4$ (From A.M. 19386)	16.0	15.0
M$_1$ (From A.M. 19386)	14.0	11.7
M$_2$	18.0	15.0
M$_3$	26.0	14.7

Length of premolar series 65.0 mm.
Length of molar series . . 58.0
Ratio . . . . ∴ . 112 : 100

Conohyus AND *Hyotherium*

Pilgrim in 1926 created a new genus *Conohyus*, to include certain pigs close to *Hyotherium*, but distinguished from this latter genus by the enlargement of the premolars and by the shortness of the last molars. Although Pilgrim's general system of a phylogenetic classification of the Suidae on the basis of the last premolars may be questioned, it would seem that his separation of the two above mentioned genera is essentially sound. A study of Siwalik and European suids formerly classified in the genus *Hyotherium*, shows that there are naturally two series. In one group, the true *Hyotherium*, the premolars are of normal size in comparison to the molars, and the third molar is usually elongated. Moreover, in the last upper premolar the outer and inner moieties of the tooth are distinct, while in the corresponding lower tooth the principal cone terminates in two separate cusps placed obliquely, the anterior one being buccal. In the other group, making up the genus *Conohyus*, the premolar teeth are enlarged in comparison to the molars, and the third molar is short. In addition, the outer and inner portions of the last upper premolar are more or less confluent, while in the last lower premolar the principal cone is simple.

Analyzing these differences it would seem that *Hyotherium* and *Conohyus* represent two branches from the original primitive suid stem (as typified by *Palaeochoerus*), the former being ancestral to a great mass of later Tertiary and Quaternary pigs, and the latter being ancestral to the gigantic and very peculiar genus, *Tetraconodon*.

Thus we do not visualize the suids as being polyphyletic, with the dichotomy of the great groups extending far back into the Oligocene, as was advocated by Pilgrim (1926, Pl. I) but rather that from the primitive suid stem an aberrant line developed, its first appearance being in the genus *Conohyus*. This line, characterized by the enlarged premolars, spread over Europe and Asia and soon became extinct, reaching the culmination of its development in the Indian genus *Tetraconodon*, a form of great size, having enormously developed premolars.

TABLES OF MEASUREMENTS AND INDICES

Analysis of Tables

The tables of indices for the grinding teeth of various Tertiary and recent Suidae show certain phylogenetic trends in the evolution of this group. In the first place, the close relations existing between *Palaeochoerus* and the typical *Hyotherium* are at once apparent, proving that the latter genus is most certainly a descendant of the former. We see that in the central forms, springing directly from the primitive stock, the premolars are relatively narrow and the third molars relatively long.

An analysis of *Conohyus* shows that it is trending *away* from the typical *Hyotherium* group, with an undoubted orthogenetic trend towards the specialized genus *Tetraconodon*. In fact, *Tetraconodon* is essentially a large edition of *Conohyus*, the proportional indices remaining remarkably similar in these two genera.

Applying the method of proportional indices to the genera *Potamochoerus* and *Sus*, it may be seen at once that they are close to each other. *Potamochoerus* can not be correlated with the *Conohyus–Tetraconodon* line by any particularly significant criteria, but would seem on the other hand (along with *Sus*) to be close to the true *Hyotherium* group. Consequently, Pilgrim's contention that *Potamochoerus* is an early development related to the *Conohyus* line of evolution, while *Sus* is a separate derivation connected with *Palaeochoerus*,

Upper dentition

Measurements in millimeters	P^3 Length	P^3 Width	P^3 Index	P^4 Length	P^4 Width	P^4 Index	M^1 Length	M^1 Width	M^1 Index	M^2 Length	M^2 Width	M^2 Index	M^3 Length	M^3 Width	M^3 Index	Length of $P^3 + P^4$	Length of molar series	Index $\frac{P^3+P^4}{\text{molar series}}$	Index $\frac{W\ P^3}{W\ M^1}$	Index $\frac{W\ P^4}{W\ M^1}$	Index $\frac{L\ M^3}{\text{molar series}}$
Palaeochoerus meissneri, A.M. 10251	14.2	9.0	63	10.2	12.7	124	12.8	13.0	101	14.8	15.2	103	16.5	14.8	90	24.4	44.1	55	64	98	37
Hyotherium sömmeringi, A.M. 15581							16.5	16.3	99	17.7	15.4	87	20.0	16.0	80		54.2				37
Conohyus chinjiensis (from fig.), Ind. Mus. B. 536	15.5	15.0	97	10.5	16.0	152	12.5	14.5	116	16.5	15.5	94	21.0	15.5	74	26.0	51.0	51	103	110	41
Conohyus sindiense, A.M. 19616	16.0	16.3	102	11.5	17.5	152	12.8	16.5	129	18.0	18.0	100	20.0	17.0	85	28.0	52.5	53	99	106	38
Tetraconodon minor (from fig.), Ind. Mus. B. 676				19.5	30.5	156	20.0	22.0	110	28.0	28.0	100	40.5	32.0	79		88.0			138	46
Tetraconodon magnus (from fig.), Ind. Mus. B. 675	47.0	40.0	85	34.0	49.0	145	26.0	31.0	119	30.0	33.0	110	40.0	32.0	80	77.5	95.5	81	129	158	42

Lower dentition

Measurements in millimeters	P_3 Length	P_3 Width	P_3 Index	P_4 Length	P_4 Width	P_4 Index	M_1 Length	M_1 Width	M_1 Index	M_2 Length	M_2 Width	M_2 Index	M_3 Length	M_3 Width	M_3 Index	Length of $P_3 + P_4$	Length of molar series	Index $\frac{P_3+P_4}{\text{molar series}}$	Index $\frac{W\ P_3}{W\ M_1}$	Index $\frac{W\ P_4}{W\ M_1}$	Index $\frac{L\ M_3}{\text{molar series}}$
Palaeochoerus meissneri, A.M. 10251	11.5	6.3	55	11.2	8.0	71	13.7	11.0	80	15.0	12.0	80	21.0	12.0	57	22.7	35.0	65	57	73	60
Hyotherium sömmeringi, A.M. 15581	15.0	7.0	47	15.0	8.5	57	15.5	11.7	75	17.2	13.0	75	23.5	13.3	58	30.0	38.0	79	60	73	62
Conohyus sindiense, A.M. 19386	18.0	14.0*	78	16.0	15.0	94	14.0	12.5	89	16.5	13.5	82	22.5	13.5	60	33.5	53.5	63	112	120	42
Conohyus sindiense, A.M. 19511	20.0	12.5	63	17.0	13.0	76	14.0	11.7	84	16.0	13.0	81	21.0	13.0	62	36.0	51.0	70	107	111	41

*Estimated.

Measurements in millimeters

	P^3 Length	P^3 Width	P^3 Index	P^4 Length	P^4 Width	P^4 Index	M^1 Length	M^1 Width	M^1 Index	M^2 Length	M^2 Width	M^2 Index	M^3 Length	M^3 Width	M^3 Index	Length of $P^3 + P^4$	Length of molar series	Index $\dfrac{\text{Length } P^3 + P^4}{\text{Length molar series}}$	Index $\dfrac{\text{Width of } P^3}{\text{Width of } M^1}$	Index $\dfrac{\text{Width of } P^4}{\text{Width of } M^1}$	Index $\dfrac{\text{Length of } M^3}{\text{Length molar series}}$
Conohyus sindiense, A.M. 19387	21.2	14.2	67	16.0	14.5	91										37.2					
Conohyus sindiense, A.M. 9929				17.0	14.0	82	12.0	11.0	92	16.0	13.7	85	24.5	13.7	56		54.0			127	45
Tetraconodon minor (from fig.), Ind. Mus. B 677	39.0	26.0	67	33.0	30.0	91										74.0					
Tetraconodon magnus, A.M. 9937	58.0	49.0	85	49.5	56.5	114	30.0	29.0	97	36.5	32.0	88	48.5	32.0	66	107.5	114.5	94	169	195	42

Measurements in millimeters

	P^3 Length	P^3 Width	P^3 Index	P^4 Length	P^4 Width	P^4 Index	M^1 Length	M^1 Width	M^1 Index	M^2 Length	M^2 Width	M^2 Index	M^3 Length	M^3 Width	M^3 Index	Length of $P^3 + P^4$	Length of molar series	Index $\dfrac{\text{Length } P^3 + P^4}{\text{Length molar series}}$	Index $\dfrac{\text{Width of } P^3}{\text{Width of } M^1}$	Index $\dfrac{\text{Width of } P^4}{\text{Width of } M^1}$	Index $\dfrac{\text{Length of } M^3}{\text{Length molar series}}$
Potamochoerus	14.0	10.5	75	13.0	14.5	111	19.0	16.5	87	23.0	21.5	93	28.0	21.0	75	27.0	72.5	37	64	88	39
Sus	13.5	11.0	81	12.5	15.5	124	17.5	16.0	91	24.5	20.0	82	35.0	22.0	63	27.0	76.5	35	69	97	46

Measurements in millimeters

	P_3 Length	P_3 Width	P_3 Index	P_4 Length	P_4 Width	P_4 Index	M_1 Length	M_1 Width	M_1 Index	M_2 Length	M_2 Width	M_2 Index	M_3 Length	M_3 Width	M_3 Index	Length of $P_3 + P_4$	Length of molar series	Index $\dfrac{\text{Length } P_3 + P_4}{\text{Length molar series}}$	Index $\dfrac{\text{Width of } P_3}{\text{Width of } M_1}$	Index $\dfrac{\text{Width of } P_4}{\text{Width of } M_1}$	Index $\dfrac{\text{Length of } M_3}{\text{Length molar series}}$
Potamochoerus	15.5	8.5	55	16.5	11.5	70	17.0	13.5	79	23.0	17.5	76	32.0	17.0	53	32.0	72.5	44	63	85	44
Sus	13.5	7.0	52	15.5	10.5	68	17.0	14.0	82	24.0	17.0	71	40.0	18.5	46	29.0	79.0	37	50	73	51

is not particularly upheld by a critical study of dental indices.   Rather it would seem that *Potamochoerus* is a relatively late offshoot from the *Sus* line, and *Conohyus* must be considered as representative of a side branch in suid phylogeny.   Therefore it would seem logical that the pigs are essentially monophyletic, and that the stem form is an animal directly ancestral to *Palaeochoerus*.

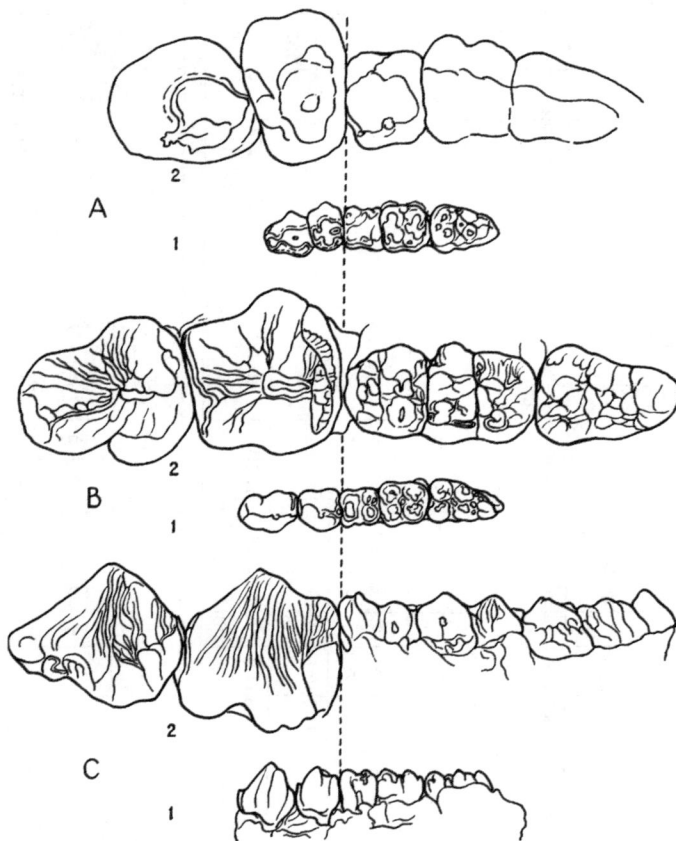

Fig. 107.   A comparison of the cheek teeth in *Conohyus* and *Tetraconodon*.
A 1. Conohyus chinjiensis Pilgrim.   Left P²–M³, crown view.   After Pilgrim.
A 2. Tetraconodon magnus Falconer.   Left P²–M³, crown view.   After Pilgrim.
B 1. Conohyus chinjiensis Pilgrim.   Left P₂–M₃, crown view.   After Pilgrim.
B 2. Tetraconodon magnus Falconer.   Right P₂–M₃, crown view.   After Lydekker.
C 1. Conohyus chinjiensis Pilgrim.   Left P₂–M₃, side view.   After Pilgrim.
C 2. Tetraconodon magnus Falconer.   Right P₂–M₃, side view.   After Lydekker.
All figures one half natural size.

Conohyus chinjiensis Pilgrim

Conohyus chinjiensis, Pilgrim, 1926, Pal. Indica, N.S., VIII, No. 4, pp. 13–14, Pl. II, figs. 10, 11.

Additional Reference.—
Matthew, W. D., 1929, p. 459.

Type.—(Lectotype.) G.S.I. No. B 536, maxilla with left P^2–M^3.
Cotype.—G.S.I. No. B 537, mandible with right P_2–M_2.
Horizon.—Lower Siwaliks, Chinji zone.
Locality.—Chinji, Salt Range, Attock District, Punjab.
Diagnosis.—Like *Conohyus sindiense*, but with a relatively narrower M^3 having a longer talon; the lower premolars are not quite so large with respect to the molars, and P^3 is not quite so wide at the base.

The characters which Pilgrim gives as distinctive between *Conohyus sindiense* and *Conohyus chinjiensis* are of doubtful value. The talon of the third molar is quite variable in any one species of suid. Therefore, in an individual with a relatively short talon, the breadth of the tooth will of necessity be relatively great. The other differences cited may be attributed to individual or subspecific variations as readily as to specific differentiations.

Conohyus chinjiensis is thus regarded here as a synonym of *Conohyus sindiense*.

Conohyus indicus (Lydekker)

Hyaenodon indicum, Lydekker, 1884, Pal. Indica (X), II, p. 349, fig. 29.
Hyotherium indicum, Pilgrim, 1910, Rec. Geol. Surv. India, XL, p. 190.
Conohyus indicus, Pilgrim, 1926, Pal. Indica, N.S., VIII, No. 4, p. 14, Pl. II, figs. 7–9.

Additional References.—
Lydekker, R., 1885B, p. 12.
Pilgrim, G. E., 1913B, p. 319; 1914B, p. 266.
Matthew, W. D., 1929, p. 459.

Type.—G.S.I. No. D 57, a lower premolar.
Paratypes.—None.
Horizon.—Lower to Middle Siwaliks.
Locality.—Kushalgar, below Attock, Punjab.
Diagnosis.—Larger than *Conohyus sindiense*.

This species, for lack of better evidence, may be considered as separate from *Conohyus sindiense*. It is seemingly a larger form, in which the premolars are relatively greater. Consequently it would seem to be a trifle nearer to *Tetraconodon*. As Pilgrim has pointed out, this species may belong to the genus *Sivachoerus*, but more adequate material is needed to settle this point.

Sivachoerus Pilgrim, 1926

Generic type, *Sivachoerus prior* Pilgrim

Sivachoerus prior Pilgrim

Sivachoerus prior, Pilgrim, 1926, Pal. Indica, N.S., VIII, No. 4, p. 20, Pl. III, fig. 3, Pl. XX, fig. 5.

Additional Reference.—
Matthew, W. D., 1929, p. 459.

Type.—G.S.I. No. B 1, a mandible. (Figured by Lydekker, 1884, Pal. Indica (X), III, p. 55, Pl. XI, fig. 1, as *Sus* cf. *giganteus*.)

Paratypes.—G.S.I. No. B 678, right M^3; G.S.I. No. B 552, a mandible; G.S.I. No. B 679, a mandible with P_4–M_3.

Horizon.—Middle Siwaliks, Dhok Pathan zone.

Locality.—From Hasnot, Salt Range, Punjab. Also from Pakokku District, Burma.

Diagnosis.—A large species of suid, with hypsodont teeth, wrinkled enamel, and enlarged posterior premolars. The orbit is relatively small. These are characters relating *Sivachoerus* to *Conohyus* and *Tetraconodon*. This species is distinguished from *S. giganteus* by its relatively short, broad third molar.

Pilgrim set *Sivachoerus* apart as a separate genus, by virtue of its enlarged premolars, hypsodont teeth and wrinkled enamel, characters that would seem to point to a relationship with *Conohyus*. That *Sivachoerus* is not directly in line between *Conohyus* and *Tetraconodon* is shown by the fact that its premolar teeth are relatively smaller than in either of these other two genera.

Sivachoerus giganteus (Falconer and Cautley)

Sus giganteus, Falconer and Cautley, 1847, Fauna Antiqua Sivalensis, Pl. LXIX, Pl. LXX, figs. 4–8, Pl. LXXI, figs. 12–19.

Sivachoerus giganteus, Pilgrim, 1926, Pal. Indica, N.S., VIII, No. 4, pp. 21–22, Pl. III, fig. 2, Pl. IV, fig. 1, Pl. XX, fig. 4.

Additional References.—

Falconer, H., 1868A, pp. 508–512.

Lydekker, R., 1880B, p. 31; 1883C, pp. 83, 93; 1884C, p. 52, Pl. XI; 1885D, pp. 268–270.

Pilgrim, G. E., 1910B, p. 202.

Matthew, W. D., 1929, p. 444.

Type.—(Lectotype of Lydekker.) Brit. Mus. No. 15385, a well preserved skull.

Cotypes.—Brit. Mus. Nos. 16166, anterior portion of a skull; 16592, a right maxilla; 16590, a left maxilla; 16592a, a canine tooth.

Horizon.—Upper Siwaliks.

Locality.—Siwalik Hills.

Diagnosis.—Like *S. prior* but with an elongated third molar.

Tetraconodon Falconer, 1868

Generic type, *Tetraconodon magnus* Falconer

Tetraconodon magnus Falconer

Tetraconodon magnum, Falconer, 1868, Pal. Memoirs, I, pp. 149–156, fig. 5.

Additional References.—

Lydekker, R., 1876, p. 79, Pl. X; 1880B, p. 31; 1883C, pp. 83, 93; 1884C, p. 99; 1885B, p. 43.

Stehlin, H., 1899, p. 52.

Pilgrim, G. E., 1910B, p. 202; 1913B, p. 300.

Matthew, W. D., 1929, pp. 449, 459.

Type.—A fragment of the maxilla with right M$^{2-3}$. This specimen, which was in the Baker and Durand collection, was lost subsequent to its description by Falconer.

Neotype.—G.S.I. No. B 71, a mandible with a greater portion of the dentition on either side.

Horizon.—The exact horizon is a matter of conjecture. It is either Middle or Upper Siwaliks, probably Upper Siwaliks. The type was found "in the Tertiary hills between the Murkunda pass and Pinjore."

Locality.—Siwalik Hills. The specimen described by Lydekker came from near Hasnot in the Salt Range.

Diagnosis.—A gigantic suid distinguished by the short, broad molars and the greatly enlarged and crenulated posterior premolars.

The status of the genus *Tetraconodon* and of its type species, *T. magnus*, presents a perplexing problem in nomenclature. The original description was made by Falconer on the basis of two upper molar teeth, and the designation *Tetraconodon magnum* was applied by this author. Subsequent to Falconer's description the type was lost, nor has it been found up to the present time. Many years after the original description appeared, the genus and species were further described by Lydekker on the basis of a lower jaw. In this description, Lydekker pointed out that the termination of the specific name as given by Falconer was incorrect, and he changed it to agree with the generic designation. Consequently the name became *Tetraconodon magnus*.

More recently Pilgrim made a study of this species, and on the grounds of his knowledge of Siwalik stratigraphy and fossil localities, he decided that the original specimen described by Falconer, since it came from "between Murkunda Pass and Pinjore," was probably of Upper Siwalik age, while the specimen described by Lydekker, having been found near Hasnot, was presumably from a Middle Siwalik horizon. Reasoning thus, he assumed that the two specimens represented two different species, one of upper Siwalik age and one of Middle Siwalik age, and to the latter one he assigned the name *Tetraconodon mirabilis*, making it a new genotype.

Of course Pilgrim's designation of Lydekker's specimen as a *new genotype* is quite wrong. The genotype of the genus *Tetraconodon* must be the original species described by Falconer. On the other hand, Pilgrim's assumption that the second specimen is a species different from that of the original one is more than likely to be correct. For that matter, the genotype specimen (judging from Falconer's figure) could be a *Potamochoerus*, such as *P. titan*, quite as easily as it could be a species of *Tetraconodon*.

Since the genotype of *Tetraconodon* is lost and the chances are very strong against its ever being found again, it seems rather futile to draw differences between it and the later specimen on the basis of probable but unproven stratigraphic occurrences. Moreover, it is quite possible that a single species persisted through the Middle Siwaliks into Upper Siwalik beds.

Therefore, for the dual purpose of simplicity of thought and for the retention of a well established name, the second specimen of *Tetraconodon*, namely G.S.I. No. B 71, is here regarded as belonging to the same species as the specimen described by Falconer. This second specimen thus assumes the status of a neotype, and as such it gives extant material on which to base the genus. Pilgrim's name, *T. mirabilis*, applied to this specimen, is considered as synonymous with Falconer's name, *T. magnus*.

While the above procedure may not be any nearer the truth than the procedure followed by Pilgrim in regard to this question, it has the advantage of establishing definite material representative of the species *Tetraconodon magnus*. And since, by the provisions of the International Rules of Zoological Nomenclature, *T. magnus* must be considered as the generic type, this procedure establishes definite material on which to base the genus. Otherwise, we would be forced into the predicament of dealing with a genus having no type material.

Tetraconodon mirabilis Pilgrim

Tetraconodon mirabilis, Pilgrim, 1926, Pal. Indica, VIII, No. 4, pp. 15–17, Pl. III, fig. 4, Pls. V, VI.

Type.—G.S.I. No. B 71, greater portion of the lower dentition on either side.

Paratypes.—G.S.I. No. B 674, a single third upper premolar; G.S.I. No. B 675, a fragmentary skull.

Horizon.—Conjectural, probably Middle Siwaliks. No. B 674 came from the Middle Siwaliks at Hari Talyangar.

Locality.—From near Hasnot, Salt Range.

Diagnosis.—See the diagnosis for *Tetraconodon magnus*.

Considered as synonymous with *Tetraconodon magnus*. See the discussion under that species.

Tetraconodon minor Pilgrim

Tetraconodon minor, Pilgrim, 1910, Rec. Geol. Surv. India, XL, p. 67.

Additional References.—

Pilgrim, G. E., 1926, p. 17.

Matthew, W. D., 1929, p. 459.

Type.—G.S.I. No. B 677, portion of a mandible, containing last two left premolars.

Cotypes.—G.S.I. No. B 676, maxilla with right P^4–M^3; G.S.I. No. B 534, mandibular ramus with right M_{2-3}.

Horizon.—Near the base of the Irrawaddy series, correlative with the lower portion of the Middle Siwaliks.

Locality.—Below Yenangyoung, Burma.

Specimens in the American Museum.—Amer. Mus. No. 19750. Fragment of a maxilla with broken right P^4–M^1. From the Middle Siwaliks, 200 feet below the Bhandar Bone Bed at a locality one mile east of Hasnot, Punjab.

This specimen, which is very fragmentary, is merely referred to the species under consideration.

Diagnosis.—Smaller than *T. magnus*, but much larger than *Conohyus*. The premolars are relatively much smaller than in *T. magnus*, but much larger than in *Conohyus*. Teeth hypsodont, palate wide. Skull similar to skull of *T. magnus*.

This species is rather intermediate between *Conohyus* and *T. magnus*.

Matthew (1929, p. 459) has expressed some doubt as to whether the upper and lower teeth figured by Pilgrim could belong to the same species. At first sight they would seem to show a considerable disparity, especially in the sizes of the fourth premolars. On the other hand, the ratio between the lengths of P^4 and P_4 of the two specimens attributed by Pilgrim to *T. minor* is almost the same as between the corresponding teeth of *T. magnus*, while the ratios between the widths in the two species shows that the P_4 of *T. magnus* is even relatively wider as compared to P^4 than is the case in *T. minor*. Thus it would seem logical to regard the upper and lower dentitions referred to *T. minor* as representative of that one species. The relatively smaller P^4 in *T. minor*, as well as the relatively narrower lower premolars of this species, are indicative of the somewhat more primitive character of the Burmese form.

Listriodon von Meyer, 1846

Generic type, *Listriodon splendens* von Meyer

Listriodon pentapotamiae (Falconer)

Tapirus pentapotamiae, Falconer, 1868, Pal. Memoirs, I, p. 415.
Listriodon pentapotamiae, Lydekker, 1876, Pal. Indica (X) I, p. 70, Pl. VIII, figs. 8, 9.
  Additional References.—
        Lydekker, R., 1880B, p. 31; 1883C, pp. 83, 93; 1884C, p. 101, Pl. VIII, figs. 13, 17; 1885D, p. 276.
        Pilgrim, G. E., 1910B, p. 202; 1926, p. 31, Pl. XI, figs. 11–14, Pl. XII, Pl. XX, fig. 3.
        Matthew, W. D., 1929, pp. 453, 455.
  Type.—G.S.I. No. B 107, a complete right M^2 and a fragment of a right M^3. Also right and left P^4.
  Paratypes.—None.
  Horizon.—Lower Siwaliks and lower portion of the Middle Siwaliks.
  Locality.—Khushalghar below Attock, Punjab, for the type.
  Specimens in the American Museum.—
From the lower portion of the Middle Siwaliks:
  Amer. Mus. No. 19382. Miscellaneous teeth. One mile south of Nathot.
  19456. Jaw fragments. South of Nathot.
  19457. Mandibular fragment with M_{1-2}. One half mile west of Phadial.
  19519. Mandible with P_4–M_3. One half mile west of Phadial.
  19540. Miscellaneous teeth. One mile south of Nathot.
  19541. Mandibular fragment. One mile south of Nathot.
  19542. Left M_3. One mile south of Nathot.
  19543. Miscellaneous teeth. One mile south of Nathot.
From the Lower Siwaliks:
  Amer. Mus. No. 19373. Miscellaneous teeth. One hundred feet above the level of Chinji Rest House, one mile west of Chinji Rest House.
  19376. Fragmentary mandible with teeth. 1600 feet above the level of Chinji Rest House, two miles northwest of Chinji Rest House.
  19392. Miscellaneous teeth. About the level of Chinji Rest House, four miles northeast of Chinji Rest House.
  19394. Palatal fragment with teeth. 100 feet above the level of Chinji Rest House, one half mile north of Chinji Rest House.
  19395. Mandibular symphysis with incisors and right canine. 600 feet above the level of Chinji Rest House, one half mile north of Chinji Rest House.
  19396. Mandibular fragments. 200 feet above the level of Chinji Rest House, one half mile north of Chinji Rest House.

19424. Teeth and jaw fragments. About the level of Chinji Rest House, four miles west of Chinji Rest House.

19425. Miscellaneous teeth. Same locality and level as 19424.

19429. Mandibular fragment with second molar. 1600 feet above the level of Chinji Rest House, 12 miles east of Chinji Rest House.

19432. Mandible with left M_{2-3}. 100 feet above the level of Chinji Rest House, two miles west of Chinji Rest House.

19451. Teeth and jaw fragments. Near Chinji Rest House.

19453. Molar and canine teeth. Near Chinji Rest House.

19559. Miscellaneous teeth. About 100 feet above the level of Chinji Rest House, two miles west of Chinji Rest House.

19565. Maxillary fragment with canine. 500 feet above the level of Chinji Rest House. One and one half miles west of Chinji Rest House.

19586. Incisor and milk molar. About level of Chinji Rest House, five miles east of Chinji Rest House.

19599. Miscellaneous teeth. 400 feet above the level of Chinji Rest House, one mile west of Chinji Rest House.

19600. Palate with right M^{1-2}. 200 feet above the level of Chinji Rest House, one half mile north of Chinji Rest House.

19603. Mandibular fragment with molars. 400 feet above the level of Chinji Rest House, one and one half miles west of Chinji Rest House.

19604. Palate with teeth. Same level and locality as 19603.

19610. Miscellaneous teeth. 1600 feet above the level of Chinji Rest House, one half mile north of Chinji Rest House.

19612. Canine tooth. 100 feet above the level of Chinji Rest House, two miles west of Chinji Rest House.

19615. Left M^3 and premolars. 1600 feet above the level of Chinji Rest House, one and one half miles northeast of Chinji Rest House.

19624. Mandible with left M_{2-3}. About the level of Chinji Rest House, four miles northeast of Chinji Rest House.

19625. Mandible with right M_{2-3}. Same level and locality as 19624.

19627. Fragments of maxilla and mandible. Lower Siwaliks, about level of Chinji Rest House, two miles west of Chinji Rest House.

19629. Mandibular fragment and miscellaneous teeth. About level of Chinji Rest House, two miles west of Chinji Rest House.

19631. Miscellaneous teeth. Lower Siwaliks, five miles west of Chinji Rest House.

19639. Palate with left P_{2-4}. About level of Chinji Rest House, four miles northeast of Chinji Rest House.

19641. Palate with left M^{2-3}, and mandible with left P_4-M_2. About level of Chinji Rest House, five miles east of Chinji Rest House.

19642. Palate with left P^3-M^2. Same level and locality as 19641.

19644. Left M^2. Same level and locality as 19641.

19650. Miscellaneous teeth. About level of Chinji Rest House, ten miles east of Chinji Rest House.

19651. Miscellaneous teeth. About level of Chinji Rest House, ten miles east of Chinji Rest House.

19929. Miscellaneous teeth.   Three miles east of Rammagar.
29813. Canine tooth.   Lower Siwaliks, locality lost.
29836. Palate with right M³.   400 feet above the level of Chinji Rest House, one
        and one half miles east of Chinji Rest House.

Diagnosis.—Like *Listriodon splendens* of Europe, but with a larger talon on the third
molar, a stronger cingulum in the fourth premolar, a shorter and more slender symphysis.

Listriodon pentapotamiae is very close to *Listriodon splendens*, from the Miocene of
southeastern Europe, a fact that was pointed out in detail by Pilgrim in 1926.   Except
for their geographic separation, the two species might almost be classed as one.   Pilgrim
has cited certain features which mark the Indian species as slightly more primitive than
the European form, and these are:

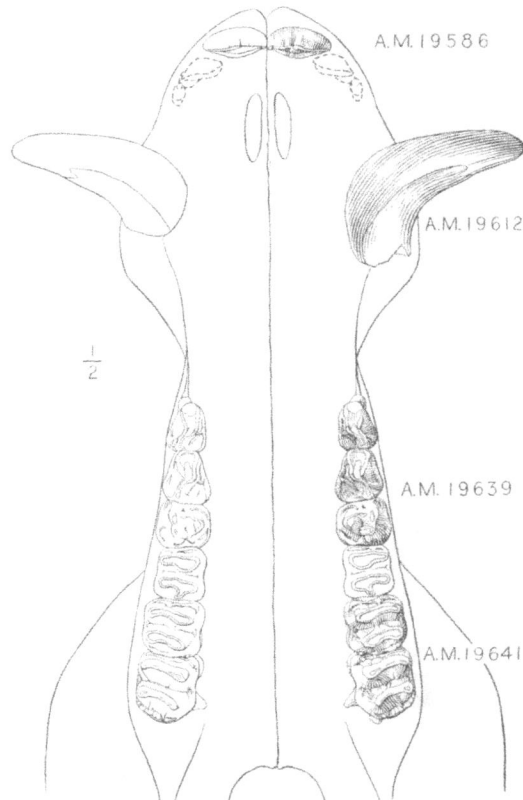

Fig. 108.   *Listriodon pentapotamiae* (Falconer).   Amer. Mus. Nos. 19586, 19612, 19639, 19641, upper incisors
canine, premolars and molars, placed in their approximate positions, relative to each other.   Crown views.   One half
natural size.

1. A more pronounced talon in M^3.
2. A stronger cingulum in P^4.
3. The structure of P_3.
4. The presence of P_1.
5. The shortness and slenderness of the symphysis.

Yet as Pilgrim admits (1926, p. 31) many of these characters are quite variable in *Listriodon*. Therefore it becomes difficult to know just how much value should be attached to the above listed characters in any comparison of the European and the Indian species.

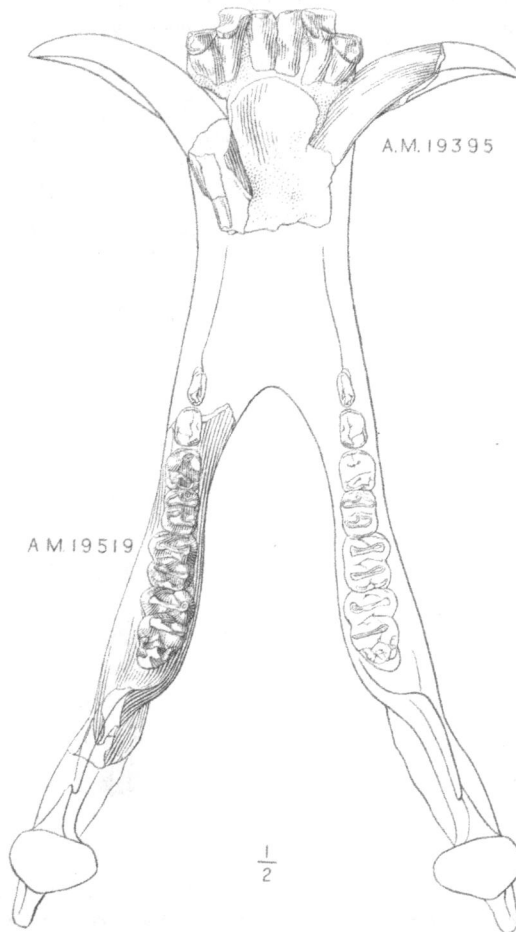

FIG. 109. *Listriodon pentapotamiae* (Falconer). Amer. Mus. Nos. 19395 and 19519, mandibular symphysis and lower cheek teeth, placed in their approximate positions relative to each other. Crown views. One half natural size.

Listriodon pentapotamiae is a fairly long ranging species, extending from the base of the Lower Siwaliks well up into the Middle Siwalik beds. Several specimens in the American Museum collection from the lower portion of the Middle Siwaliks should definitely establish the persistence of this genus beyond its typical Chinji development, a fact that was of some doubt to Matthew (1929, p. 451).

This persistence of *Listriodon* into the Middle Siwaliks, together with the holdover of other typically Lower Siwalik forms, such as *Dissopsalis*, *Giraffokeryx*, etc., demonstrates the really close relationship of these horizons to each other. One sees here a faunal sequence extending through the entire Chinji zone and into the lower portion of the Middle Siwaliks, a sequence quite distinct from the typical upper Middle Siwalik, or Dhok Pathan fauna.

Varities of *Listriodon*

Two varieties of *Listriodon* teeth are found in the American Museum collection. In one, the more common of the two, the enamel is very smooth, while in the other it is rugose. These variations in the enamel seem in no way to be connected with size differences, or with changes in stratigraphic levels, but run indiscriminately through the series at hand. Thus it would seem probable that they represent individual or at the most, geographic variations, in one species, and as such they must not be regarded as especially significant.

The Upper Incisors of *Listriodon*

In the Siwalik *Listriodon* the central upper incisor is very broad, so that it bites with the central and median lower incisors. Its outer surface is divided by a deep vertical sulcus or groove into two halves or moieties—the portions of the tooth fitting over the central and median lower incisors respectively. There are no teeth in the American Museum collection that can be definitely referred to the median or lateral positions in the upper incisor series of this genus under consideration. Presumably the median incisor was much narrower than the central one, and its occlusion was with the second and the third lower incisors. A lateral incisor, if present, must have been very small. The reader is referred to the accompanying figures of *Listriodon* for a portrayal of the probable occlusal relationships of these teeth.

Measurements (in mm.) of *Listriodon pentapotamiae*

Amer. Mus. Number	M¹			M²			M³		
	Length	Width	$\frac{W}{L} \times 100$	Length	Width	$\frac{W}{L} \times 100$	Length	Width	$\frac{W}{L} \times 100$
19642	16.5	15.0	91						
19651				24.0	21.5	90	27.0	21.5	79
19600	16.0	14.0	87	21.0	18.5	88			
19615							24.0	21.0	87
19650				21.0	20.0	95	27.0	21.0	78
19394	16.0	15.0	94						
19604	17.0	16.0	94						
19565	14.5	11.0	76	19.0	15.0	79			
29836							23.0	20.0	87
19644				18.0	17.5	97			

Amer. Mus. Number	M_1			M_2			M_3		
	Length	Width	$\frac{W}{L} \times 100$	Length	Width	$\frac{W}{L} \times 100$	Length	Width	$\frac{W}{L} \times 100$
19519	16.0	12.0	75	19.0	16.0	84	27.5	18.0	65
19457	17.0	13.5	77						
19629				18.0	14.0	78			
19599				18.0	14.0	78			
19624				23±	18.0	78			
19641	16.0	12.5	78	19.0	15.5	82			
19432				21.0	15.0	72	28.0	16.0	58
19625				19.5	14.5	74	28.0	16.0	58
19603							30.0	19.0	63
19424							31.0	19.0	61
19540				17.0	15.0	88			
19456				22.5	17.0	76			
19542							28.5	17.0	60
19541				19.0	13.0	69			
19543	21.5	16.0	75	24.0	19.0	79			
19429				21.5	16.5	77			
19453				21.0	15.5	74			
19929	16.0	12.0	75						

Listriodon theobaldi Lydekker

Listriodon theobaldi, Lydekker, 1878, Rec. Geol. Surv. India, XL, p. 98.

  Additional References.—

  Lydekker, R., 1880B, p. 31; 1883C, pp. 83, 93; 1884C, p. 102, Pl. VIII, fig. 12. Stehlin, H., 1899, p. 13.

  Pilgrim, G. E., 1910B, p. 202; 1926, p. 34, Pl. XI, figs. 7–10.

Type.—G.S.I. No. B 109, a molar, regarded by Lydekker as right M^1 or M^2.

Paratypes.—None.

Horizon.—Lower or Middle Siwaliks; not definitely known.

Locality.—Near the village of Jabi, Punjab.

Diagnosis.—(After Lydekker, R., 1878A, p. 98.)

Smaller than *Listriodon pentapotamiae*; transverse valley of upper molar wider and more open; oblique ridges running from fore and aft cingula to the summits of the main ridges are absent.

Lydekker's distinction of *L. theobaldi* from *L. pentapotamiae* was based mainly on the differences in size which existed between certain specimens under his observation. He admitted that structurally no constant distinctions could be drawn between the smaller teeth, which were classified as *L. theobaldi*, and the larger teeth of *L. pentapotamiae*, a conclusion that was subsequently borne out by Pilgrim.

After a careful study of numerous *Listriodon* teeth in the American Museum collection, it seems inadvisable to recognize a second species, smaller than *L. pentapotamiae*. A gradual gradation in size exists between the smaller and larger teeth in the collection, so that any definite lines of demarcation can be established only on very arbitrary or artificial criteria. It might be well to add that size variations run indiscriminately through specimens ranging from the base of the Chinji well up into the Middle Siwaliks. Therefore it seems best to consider *L. theobaldi* as being synonymous with *L. pentapotamiae*.

Listriodon guptai Pilgrim

Listriodon guptai, Pilgrim, 1926, Pal. Indica, N.S., VIII, No. 4, pp. 34–35, Pl. XI, figs. 2–6.
  Type.—G.S.I. No. B 701, a last upper molar.
  Paratypes.—G.S.I. No. B 702, right M_3; B 703, left M_3; B 704, upper incisor; B 705, incisor.
  Horizon.—From the Lower Manchars, correlative with the Kamlial horizon of the Punjab.
  Locality.—Bhagothoro, Sind.
  Diagnosis.—A small, bunodont species, related to *L. lockharti* of Europe.

There would seem to be in the Siwaliks a true bunodont *Listriodon*, a form more primitive than the typical lophodont *Listriodon*.

Lophochoerus Pilgrim, 1926

Generic type, *Lophocohoerus nagrii* Pilgrim

Lophochoerus nagrii Pilgrim

Lophochoerus nagrii, Pilgrim, 1926, Pal. Indica, N.S., VIII, No. 4, p. 28, Pl. X, fig. 4.
  Additional Reference.—
      Matthew, W. D., 1929, p. 459.
  Type.—G.S.I. No. B 692, a left mandibular ramus with a partial milk dentition.
  Paratypes.—None.
  Horizon.—Middle Siwaliks, Nagri zone.
  Locality.—Hari Talyangar, in the Simla Hills.
  Diagnosis.—Cusps semi-selenodont, without accessory cuspules. A tendency towards transverse crests in the molars; talonid of M_3 very small.

Lophochoerus himalayensis Pilgrim

Lophochoerus himalayensis, Pilgrim, 1926, Pal. Indica, N.S., VIII, No. 4, p. 29, Pl. X, fig. 5.
  Type.—G.S.I. No. B 693, a right mandibular ramus.
  Paratypes.—None.
  Horizon.—Middle Siwaliks, Nagri zone.
  Locality.—Hari Talyangar, in the Simla Hills.
  Diagnosis.—Like *L. nagrii*, but larger.

Lophochoerus exiguus Pilgrim

Lophochoerus exiguus, Pilgrim, 1926, Pal. Indica, N.S., VIII, No. 4, p. 29, Pl. X, fig. 6.
  Type.—G.S.I. No. B 694, a last lower molar.
  Paratypes.—None.
  Horizon.—Lower Siwaliks, Chinji zone.
  Locality.—Near Chinji, Attock District, Punjab.
  Diagnosis.—Smaller than *L. nagrii*, with closely connected cusps.

The genus *Lophochoerus* was erected by Pilgrim to contain certain very small suids characterized by their semi-selenodont molars. Since the genus is founded on very fragmentary material, it is as yet not well known.

Pilgrim has compared *Lophochoerus* to *Choerotherium*, and if a relation does exist between these two genera, the Siwalik form would then be allied to the peccaries, rather than to the true pigs.[62]

On the other hand, the last molar of *L. exiguus* is typically suilline in its characters, and because of its size it would seem to place this genus in line with *Palaeochoerus*. The teeth of *L. himalayensis* also have a typical suilline appearance. The relationships of this genus can be settled only when more complete material is known.

Propotamochoerus Pilgrim, 1926

Generic type, *Propotamochoerus salinus* Pilgrim

Propotamochoerus salinus Pilgrim

Propotamochoerus salinus, Pilgrim, 1926, Pal. Indica, N.S., VIII, No. 4, pp. 23–24, Pl. VII, fig. 1.

Additional Reference.—

Matthew, W. D., 1929, p. 459.

Type.—G.S.I. No. B 680, a right mandibular ramus.

Paratypes.—Brit. Mus. No. 37267, a skull; Brit. Mus. No. M 3438, a palate. G.S.I. No. B 681, a skull.

Horizon.—Middle Siwaliks, Nagri zone, for the type. Other specimens also come from the Lower Siwaliks, Chinji zone.

Locality.—Nagri, and adjacent regions, Salt Range, Punjab.

Specimens in the American Museum.—Amer. Mus. No. 19637, a left M_2. From the Lower Siwaliks, about 600 feet above the level of Chinji Rest House, one mile northwest of Chinji Rest House.

Diagnosis.—A small suid, with rugose enamel on the molars and a small heel on the last tooth. Lower premolars compressed.

The genus *Propotamochoerus* was erected by Pilgrim to contain certain species of Siwalik suids seemingly ancestral to *Potamochoerus*. He based his distinctions between the ancient and the modern genera not so much on dental characters as on differences in skull form, pointing out the fact that *Potamochoerus* is characterized by its advanced traits, such as the lengthening of the preorbital portion of the face in relation to the postorbital region, the reduction in size of the orbit and the expansion of the cranial cavity.

In *Propotamochoerus* there is a first premolar both above and below, and this tooth has disappeared in *Potamochoerus*.

Just how great an advance is shown by the recent genus *Potamochoerus* over the Siwalik genus *Propotamochoerus* is a question open to some doubt. Certainly the changes in proportions cited by Pilgrim do not reach any great degrees of difference. And these changes do not affect the teeth in any marked manner. Thus it would seem that the creation of a new genus to include the Siwalik forms is a step that may not be subsequently validated. On the other hand, since this new genus has been set apart, and since it does express the

---

[62] See Pearson, H. S., 1927, pp. 405–410, as to the relationships of *Choerotherium*.

difference between Pliocene and Pleistocene to recent forms, it may be best to retain the two genera in the way that they have been used by Pilgrim.

Propotamochoerus hysudricus (Stehlin)

Potamochoerus hysudricus, Stehlin, 1899, Abhandl. Schw. Pal. Gesell., XXVI, pp. 18, 164.
Propotamochoerus hysudricus, Pilgrim, 1926, Pal. Indica, N.S., VIII, No. 4, pp. 25–26.

Additional References.—
> Pilgrim, G. E., 1910B, p. 202.
> Matthew, W. D., 1929, p. 444.

Type.—G.S.I. No. B 30, a mandibular ramus.
Paratypes.—None.
Horizon.—Middle Siwaliks, Dhok Pathan zone.
Locality.—From the Potwar District, Punjab, for the type. Other specimens from various localities in the Punjab, also from Perim Island.

Specimens in the American Museum.—Amer. Mus. No. 19502. Left M³. Upper Middle Siwaliks, one mile north of Nathot.

19516. Fragment of palate, with right P⁴–M³. Also some left molars. Middle Siwaliks, one mile northeast of Hasnot.
19521. Fragment of left M₃. Middle Siwaliks, two miles east of Hasnot.
19534. Left ramus with P₄–M₃. Middle Siwaliks, 500 feet below the Bhandar bone bed, one and one half miles west of Hasnot.
19698. Miscellaneous teeth. Middle Siwaliks, four miles west of Dhok Pathan.
19703. Left M³ and M₃. Middle Siwaliks, one mile west of Dhok Pathan.
19707. Ramus with left M₃. Middle Siwaliks, four miles east of Dhok Pathan.
19713. Left M₃. Middle Siwaliks, four miles west of Dhok Pathan.
19724. A skull and mandible. Middle Siwaliks, one and one half miles northeast of Hasnot.
19728. Mandible. Middle Siwaliks, 100 feet above the Bhandar bone bed, one and one half miles northeast of Hasnot.
19765. Teeth. Middle Siwaliks, near Karsai.
19896. Fragment of ramus with left P₃₋₄. Middle Siwaliks, one mile north of Hasnot.
19936. Left M₃. Middle Siwaliks, two miles northwest of Hasnot.

Diagnosis.—Like *P. salinus*, but larger, and with a longer and more complex third molar. This species is of Dhok Pathan age, whereas *P. salinus* is found in the earlier Chinji and Nagri beds.

Propotamochoerus hysudricus may be readily compared with the modern African *Potamochoerus*. The skulls of the Siwalik and the modern form are of about the same size and they are essentially similar to each other. They differ in the characters pointed out by Pilgrim, namely the more centrally placed and smaller orbit of the fossil genus, and the smaller brain case as compared to the recent genus.

The skull of *P. hysudricus* (A.M. No. 19724) in the American Museum collection is that of a female, as shown by the small canine teeth. It shows certain primitive characters of minor importance that distinguish it from the modern form. For instance, all of the upper premolars are retained, whereas in the recent *Potamochoerus* the first premolar has disappeared. It would seem also, that there is no premolar reduction in the mandible of the fossil. Premolar reduction is, however, a variable feature within species, and consequently should not be given too much weight in taxonomic differentiations.

FIG. 110. *Propotamochoerus hysudricus* (Stehlin). Amer. Mus. No. 19724, associated skull and mandible. Lateral view. One half natural size.

MEASUREMENTS

	Propotamochoerus hysudricus (A.M. 19724)	*Potamochoerus porcus* (A.M. 53727)
Length of P^2–P^4	42.5 mm.	36.0 mm.
Length of M^1–M^3	68.0	74.0
Length of P_3–P_4	33.0	31.0
Length of M_1–M_3	69.0	75.0
Ratio P^{2-4}/M^{1-3}	62	49
Ratio P_{3-4}/M_{1-3}	48	41

MEASUREMENTS (IN MM.) OF *Propotamochoerus hysudricus*

Amer. Mus. Number	M^1	M^2	M^3	Length $P_2 + P_4$ (A)	Length Molar series (B)	$\frac{A}{B}$	Ratio: $\frac{M^1 + M^2}{M^3}$	Index M^3
19724	15 × 16	21 × 20	34 × 22	27	68	40	106	65
19703			34.0 × 22.5					66
No. num.	16.0 × 15.5	22.0 × 20.0	29.0 × 20.0		67.0		130	69

Amer. Mus. Number	M_1	M_2	M_3	Length $P_2 + P_4$ (A')	Length Molar series (B')	$\frac{A'}{B'}$	Ratio: $\frac{M_1 + M_2}{M_3}$	Index M_3
19724	13 × —	22 × 17	34 × 19	34	69	49	97	57
19728	15.0 × 12.5	22.0 × 18.0	40.0 × 20.0	40.0	79.0	50	92	50
19703			30.0 × 15.5					52
19534	15.0 × 12.5	21.0 × 17.0	36.0 × 18.5		72.0		100	52
19698			33.0 × 18.0					55
19707			31.5 × 16.5					52

Sus hysudricus AND *Potamochoerus hysudricus*

The name *Sus hysudricus* first appeared on certain plates (70, 71) of the Fauna Antiqua Sivalensis, to accompany figures of that species. At that time no description of the species was given, although later short descriptions of the figured specimens were published (Falconer, H., 1868A). In 1884 Lydekker gave a full description and discussion of the specimens figured in the Fauna Antiqua Sivalensis, and in addition he described and figured some additional material, all of which he referred to *Sus hysudricus*.

In 1899, Stehlin, in his "Geschichte des Suiden Gebisses," designated *Sus hysudricus* as Falconer's species, on the basis of the material figured in the Fauna Antiqua Sivalensis. Then he listed *Potamochoerus hysudricus* Falconer on the basis of two of the additional specimens figured by Lydekker. Then, notwithstanding the fact that Stehlin attributed *Sus hysudricus* to Falconer (on page 16 of his monograph), which was only partly correct, since Falconer and Cautley are the authors, on a subsequent page of the same work (page 18) he attributed this species to Lydekker. Finally, to make things even more complicated, in another section of his work he assigned *Sus hysudricus* to Falconer and *Potamochoerus hysudricus* to Lydekker.

Pilgrim, in 1910, listed two species, *Potamochoerus hysudricus* Falconer and Cautley, and *Sus hysudricus* Falconer and Cautley. In 1926 this same author designated these species as follows:

Sus hysudricus Falconer and Cautley.

Propotamochoerus hysudricus (Lydekker).

In 1929 Matthew listed *Potamochoerus hysudricus* without designating an author.

To summarize the complicated histories of these species, we may present them in a tabular form.

Species	Attributed to	By
Sus hysudricus	Falconer and Cautley	Falconer, 1868
	Falconer and Cautley	Lydekker, 1884
	Falconer	Stehlin, 1899
	Lydekker	Stehlin, 1899
	Falconer and Cautley	Pilgrim, 1910
		1926
Potamochoerus hysudricus	Falconer	Stehlin, 1899
	Lydekker	Stehlin, 1899
	Falconer and Cautley	Pilgrim, 1926
Propotamochoerus hysudricus	Lydekker	Pilgrim, 1926

As a matter of fact, Stehlin is the author of *Potamochoerus hysudricus*, and his description on page 164 of "Geschichte des Suiden Gebisses" is the type description of the species. Neither Falconer nor Lydekker showed any intentions of placing the Siwalik material in the genus *Potamochoerus*. Lydekker, although he recognized differences in the *Sus hysudricus* material, did not propose to separate any part of this out as representative of a new genus, but rather he attributed these differences to sexual disparities.

Propotamochoerus uliginosus Pilgrim

Propotamochoerus uliginosus, Pilgrim, 1926, Pal. Indica, N.S., VIII, No. 4, p. 25, Pl. VII, figs. 2–5, Pl. VIII.

Type.—G.S.I. No. B 683, a last lower premolar.

Paratypes.—G.S.I. Nos. B 685, right mandibular ramus, with P_3; B 686, a palate; B 29, a skull; B 684, a fourth upper premolar. Brit. Mus. No. M 2050, palate.

Horizon.—Lower Siwaliks, Chinji zone. Some of the paratypes come from Hari Talyangar and Perim Island, which would indicate that this form extends up into the lower portion of the Middle Siwaliks.

Locality.—Chinji. Also Hari Talyangar and Perim Island.

Diagnosis.—Like *P. salinus*, but smaller. Minor differences in the teeth.

The validity of this species may be doubted. It may very likely be synonymous with *P. salinus*.

Propotamochoerus ingens Pilgrim

Propotamochoerus ingens, Pilgrim, 1926, Pal. Indica, N.S., VIII, No. 4, p. 26, Pl. VIII, fig. 8, Pl. XX, fig. 6.

Type.—G.S.I. No. B 10, a left mandibular ramus.

Paratype.—G.S.I. No. B 690, a right maxilla, with P^3–M^1.

Horizon.—Middle Siwaliks, Dhok Pathan zone.

Locality.—From the Salt Range, Punjab. The paratype is from Chakwal.

Diagnosis.—A very large species of the genus. Except for size, like *P. hysudricus*, of which the species under discussion may be a variant.

Potamochoerus Gray, 1854

Generic type, *Sus koiropotamus* Desmoulins

Potamochoerus palaeindicus Pilgrim

Potamochoerus palaeindicus, Pilgrim, 1926, Pal. Indica, N.S., VIII, No. 4, p. 27, Pl. X, fig. 2.

Type.—G.S.I. No. B 691, a right mandibular ramus.

Paratypes.—None.

Horizon.—Presumably from the Pinjor horizon, Upper Siwaliks.

Locality.—Collected on the Tatrot plateau.

Specimens in the American Museum.—Amer. Mus. No. 19878, a palatal fragment with left P^2–M^2. Upper Siwaliks, upper clays below the conglomerate. Four miles west of Mirzapur.

Diagnosis.—A large species with an elongated third molar.

Potamochoerus appears in the Upper Siwalik beds, and it may be considered as a direct descendant of the earlier and closely related *Propotamochoerus*.

The palate in the American Museum collection, listed above, is here referred to this species. Thus it augments our knowledge of the species, since Pilgrim's description is based on a mandible. The American Museum palate is referred to this species because it would seem to be of a size proper for inclusion here, and it comes from the Upper Siwaliks. This procedure is followed, rather than naming a new species.

The American Museum specimen contains only P^3 to M^2 on the left side. The teeth are slightly larger than those of a modern *Potamochoerus*. As to the form and disposition of cusps, the cheek teeth of the fossil form are exceedingly similar to those of the modern species. The only notable differences are the somewhat larger postero-internal angle on the third premolar, and the presence of an external conule in the transverse valley of the second molar. These are, however, characters of considerable variability, and consequently they are likely to be of little importance. The American Museum specimen is illustrated by the accompanying figure.

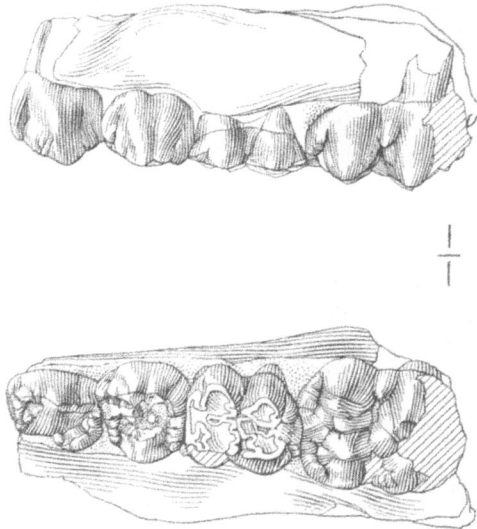

A. M. 19878

Fig. 111. *Potamochoerus palaeindicus* Pilgrim. Amer. Mus. No. 19878, left P^3–M^2. Side view above, crown view below. Natural size.

MEASUREMENTS

		Potamochoerus palaeindicus A.M. 19878	*Potamochoerus porcus* A.M. 53727
P³	Length	16.0 mm.	14.0 mm.
	Width	13.0	10.5
P⁴	Length	15.0	13.0
	Width	17.0	15.0
M¹	Length	19.0	19.0
	Width	17.0	17.0
M²	Length	26.0	23.0
	Width	23.0	22.0

Indices

		P⁴	M¹	M²
P. palaeindicus, A.M. 19878		113,	90,	88
P. porcus, A.M. 53727		115,	90,	95

Potamochoerus theobaldi Pilgrim

Potamochoerus theobaldi, Pilgrim, 1926, Pal. Indica, N.S., VIII, No. 4, pp. 27–28, Pl. X, fig. 3.

Type.—G.S.I. No. B 11, a right mandibular ramus.

Paratypes.—None.

Horizon.—Upper Siwaliks, Pinjor zone.

Locality.—Kangra District.

Diagnosis.—Differs from *P. palaeindicus* in that the premolars are reduced in size as compared to the molars, and the heel of the last molar is longer and more complex.

Dicoryphochoerus Pilgrim, 1926

Generic type, *Sus titan* Lydekker

Dicoryphochoerus titan (Lydekker)

Sus titan, Lydekker, 1884, Pal. Indica (X), III, pp. 59–66, Pls. VII–XII.

Potamochoerus titan, Stehlin, 1899, Abhandl. Schweiz. Pal. Gesell., XXVI, pp. 18, 74, 163, 258, 390.

Dicoryphochoerus titan, Pilgrim, 1926, Pal. Indica, N.S., VIII, No. 4, pp. 38–40, Pl. XIII, figs. 1, 2, Pl. XX, figs. 10, 11.

Additional References.—

Lydekker, R., 1885B, pp. 45–46; 1885D, p. 270.

Pilgrim, G.E., 1910B, p. 202.

Matthew, W. D., 1929, p. 449.

Type.—(Lectotype).—G.S.I. No. B 26, a skull and mandible.

Cotypes.—G.S.I. Nos. B 4, a mandible; B 15 a molar tooth; B 19, a right M³; B 358, a right mandibular ramus; B 435, a right mandibular ramus. Royal College of Surgeons, No. 1805, a ramus.

Horizon.—Middle Siwaliks, Dhok Pathan zone.

Locality.—Niki, Punjab for the lectotype; vicinity of Hasnot for the cotypes. Also other localities in the Punjab.

Specimens in the American Museum.—Amer. Mus. No. 19380. Fragmentary material, with several upper and lower teeth. Middle Siwaliks, about 1000 feet below the Bhandar bone bed. Locality doubtful.

19381. A left P⁴. Middle Siwaliks, one mile west of Dhok Pathan.

19507. Fragmentary upper teeth. Middle Siwaliks, one mile northeast of Hasnot.

19546. Back portion of right M₃. Middle Siwaliks, three miles north of Hasnot.

19579. Mandible with M₂₋₃. Lower Siwaliks, 1600 feet above the level of Chinji Rest House, two miles west of Chinji Rest House. (Tentatively referred to this species.)

19679. Fragment of ramus with left M₂₋₃. Middle Siwaliks, four miles west of Dhok Pathan.

19699. Right M³. Middle Siwaliks, four miles west of Dhok Pathan.

19742. Right M³. Middle Siwaliks, two and one half miles northeast of Hasnot.

Diagnosis.—An extremely large suid, with broad, complex molars. The skull is deep with a short premaxillary region. Fourth lower premolar with an accessory cusp, internal to the main cusp.

Potamochoerus and *Sus* are the forms most closely related to the genus *Dicoryphochoerus* as defined by Pilgrim. The constant distinction between *Potamochoerus* and the genus under consideration is that in *Potamochoerus* the fourth lower premolar consists of a simple central cone, whereas in *Dicoryphochoerus* there are two cusps in this tooth, there being an accessory cusp internal to the main cusp. In this last respect, *Dicoryphochoerus* is similar to *Sus*. In many respects the smaller and more primitive species of *Dicoryphochoerus* show characters ancestral in nature to the characters of *Sus*, so that we may imagine that the fossil genus is rather close to the direct antecedents of *Sus*. *Dicoryphochoerus*, like *Potamochoerus* and *Sus* is typified by the strong tendencies towards polybuny in the molars.

The material representative of *D. titan* in the American Museum collections does not furnish any essential information additional to that already supplied in the extended descriptions by Lydekker and by Pilgrim. Consequently it will not be necessary to describe the new material here. Some measurements are given below.

MEASUREMENTS
Dicoryphochoerus titan

Amer. Mus. No.	M³			M₃		
	Length	Width	Index	Length	Width	Index
19380	52 mm.	31 mm.	60			
19679				56 mm.	29 mm.	52
19699	52	37	71			
19742	57	38	67			

Dicoryphochoerus titanoides Pilgrim

Dicoryphochoerus titanoides, Pilgrim, 1926, Pal. Indica, N.S., VIII, No. 4, pp. 40–41, Pls. XIII, fig. 3.

Type.—G.S.I. No. B 354, a right maxilla with P³–M³.

Paratypes.—None.

Horizon.—Middle Siwaliks, Dhok Pathan zone.

Locality.—From the Punjab, definite locality not given.

Diagnosis.—Like *D. titan*, but smaller.

This species is very probably synonymous with *D. titan*, the specimen referred to by Pilgrim in his description being a small variant of the larger Middle Siwalik suid.

"There appears to be some variation in size in both *D. titan* and *D. titanoides*, and as the material is insufficient to show any precise differences of structure between them, their specific separation was not exactly a foregone conclusion." [63]

According to Pilgrim *D. titanoides* is allied to *D. chisholmi*.

Dicoryphochoerus chisholmi Pilgrim

Dicoryphochoerus chisholmi, Pilgrim, 1926, Pal. Indica, N.S., VIII, No. 4, p. 41, Pl. XIV, figs. 1, 2.

Type.—G.S.I. No. B 707, an isolated left P_4.

Paratype.—G.S.I. No. B 708, a P_3.

Horizon.—Lower Siwaliks, Chinji zone.

Locality.—Near Chinji, Salt Range, Punjab.

Diagnosis.—Like *D. titan* and *D. titanoides*, but with a weaker anterior cusp and cingulum and a feebler development of the third cusp in the fourth premolar.

D. chisholmi is based on such really insufficient material that one is inclined to regard the species as of little value.

Dicoryphochoerus vagus Pilgrim

Dicoryphochoerus vagus, Pilgrim, 1926, Pal. Indica, N.S., VIII, No. 4, pp. 43–45, Pl. XV, figs. 1, 2, 7, 8, Pl. XX, figs. 7, 8.

Type.—G.S.I. No. B 39, a right mandibular ramus, with P_3–M_3.

Paratypes.—G.S.I. Nos. B 742, left ramus with P_4–M_3; B 57, maxilla with right P^3–M^1. Brit. Mus. No. 15363–4, right mandibular ramus with P_3–M_3.

Horizon.—Middle Siwaliks, Dhok Pathan zone.   A variety seems to occur in the Upper Siwaliks, Pinjor zone.

Locality.—Hasnot for the type.   Hasnot, Nila, Diwal and other localities in the Punjab for the paratypes.

Diagnosis.—Upper premolars stout, with weak antero-internal cingulum; inner cusp of lower fourth premolar weak.   Otherwise, except for size, quite similar to *Sus hysudricus*.

This species may very possibly be synonymous with *Sus hysudricus*, a possibility recognized by Pilgrim.   If so, it would represent a small variation of that species.

Dicoryphochoerus durandi Pilgrim

Dicoryphochoerus durandi, Pilgrim, 1926, Pal. Indica, N.S., VIII, No. 4, p. 45, Pl. XX, fig. 9.

Type.—Brit. Mus. No. 37130, a mandibular symphysis.

Paratypes.—Brit. Mus. No. 16605, a right mandibular ramus.   G.S.I. No. B 716 (this species?), a canine.

Horizon.—Upper Siwaliks, Pinjor zone.

Locality.—Siwalik Hills.

Diagnosis.—Distinguished by the stout premolars and the elongated muzzle.   Smaller than *D. vagus*.   According to Pilgrim, this species and *D. vagus* approach more closely than any of the other species of *Dicoryphochoerus*, the genus *Sus*.

[63] Pilgrim, G.E., 1926, p. 40.

Dicoryphochoerus instabilis Pilgrim

Dicoryphochoerus instabilis, Pilgrim, 1926, Pal. Indica, N.S., VIII, No. 4, p. 43, Pl. XIV, figs. 7, 8.

Type.—(Lectotype).—G.S.I. No. B 712, a left P_3.

Cotype.—G.S.I. No. B 713, a left P_4.

Horizon.—Lower Siwaliks, Chinji zone.

Locality.—Near Chinji.

Diagnosis.—A small species, distinguished by the close appression of the accessory cusp to the main cusp in P_4. Heel of tooth simple.

This species is founded on very scanty and insufficient material.

Dicoryphochoerus vinayaki Pilgrim

Dicoryphochoerus vinayaki, Pilgrim, 1926, Pal. Indica, N.S., VIII, No. 4, p. 43, Pl. XV, fig. 3.

Type.—G.S.I. No. B 715, a left mandibular ramus with P_4–M_3.

Paratypes.—None.

Horizon.—Dhok Pathan zone, Middle Siwaliks.

Locality.—Not definitely stated, Punjab.

Diagnosis.—Like *D. chisholmi* but larger and with a more robust P_4.

According to Pilgrim this species is fairly abundant in the Dhok Pathan zone, and it occurs in the Nagri zone as well.

Dicoryphochoerus robustus Pilgrim

Dicoryphochoerus robustus, Pilgrim, 1926, Pal. Indica, N.S., VIII, No. 4, pp. 42–43, Pl. XIV, figs. 9–11.

Type.—G.S.I. No. B 714, a series of teeth, possibly representing a single individual and consisting of left P_{3-4}, right M_{1-2}.

Paratypes.—None.

Horizon.—Found at a level approximate to the boundary between the Lower and the Middle Siwaliks.

Locality.—Hari Talyangar, Bilaspur State.

Diagnosis.—Intermediate between *D. haydeni* and *D. titan,* and closely related to the former. First molar small, as compared to second molar. P_3 a slender single cusped tooth; P_4 with a strong intermediate cusp.

This species is founded on rather fragmentary material. However, it may be a separate form, directly ancestral to *D. titan.*

Dicoryphochoerus haydeni Pilgrim

Dicoryphochoerus haydeni, Pilgrim, 1926, Pal. Indica, N.S., VIII, No. 4, pp. 41–42, Pl. XIV, figs. 3–6.

Type.—G.S.I. No. B 709, a left P_4.

Paratypes.—G.S.I. No. B 710, a mandibular ramus with left P_3–M_2; G.S.I. No. B 711, a palate with milk and permanent molars.

Horizon.—Lower Siwaliks, Chinji zone.

Locality.—Near Chinji, Salt Range, Attock District, Punjab.

Specimens in the American Museum.—Amer. Mus. No. 19377. A left P_4. Lower Siwaliks, about 1600 feet above the level of Chinji Rest House, thirteen miles east of Chinji Rest House.

19391. Left M^3. Lower Siwaliks, about level of Chinji Rest House, five miles east of Chinji Rest House.

19423. Mandible with left P_4–M_3. About level of Chinji Rest House, two miles west of Chinji Rest House.

19434. Mandibular fragments with right M_{1-2}, left M_3, left M^{1-2}. Lower Siwaliks, about 100 feet above the level of Chinji Rest House, two miles west of Chinji Rest House.

19597. Right M_3. Lower Siwaliks, 1600 feet above the level of Chinji Rest House, eight miles west of Chinji Rest House.

19643. Left M^{1-3} and right M^3. Lower Siwaliks, about level of Chinji Rest House, five miles east of Chinji Rest House.

19960. Palatal fragment with right P^4–M^3. Lower Siwaliks, 1600 feet above the level of Chinji Rest House, three miles west of Chinji Rest House.

29814. Maxillary fragment with left M^2. Lower Siwaliks, 1600 feet above the level of Chinji Rest House, twelve miles east of Chinji Rest House.

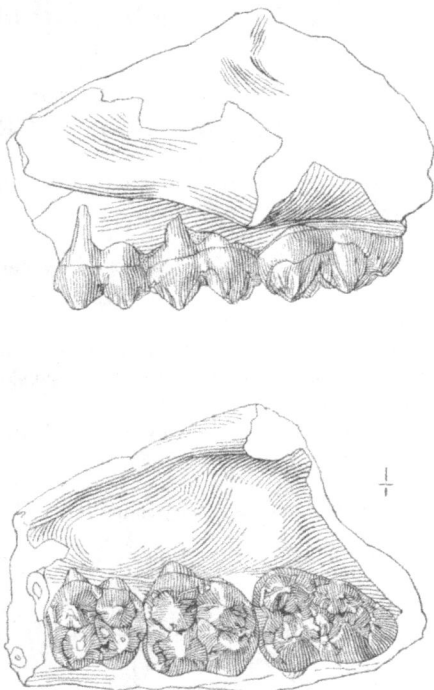

FIG. 112. *Dicoryphochoerus haydeni* Pilgrim. Amer. Mus. No. 19643, fragment of maxilla with left M^{1-3}. Lateral view above, crown view below. Natural size.

Diagnosis.—A primitive species of the genus. Skull primitive and low, with broad zygomatic arches. P_4 distinguished by the close appression of the accessory cusp to the main cusp. M^3 with an elongated talon, as in other species of the genus, but narrow.

A number of specimens in the American Museum collection from the lower Siwalik beds at Chinji, which obviously do not belong to the genus *Conohyus* are here assigned to the genus and species under discussion. In size they agree very closely with the specimens

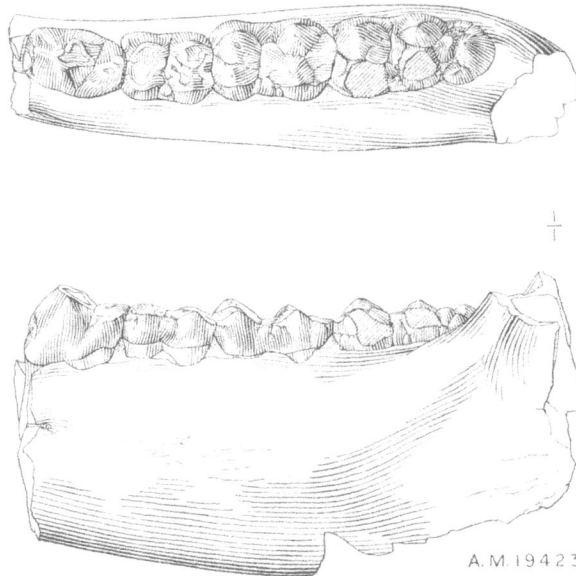

FIG. 113. *Dicoryphochoerus haydeni* Pilgrim. Amer. Mus. No. 19423, ramus with left P_4–M_3. Crown view above, lateral view below. Natural size.

figured by Pilgrim. Characteristic features are the tapering heel of the third upper molar, the long typically suilline heel of the third lower molar, and the double cone on the last lower premolar.

That this is a rather primitive form is shown by the lack of development of supernumerary cusps in either upper or lower dentitions. In each molar tooth there is a median conule developed in the valley between the anterior and posterior pairs of cusps. The molar cusps show small ridges along their surfaces in the unworn tooth, but these are largely obliterated by wear.

Measurements of some of the American Museum specimens are presented in the following table.

MEASUREMENTS (IN MM.) OF *Dicoryphochoerus haydeni*

Amer. Mus. Number	P⁴	Index	M¹	Index	M²	Index	M³	Index	Ratio: $\frac{M^1 + M^2}{M^3}$
19960	14.0 × 18.0	128	17.5 × 16.5	94	22.0 × 19.0	86	27.0 × 17.5	65	146
19643			16.0 × 15.0	94	18.5 × 17.0	92	24.0 × 17.5	73	144
29814					18.5 × 16.5	89			
19391							27.0 × 19.5	72	

Amer. Mus. Number	P₄	Index	M₁	Index	M₂	Index	M₃	Index	Ratio: $\frac{M_1 + M_2}{M_3}$
19423	17.0 × 12.0	71	15.0 × 12.0	80	21.0 × 14.0	67	26.0 × 14.5	56	138
19434			17.0 × 14.0	82	24.0 × 17.5	73			
19597							28.5 × 15.0	53	
19377	18.5 × 12.5	68							

Hyosus Pilgrim, 1926

Generic type, *Hyosus tenuis* Pilgrim

Hyosus tenuis Pilgrim

Hyosus tenuis, Pilgrim, 1926, Pal. Indica, N.S., VIII, No. 4, pp. 57–58, Pl. XIX, fig. 1.
 Type.—G.S.I. No. B 729, a right mandibular ramus.
 Paratypes.—None.
 Horizon.—Middle Siwaliks, Dhok Pathan zone.
 Locality.—Hasnot, Punjab.
 Diagnosis.—A very small suid, with complex molars and progressive premolars. The molars are distinguished by the folding of the enamel, which however, has not reached the degree of complexity to be found in *Hippohyus*. The third molar is small. The lower premolars are distinguished by the fact that the cusps are of subequal height. The last lower premolar is broad.

Hyosus punjabiensis (Lydekker)

Sus punjabiensis, Lydekker, 1878, Rec. Geol. Surv. India, XI, p. 81.
Hyosus punjabiensis, Pilgrim, 1926, Pal. Indica, N.S., VIII, No. 4, p. 58, Pl. XIX, figs. 2, 3.
 Additional References.—
   Lydekker, R., 1880B, p. 31; 1883C, pp. 83, 93; 1884C, p. 82, Pl. VIII, fig. 9.
   Stehlin, H., 1899, p. 75.
   Pilgrim, G. E., 1910B, p. 202.
   Matthew, W.D., 1929, p. 449.
 Type.—G.S.I. No. B 61, a mandibular ramus with left M₁₋₃.
 Paratypes.—None.
 Horizon.—Middle Siwaliks, Dhok Pathan zone.
 Locality.—Hasnot, Punjab.
 Diagnosis.—Like *H. tenuis*, but larger. According to Pilgrim, this species is less specialized than *H. tenuis*, in that the cusps are more rounded and the enamel folding is less complex.

This form was first described by Lydekker as *Sus punjabiensis*, on the basis of a fragmentary ramus with the molars present. In 1926 Pilgrim created a new genus, on very good grounds, to contain this species, and in addition he named another species, *Hyosus tenuis*, making it the generic type. A new genus and species were created, *Sivahyus hollandi*, for a palate and jaw very similar to *H. punjabiensis*, and coming from the same horizon and locality.

The differentiations of these three species, as worked out by Pilgrim, are based on:

1. Slight differences in proportions.
2. Slight differences in size.
3. Slight differences in the amount of crenulation of the enamel.
4. A somewhat different development of the heel in the third molar.

These characters are all quite variable among the Suidae, so it becomes rather doubtful whether some of the distinctions drawn by Pilgrim are of specific or generic calibre. It may very well be that the species listed above are all variants of *H. punjabiensis*, or at least closely related to it.

Sivahyus Pilgrim, 1926

Generic type, *Sivahyus hollandi* Pilgrim

Sivahyus hollandi Pilgrim

Sivahyus hollandi, Pilgrim, 1926, Pal. Indica, N.S., VIII, No. 4, pp. 53–54, Pl. XVIII, figs. 2–4.

Additional Reference.—

Matthew, W.D., 1929, p. 459.

Type.—(Lectotype.)—G.S.I. No. B 726, a left ramus with M_{1-3}.

Cotype.—G.S.I. No. B 725, a left ramus with DM_4, M_{1-2}.

Horizon.—Middle Siwaliks, Dhok Pathan zone.

Locality.—Hasnot, Punjab.

Diagnosis.—Molar teeth with complexly folded enamel, like *Hippohyus*, but they are brachyodont. Premolars relatively large as compared to molars.

This form may be regarded as representative of the direct ancestor of *Hippohyus*.

Sanitherium von Meyer, 1866

Generic type, *Sanitherium schlagentweitii* von Meyer

Sanitherium schlagentweitii von Meyer

Sanitherium schlagentweitii, von Meyer, 1866, Palaeontographica, XV, p. 15, Pl. II, figs. 9–12.

Additional References.—

Lydekker, R., 1876, p. 58, Pl. IX, figs. 6–9; 1880B, p. 31; 1883C, pp. 83, 92; 1884C, p. 91, Pl. VIII, fig. 7; 1885B, p. 44.

Pilgrim, G. E., 1910B, p. 202; 1926, p. 55, Pl. I.

Matthew, W. D., 1929, p. 453.

Type.—The second, and a portion of the third lower molars of the left side, described and figured by von Meyer.

Paratypes.—None.

Horizon.—Lower Siwaliks, probably from a level equivalent to the Chinji zone.
Locality.—Kushalgar, Punjab.
Diagnosis.—A very small suid, distinguished by its narrow teeth, folded enamel and by the strong beaded cingulum of the molar teeth.

Sanitherium cingulatum Pilgrim

Sanitherium cingulatum, Pilgrim, 1926, Pal. Indica, N.S., VIII, No. 4, pp. 54–55, Pl. XVIII, fig. 5.
Type.—G.S.I. No. B 728, fragment of a right mandibular ramus, with M_{1-3}.
Paratypes.—None.
Horizon.—Lower Siwaliks, Chinji zone.
Locality.—From near Chinji, Salt Range, Punjab.
Specimens in the American Museum.—Amer. Mus. No. 29842. Mandible, with milk dentition and first two permanent molars. Both rami present. Lower Siwaliks, 100 feet below the level of Chinji Rest House, two miles west of Chinji Rest House, Punjab.
19385 (referred). Fragment of ramus, with right M_{1-3}, badly broken. Lower Siwaliks, about level of Chinji Rest House, four miles west of Chinji Rest House.
Diagnosis.—Like *Sanitherium schlagentweitii*, but smaller, with more compressed teeth, and a more slender mandible.

Sanitherium is a very small upper Tertiary suid, characterized by the complexity of the furrowings in the cusps, which in turn causes the enamel in a worn tooth to assume a complicated pattern, and by the presence of a beaded cingulum in the molars. Von Meyer's original description was based on very fragmentary material, but our knowledge of the genus was supplemented by descriptions given by Lydekker and Pilgrim, on the basis of new material under their observation.

A mandible in the American Museum collection (A.M. No. 29842), which is the subject of this description, would seem to belong to the species *S. cingulatum*, as described by Pilgrim, since it has narrow molars and a relatively slender ramus. The milk dentition is present, as well as the first two permanent molars. Two permanent premolars, presumably the second and third, have appeared beneath the milk molars.

The specimen is badly crushed, and the teeth are shattered, making a detailed study a rather tantalizing experience. It would seem, however, that there are but two deciduous incisors, the outer one of which is closely pressed against the deciduous canine. It would seem as if the third deciduous incisor has been suppressed in this genus, possibly it was crowded out by a forward migration of the deciduous canine, or by a backward or lateral growth of the incisor series.

The deciduous canine is triangular in cross section, the outer and inner facets being long and the posterior one narrow. Thus this tooth is of the primitive verrucose type, as might be expected in a relatively primitive suid.

The tooth immediately following the deciduous canine would seem to be the first member of the permanent premolar series (or as certain authorities would have it, a retained milk molar). It is long, slender and upright, and is peculiar in having assumed a caniniform shape. That this is actually a premolar tooth was proven by an examination of the cross section near the base, which showed two distinct roots.

Following this tooth are two milk molars, the third and the fourth. The third milk molar is two rooted, long and narrow. The last milk molar, as is usually the case in artiodactyls, is very long, with three roots and with a complex crown pattern. This tooth is so badly broken that a detailed description or figure of its coronal surface is impossible.

The first and second molars are essentially four-cusped, but with numerous secondary features that make their crowns rather complicated. The right second molar, which is the only one of the series preserved sufficiently well to make a description possible, is long in comparison to its width, and compressed. The four cusps are equally disposed, that is the outer cusps are not offset in relation to the inner cusps. From each of the two anterior cusps, the protoconid and metaconid, two ridges run posteriorly to the median valley of the tooth, the outer ridge being parallel to the edge of the tooth, and the inner one projecting medially to meet its fellow from the neighboring cusp. The hypoconid has a ridge projecting forward to the median valley of the tooth, and another one projecting postero-internally. The postero-internal cusp, the entoconid, is characterized by two forwardly projecting ridges, one running parallel to the internal edge of the tooth and the other projecting medially.

There is a beaded cingulum, a diagnostic character in this genus, running along the external side of the tooth and projecting inwardly on each end. The enamel surface is corrugated.

Below the third and fourth milk molars may be seen two erupting premolars, presumably the second and third of the permanent series. The so-called third premolar is so identified because it occupies a place beneath the anterior end of the last milk molar. Without question a fourth permanent premolar came up between it and the first molar.

The rami are characterized by their curved lower borders and the lack of produced angles. These are probably adolescent characters. The external openings of the mental canal are beneath the third and fourth premolars.

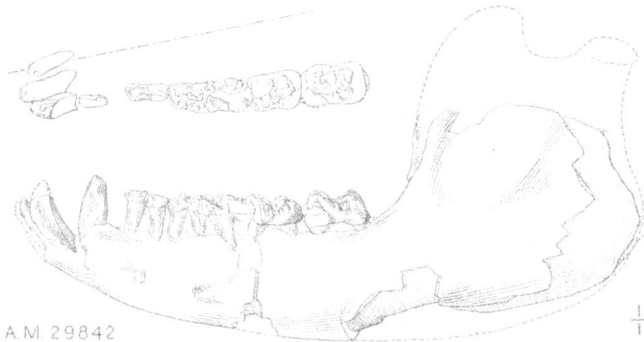

A.M. 29842

FIG. 114. *Sanitherium cingulatum* Pilgrim. Amer. Mus. No. 29842, left mandibular ramus with milk dentition and permanent molars. Crown view above, side view below. Natural size.

From the above description it may be seen that *Sanitherium* had a rather peculiar dental formula. That is, one deciduous incisor and a deciduous molar would seem to have been lacking.

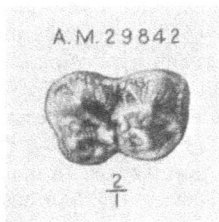

FIG. 115. *Sanitherium cingulatum* Pilgrim. Amer. Mus. No. 29842, right M₁. Crown view. Twice natural size.

MEASUREMENTS

Sanitherium cingulatum, Amer. Mus. No. 29842

	Antero-posterior Diameter	Transverse Diameter.
C.	5.0 mm.	3.5 mm.
P₁.	4.5	2.0
DM₃.	7.5	3.0
DM₄.	13.0	6.0
M₁.	9.0	6.5
M₂.	12.5	7.5

Hippohyus Falconer and Cautley

Generic type, *Hippohyus sivalensis* Falconer and Cautley

Hippohyus sivalensis Falconer and Cautley

Hippohyus sivalensis, Falconer and Cautley, 1840–5. In Owen, R., Odontography, p. 562, Pl. CXL, fig. 7.

Additional References.—

Falconer, H., and Cautley, P. T., 1847G, Pl. LXX, figs. 1, 1b, Pl. LXXI, figs. 1–4.

Falconer, H., 1868A, pp. 22, 509, 511.

Lydekker, R., 1880B, p. 31; 1883C, pp. 83, 91; 1884C, pp. 85–91, Pl. XII, figs. 17–21; 1885B, p. 44; 1885D, p. 259.

Pilgrim, G. E., 1910B, p. 202; 1926, p. 50.

Stehlin, 1899, various pages.

*Type.—*The molar tooth figured in Owen's Odontography, Pl. CXL, fig. 7.

*Paratypes.—*None.

*Horizon.—*Upper Siwaliks.

*Locality.—*Siwalik Hills.

*Specimen in the American Museum.—*Amer. Mus. No. 19497. Right M³, also jaw fragment with a portion of right M₁₋₂. Upper Siwaliks, one half mile east of Kotal Kund.

*Diagnosis.—*A very specialized suid with hypsodont teeth; with the molars distinguished by the complexity of the folding of their enamel. The skull is rather heavy, with the orbit centrally placed, the preorbital portion of the skull is relatively short, and the zygomatic arches heavy.

The name *Hippohyus sivalensis* first appears in Owens Odontography, two years before the publication of the Fauna Antiqua Sivalensis. Thus there arises the possibility that

Owen may have been the author of this genus and species, though in any case, an unwitting author. However, on page 544, where he first mentions *Hippohyus* (without using the specific name), he says in a footnote; "An extinct genus so called by its discoverers, Captain Cautley and Dr. Falconer." This footnote by Owen may be considered as a sufficient indication, according to the International Rules, that he is not the author of the genus. Consequently *Hippohyus sivalensis* may be logically attributed to Falconer and Cautley.

The type is the tooth figured in Owen's Odontography, not a skull as indicated by Lydekker.

Hippohyus lydekkeri Pilgrim

Hippohyus lydekkeri, Pilgrim, 1910, Rec. Geol. Surv. India, XL, p. 68.

Additional Reference.—

Pilgrim, G. E., 1926, p. 52, Pl. XVII, fig. 2, Pl. XVIII, fig. 6.

Type.—(Lectotype.)—G.S.I. No. B 68, mandibular ramus with left P_3–M_3.

Cotypes.—G.S.I. No. B 69a, mandible with left M_{2-3}; G.S.I. No. B 63, fragmentary skull with palate; G.S.I. No. B 724, right maxilla with P^3–M^3.

Horizon.—Middle Siwaliks, Dhok Pathan zone.

Locality.—From Hasnot and the adjacent region, Salt Range, Punjab.

Specimen in the American Museum.—Amer. Mus. No. 19947, jaw fragment with left M_{2-3}. Upper portion of Middle Siwaliks, near Tatrot.

Diagnosis.—According to Pilgrim this species is distinguished from the other forms of *Hippohyus* by its small size.

Hippohyus grandis Pilgrim

Hippohyus grandis, Pilgrim, 1926, Pal. Indica, N.S., VIII, No. 4, pp. 50–51, Pl. XVI, figs. 1–5, Pl. XVII, fig. 1.

Type.—G.S.I. No. B 720, a well preserved skull, with the crowns of the cheek teeth badly battered.

Paratypes.—G.S.I. No. B 722, a palate; G.S.I. No. B 64, ramus with symphysis; G.S.I. No. B 721, right ramus with P_4–M_3.

Horizon.—"Tatrot stage of the Upper Siwalik, and possibly also Middle Siwalik."

Locality.—From around Tatrot and Kotal Kund, Salt Range, Jhelum District, Punjab.

Diagnosis.—Like *Hippohyus sivalensis*, but larger.

Hippohyus tatroti Pilgrim

Hippohyus tatroti, Pilgrim, 1926, Pal. Indica, N.S., VIII, No. 4, pp. 51–52, Pl. XVII, fig. 3, Pl. XVIII, fig. 1, Pl. XX, fig. 13.

Type.—(Lectotype.)—G.S.I. No. B 62, a skull.

Cotypes.—G.S.I. No. B 65, maxilla with right M^{2-3}; G.S.I. No. B 66, maxilla with left M^{2-3}.

Horizon.—Upper Siwaliks, Tatrot zone.

Locality.—Salt Range, Punjab; also Potwar district.

Diagnosis.—Molar series relatively short, the third molar is especially short; and the premolars are simpler than in the other species.

Hippohyus deterrai Lewis

Hippohyus deterrai, Lewis, 1934, Amer. Jour. Sci., Ser. V, No. 162, XXVII, pp. 457–459, Pl. I.

Type.—Yale Peabody Museum, No. 13812, an incomplete skull with right and left P³–M³.

Paratypes.—None.

Horizon.—Middle Siwaliks, Nagri zone.

Locality.—Three and three fourths miles west southwest of Hasnot, Punjab.

Diagnosis.—(See Lewis, 1934.) "The size is exceptionally small. The molars are relatively short and broad, with typical *Hippohyus* enamel folds."

This species is a typical *Hippohyus*, but it is remarkable because of its very small size.

It would seem advisable to recognize three species of *Hippohyus* in the Siwalik series, a large form, *H. sivalensis* from the Upper Siwaliks, a smaller and slightly more primitive species from the Middle Siwaliks, namely *H. lydekkeri*, and a very small form from the base of the Middle Siwaliks, *H. deterrai*.

It seems very likely the *H. grandis* of Pilgrim is a large variety of *H. sivalensis*, and that *H. tatroti* is a variant of *H. lydekkeri* persisting on up into the base of the Upper Siwaliks.

Sus Linnaeus, 1758

Generic type, *Sus scrofa* Linnaeus

Sus hysudricus Falconer and Cautley

Sus hysudricus, Falconer and Cautley, 1847, Fauna Antiqua Sivalensis, Pl. LXX, fig. 2, 3, Pl. LXXI, figs. 5–11.

Additional References.—

Falconer, H., 1868A, pp. 510–512.

Lydekker, R., 1880B, p. 31; 1883C, pp. 83, 92; 1884C, pp. 77–84, Pl. VIII; 1885B, pp. 48–50; 1885D, pp. 271–272.

Stehlin, H., 1899, pp. 16, 18, 164.

Pilgrim, G. E., 1910B, p. 202; 1913B, p. 276; 1926, pp. 61–62, Pl. I.

Matthew, W. D., 1929, p. 444.

Type.—(Lectotype.)—Brit. Mus. No. 15362, a skull.

Cotypes.—Brit. Mus. Nos. M 2050, a right maxilla; M 2051, a left maxilla; M 2052, a nearly complete mandible; 16603, a right mandibular ramus; 16605, a right ramus; 15373, a left ramus; 37130, anterior portion of a mandible.

Horizon.—Upper Siwaliks.

Locality.—Siwalik Hills.

Diagnosis.—A medium sized *Sus*, very similar to the modern species *Sus cristatus*. It may be distinguished from the latter by its relatively larger premolars and the shorter heel in the third molar. Diastema between first and second premolars short.

This species has been discussed to some extent with *Propotamochoerus hysudricus*.

Lydekker gave very full descriptions of *Sus hysudricus*, and nothing need be added to his account of the species. None of the material in the American Museum collection has been definitely identified as belonging to this genus and species.

Sus advena Pilgrim

Sus advena, Pilgrim, 1926, Pal. Indica, N.S., VIII, No. 4, pp. 59–60, Pl. XX, figs. 1, 14.

Type.—G.S.I. No. B 732, front portion of a right mandibular ramus.

Paratypes.—None.

Horizon.—Middle Siwaliks, Nagri zone.

Locality.—Hari Talyangar, Bilaspur State.

Diagnosis.—Like *Sus falconeri*, but smaller and more primitive.

Sus falconeri Lydekker

Sus falconeri, Lydekker, 1884, Pal. Indica (X), III, pp. 66–77, Pl. VII, figs. 1, 2, 3–9, Pl. X.

Additional References.—

Lydekker, R., 1885D, pp. 263–266.

Stehlin, H., 1899, pp. 72, 155, 265, 373.

Pilgrim, G. E., 1910B, p. 202; 1913B, p. 324; 1926, p. 63, Pl. XX, fig. 15.

Matthew, W. D., 1929, pp. 444, 459.

Type.—Brit. Mus. No. 16386, a skull. (Catalogued by Lydekker, 1885D, as No. 15386.)

Paratypes.—Sci. and Art Mus., Dublin, Nos. C 27, a skull; C 26, a mandible. G.S.I. Nos. B 16, fragment of a mandibular ramus; B 18 fragment of a maxilla. Brit. Mus. Nos. M 2044, a skull; M 2042, a skull; M 2013, a skull; M 2043, a palate; M 2012, a mandible; 16614, a mandible; 15387, a mandible; 16612, a mandible.

Horizon.—Upper Siwaliks.

Locality.—Siwalik Hills.

Specimens in the American Museum.—Amer. Mus. No. 19825. A left M^3. Upper Siwaliks, upper clays below conglomerate, two miles north of Siswan.

19887. A left M^3. Upper Siwaliks, upper clays below conglomerate, three miles north of Siswan.

19913. A right M^3. Upper Siwaliks, upper clays below conglomerate, three miles west of Chandigarh.

Diagnosis.—An extremely large suid, with hypsodont cheek teeth, and a much elongated third molar. The skull is distinguished by the length of the preorbital portion, and the posterior position of the orbit; the mandible has a high ascending ramus and a long, heavy symphysis.

Sus falconeri AND THE ORIGIN OF THE WART HOGS

Sus falconeri is distinguished not only by its great size, but also by the length and complexity of the third molar, both above and below. This tooth, when worn, presents a very complex enamel pattern, due to the numerous accessory conules. When this enamel pattern is analyzed, and reduced to its simplest form, that is, to the original arrangement of cones and conules, it is seen to be made up of an orderly arrangement of three longitudinal rows of cusps or cones. Moreover, these cones are formed from the elongated talon or talonid of the tooth. This at once suggests a condition similar to that of the third molar of *Phacochoerus*, and it would thereby seem reasonable to assume that the African wart hog sprang from the suid stem at a point not far distant from *Sus falconeri*.

Modern species of *Sus*   *Phacochoerus*

— *Sus falconeri* ——————

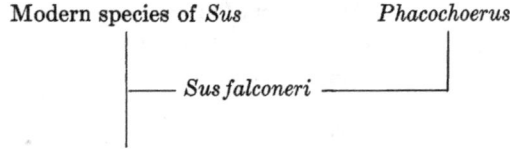

The third molar, however, is not the only character suggestive of a link between *Sus falconeri* and *Phacochoerus*. A study of the skulls in these two forms shows that characters which are usually taken as being very typical of *Phacochoerus*, are certainly developed in a moderate way in the Siwalik species.

Perhaps it may be well to list the specialized habitus characters that set *Phacochoerus* apart from the typical suids, and then with these characters in mind we may proceed to an examination of *Sus falconeri*, to see how far it has progressed along the phacochoerid branch of phylogenetic development. Putting our knowledge in tabular form, we may say that *Phacochoerus* displays the following specialized characters.

1. The dentition is reduced, the formula being $\frac{1\text{-}1\text{-}\ \ 1\ \ \text{-}3}{3\text{-}1\text{-}(0\text{-}1)\text{-}3}$.

2. The third molar is extremely elongated at the expense of the anterior cheek teeth. This has been accomplished by an elongation of the heel of the tooth. The cusps are arranged in three longitudinal rows.

3. The canines are enlarged.

4. The preorbital portion of the skull is greatly elongated, while the postorbital portion is compressed.

5. The orbit has migrated upward and backward, so that it projects above the level of the frontal bones. The front of the orbit is considerably behind the third molar.

6. There has been an accompanying upward pushing of the whole basicranial region, causing the glenoid-condylar articulation to occupy a very high position.

7. The compression of the postorbital region has occasioned the change in proportion of nearly all of the skull bones.

9. The jugal is very heavy, especially in the male.

10. The frontals are very wide, making the eyes far apart.

11. The bullae are quite compressed.

12. There is a peculiar development of two deep pockets, one on either side of the vomer, and placed just anteriorly to the basisphenoid. These are formed by a median outgrowth from the alisphenoid, and they face towards the opening of the posterior nasal choanae.

13. The ascending ramus of the mandible has become very high, as an accompaniment of the raising of the glenoid.

14. The mandibular symphysis is long, heavy and wide, making a sort of shovel of the front of the jaw.

Turning now to *Sus falconeri*, we see that it shows the following traits that are similar to or prophetic of the above listed specializations of the wart hog.

1. Although the dentition is not reduced, the third molar is elongated and complicated as pointed out above.

2. The preorbital portion of the skull is greatly elongated.

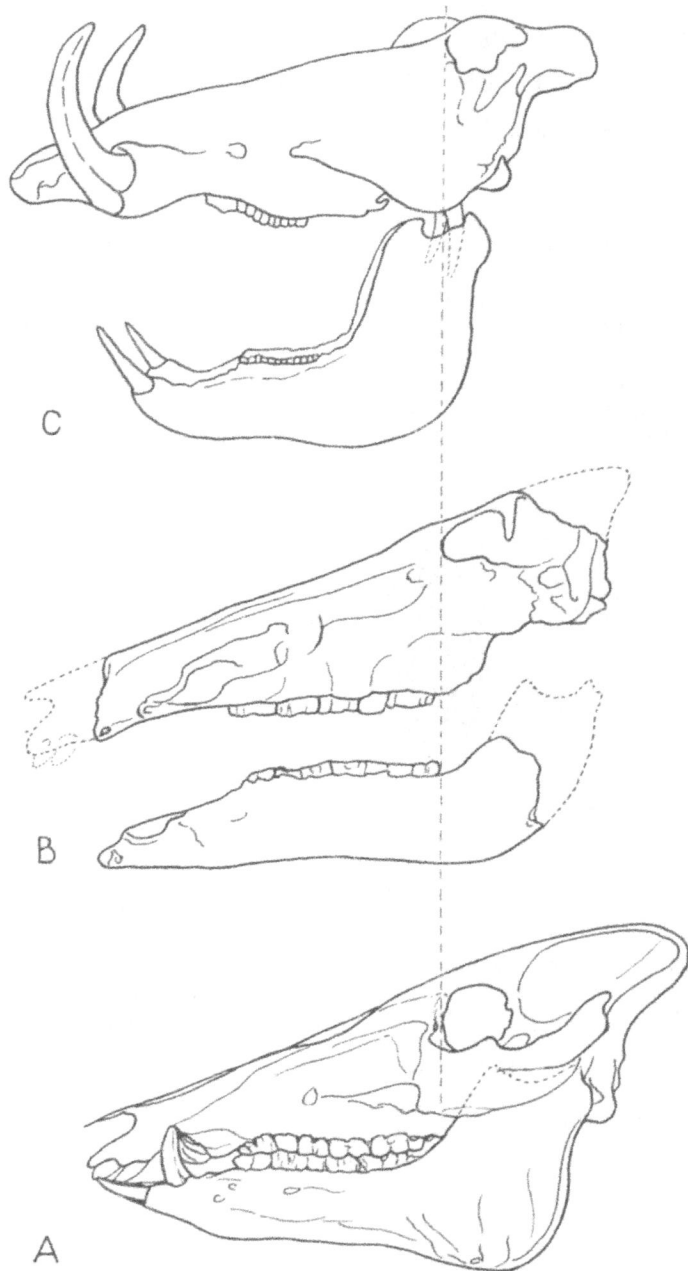

Fig. 116. Comparison of the skulls and mandibles of (A) *Sus scrofa*, (B) *Sus falconeri* and (C) *Phacochoerus aethiopicus*. Anterior borders of the orbits aligned vertically. Figures one fourth natural size. After Falconer and Cautley, and Sclater.

3. The canines are moderately large.

4. The eye has migrated upward and backward to some extent. The anterior border of the orbit is back of M³.

5. There is a slight upward growth of the glenoid.

6. The jugal is rather heavy, much heavier than in the typical *Sus*, but not so much so as in *Phacochoerus*.

7. The frontals are moderately wide.

8. The mandibular symphysis is heavy and long, much as in *Phacochoerus*.

These phacochoerid traits are much more pronounced in the type skull, which is that of a male, than in the female skull figured by Lydekker.

Thus it is seen that *Sus falconeri*, though a typical suid in many of its features, does show a decided trend towards the *Phacochoerus* line of development.

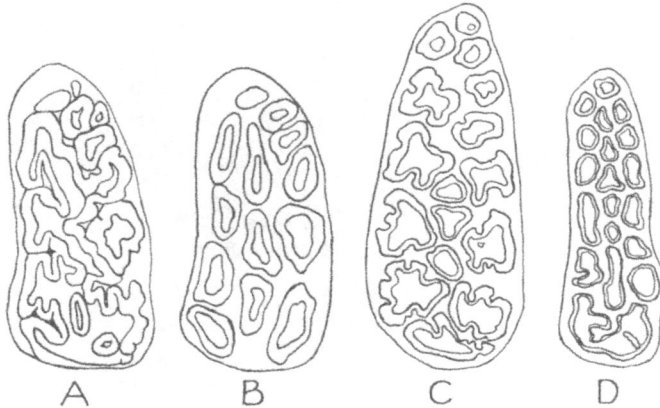

FIG. 117. Comparison of the right third upper molar in (*A*) *Sus falconeri* [worn], (*B*) *Sus falconeri* [unworn], (*C*) *Kolpochoerus sinuosus* and (*D*) *Phacochoerus aethiopicus*. Natural size. A and B after Lydekker, C and D after van Hoepen.

Some phacochoerid teeth recently described from South Africa by van Hoepen and van Hoepen (1932) show stages that are really intermediate between *Sus falconeri* and *Phacochoerus*. One of these, that of *Kolpochoerus*, is shown in the accompanying figure (Fig. 117).

MEASUREMENTS

Amer. Mus. No. 19825	R M³	Length... 62.0 mm.	Width... 27.0 mm.	Index... 43.5
19887	R M³	65.0	30.0	46.0

PROPORTIONS

	Sus scrofa	*Sus falconeri*	*Phacochoerus aethiopicus*
	A.M. 57326	B.M. 15386	A.M. 54174
			Dub. Mus. C 27
Preorbital length	207 mm.	255 mm.	228 mm.	250 mm.
Postorbital length	106	108	124	70
Ratio $-\dfrac{\text{Preorb.}}{\text{Postorb.}} :: \dfrac{x}{1}$	1.95	2.36	1.84	3.58

Sus comes Pilgrim

Sus comes, Pilgrim, 1926, Pal. Indica, N.S., VIII, No. 4, p. 60, Pl. XIX, fig. 8.

    Type.—G.S.I. No. B 733, a left mandibular ramus with DM_4, M_{1-2}.

    Paratypes.—None.

    Horizon.—Middle Siwaliks, Dhok Pathan zone.

    Locality.—Hasnot, Punjab.

    Specimen in the American Museum.—Amer. Mus. No. 19709, fragment of the left mandibular ramus, with DM_4, M_{1-2}.   From the Middle Siwaliks, three miles east of Dhok Pathan.

    Diagnosis.—A medium sized suid, with relatively narrow molars.

Pilgrim's description of this species is not very detailed, and one must needs base comparisons on his figures.   There is, however, a specimen in the American Museum collection that would seem to compare more closely with Pilgrim's figure of *Sus comes* than with any other Siwalik suid.   This specimen is slightly larger than the type, but this discrepancy in size may readily be attributed to individual variation.

The American Museum specimen is characterized by the development of accessory conules, thus adding to the complexity of the molar pattern, and giving it a distinct "modern" appearance.   The molars are rather narrow in comparison to their length.

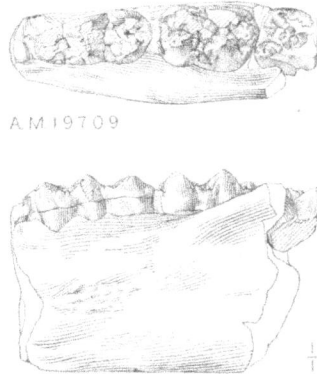

FIG. 118.   *Sus comes* Pilgrim.   Amer. Mus. No. 19709, left ramus with DM_4, M_{1-2}.   Crown view above, lateral view below.   Natural size.

MEASUREMENTS

Sus comes M_1	Length	Width	Index
G.S.I. No. B 733	14.5 mm.	10.0 mm.	69
A.M. No. 19709	16.5	12.0	73

Sus adolescens Pilgrim

Sus adolescens, Pilgrim, 1926, Pal. Indica, N.S., VIII, No. 4, p. 60, Pl. XX, fig. 2.

Type.—G.S.I. No. B 734, a left mandibular ramus with P_4, M_{1-3}.

Paratypes.—None.

Horizon.—Middle Siwaliks, Dhok Pathan zone.

Locality.—Hasnot, Punjab.

Diagnosis.—Like *Sus falconeri*, but smaller, with a simpler talon on the last premolar, and a proportionately shorter third molar.

Sus praecox Pilgrim

Sus praecox, Pilgrim, 1926, Pal. Indica, N.S., VIII, No. 4, p. 60, Pl. XIX, figs. 4–7.

Type.—G.S.I. No. B 735, a last left lower premolar.

Paratypes.—G.S.I. Nos. B 736, a second lower premolar; B 737, a third upper premolar; B 738, a first upper premolar. Brit. Mus. No. M 12692, a last lower premolar.

Horizon.—Middle Siwaliks, Dhok Pathan zone.

Locality.—Hasnot, Punjab.

Diagnosis.—Like the modern *Sus verrucosus*, but larger and with heel of fourth lower premolar less elevated.

Sus peregrinus Pilgrim

Sus peregrinus, Pilgrim, 1926, Pal. Indica, N.S., VIII, No. 4, p. 61, Pl. XIX, fig. 9.

Type.—G.S.I. No. B 739, a left mandibular ramus.

Paratypes.—None.

Horizon.—Base of Upper Siwaliks, Tatrot zone.

Locality.—Tatrot, Jhelum District, Punjab.

Diagnosis.—A small *Sus* having simple, slender cheek teeth; third molar with a long talon.

Sus bakeri Pilgrim

Sus bakeri, Pilgrim, 1926, Pal. Indica, N.S., VIII, No. 4, pp. 62–63.

Type.—Brit. Mus. No. 15373, a left mandibular ramus. Originally designated as *Sus hysudricus* by Falconer and Cautley, and by Lydekker.

Paratypes.—None.

Horizon.—Upper Siwaliks.

Locality.—Siwalik Hills.

Diagnosis.—Distinguished by the hypsodont molars, having simple cusps. Talon of third molar simple.

Sus cautleyi Pilgrim

Sus cautleyi, Pilgrim, 1926, Pal. Indica, N.S., VIII, No. 4, pp. 63–64.

Type.—Brit. Mus. No. M 2043, a palate, with right and left M^{2-3}. Originally designated as *Sus giganteus* by Falconer and Cautley, and as *Sus falconeri* by Lydekker.

Paratypes.—None.

Horizon.—Upper Siwaliks.

Locality.—Siwalik Hills.

Diagnosis.—Larger than *Sus falconeri* and with longer and more complicated talon in the third molar.

This form may very well be a large variant of *Sus falconeri*.

Various suid bones, especially astragali and phalanges, are present in the American Museum collection, but since they are not definitely associated with teeth it is difficult to assign them to particular genera.

Amer. Mus. No. 19622. Teeth. Lower Siwaliks, 200 feet above the level of Chinji Rest House, four miles west of Chinji Rest House.

19648. Mandibular ramus. Lower Siwaliks, 1100 feet above the level of Chinji Rest House, one mile northwest of Chinji Rest House.

19890. Mandibular ramus. Middle Siwaliks, one half mile east of Pati.

29813. Upper Canine. Lower Siwaliks, locality unknown.

THE PHYLOGENY OF THE INDIAN SUIDAE [63]

From the above systematic review it will be seen that fossil Suidae are abundant in the Siwalik beds. The Indian region would seem, during Upper Tertiary times, to have been a center for the adaptive radiation of the suid group, and here, in one relatively restricted area, there may be found almost the complete phylogenetic history of this family, from primitive undifferentiated types to advanced, highly specialized genera.

The Indian Suidae may be divided into several phylogenetic groups or branches, each of which is an evolutionary unit. These groups are listed below.

Group   I.   *Palaeochoerus.*
Group   II.   *Listriodon.*
Group III.   *Conohyus—Sivachoerus—Tetraconodon.*
Group   IV.   *Dicoryphochoerus—Sus—[Phacochoerus].*
              Propotamochoerus—Potamochoerus.
              Hyosus—Sivahyus—Hippohyus.
Group   V.   *Lophochoerus.*
Group   VI.   *Sanitherium.*

These phylogenetic groups naturally are not of equal rank or importance, and it may be that some of them might be advantageously combined or separated. They will be briefly discussed in the following paragraphs.

GROUP I

Palaeochoerus

Palaeochoerus in the Siwaliks is an excellent example of a primitive, ancestral form persisting on in time to a period contemporaneous with its specialized descendants. The species of *Palaeochoerus* in the Siwaliks are similar to the typical *Palaeochoerus* of the European Oligocene, so we are justified in thinking that this genus held on in India into the Pliocene without undergoing any appreciable evolutionary changes. Thus *Palaeochoerus* is structurally ancestral, in a general way, to the advanced genera with which it is contemporaneous.

[63] The reader is referred to Colbert, Edwin H., 1935, Amer. Mus. Novitates, No. 799, for a more detailed treatment of this subject.

Group II

Listriodon

This genus must have split off at a very early date from the primitive *Palaeochoerus*—type of ancestor, for it is specialized in an aberrant way, quite separated from the more normal kinds of pigs. *Listriodon guptai* of the Kamlial would seem to be a species somewhat transitional in structure between the normal bunodont pigs and the specialized listriodont forms.

Group III

Conohyus—Sivachoerus—Tetraconodon

Conohyus, a genus closely related to the European form, *Hyotherium*, is distinguished by its enlarged posterior premolars. Thus it would seem to be on a line ancestral to *Sivachoerus* and *Tetraconodon* of the Middle and Upper Siwaliks, characterized by their large size and their tremendously enlarged premolars. We may imagine that *Sivachoerus* and *Tetraconodon* were contemporaneous developments from *Conohyus*. In *Tetraconodon*, however, evolution was more rapid than in *Sivachoerus* so that it became much the larger and more specialized of the two genera.

Group IV

Dicoryphochoerus—Sus—Phacochoerus

Propotamochoerus—Potamochoerus

Hyosus—Sivahyus—Hippohyus

This group is composed of a variety of related genera, which represent a fairly wide range of adaptive radiation. These are what we might call the advanced and "normal" suids, typified by the elongation of the skull, the complication of the cheek teeth, and the various development of fighting tusks.

Dicoryphochoerus, described by Pilgrim, may be considered as approximating the stem, or the central branch of this group. *Dicoryphochoerus haydeni* is a primitive species that would seem to approximate a structural ancestor for the entire phylogenetic group. The later species of *Dicoryphochoerus* are considerably advanced in size and in complexity of the cheek teeth over *D. haydeni*, and the culmination of the genus is reached in the form *D. titan*, a species of truly gigantic proportions. In the phylogenetic advance of this genus the tendency has been towards a retention of the primitive shortness of the cheek teeth, which may be compared to the elongation of the cheek teeth in the genus *Sus*.

Sus seemingly branched out from the relatively primitive stem of this group, from a species like *Dicoryphochoerus haydeni*, and paralleled *Dicoryphochoerus* in its evolutionary development. As mentioned above, *Sus* was characterized by the elongation of the cheek teeth.

In a preceding section of this present work (pp. 257–260), it has been shown how *Phacochoerus* might have been derived from *Sus* through a form like *Sus falconeri*. The reader is referred to the pages cited.

The hypsodont branch of this suid group is represented by the genera *Hyosus*, *Sivahyus* (possibly synonymous) and *Hippohyus*. In these forms the cheek teeth became quite tall, and the enamel pattern became complicated, due to the intergrading upgrowth of cones and accessory conules. The skull of *Hippohyus* is quite distinct from the *Sus* skull, being typified by its heaviness, the rather centrally placed orbit of small size, and the stout zygomatic processes.

I cannot agree to Pilgrim's interpretation that the genera *Propotamochoerus* and *Potamochoerus* are widely divergent from the *Sus—Dicoryphochoerus* branch of suid development. It would seem to me that any differences between them are of a rather minor character. In the *Potamochoerus* line the last lower premolar consists of single cone, with sometimes a minute secondary cone behind it, whereas in the *Sus* line the last lower premolar has the form of a double cone, with the two apices closely appressed, and either in line, antero-posteriorly, or slightly oblique. The cheek teeth, too, in the *Potamochoerus* line are slightly less complex than they are in the *Sus* line. On the other hand, the essential

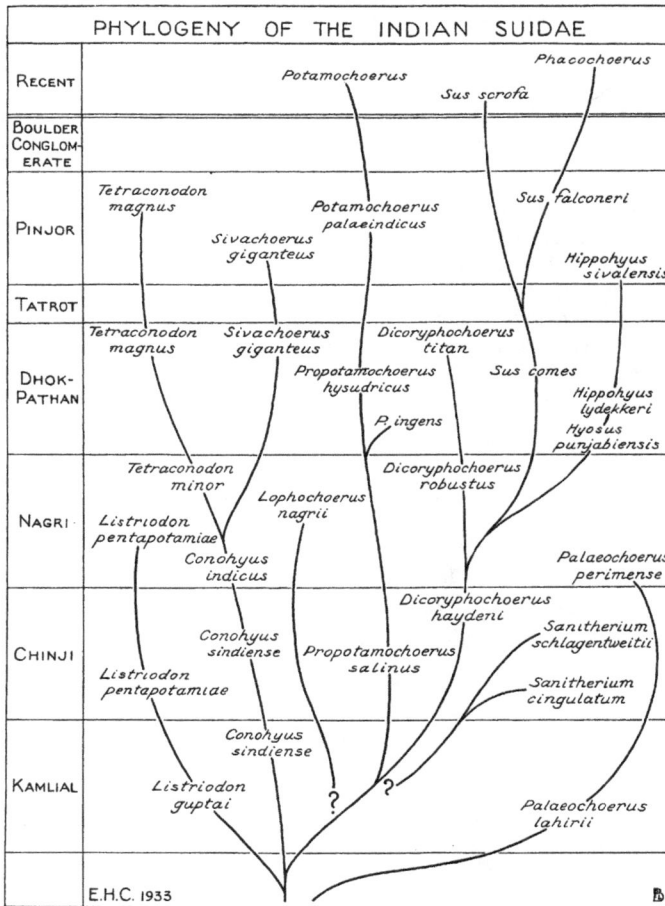

FIG. 119. Phylogeny of the Indian Suidae.

features of skull and tooth structure are so much alike in these two phylogenetic branches, that it would seem as if they must be closely related to each other. Thus we imagine that they split late in the Tertiary period, retaining their similar heritage characters, but developing habitus characters that set them slightly apart one from the other. Just as *Dicoryphochoerus haydeni* would seem to approximate the central stem for the *Sus* line, so would *Propotamochoerus salinus* seem to be near the central stem for the group under consideration.

Group V

Lophochoerus

Lophochoerus would seem to be an aberrant genus, related to *Choerotherium*. Just how it is related to the other suids is a difficult question. Pilgrim has suggested that it is an offshoot from the *Propotamochoerus—Potamochoerus* line.

Group VI

Sanitherium

Sanitherium would seem to be an aberrant genus, possibly derived from the *Dicoryphochoerus—Sus* branch of phylogenetic development.

The accompanying chart is a graphic presentation of the foregoing discussion. The several phylogenetic groups as outlined above are indicated by branching lines, and the species most typical of them are placed at their proper stratigraphic levels.

It will be noticed that the present phylogenetic chart differs from the chart outlined by Pilgrim in 1926 by its greater simplicity. This difference is due to certain fundamental differences of thought. Dr. Pilgrim considers the Suidae to be polyphyletic down to the base of the Eocene, while I consider the Suidae to be monophyletic through the earlier portions of the Tertiary period.

ANCODONTA

ANTHRACOTHERIOIDEA

ANTHRACOTHERIIDAE

Choeromeryx Pomel, 1848

Generic type, *Anthracotherium silistrense* Pentland

Choeromeryx silistrense (Pentland)

Anthracotherium silistrense, Pentland, 1828, Trans. Geol. Soc., London, Ser. 2, II, p. 393, Pl. XLV, figs. 2–5.

Choeromeryx silistrense, Pomel, 1848, Comptes Rendus, Acad. Sci., Paris, XXVI, No. 25, p. 687.

Additional References.—

Falconer, H., and Cautley, P. T., 1847F, Pl. LXVIII, figs. 22, 22a.

Falconer, H., 1868A, p. 508.

Lydekker, R., 1877B, p. 77; 1878A, pp. 77–78; 1880B, p. 32; 1883A, pp. 166–167; 1883C, p. 90; 1885D, p. 165; 1885E, pp. 72–73; 1886F, p. xiii.

Pilgrim, G. E., 1910B, p. 202; 1915C, p. 227.

Matthew, W. D., 1929, p. 463.

Type.—(Lectotype.)—Brit. Mus. No. 19040, a right maxilla with DM$^{3-4}$.

Cotype.—Brit. Mus. No. 19041, a right M^3. (Not this species.)

Horizon.—Lower or Middle Siwaliks.

Locality.—"The Siwaliks of Káribári, Gáro Hills, N. E. Bengal."

Diagnosis.—A small, selenodont anthracothere, with four cusps in the upper molars.

A great deal of confusion exists in the literature regarding the genus and species *Choeromeryx silistrense.*

In 1828, Pentland described two specimens from the Siwaliks as *Anthracotherium silistrense.* His description was based on a right maxilla, containing the last two milk molars, and a right upper third molar. Now it happens that the maxillary fragment, described by Pentland, contains a four cusped milk molar, whereas the single upper molar referred by him to the same species is a five cusped tooth.

In 1848, Pomel renamed the genus, calling it *Choeromeryx,* on the basis of the four cusped tooth.

"Quatrième tribu: *Dichodiens.*—Molaires supérieures à quatre mamelons suelement, au lieu de cinq; canines et incisives des Anoplothériums; face externe des molaires supérieures dépourvue d'arêtes en double U; à leur place un tubercule comme dans les deux précédentes tribus.—Dichodon, Choeromeryx (I), Merycopotamus. (I) *Anthracotherium silistrense,* Pentl."[65]

Thus Pomel unconsciously selected a lectotype for the species, and this lectotype was the four cusped tooth. Subsequent authors confused the problem by thinking that Pomel included both the four cusped and the five cusped tooth in his genus *Choeromeryx.* This lead to a great deal of muddled thinking regarding the question of the taxonomy of *Choeromeryx silistrense* and *Anthracotherium silistrense.*

It is quite clear, however, that Pomel intended *Choeromeryx* to be a four cusped genus of anthracotheres, and his description, quoted above, restricts the genus to the four cusped tooth of the two specimens described by Pentland. Consequently the five cusped tooth inadvertently was left by Pomel without a name.

A discussion of this five cusped tooth will be given in the consideration of the species *Anthracotherium punjabiense.*

Choeromeryx mihi, listed by Pilgrim in his faunal list of 1910 (Pilgrim, G. E., 1910B, p. 202), is a nomen nudum.

Anthracotherium Cuvier, 1822

Generic type, *Anthracotherium magnum* Cuvier

Anthracotherium punjabiense Lydekker

Anthracotherium punjabiense, Lydekker, 1877, Rec. Geol. Surv. India, X, p. 78.

Rhagatherium sindiense, Lydekker, 1877, Rec. Geol. Surv. India, X, p. 225.

Anthracotherium silistrense, Lydekker, 1878, Rec. Geol. Surv. India, XI, p. 78.

Microselenodon silistrense, Pilgrim, 1910, Rec. Geol. Surv. India, XL, p. 201.

Microbunodon silistrense, Pilgrim, 1913, Rec. Geol. Surv. India, XLIII, p. 317.

[65] Pomel, A., 1848, p. 687.

Additional References.—

Falconer, H., and Cautley, P. T., 1847F, Pl. LXVIII, figs. 23–23a.
Falconer, H., 1868A, p. 508.
Lydekker, R., 1880B, p. 31; 1883A, pp. 149–152, Pl. XXIII, fig. 12, Pl. XXIV, fig. 1; 1883C, p. 89; 1885D, p. 243.
Pilgrim, G. E., 1913B, p. 285.
Cooper, C. Forster, 1924, pp. 18–20.
Matthew, W. D., 1929, pp. 449, 453.

In all of the above references this species is listed as *Anthracotherium silistrense* or *Microbunodon silistrense*. In Forster Cooper, 1924, both *Anthracotherium silistrense* and *Anthracotherium punjabiense* are listed, but are considered as separate species.

Type.—G.S.I. No. B 104, a right mandibular ramus with $M_{2–3}$.

Paratypes.—None.

Horizon.—Middle Siwaliks(?) for the type. Also from the Lower Siwaliks.

Locality.—Hasnot, Punjab. Also Chinji, and other localities in the Punjab.

Specimen in the American Museum.—Amer. Mus. No. 19444, a mandible, complete except for the anterior portion of the symphysis. From the Lower Siwaliks, one mile northwest of Chinji Rest House, Punjab.

Diagnosis.—A small primitive anthracothere, with brachyodont teeth. Upper molars with protoconule well developed. Lower premolars simple.

It has been shown on a foregoing page that *Choeromeryx silistrense* (Pentland) is a four-cusped form of anthracotherine. Therefore the five-cusped tooth, originally described by Pentland as belonging to the same species, i.e., *Anthracotherium silistrense*, as the four cusped one, becomes, according to Pomel's limitation of the genus *Choeromeryx*, a specimen without a name. Lydekker had no right, in 1878 to designate two species and attribute them to two different authors, thus:

Anthracotherium silistrense Pentland.
Choeromeryx silistrense Pomel.

There is only one species, and that is *Choeromeryx silistrense* (Pentland), emend Pomel.

Thus, unfortunate as it may be, it so happens that the small, five cusped anthracothere described and figured by Pentland, hitherto known as *"Anthracotherium silistrense"* is a specimen without a name.

In 1877, Lydekker named a new species of small anthracothere, on the basis of a fragmentary mandible. This he called *Anthracotherium punjabiense*. Soon after, in 1878, he combined this species with *"Anthracotherium silistrense,"* regarding it as a synonym of the latter. Lydekker was justified in considering *A. punjabiense* and *"A. silistrense,"* as belonging to a single species. A comparison of the mandible of *A. punjabiense* with various mandibles of *"A. silistrense"* readily demonstrates their identity with each other. And there can be very little doubt but that the five cusped upper tooth is to be associated with these lower dentitions in one species.

In 1924 Forster Cooper resurrected Lydekker's species, *Anthracotherium punjabiense*, but he considered it as different from *"A. silistrense."*

There seems to be but one logical and comparatively easy way out of this dilemma.

It is hereby proposed to recognize Lydekker's species, *Anthracotherium punjabiense*, based on a lower jaw, and to refer all of the small, Lower and Middle Siwalik five cusped anthracotheres to this species. Thus the forms heretofore known as *Anthracotherium silistrense* in the literature, become *Anthracotherium punjabiense*. Unfortunate as it may seem, to supplant a specific name that has become more or less intrenched, the above course of action seems to me the only one that really solves the difficult problem.

The name *Choeromeryx silistrense* (Pentland) applies to the milk molar with four cusps.

It may be well here to consider briefly the suitability of the name *Microbunodon* for this species.

Depéret, in his original designation of the genus *Microbunodon*, characterized it in the following way.

"*Microbunodon* n.g.; type *Anthracotherium Laharpei* Renevier. Ce rameau, encore mal suivi dans son évolution, comprend de petites formes à dentition *brachybunodonte*, comme celles du premier rameau. L'espèce type de Rochette, connue seulement par ses molaires supérieures (m³ = 20 mm) portait de longues canines aplaties, traguliformes, qui distinguent bien les *Microbunodon* de tous les autres Anthracothéridés." [66]

Forster Cooper, in 1924, evidently misinterpreted Depéret's analysis of the genus, because he makes the following statement.

"As the genus *Microbunodon* according to Depéret's diagnosis is very sharply marked off from the others by the triangular upper canines and premolariform lower canines. . . ." [67]

Evidently Forster Cooper mistook the word *traguliform* for *triangular*. Then supposing *Microbunodon* to be distinguished by triangular upper canines, he excluded *Anthracotherium silistrense* from it.

So we see that *Anthracotherium punjabiense* (called *A. silistrense* by Forster Cooper) can not be excluded from the genus *Microbunodon* on the grounds that it does not have a triangular canine. There are, however, good reasons for placing *A. punjabiense* in a genus different from *Microbunodon*. In speaking of the species under discussion, Matthew says:

"Pilgrim refers this to *Microbunodon*, but it is remarkably selenodont for an anthracotherine, as much so as *Brachyodus*, etc." [68]

In the original definition of *Microbunodon*, Depéret stipulates that it is a small *brachybunodont* anthracothere. Consequently it would seem logical to exclude the Siwalik species from Depéret's genus, in the light of Matthew's positive observations of its very selenodont teeth.

Of course, the Siwalik form might be included in Depéret's genus *Microselenodon*, as it once was by Pilgrim. But in view of the fact that the definition of this latter genus is not very clear, and that the necessary information for such an inclusion is not just now available, it seems best to place the form under discussion in the genus *Anthracotherium*, recognizing the desirability of a future transference to a more appropriate genus.

A Mandible of *Anthracotherium punjabiense*

This species, discussed at some length in preceding paragraphs has been described by several authors from fragmentary remains of the upper and lower dentitions. The jaw

[66] Depéret, C., 1908, p. 2.
[67] Forster Cooper, C., 1924, p. 18.
[68] Matthew, W. D., 1929, p. 463.

described in the following paragraphs, would seem to be the first complete mandible recorded, and for this reason it is treated in some detail.

Both rami are present and complete. A portion of the symphysis is broken, anterior to the first premolar, so that the canines and incisors are missing. These missing parts may, however, be restored from comparable anthracotheres.

The horizontal ramus is long and shallow, and the lower border is comparatively straight. The angle, though partially restored, would seem to have projected slightly below the lower border line of the ramus. The ascending ramus is low, so that the condyle is only a short distance above the level of the occlusal line of the molars. The coronoid projects somewhat above the condyle, and it is curved strongly back.

There is a full set of cheek teeth. The premolars are distinguished by their simplicity, each tooth consisting of a single triangular blade, the anterior edge of which is convex, and the posterior edge slightly concave. The posterior edge of the fourth premolar is broad, being made up of two ridges that run down from the tip of the tooth and diverge laterally at the base.

The molars are typically anthracotherine, each tooth consisting of four cusps, while in the last molar there is a well developed talonid. The teeth are broad and brachyodont.

MEASUREMENTS

Anthracotherium punjabiense Amer. Mus. No. 19444

Length of mandible, condyle to first incisor (estimated)........ 200 mm.
Height of condyle above lower border of mandible............ 47
Length P_1–M_3.. 112
Length of premolar series................................... 58
Length of molar series...................................... 55

	Length	Width
Right P_1...............................	8.0 mm.	4.0 mm.
P_2....................................		
P_3....................................	16.5	8.0
P_4....................................	14.5	9.0
M_1....................................	14.0	10.0
M_2....................................	17.0	12.5
M_3....................................	24.5	14.0

Pilgrim has shown how the lower molars of *Anthracotherium punjabiense* may be distinguished from those of *Dorcabune anthracotherioides*, the primitive tragulid from the Lower Siwaliks, by the fact that in the former the anterior spur of the hypoconid runs inwardly to meet the posterior spur of the metaconid. To put it in a more general way, the lower molars of the anthracotheres are distinguished by the dominance of the spurs from the external cusps, while the primitive tragulid molars show their true relations by the fact that the external cusp spurs are restricted as to their lingual extension. Another difference may be noted; in the anthracotheres the external and internal cusps are laterally opposite to each other, whereas in the primitive tragulids these cusps are always slightly oblique. The differences pointed out above are basic distinctions, outlining the trend in separate evolutionary lines, and though it would seem at first sight that a fine division is being drawn between these teeth, it must be remembered that the artiodactyls make up a group in which

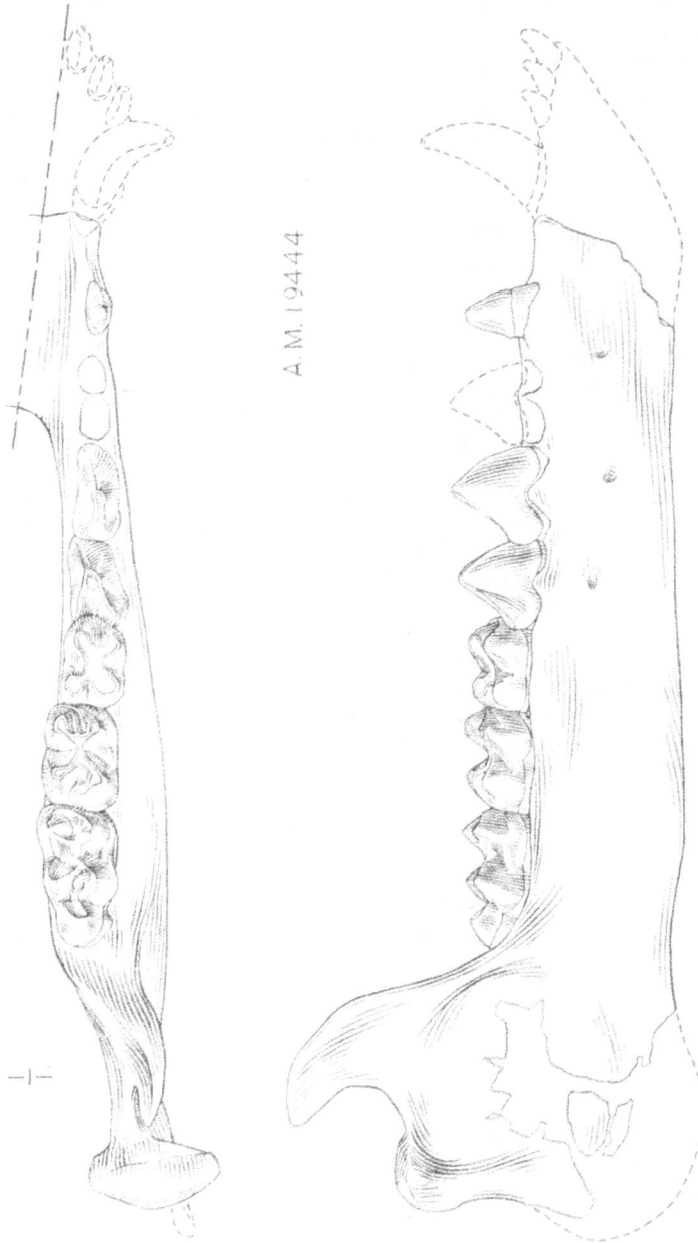

A.M. 19444

FIG. 120. *Anthracotherium punjabiense* Lydekker. Amer. Mus. No. 19444, mandible. Crown view of right ramus above, side view below. Natural size.

evolutionary changes have been pronounced as to skull and body form, while the teeth have followed more conservative lines of development. The fact must then be accepted that the student of this group is required to search out small constant distinctions in the teeth as guides of taxonomic and phylogenetic value. The recognition of these seemingly minute differences is especially important when dealing with fossil forms in which most of our information is necessarily based on the dentition.

Hyoboops Trouessart, 1904

Generic type, *Hyopotamus palaeindicus* Lydekker

Hyoboops palaeindicus (Lydekker)

Hyopotamus palaeindicus, Lydekker, 1877, Rec. Geol. Surv. India, X, p. 77.

Hyoboops palaeindicus, Trouessart, 1904, Catalogus Mammalium, I quinquennale sup-
plementum, p. 651.

> *Additional References.—*
>
> > Lydekker, R., 1878A, p. 80; 1880B, p. 31; 1883A, pp. 158–160, Pl. XXIII, figs.
> > 4, 6, 7, 9, Pl. XXIV, fig. 4; 1883C, p. 91; 1885B, p. 40.
> >
> > Pilgrim, G. E., 1910B, p. 201.
> >
> > Cooper, C. Forster, 1924, p. 47, Pl. IV, fig. 6.

Type.—(Lectotype.)—G.S.I. No. B 82, a left upper molar. This is presumably one of the two teeth mentioned by Lydekker in his original description of 1877. This is a left upper molar, not a right, as stated by Lydekker in the Catalogue of Siwalik Vertebrates in the Indian Museum, 1885, p. 40.

Cotypes.—Lydekker mentions a second specimen, an upper molar in his original description of the species. He does not, however, give any indication of the identity of this tooth, except to say that it is "much worn."

Horizon.—Lower Siwaliks (Lower Manchar).

Locality.—From the Laki Hills, Sind.

Diagnosis.—Brachyodont, strongly selenodont molars, with a heavy cingulum.

Hemimeryx Lydekker, 1883

Generic type, *Hemimeryx blanfordi* Lydekker

Hemimeryx blanfordi Lydekker

Hemimeryx blanfordi, Lydekker, 1883, Pal. Indica (X), II, pp. 167–169, Pl. XXIII, figs. 5, 8.

> *Additional References.—*
>
> > Lydekker, R., 1885B, p. 39.
> >
> > Pilgrim, G. E., 1910B, p. 201.

Type.—(Lectotype.)—G.S.I. No. B 89, a left upper molar.

Cotype.—G.S.I. No. B 90, two lower molars.

Horizon.—Lower Siwaliks (Lower Manchars).

Locality.—From Sind. Exact locality not given.

Diagnosis.—A medium sized anthracothere with selenodont molars. Molars with four cusps. Protocone crescent incomplete posteriorly.

The genus *Hemimeryx* was founded by Lydekker in 1883. In 1877 and 1878, in the Records of the Geological Survey of India, he discussed some teeth, which he considered as representative of a genus different from *Merycopotamus*, but in neither of these previous records did he give any designation for the specimens.

Hemimeryx pusillus (Lydekker)

Merycopotamus pusillus, Lydekker, 1885, Rec. Geol. Surv. India, XVIII, p. 146.
Hemimeryx pusillus, Pilgrim, 1910, Rec. Geol. Surv. India, XL, p. 202.

Additional References.—

Lydekker, R., 1885D, pp. 214, 215, fig. 28; 1886F, p. xii, fig. 3.
Matthew, W. D., 1929, p. 453.
Pilgrim, G. E., 1913B, p. 285.

*Type.—*G.S.I. No. B 324, a right maxillary fragment with the third molar.
*Paratypes.—*None.
*Horizon.—*Lower Siwaliks. Also lower Middle Siwaliks.
Locality.—"From the Siwaliks of Khushalghar, below Attock, Punjab."
*Specimens in the American Museum.—*Amer. Mus. No. 19357, a maxilla with left M^{1-3}. One half mile north of Chinji Rest House, about 200 feet above the level of Chinji Rest House.

19358, a right M^3. 150 feet below the level of Chinji Rest House, five miles west of Chinji Rest House.

19359, a maxilla with left M^{2-3}. About 400 feet above the level of Chinji Rest House, one and one half miles northeast of Chinji Rest House.

19360, maxilla with right M^{1-2}. About the level of Chinji Rest House, three miles west of Chinji Rest House.

19361, fragment of mandibular ramus with right M_3. About the level of Chinji Rest House, two miles west of Chinji Rest House.

19362, mandible with left M_3. 100 feet above the level of Chinji Rest House, two miles west of Chinji Rest House.

19363, miscellaneous teeth. About the level of Chinji Rest House, four miles northeast of Chinji Rest House.

19364, ramus with right P_{2-4}. 500 feet above the level of Chinji Rest House, one and one half miles west of Chinji Rest House.

19370, mandible with worn cheek teeth. Four and one half miles west of Hasnot. From the lower portion of the Middle Siwaliks.

19606, a left P^3, right M_2. One and one half miles east of Chinji Rest House.

*Diagnosis.—*Considerably smaller than *H. blanfordi*, with a cingulum completely encircling the tooth. Molars with four crescentic cusps.

This species was founded by Lydekker on the basis of a right upper molar. Since the first designation of the species, various specimens have been attributed to it, but all of the published descriptions have been based entirely on teeth.

Two closely related genera, namely *Hemimeryx* and *Merycopotamus* are known from the Siwalik series, the former occurring in the Lower Siwalik beds and extending into the lower portion of the Middle Siwaliks, and the latter ranging through the Middle and Upper

Siwaliks. *Hemimeryx* is more primitive than is *Merycopotamus*, not only in regard to tooth structures, as will be pointed out below, but also as to the form of the mandible, which will now be considered.

The mandible to be considered (Amer. Mus. No. 19370) is attributed to *Hemimeryx*. It is primitive as compared to *Merycopotamus*, and it is primitive to the degree that might be expected in a form directly ancestral to the latter genus. The symphysis is less massive than is the case in *Merycopotamus*, and its front border is less steeply inclined—a condition approaching that of the earliest anthracotheres. The horizontal ramus is relatively slender and the angle is very much less downwardly produced than it is in *Merycopotamus*. This latter feature is an important distinction between the two genera. A fragment of the ascending ramus shows that the coronoid was tall and vertical, much as in *Merycopotamus*.

The anterior outlet of the mental canal is beneath the front edge of the second premolar.

In the specimen under consideration the right canine alveolus is filled with cancellous tissue, showing that this tooth was lost during the life of the individual. This has occasioned a peculiar lop-sided growth of the front part of the jaw, due to the concentration of kinetic stresses on one side of the mandible. The left alveolus shows that the canine was compressed laterally, and that it curved strongly outward. Undoubtedly there were six incisors, although the bone carrying them has been broken away.

The cheek teeth are similar to those of *Merycopotamus*, but they are smaller. In the specimen at hand they are so worn as to have had many of the diagnostic characters obliterated. Each premolar is of simple form, consisting of a high central cusp, with two small cusps rising from its internal slope, and a broad talonid enclosed by a posterior cingulum, and divided by a longitudinal ridge which abuts against the back surface of the protoconid. The molars are too worn to furnish the basis for a description.

Fig. 121. *Hemimeryx pusillus* (Lydekker). Amer. Mus. No. 19370, right mandibular ramus. Lateral view. One half natural size.

MEASUREMENTS

Hemimeryx pusillus

Amer. Mus. No. 19357, maxilla.
Length of molar series.................................. 64.0 mm.

	Length	Width
M^1...................................	19.0 mm.	22.0 mm.
M^2...................................	24.5	26.0
M^3...................................	27.0	29.0

Amer. Mus. No. 19358, right M^3
Length.. 22.5 mm.
Width... 24.5

Amer. Mus. No. 19361, right M$_3$
Length.. 34.0
Width... 17.5

Amer. Mus. No. 19370, mandible.
Length of premolar—molar series...................... 131.0 mm.
Length of premolar series........................... 60.0
Length of molar series............................. 72.5
Depth of ramus below M$_3$........................... 43.0

	Length	Width
P$_1$...................................	12.0 mm.	6.5 mm.
P$_2$...................................	16.5	9.0
P$_3$...................................	18.0	11.0
P$_4$...................................	19.0	12.0
M$_1$...................................	17.0	13.0
M$_2$...................................	18.0	15.5
M$_3$...................................	35.0	17.5

Merycopotamus Falconer and Cautley, 1847

Generic type, *Hippopotamus dissimilis* Falconer and Cautley

Merycopotamus dissimilis (Falconer and Cautley)

Hippopotamus dissimilis, Falconer and Cautley, 1836, Asiatic Researches, XIX, pp. 49–51.
Merycopotamus dissimilis, Falconer and Cautley, 1847, Fauna Antiqua Sivalensis, Pls.
    LXVII–LXVIII.

Additional References.—The name *Merycopotamus* was published in 1845, in Owen's
'Odontography,' but without any specific name accompanying it. Consequently it was
not published according to the principles of binary nomenclature.

    Falconer, H., 1868A, pp. 505–507.
    Lydekker, R., 1876B, pp. 144; 1877A, p. 34; 1880B, p. 32; 1883C, p. 92; 1883A,
        pp. 164–165; 1885D, pp. 209, 214.
    Pilgrim, G. E., 1910B, pp. 202; 1913B, p. 322.
    Matthew, W. D., 1929, p. 449.

Type.—(Lectotype.)—Brit. Mus. No. 18441, a skull.
Cotype.—Brit. Mus. No. 18442, a right mandibular ramus.
Horizon.—Upper Siwaliks. Also Middle Siwaliks.
Locality.—Siwalik Hills. Also from the Salt Range, Punjab.

Specimens in the American Museum.—Amer. Mus. No. 19371, a maxilla with left M^{1-2}. Middle Siwaliks, two miles northwest of Hasnot.

19372, a right M^3. Middle Siwaliks, three miles east of Dhok Pathan.

19504, maxilla with right and left M^{1-2}. One half mile northeast of Hasnot.

19574, fragment of ramus with left M_{2-3}. Middle Siwaliks, three miles northwest of Chinji Rest House.

19702, fragment of ramus with left DM_4. Middle Siwaliks, one mile west of Dhok Pathan.

19726, rami with milk molars. Middle Siwaliks, locality unknown.

19756, fragment of ramus with right M_{1-2}. Middle Siwaliks, one half mile north of Hasnot.

Diagnosis.—A large, advanced anthracothere, distinguished by its selenodont, four cusped molars. The skull has a heavy muzzle, and elevated orbits. The mandible has a broad symphysis and a deep angle. The skeleton is comparatively heavy.

It seems hardly worth while to present any further description of *Merycopotamus* here, as this genus has been fully dealt with by Falconer and by Lydekker.

Lydekker's species *Merycopotamus nanus* is here considered as a synonym of Falconer's *M. dissimilis*. The specimens on which the former species was founded are merely smaller individuals of the original species.

Measurements of *M. dissimilis* are tabulated below.

MEASUREMENTS

Merycopotamus dissimilis

Amer. Mus. No. 19372, R M^3 .... Length... 25.0 mm.	Width...	24.5 mm.
Amer. Mus. No. 19756, R M_1 .... 21.0		13.0
R M_2 .... 27.0		17.0
Amer. Mus. No. 19574, L M_2 .... 26.0		16.5
L M_3 .... 41.0		20.0
Depth of ramus below M_3 .... 55.0		

Merycopotamus nanus Lydekker

Merycopotamus nanus, Lydekker, 1884, Geol. Mag., Dec. I, I, p. 545.

Additional References.—

Lydekker, R., 1885B, pp. 38–39; 1885D, pp. 211–212.

Pilgrim, G. E., 1910B, p. 201.

Type.—Brit. Mus. No. 16551, a skull.

Paratypes.—Brit. Mus. Nos. 16552, a skull; 15349, a right mandibular ramus; 18407, a left mandibular ramus.

Horizon.—Presumably Upper Siwaliks.

Locality.—Siwalik Hills.

Diagnosis.—See the diagnosis for *Merycapotamus dissimilis*.

This species is considered as synonymous with *Merycopotamus dissimilis*.

NOT SPECIFICALLY IDENTIFIED.

Amer. Mus. No. 19378, fragmentary mandible of an anthracothere. Upper Siwaliks, one half mile east of Kotal Kund.

Hemimeryx AND *Merycopotamus*

The two genera *Hemimeryx* and *Merycopotamus* are very close to each other, and their resemblances are so striking that it seems rather safe to suppose that the former, which is typically found in Lower Siwalik beds, is directly ancestral to the latter, which appears first in the Middle Siwaliks. That *Hemimeryx* is actually more primitive than *Merycopotamus* is shown by the greater hypsodonty of the teeth and the more acutely pointed cusps in the latter genus, as well as by the mandibular characters, elucidated on a preceding page.

The differences between the teeth of *Hemimeryx* and *Merycopotamus* are as follows.

1. In *Hemimeryx* the cheek teeth are more brachyodont than in *Merycopotamus*.

2. In *Hemimeryx* the individual cusps are less acute than in *Merycopotamus*.

3. In the upper molars of *Hemimeryx* the mesostyle is developed in a continuous wall joining the spurs from the paracone and metacone, whereas in *Merycopotamus* this wall, though present, is greatly reduced, so that the paracone and metacone spurs terminate medially in rather independent styles.

4. In the last lower molar of *Hemimeryx* there is an external ridge from the entoconid, running anteriorly. In *Merycopotamus* this ridge has disappeared. Moreover, the talonid is less compressed in *Merycopotamus*.

Comparing *Hemimeryx pusillus* with *Hemimeryx blanfordi*, on the basis of the criteria outlined above, it is seen that the former is more advanced than the latter, standing indeed in a position intermediate between *H. blanfordi* and *Merycopotamus*. Thus, the phylogenetic development of these forms in India, would seem to have followed a course such as that outlined below.

Merycopotamus dissimilis Middle and Upper Siwaliks
Hemimeryx pusillus   Chinji
Hemimeryx blanfordi   Kamlial

EVOLUTION OF THE SIWALIK ANTHRACOTHERES

The several genera and species of anthracotheres in the Siwalik deposits are representative of two separate phylogenetic lines, one a bunodont, conservative branch, and the other a selenodont, specialized branch. The first group is conservative in that the teeth remain bunodont, and in the upper molars a fifth cusp, the protoconule, is retained. The other, specialized group is characterized by the sharp, selenodont cusps and by a loss of the protoconule. *Hyoboops* is a genus of a somewhat intermediate position, in that it retains the protoconule in a much reduced state, and is selenodont.

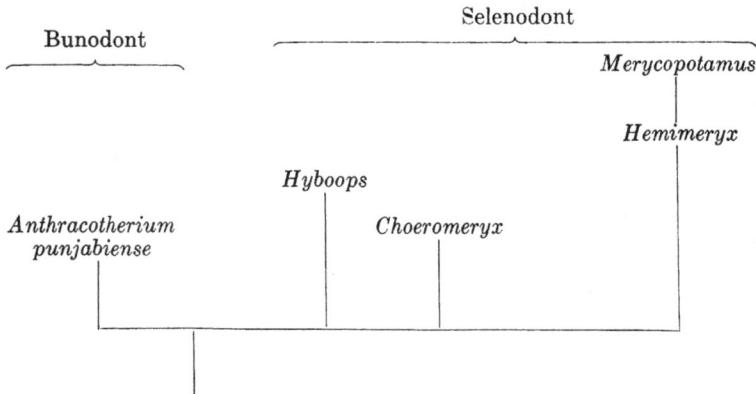

HIPPOPOTAMOIDEA

HIPPOPOTAMIDAE

Hexaprotodon Falconer and Cautley, 1836

Generic type, *Hexaprotodon sivalensis* Falconer and Cautley

Hexaprotodon sivalensis Falconer and Cautley

Hippopotamus (Hexaprotodon) sivalensis, Falconer and Cautley, 1836, Asiatic Researches, XIX, pp. 39–53.

Additional References.—

Falconer, H., 1835, p. 706. (Nomen nudum.)

McClelland, J., 1838, pp. 1038–1047, Pl. LIX.

Falconer, H., and Cautley, P. T., 1847F, Pls. LIX–LX.

Owen, R., 1845, p. 566, Pl. CXLIII.

Falconer, H., 1868A, pp. 130–148, Pls. XI–XIII; 1868B, pp. 404–407.

Lydekker, R., 1880B, p. 31; 1883C, pp. 83, 91; 1884C, pp. 36–42, Fig. 1, Pl. VI; 1885B, pp. 51–53; 1885D, pp. 297–309.

Pilgrim, G. E., 1910B, p. 202; 1913B, p. 324.

Type.—(Lectotype.)—Brit. Mus. No. M 2269, a skull.

Cotypes.—A skull mentioned by Falconer and Cautley in the original description, the identity of which is not indicated. These authors also described the mandible, but did not designate any particular specimen on which their description was based. The numerous specimens on which the original description of *H. sivalensis* was founded, are figured in the Fauna Antiqua Sivalensis.

Horizon.—Upper Siwaliks.

Locality.—Siwalik Hills.

Specimens in the American Museum.—Amer. Mus. No. 19741, a left maxilla with M_{1-3}. Upper portion of the Middle Siwaliks, three miles northwest of Hasnot.

19775, posterior portion of a skull. Upper Siwaliks, below conglomerate, three miles west of Chandigarh.

19776, a complete skull. Upper Siwaliks, below conglomerate, three miles northwest of Chandigarh. Also an associated mandibular symphysis.

19781, a complete skull. Upper Siwaliks, near Siswan.

19784, a left mandibular ramus with P_3–M_3. Upper Siwaliks, below conglomerate, three miles northwest of Chandigarh.

19789, maxillary fragment with right DM^{3-4}. Upper Siwaliks, twenty-five miles east of Chandigarh.

19794, back portion of a skull. Upper Siwaliks, near Kotal Kund.

19817, a partial skull. Upper Siwaliks, three miles west of Chandigarh.

19820, posterior portion of a skull of a young individual. Upper Siwaliks, near Siswan.

19839, partial skull of a young individual with the milk dentition. Upper Siwaliks, six miles west of Kalka.

19868, skull. Upper Siwaliks, near Siswan.

19910, fragmentary ramus with left M_{2-3}. Upper Siwaliks, below conglomerate, three miles west of Chandigarh.

19914, fragment of a maxilla with left M^{1-3}. Upper Siwaliks, below conglomerate, three miles west of Chandigarh.

19918, fragmentary ramus with right M_3. Upper Siwaliks, below conglomerate, three miles west of Chandigarh.

19920, mandibular symphysis of a juvenile. Upper Siwaliks, below conglomerate, three miles west of Chandigarh.

19926, right DM_4. Upper portion of Middle Siwaliks, two miles south of Hasnot.

19972, canine tooth. Upper Siwaliks, three miles west of Chandigarh.

Diagnosis.—Like *Hippopotamus amphibius*, but with six incisors of subequal size. First premolar large, brain case relatively small, well developed sagittal crest, lacrymal in contact with orbit but separated from nasal by an extension of the frontal.

The name *Hexaprotodon* was established by Falconer and Cautley in a subgeneric sense, in reference to the large Hippopotamus of the Upper Siwalik beds. The name was given in allusion to the fact that six incisor teeth are present, whereas in *Hippopotamus* s.s. there are but four incisors.

Owen elevated *Hexaprotodon* to full generic rank in 1845, but subsequent authors have been inclined to regard it as of subgeneric value or as a term synonymous with *Hippopotamus*. Matthew, in 1929, distinguished *Hexaprotodon* as a genus of rather doubtful status, but he did not give a full list of its distinguishing characters.

"The only character that I can find to separate this genus is the one specified by the describer, the presence of six subequal incisors in a transverse row, three on each side of the jaw." [69]

A careful study of various hippopotami has convinced me that *Hexaprotodon* is a distinct genus with a good number of distinguishing characters. These are as follows:

1. There are six subequal incisors, as compared to four in *Hippopotamus*. The central incisors are not enlarged. (There is good reason to think that the incisors in *Hippopotamus* may be homologized with the first and third incisors of *Hexaprotodon*.)

2. The anterior premolars diverge from each other on the opposite sides of the skull and mandible. In *Hippopotamus* the tooth rows are approximately parallel.

3. The brain case is relatively small, and the sagittal crest is consequently high.

4. The maxilla is in contact with the frontal and the lacrymal is not in contact with the nasal. In *Hippopotamus* the lacrymal is situated between the maxilla and the frontal, thereby excluding these bones from any contact with each other.

5. The nasal is unexpanded posteriorly, whereas in *Hippopotamus* it is greatly broadened at the back. The anterior tip is not retracted as in *Hippopotamus*.

6. The premaxillae are in contact with each other along their entire length.

7. The mandibular symphysis is somewhat longer in comparison to its width than is the case in *Hippopotamus*.

8. Lydekker has pointed out the relative length of the astragalus in *Hexaprotodon* as compared to that in *Hippopotamus*. This would seem to be a valid distinction. "The astragalus (of which also there seems to be more than one form) is decidedly longer than the corresponding bone of *H. amphibius*, and thereby makes a marked step in the direction of the pigs." [70]

[69] Matthew, W. D., 1929, p. 556.
[70] Lydekker, R., 1884C, p. 41.

The characters listed above as characteristic of *Hexaprotodon* are primitive, thus distinguishing this genus as a form in which certain heritage characters have persisted. Moreover, these characters all seem to be definitely and constantly different from the comparable characters in *Hippopotamus*, and therefore they would seem to be of true generic significance.

Hexaprotodon iravaticus Falconer and Cautley

Hexaprotodon iravaticus, Falconer and Cautley, Fauna Antiqua Sivalensis, Pl. LVII, figs. 10, 11.

Additional References.—

Falconer, H., 1868A, pp. 142, 498.

Lydekker, R., 1880B, p. 31; 1882B, p. 31; 1883C, pp. 83, 91; 1884C, pp. 42–43; 1885B, p. 53; 1885D, pp. 309–310.

Pilgrim, G. E., 1910B, p. 203.

Matthew, W. D., 1929, pp. 449, 557.

Type.—(Lectotype.)—Brit. Mus. No. 14771, a mandibular symphysis.

Cotype.—A fragmentary mandibular symphysis, figured in the F.A.S., Pl. LVII, fig. 11.

Horizon.—From the Irrawaddy Beds, equivalent to the Middle Siwaliks.

Locality.—Irrawaddy River, Burma.

Diagnosis.—Like *H. sivalensis*, but very much smaller.

Having discussed in the foregoing paragraphs the differences existing between *Hexaprotodon* and *Hippopotamus*, we may now turn to a fuller and more detailed consideration of the former genus.

Falconer in the original description of *Hexaprotodon sivalensis*, discussed at some length the differences which he observed existing between the skull in this species and the skull in the modern form, *Hippopotamus amphibius*. His description was supplemented in 1884 by Lydekker, who described additional differences between the fossil and the recent species. Some of Lydekker's conclusions were based on certain "aberrant crania" in the British Museum, which specimens he supposed to be representative of a variety different from the type cranium. More recently, Matthew studied the skulls of the Siwalik hippopotami, and he realized that among the fossil forms, as well as in the recent species, a great deal of individual variation is expressed in the anatomical structures of the skull. Thus his studies invalidated some of Lydekker's conclusions regarding subspecific differences in this group.

In view of the variability of the skull characters in *H. sivalensis*, it seems quite likely that certain of these characters which both Falconer and Lydekker considered as of specific rank, are actually of varietal magnitude, or even more likely, are but individual variations. Naturally, with only a limited series of fossil specimens to study, certain variations that were probably quite typical within the species would not appear. It is certain that the several skulls in the American Museum collection show some differences from the specimens described by Falconer and by Lydekker. Therefore it may be well to present a list of the distinctive dental and cranial characters of *H. sivalensis* as shown by the specimens in the American Museum, not only to review and to give additional information about this species, but also to evaluate, if possible, what characters may be considered as of real worth in the establishment of taxonomic distinctions.

1. There are six incisors, as noted above, a fact that caused Falconer to create the new genus *Hexaprotodon*. In this connection it may be well to point out the fact that Lydekker described and figured a mandible of *Hexaprotodon palaeindicus* in which the second incisor on either side is quite small. This tooth, as figured by Lydekker, is crowded up, out of place, by the large first and third incisors on either side of it. Thus the evidence would seem to point to the fact that tetraprotodonty in the *Hippopotamus* was attained by the unusual method of eliminating the second incisor.

Lydekker states that the lower incisors of *Hexaprotodon sivalensis* are more oblique than in *Hippopotamus amphibius*, thereby "indicating relations to the pigs." The point does not seem to be well taken.

2. In *Hexaprotodon sivalensis* the incisors are all subequal in size, whereas in *Hippopotamus amphibius* the central incisors are greatly enlarged.

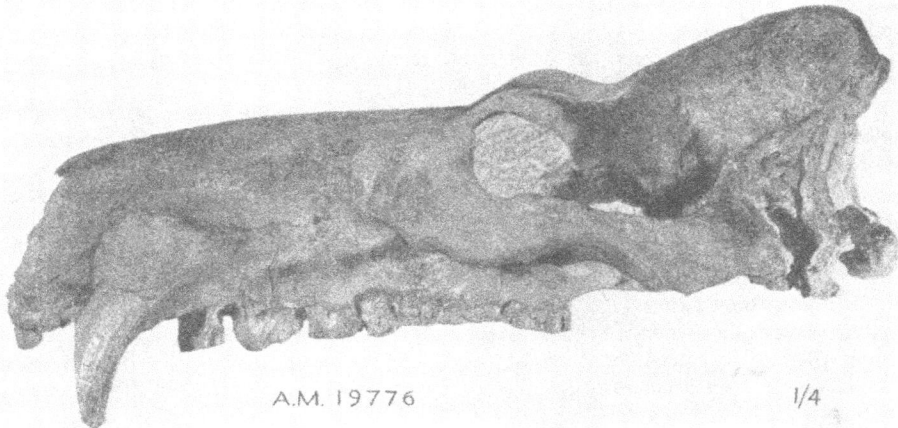

FIG. 122. *Hexaprotodon sivalensis* Falconer and Cautley. Amer. Mus. No. 19776, skull. Lateral view, one fourth natural size.

3. The first premolar in *Hexaprotodon sivalensis* is large; in *H. amphibius* this tooth is reduced, or often wanting.

4. The canine premolar diastema is very short.

5. Any differences between the two species in the molar teeth are very slight, in fact, so slight as to be practically non-existent.

6. A difference is to be seen in the alignment of the cheek teeth. In *H. amphibius* the distance between the last molars is somewhat greater than that between the second premolars, while in *H. sivalensis* the molars are about parallel to each other and the two premolar series diverge so that the distance between the first premolars is greater than between any of the other opposite teeth. This is a function of the posteriorly placed canine in the Siwalik form.

7. The ratios of preorbital to postorbital length is practically the same in the two species. Falconer placed special emphasis on his observation that in the Siwalik form the

eye is further forward than in the modern species, thereby causing a proportional shortness of the facial portion and a greater length for the temporal region. An analysis of the British Museum material, as well as of that in the American Museum, indicates that the supposed differences seen by Falconer are more apparent than real. It is true that the temporal region is slightly longer in the fossil form than it is in the recent species, but the difference is very small. Reference should be made to the appended table.

8. The orbits are perhaps slightly less elevated in the fossil form than they are in the recent species. The difference is small.

9. The brain case in *Hexaprotodon sivalensis* is relatively small. This is a primitive character that would be expected in a fossil animal. The difference is so great between the Siwalik species and the modern *Hippopotamus*, that it becomes evident there has been a very considerable expansion of the cerebrum in the Hippopotamidae during the relatively short duration of Pleistocene times.

10. In connection with the small brain case in *H. sivalensis* there are the high sagittal and lambdoidal crests, for the purpose of adequate muscular attachments. Naturally, as a corollary to the high crests, the occiput of *H. sivalensis* is narrower and higher than in *H. amphibius*.

11. In *H. sivalensis* there is a rather deep pit in the maxilla above the first molar and beneath the anterior portion of the zygomatic arch. This probably served as an attachment for the buccinator muscle.

12. The tips of the nasals in *H. sivalensis* extend forward, so that they project slightly beyond the border of the nasal opening. In *H. amphibius* the nasals are somewhat retracted.

13. The nasals are shaped differently in the two species under consideration. In *H. sivalensis* they are rounded at the posterior end, as is common among mammals, but in *H. amphibius* they flare laterally to points opposite the front borders of the orbits. This development of the posterior portions of the nasals, accompanies the position of the lacrymal, which will next be considered.

14. In *H. sivalensis* the lacrymal is in contact with the orbit and on the upper surface of the skull it is separated entirely from the nasal by an anteriorly projecting tongue of the frontal bone. The facial portion of the lacrymal is long, extending forward to above the anterior border of the second molar. In *H. amphibius*, on the other hand, the lacrymal has migrated forward so that it has lost entirely the contact with the orbit on the upper surface of the skull. This has occasioned the disappearance of the lacrymal foramina from the front border of the orbit. Moreover, the lacrymal has established a contact with the nasal, which in turn has caused the anteriorly projecting tongue of the frontal (typical of *H. sivalensis*) to disappear. On the inside of the orbit, the lacrymal extends forward to almost the front border.

Obviously the position and shape of the lacrymal in *H. sivalensis* is primitive, and generalized, and it probably represents an heritage character retained from an earlier ancestral form. On the other hand, the position of the lacrymal in *H. amphibius* is an habitus character, which may be an adaptation to the elevated orbits.

15. In *H. amphibius* the maxilla is expanded above the third molar, while this expansion is not present in the fossil form.

16. The modern *Hippopotamus* shows a striking development of the premaxillae, in that they are broadly separated along a portion of the median line. In *H. sivalensis* a

more primitive condition still holds, that is the premaxillae are united along the entire length of their median symphysis. This is a distinct contradiction to Falconer's statement that the premaxillae were separate, as in the modern form.

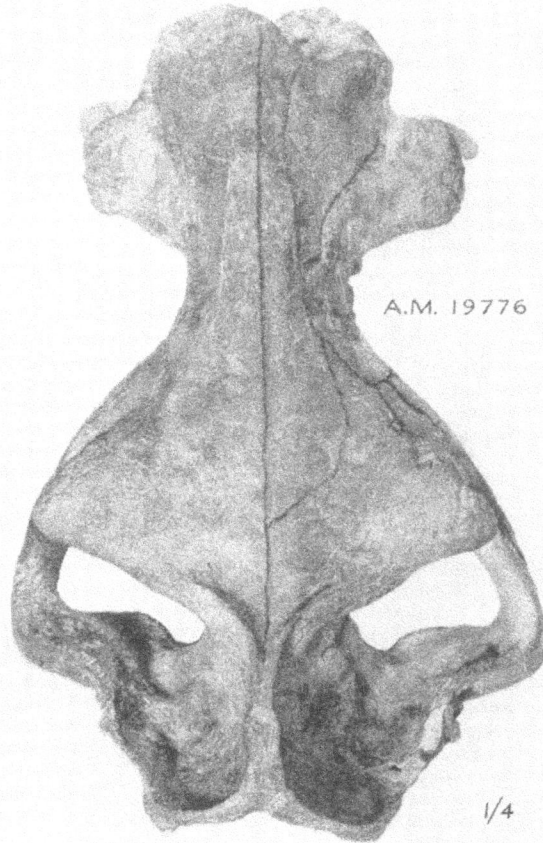

Fig. 123. *Hexaprotodon sivalensis* Falconer and Cautley. Amer. Mus. No. 19776, skull. Top view, one fourth natural size.

17. There is a great deal of variation shown in the development of the posterior nasal choanae. Lydekker cited the posterior position of the opening as indicative of a specific difference in the fossil form. An examination of the several skulls in the American Museum collection shows that while for the most part the choanae are situated at a considerable distance back of the third molar, they do at times (as in Amer. Mus. No. 19784) reach as

far forward as the last molar.   Thus this character would seem to be of little specific value.

On the whole, the chonae are narrower and more V shaped in the fossil form than in the recent species.

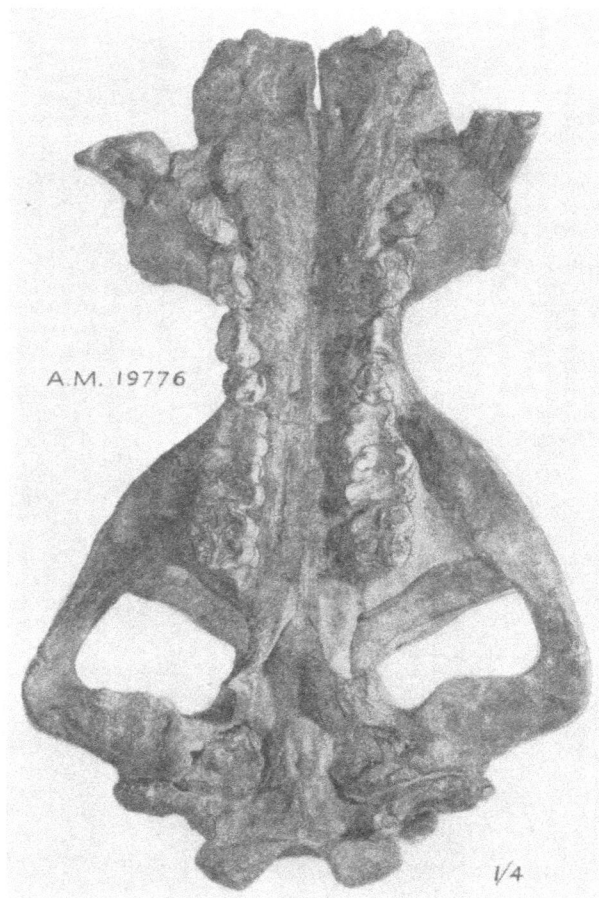

FIG. 124.   *Hexaprotodon sivalensis* Falconer and Cautley.   Amer. Mus. No. 19776, skull.   Ventral view, one fourth natural size.

18. The pterygoid processes are rather small in *H. sivalensis* as compared to the same structures in *H. amphibius*.

19. The basicranial foramina are the same in *H. sivalensis* and *H. amphibius*.   To review briefly their disposition: the ethmoid and optic foramina are large and well developed; the foramen lacerum anterius and the foramen rotundum are fused into a large

opening, while the foramen ovale, foramen lacerum medius and the foramen lacerum posterius are united to form a broad gap which entirely separates the bulla from the alisphenoid; the stylomastoid foramen is separated from the foramen lacerum posterius by a very thin bridge of bone; the condylar foramen is very large.

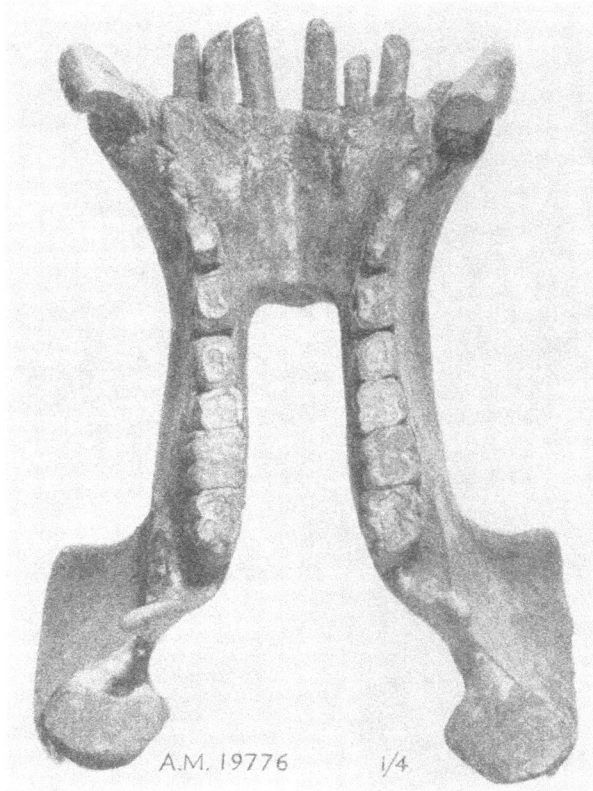

FIG. 125. *Hexaprotodon sivalensis* Falconer and Cautley. Amer. Mus. No. 19776, mandible. Crown view, one fourth natural size.

20. The bullae are somewhat rounder in *H. sivalensis* than in *H. amphibius*.

21. In the mandible, the symphysis is slightly longer and more abrupt, as seen from the side, in *H. sivalensis* than in *H. amphibius*. The lower border of the angle, though deeply expanded in the fossil form, is not produced anteriorly to a point as in the modern species.

22. Falconer states that in *H. sivalensis* the coronoid is not projected so far forward as in *H. amphibius*.

The above outline shows that there are many points whereby the hippopotamus of the Siwaliks differs from the modern hippopotamus. Of course, these differences are for the most part of a comparatively minor kind, but small differences constitute the distinctions between advanced, specialized forms such as the fossil and the recent hippopotami. Nearly

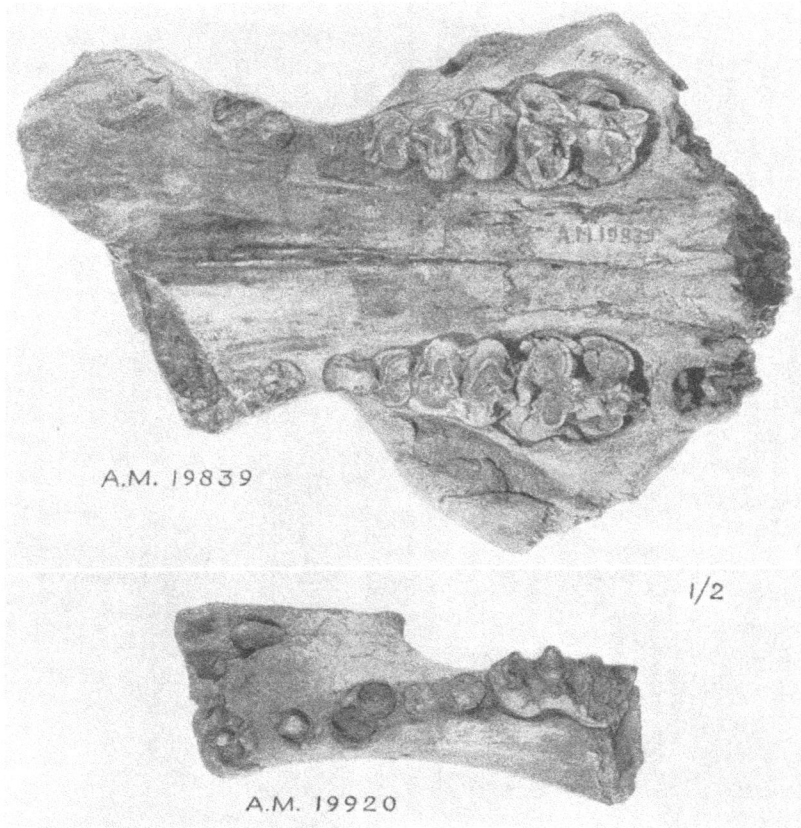

Fig. 126. *Hexaprotodon sivalensis* Falconer and Cautley. Amer. Mus. No. 19839, partial skull of an immature animal, and Amer. Mus. No. 19920, mandibular symphysis and partial left ramus of an immature animal. Skull above, mandible below. Crown views. Both figures one half natural size.

all of the differences of the Siwalik hippopotamus from the modern species show more primitive characters in the fossil form, as might be expected.

Lydekker distinguished two varieties of *Hexaprotodon sivalensis*, which he distinguished by the relative width of the molars. The "broad molar" type he distinguished as *H. sivalensis latidens*, while the "narrow molar" type was designated as *H. sivalensis angustidens*. It is difficult to say whether these varietal distinctions hold constantly in the species.

Measurements and ratios of the American Museum specimens of *Hexaprotodon sivalensis* are presented in the accompanying tables.

MEASUREMENTS (IN MM.) AND RATIOS OF *Hexaprotodon sivalensis*

	H. am-phibius	H. siva-lensis	H. siva-lensis	H. siva-lensis	H. siva-lensis	H. siva-lensis	H. siva-lensis	H. siva-lensis
	24283	19781	19776	19794	19817	19775	19784	19972
Length of skull: condyle—premaxilla.......	630	610	605	..	600	..		
Width of skull: zygomatic arch...........	413	390*	390	332	360	..		
Height, top of orbit to M³ alveolus........	180	160	150	140	160*	..		
Preorbital length......................	360	350	330	..	330*	..		
Postorbital length (condyles).............	270	260	275	280	270	..		
Width of condyles......................	146	130	140	120	120	130		
Width of occiput.......................	280	245	263	220	235	240*		
Height of occiput......................	190	183	188	165	185	170*		
Width of cranium (parietals).............	123	90	95	95	100	90		
Width of palate at M¹...................	60	47	48	53	45	60	60	
Width of palate at Pm¹.................	75	104	111	..	115	..	125	
Length of premolar series (H. amphibius, P²⁻⁴).....................................	139	129	149	..	137	..	153	
Length of molar series..................	139	130	113	108	132	128	145	
P¹ to canine diastema (P² in H. amphibius)...	90	35	32	..	40	..	28	
Transverse diameter of canine............	34	50*	45	..	50*	..	60	50

* Estimated.

		24283	19781	19776	19794	19817	19784
Ratio..	$\dfrac{\text{Width of skull}}{\text{Length of skull}} \times 100$.................	65	64	64	..	60	..
Ratio..	$\dfrac{\text{Postorbital length}}{\text{Preorbital length}} \times 100$.............	75	74	83	..	82	..
Ratio..	$\dfrac{\text{Height of occiput}}{\text{Width of occiput}} \times 100$..............	68	75	71	75	78	..
Ratio..	$\dfrac{\text{Molar series}}{\text{Premolar series}} \times 100$.................	100	100	76	..	96	95

	Preorbital length	Postorbital length	Ratio
H. amphibius, No. 24283.........................	355	270	76
H. sivalensis:			
A.M. 19781...............................	335	275	82
19776...............................	340	270	79
19817...............................	345*	260*	..
Fauna Antiqua Sivalensis, Pl. 59, fig. 1............	324	260	80
Fauna Antiqua Sivalensis, Pl. 59, fig. 2...........	300*	240	80

* Estimated.

	H. amphibius	H. sivalensis	H. sivalensis	H. sivalensis
	24283	19776	19784	19918
Length of mandible, condyle-incisor alveolus......	520	..	..	..
Breadth of symphysis.........................	220	200	..	..
Length of symphysis..........................	165	..	..	..
Height of condyle above alveolar border.........	65	..	80	75
Depth of ramus at M_3.........................	125	133	128	105
Transverse diameter of canine..................	33	35	..	..
Length of premolar series ($H.$ $amphibius$, P_{2-4})	126	..	..	..
Length of molar series.......................	173	..	173	..
P_2 — C diastema $\dfrac{19776}{70}\bigg\vert\dfrac{24283}{120}$	120	70	..	...

The Origin of the Hippopotamidae [71]

Three explanations for the origin of the Hippopotamidae have been advanced.

1. The Hippopotamidae have been derived from the Suidae, a view advocated by various authors, and recently supported by Matthew.

2. The opposite theory is that one which would derive the Hippopotamidae from the Anthracotheriidae. This idea was especially supported by Andrews, but before him Falconer and Lydekker had suggested a relationship between these two families, basing their conclusions particularly on the *Hippopotamus* like habitus of the anthracothere, *Merycopotamus*.

3. A more recent theory is that advanced by Miss H. S. Pearson, deriving the Hippopotamidae from certain Eocene bunodont artiodactyls, notably *Cebochoerus*.

Good arguments may be elicited in favor of any of these theories, for there may be found in the genus *Hippopotamus* characteristic pig, anthracothere and cebochoerid features. The problem is, how to evaluate this assemblage of characters in *Hippopotamus*, and to which ones emphasis should be given. Perhaps the problem is beyond solution at the present time, but at least it can be attacked, and the various possibilities presented by it can be carefully considered.

A Comparison of the Hippopotamidae, the Anthracotheriidae and the Suidae

A. Dentition.

If the *Hippopotamus* molar was derived from the suid molar the following steps were involved.

1. Coalescence of certain accessory conules with the primary cusps of the suid molar, to form the trefoil of the *Hippopotamus* molar.

2. Suppression of all superfluous accessory conules in the suid molar.

3. Shortening of the tooth and the elimination of the heel in the third molar.

If the *Hippopotamus* molar was derived from the anthracothere molar, this derivation followed the phylogenetic course outlined below.

1. Transformation of the anthracothere crescents into the *Hippopotamus* trefoils.

The manner in which the anthracothere or the suid molars might have been changed into a *Hippopotamus* molar is shown in the accompanying illustration (Fig. 127).

Considering the premolar teeth, we see that the *Hippopotamus*, anthracothere and suid

---

[71] A detailed treatment of this subject will be found in Colbert, Edwin H., 1935, Amer. Mus. Novitates, No. 799.

premolars are all very much alike. In the last upper premolar of the Suidae, however, there are two outer cusps and one internal cusp, whereas in the Hippopotamidae and the Anthracotheriidae there is but one outer cusp. In some of the anthracotheres, especially such genera as *Gelasmodon* or *Merycopotamus* the lower premolars are strikingly similar to those of *Hippopotamus*.

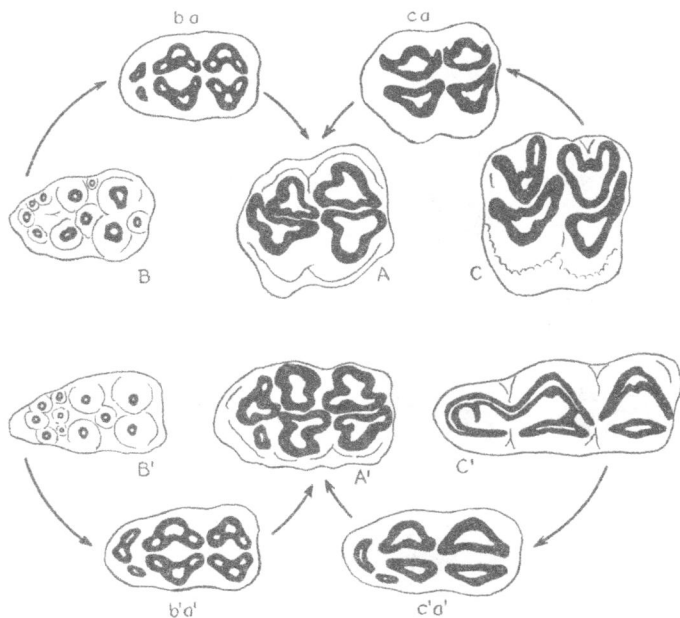

FIG. 127. The origin and evolution of the tooth pattern in the Hippopotamidae.
A, A'. Upper and lower third molars of *Hexaprotodon sivalensis*.
B, B'. Upper and lower third molars of *Conohyus sindiense*.
C, C'. Upper and lower third molars of *Merycopotamus dissimilis*.
ba, b'a'. Hypothetical intermediate stages between the Suidae and the Hippopotamidae.
ca, c'a'. Hypothetical intermediate stages between the Anthracotheriidae and the Hippopotamidae.
From Falconer and Cautley, Forster Cooper and Pilgrim. Figures not to scale.

The Hippopotamidae and the Anthracotheriidae are characterized by their vertically directed canines, whereas the Suidae have laterally directed canines.

B. Skull and Mandible.

In the skull and mandible there is a marked resemblance between the Hippopotamidae and some of the advanced Anthracotheriidae, such as *Merycopotamus*. The following characters are common to these two families (as expressed in the Anthracotheriidae by advanced genera).

1. Elevation of the orbit.
2. Position of the infraorbital foramen.

3. Shape of the zygomatic arch.
4. The broad, vertical occiput.
5. The low, wide glenoid.   (Secondarily raised in *Merycopotamus*.)
6. The short paroccipital processes.
7. The general configuration of the auditory bulla.

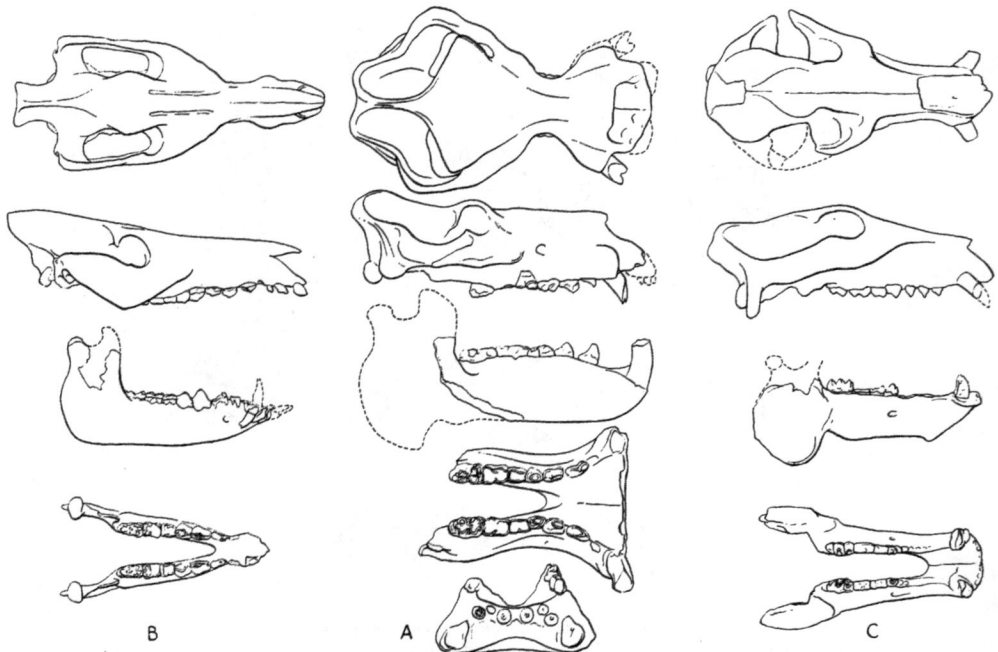

Fig. 128.   Comparison of the skull and mandible in the Hippopotamidae, the Suidae and the Anthracotheriidae.
A. *Hexaprotodon sivalensis*, Upper Siwaliks.
B. *Conohyus sindiense*, Lower Siwaliks.
C. *Merycopotamus dissimilis*, Upper Siwaliks.
The front view of the mandibular symphysis of *Hexaprotodon* is included to show the diminution in size of the second incisor.
From Falconer and Cautley, and Colbert.   Figures not to scale.

8. The post-glenoid compression.
9. The high mandibular coronoid.
10. The deep angle of the mandible.
11. The broad mandibular symphysis.

In comparing the *Hippopotamus* skull and mandible with those of the pigs, we see the following characters common to both families.

1. Elevation of the orbit (in certain specialized suids).
2. The postglenoid compression.

3. The long tube for the external auditory meatus, opening in an upward direction.
From the above it will be seen that there is a preponderance of like skull and jaw

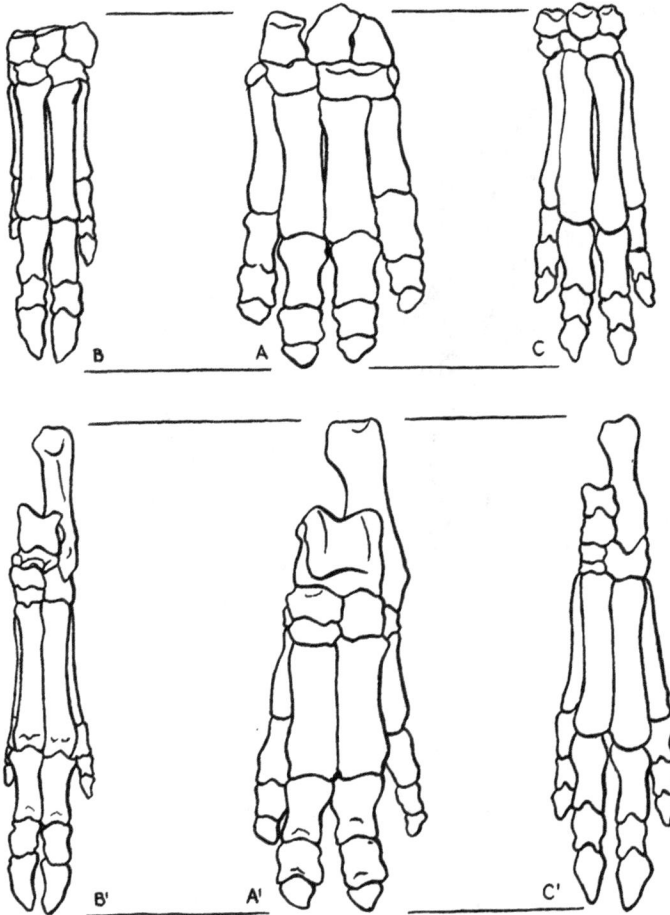

FIG. 129. Comparison of the manus and pes in the Hippopotamidae, the Suidae and the Anthracotheriidae.
A, A'. Manus and pes of *Hippopotamus amphibius*.
B, B'. Manus and pes of *Sus scrofa*.
C, C'. Manus and pes of *Ancodus brachyrhynchus*.
From de Blainville, Scott and others. Figures not to scale.

characters in the Hippopotamidae and the advanced Anthracotheriidae, as compared with
the characters common to the Hippopotamidae and the Suidae. Of course the above lists
may not be strictly diagnostic of the real resemblances and differences in these groups.

Some of the above outlined characters may be due to parallelisms, rather than to direct genetic relationships. It seems difficult, however, to account for many of the resemblances in the skulls of the Hippopotamidae and the Anthracotheriidae as due entirely to parallel evolution. The shape of the occiput, the low glenoid, the short paroccipital processes, the

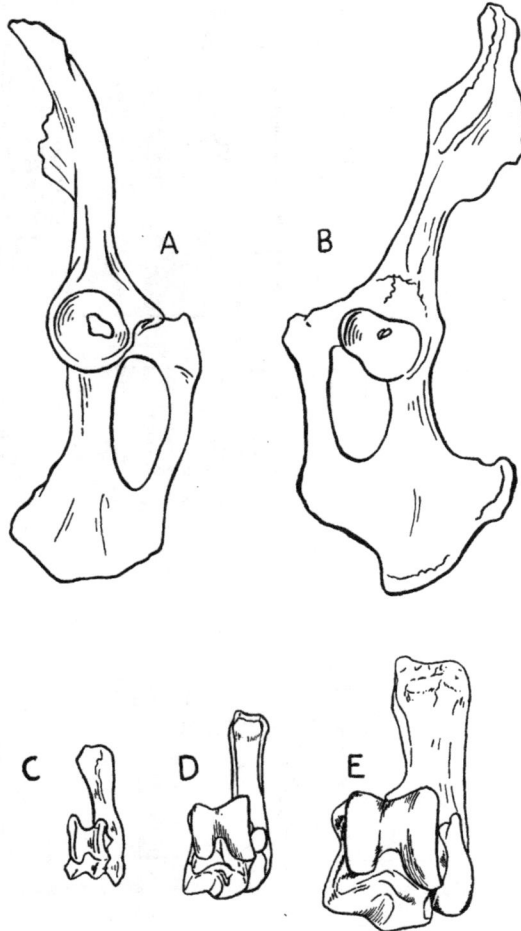

FIG. 130. Comparison of the pelvis in (A) *Ancodon* and (B) *Hippopotamus*, and of the tarsus in (C) *Sus*, (D) *Merycopotamus* and (E) *Hippopotamus*. After Andrews, de Blainville and Falconer and Cautley.

auditory bulla and the form of the mandible would seem to be characters which, when taken together, are significant of a probable connection between the Hippopotamidae and the Anthracotheriidae. The trend of evolution in the suid skull and jaw, from the Oligocene on, is *away* from the *Hippopotamus* habitus, whereas the trend of evolution in the anthracothere skull and jaw is seemingly *towards* the *Hippopotamus* habitus.

C. Skeleton.

Coming now to the postcranial skeleton, certain resemblances are to be seen between the Anthracotheriidae and the Hippopotamidae. Andrews has noted the similarity in the pelves of *Brachyodus*, an anthracothere from the Fayûm deposits of Egypt, and of *Hippopotamus*.

". . . these animals [i.e., the anthracotheres] in many points, e.g., in the pelvis, approach very nearly to the Hippopotamidae, which were probably derived from them. Remains of one of the earliest and most primitive Hippopotami known, viz. *H. hipponensis*, have already been found in the Middle Pliocene of Egypt, so that there is every prospect that annectent forms between *Hippopotamus* and the Anthracotheres may be discovered in this region in deposits between the Lower Miocene and the Pliocene." [72]

In the Suidae the pelvis is narrow as compared to the pelvis in the Anthracotheriidae and the Hippopotamidae, and the ilia are not flared. Other minor differences are to be noted.

In comparing the tarsus of the forms under consideration a close resemblance is to be seen in the broad astragalus of the anthracotheres and that of the Hippopotami, whereas the astragalus in the pigs is narrow.

Of course these skeletal characters may be due to the adaptations of the anthracotheres and the hippopotami to the increase in body weight, necessitating at least semi-graviportal structures, as compared to the cursorial adaptations in the suids. The fact remains, however, that some of the largest anthracotheres like *Merycopotamus* were not much larger than large pigs, yet in many of their skeletal characters, notably in the pelvis and the feet, they showed a distinct trend towards the *Hippopotamus* habitus.

Taking the evidence all in all it would seem that resemblances between the anthracotheres and the hippopotami are more numerous, and probably more important, than the resemblances between the pigs and the hippopotami. A definite and a final solution of this vexing problem of the origin of the Hippopotamidae is difficult to reach, because the evidence is peculiarly suited to factors of individual interpretation. The most primitive Hippopotamidae known are not appreciably different from the modern forms. Thus it becomes necessary to use many advanced, specialized habitus characters in drawing our comparisons, and these tend to mask, or rather to crowd out the more significant primitive heritage characters, the characters on which phylogenetic ties between groups must ultimately rest.

It would seem that Andrews made an unusually sagacious remark when he said that "there is every prospect that annectent forms between *Hippopotamus* and the Anthracotheres may be discovered in this region [Egypt] in deposits between the Lower Miocene and the Pliocene." We do not know any primitive Hippopotamidae from the earlier portions of the Tertiary. May not this be due to the fact that the Hippopotamidae is a family of late evolutionary development, a family that broke away from its ancestral group in the late Miocene? It may be that primitive, early Tertiary Hippopotamidae have never been found because such animals never existed.

[72] Andrews, C. W., 1906, p. xx.

If the Hippopotamidae did break away from an ancestral stem in the late Miocene, what might this stem have been? Certainly not the Suidae, which by that time have set themselves in an evolutionary trend quite away from the *Hippopotamus* type of structure. Why not, therefore, the anthracotheres, which in the Upper Tertiary show numerous structural characters strongly suggestive of the Hippopotamidae? Perhaps the great similarity of *Merycopotamus* to *Hippopotamus* is due to the fact that the former is not far removed from the *advanced* anthracothere type from which the latter may have been derived. This suggestion may be represented diagrammatically as follows:

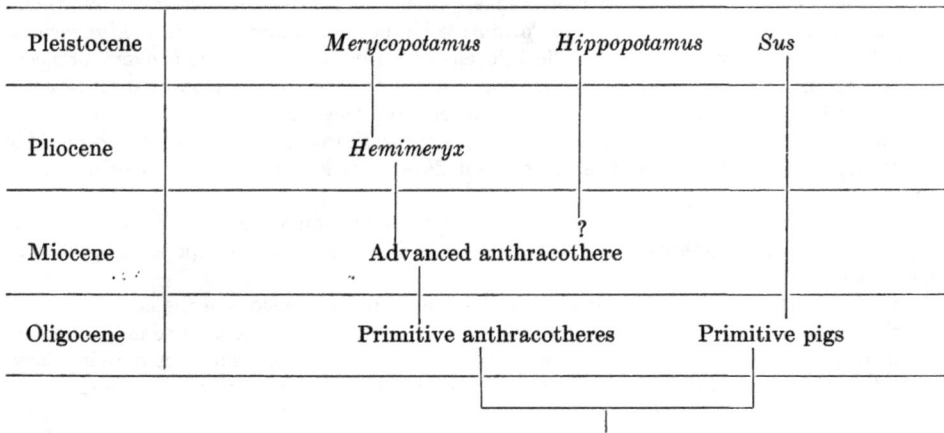

Pleistocene	*Merycopotamus*	*Hippopotamus*	*Sus*
Pliocene	*Hemimeryx*		
Miocene	Advanced anthracothere	?	
Oligocene	Primitive anthracotheres	Primitive pigs	

In the above discussions the arrangement of the basicranial foramina in the Anthracotheriidae, the Suidae and the Hippopotamidae was not utilized for comparative purposes, because of the similarities existing among the advanced forms of the first two families mentioned and the Hippopotamidae. In all of these families the foramina are coalesced around the bulla, a common character in the more specialized artiodactyls.

WAS *Cebochoerus* ANCESTRAL TO THE HIPPOPOTAMIDAE?

Miss H. S. Pearson has suggested that the Hippopotamidae may have been derived from *Cebochoerus* of the Eocene of Europe. She bases her argument on the following characters common to or similar in both groups.
1. The shape of the glenoid.
2. The position of the glenoid.
3. The position of the post-tympanic, antero-lateral to the paroccipital.
Although there are resemblances between *Cebochoerus* and the Hippopotamidae in the above mentioned characters, many differences, which it would seem are of greater importance, may be found. The skull and the dentition especially, of *Cebochoerus*, are specialized along a quite definite trend, not towards the Hippopotamidae but rather in a direction of their own.

TYLOPODA

CAMELOIDEA

CAMELIDAE

Camelinae

Camelus Linnaeus, 1758

Generic type, *Camelus dromedarius* Linnaeus

Camelus sivalensis Falconer and Cautley

Camelus sivalensis, Falconer and Cautley, 1836, Asiatic Researches, XIX, pp. 120–134, Pl. XX, figs. 1–7, Pl. XXI, figs. 8–13.
Additional References.
    Falconer, H., and Cautley, P. T., 1849, Pls. LXXXVI–XC.
    Falconer, H., 1868A, pp. 227–246, Pl. XVIII; pp. 532–537.
    Lydekker, R., 1876, pp. 61–63; 1880B, p. 32; 1883C, p. 34; 1884D, p. 129; 1885B, pp. 35–36; 1885D, pp. 140–145.
    Pilgrim, G. E., 1910B, p. 203; 1913B, p. 324.
    Matthew, W. D., 1929, pp. 444, 446–447.

Type.—(Lectotype.)—Brit. Mus. No. 15357, a fragmentary skull and mandible. [Not B.M. No. 39517, as stated by Lydekker. This specimen was neither mentioned nor figured in the original description.]

Cotypes.—Brit. Mus. Nos. 17558, a partial palate and a nearly complete mandible; 17557, posterior portion of a right mandibular ramus; 15348, a right mandibular ramus with DM_4–M_2; 40567, a fragmentary left maxilla with M^{2-3}. Also the back portion of a skull and a left mandibular ramus, figured in Asiatic Researches, XIX, Pl. XX, figs. 1, 2, and Pl. XXI, figs. 9, 10, respectively.

Horizon.—Upper Siwaliks.

Locality.—Siwalik Hills.

Specimens in the American Museum.—Amer. Mus. No. 19832. An associated skeleton from the variegated beds, just below the Boulder Conglomerate, Upper Siwaliks. From a locality about fifteen miles east of Chandigarh. The following elements comprise this skeleton.

A right maxilla with M^3; a left maxilla with P^3–M^3; both mandibular rami; the front portion of the skull, including the premaxillae, maxillae and nasals; four cervical vertebrae; eight dorsal vertebrae; six lumbar vertebrae; the sacrum; nine caudal vertebrae; numerous ribs and sternal ribs; right humerus; left radius and ulna; left trapezoid, magnum, unciform, metacarpals and phalanges; right carpals, metacarpals and phalanges; left femur, tibia, tarsals, metatarsals and phalanges, various fragments.

Amer. Mus. No. 19785, a skull from the top of the variegated beds, below the boulder Conglomerate, Upper Siwaliks. Three miles northwest of Chandigarh.

Diagnosis.—As large or larger than the modern camels. Enamel rugose; inner surface of lower molars nearly flat in the middle line; antero-external fold on the lower premolars; mandible long and shallow.

Camelus sivalensis was first named by Falconer and Cautley on the basis of some fragmentary skulls and jaws from the Siwalik Hills. Falconer later made a thorough study of the species, using his original type material, and in addition to this a large series of skulls, jaws, vertebrae and limb bones collected by P. T. Cautley at various localities in the Siwalik Hills. He compared the Siwalik form to the modern dromedary, pointing out the great similarity between the two species, and coming to the conclusion that the fossil form differed from the modern one mainly by virtue of its somewhat greater size. Furthermore, he intimated that there was a second species of camel in the Siwalik material,[73] to which he provisionally assigned the name *Camelus antiquus*, distinguishing it from *Camelus sivalensis* by its smaller size. Falconer failed, however, to publish a description of this second species, or to distinguish the various type specimens of it.

Later, Lydekker, using Falconer's manuscript as a guide, published the name *Camelus antiquus*. According to Lydekker. "The chief characters by which this species is distinguished from the last are its inferior size, the smooth enamel of the teeth, the concavity and vertical ridge in the middle of the inner surface of the lower true molars (the ridge being only observable in little-worn teeth) and the short deep mandible."[74]

Dr. Matthew suggested in 1929, that *Camelus antiquus* is a synonym of *Camelus sivalensis*, and that it is based on small individual variations in the latter species—variations that have no valid taxonomic value.[75] His view is here adopted as the correct one.

The comparison of *Camelus sivalensis* with the modern camel shows that the fossil and the recent forms are very closely related to each other, and that no strikingly constant characters separate them. A consideration of minute characters (the procedure necessarily followed in a comparative study of quite closely related species, such as the ones under discussion) has revealed that *Camelus sivalensis* is slightly more primitive in some of its osteological structures than is *Camelus bactrianus*. The differences in these two species are listed below.

1. The third upper premolar is relatively larger in *C. sivalensis* than it is in *C. bactrianus*, showing that the Siwalik camel had not progressed as far in the matter of tooth reduction, as has the modern Asiatic camel.

2. In a like manner, the second upper premolar and the third lower premolar of *C. sivalensis* are somewhat less reduced than are the corresponding teeth in *C. bactrianus*.

3. Again, the canines of the fossil species are considerably larger than is the case in the modern Asiatic camel.

4. It is to be noted that the cheek teeth of *C. sivalensis* are slightly wider in proportion to their length than is the case in *C. bactrianus*. This would indicate a departure in the modern camel from the more primitive quadrate type of tooth. (In regard to the length and breadth indices of the camel molar, careful attention must be given to the degree of attrition in the tooth, because owing to the disparity between the width of the crown and the base, the occlusal surface increases transversely as the tooth is worn down.)

The talonid of the third molar is quite large in *C. sivalensis*, while in *C. bactrianus* this structure is greatly reduced. Here again the Siwalik camel displays its slightly more primitive character.

[73] Falconer, H., 1868A, p. 234.
[74] Lydekker, R., 1885A, p. 78.
[75] Matthew, W. D., 1929, pp. 554–555.

5. In *C. sivalensis* the maxilla does not border on the nasal opening, as it does in the Bactrian camel. The encroachment of the maxilla on the nasal opening is a specialization among the ruminants, and it is always present in the modern camel.

6. The mandible of *C. sivalensis* is distinguished by its less vertical ascending ramus, a feature indicating a slightly more primitive development in the fossil species, as compared to the modern form.

There are certain other differences between the Siwalik camel and the modern species which are not especially indicative of primitive or advanced traits. They are as follows.

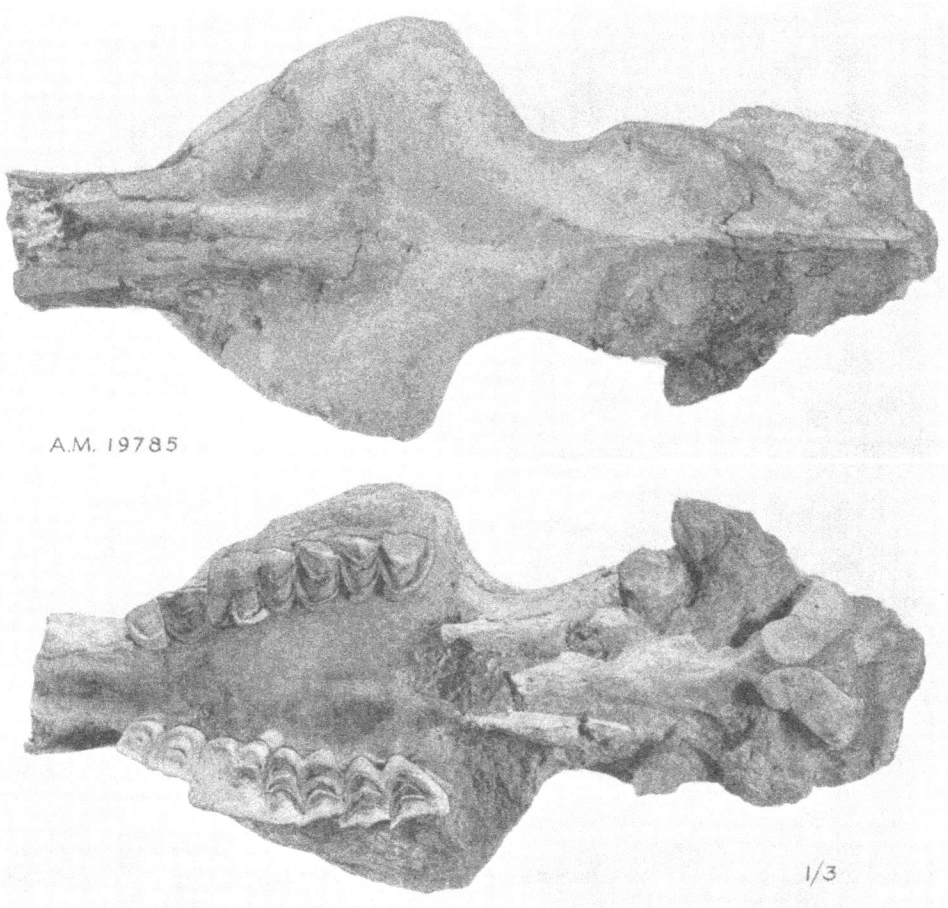

FIG. 131. *Camelus sivalensis* Falconer and Cautley. Amer. Mus. No. 19785, skull. Top view above, palatal view below. One third natural size.

1. In *C. sivalensis* the mandibular ramus is relatively deep, while in *C. bactrianus* it is shallow.

2. Minor differences occur in the basicranium.

(*a*) The anterior surfaces of the occipital condyles are less concave in the fossil species.

(*b*) The basioccipital is somewhat more constricted in *C. sivalensis*.

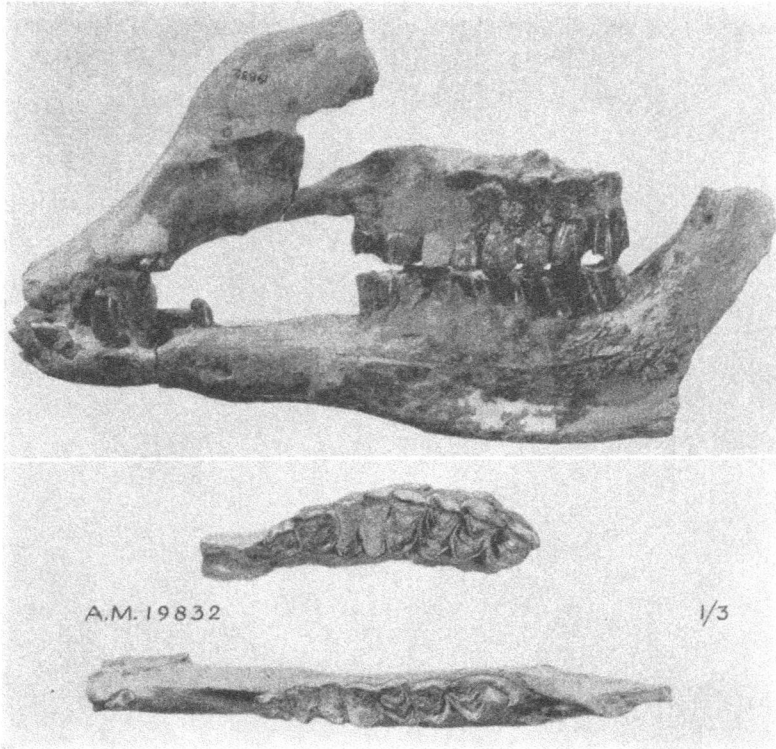

Fig. 132. *Camelus sivalensis* Falconer and Cautley. Amer. Mus. No. 19832, partial skull and mandible. Lateral view above, crown views below. One third natural size.

It might be well to say at this point, that the brain case of *Camelus sivalensis* is fully as large as in the modern camels. Evidently there has been little or no increase in brain capacity in these animals during the course of Pleistocene time.

As to any other differences, they are so slight or so variable as to be of no value. The reputed greater size of the Siwalik camel over the modern forms must be disregarded. A skull of *Camelus bactrianus* in the American Museum is fully as large as the large skull of *Camelus sivalensis* (A.M. 19785) and is somewhat larger than the other Siwalik specimen (A.M. 19832).

Little need be said about the post cranial skeleton of *Camelus sivalensis*. The various bones comprising it are all very similar to the corresponding bones in the modern camel, and thus they do not require any particular elucidation. Reference should be made to the accompanying figures and measurements.

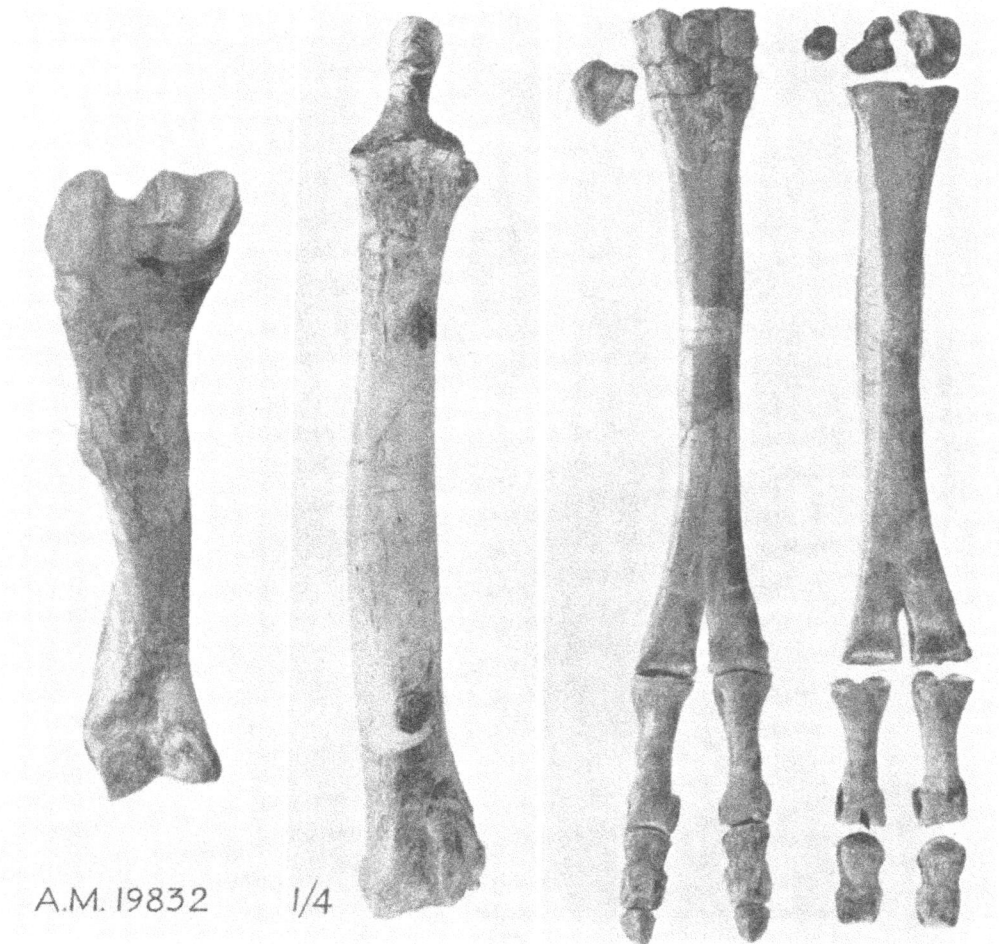

Fig. 133. *Camelus sivalensis* Falconer and Cautley. Amer. Mus. No. 19832, elements of the fore limb. Left to right, humerus, ulna-radius, right manus, left manus. One fourth natural size.

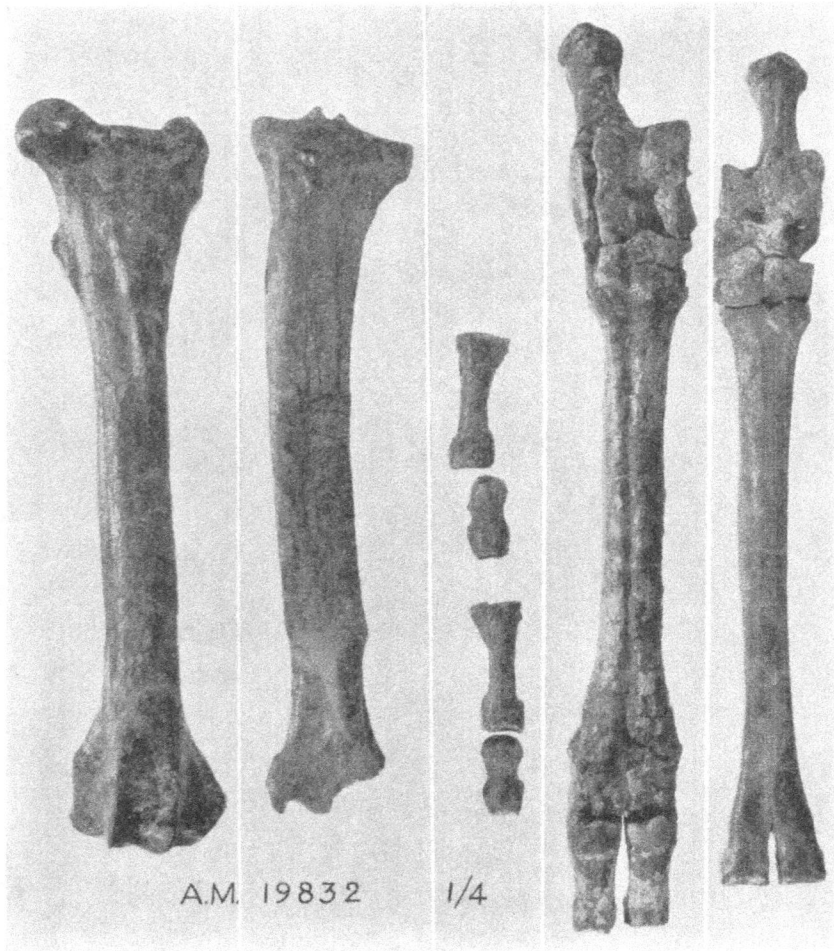

FIG. 134. *Camelus sivalensis* Falconer and Cautley. Amer. Mus. No. 19832, elements of the hind limb. Left to right, femur, tibia, proximal and median phalanges, right pes, left pes. One fourth natural size.

MEASUREMENTS (IN MM.) OF *Camelus sivalensis*

	C. sivalensis		C. bactrianus
	19832	19785	14110
Length of upper molar series....................	100.0	123.0	113.0
Length of P^{3-4}................................	40.0	46.0	43.0
Length of lower molar series....................	118.0	....	120.0
Length of P_4...................................	20.0	....	24.0
Length of M^3...................................	45.5	55.0	42.0
Width of M^3....................................	34.0	39.0	28.0
Length of skull (to condyle)....................	....	580*	515.0
Depth of ramus at M_3..........................	65.0	....	50.0

* Estimated.

Camelus antiquus Lydekker

Camelus antiquus, Lydekker, 1885, Rec. Geol. Surv. India, XVIII, p. 78.

Additional References.—

Lydekker, R., 1885D, pp. 46–47.

Pilgrim, G. E., 1910B, p. 203; 1913B, p. 324.

Matthew, W. D., 1929, pp. 444, 554–555.

Type.—(Lectotype.)—Brit. Mus. No. 16165, a left mandibular ramus.

Cotypes.—Brit. Mus. Nos. 15347, a left maxilla; 40562, a skull; 40568, a left ramus; 39599, a right ramus.

Horizon.—Upper Siwaliks.

Locality.—Siwalik Hills.

Diagnosis.—Distinguished, according to Lydekker, by its smaller size, shorter and deeper mandible, and less rugose enamel.

This species is very likely synonymous with *Camelus sivalensis*, as has been suggested by Matthew.

PECORA

TRAGULOIDEA

TRAGULIDAE

Dorcabune Pilgrim, 1910

Generic type, *Dorcabune anthracotherioides* Pilgrim

Dorcabune anthracotherioides Pilgrim

Dorcabune anthracotherioides, Pilgrim, 1910, Rec. Geol. Surv. India, XL, p. 68.

Additional References.—

Pilgrim, G. E., 1915C, pp. 228–231, Pl. XXI, figs. 1, 2, 7, 8, Pl. XXII, figs. 4, 5.

Matthew, W. D., 1929, pp. 453, 460.

Type.—(Lectotype.)—G.S.I. No. B 580, a maxilla with the molars present.

Cotypes.—G.S.I. Nos. B 581, a left M^2; B 582, a left M_3; B 583, a ramus with right M_{1-2}; B 584, a left M_2; B 588, a fragmentary right P_4.

In his original description of this species, Dr. Pilgrim did not indicate any specimens as types. It is clear, however, from his later description written in 1915, that the original description was based on the several specimens described and illustrated in his detailed account of the species. These are therefore considered as cotypes.

Horizon.—Lower Siwaliks, Chinji zone.

Locality.—Near Chinji, Salt Range, Punjab.

Specimens in the American Museum.—Amer. Mus. No. 19353. A left M_3. From the lower portion of the Middle Siwaliks, one half mile west of Phadial.

19355. Fragment of a mandible with right M_{1-2}. Lower Siwaliks, five miles west of Chinji Rest House.

19652. Maxilla with right M^{1-3}. Lower Siwaliks, one mile north of Chinji Rest House.

29998. Left M^3. Lower portion of Middle Siwaliks, one mile south of Nathot.

No numbers. Fragment of a right M^3, Lower Siwaliks, twelve miles east of Chinji Rest House.

Right M_2 and a fragment of an upper molar. Lower Siwaliks, near Nathot.

Right M_2. Lower Siwaliks, four miles west of Chinji Rest House.

Diagnosis.—A very large tragulid with bunodont teeth. The upper molars are characterized by the isolated parastyle and mesostyle, the prominent cingulum and the rugose enamel. The lower molars are broad, with a wide talonid in the third molar. Protocone pyramidal with two posteriorly directed folds.

Dorcabune, a genus founded and described by Pilgrim, is remarkable for the primitive construction of its teeth. Although this form is clearly a tragulid, it is at once distinguishable from the more normal and more advanced genera by its several primitive tooth characters. These primitive features of *Dorcabune* cause it to resemble, in certain respects, some of the primitive anthracotheres. It must be realized, however, that the resemblances of *Dorcabune* to certain anthracotheres is due to the fact that this primitive tragulid has retained various ancient artiodactyl heritage characters, and that they do not represent a direct phylogenetic connection between the two families.

The primitive anthracothere like characters in Dorcabune are as follows:

1. The molar cusps are very bunodont.
2. The enamel is heavy and rugose.
3. Well developed cingular shelves are retained.
4. There is a backwardly directed buccal ridge on the protocone.
5. The median axis of the talonid is parallel to the median axis of the third molar, rather than being at an angle to the latter, as in the more advanced tragulids.

Unfortunately the skull of *Dorcabune* is as yet unknown.

No detailed description of *Dorcabune anthracotherioides* need be given here. The reader is referred to Dr. Pilgrim's description in the Records of the Geological Survey of India for 1915.

Measurements and figures of various specimens of *Dorcabune anthracotherioides* are given on the accompanying pages.

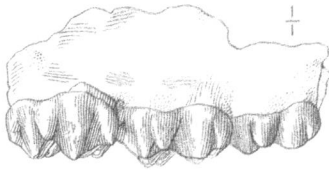

A.M. 19355

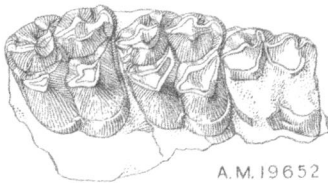

A.M. 19652

Fig. 135.

Fig. 136.

Fig. 135. *Dorcabune anthracotherioides* Pilgrim. Amer. Mus. No. 19652, maxilla with right M$^{1-3}$. Lateral view above, crown view below. Natural size.

Fig. 136. *Dorcabune anthracotherioides* Pilgrim. Amer. Mus. No. 19355, ramus with right M$_{1-2}$. Crown view above, lateral view below. Natural size.

A.M. 19353

Fig. 137. *Dorcabune anthracotherioides* Pilgrim. Amer. Mus. No. 19353, left M$_3$. Crown view above, lateral view below. Natural size.

MEASUREMENTS

Dorcabune anthracotherioides

Amer. Mus. No. 19652.

Length of molar series................ 50.0 mm

M^1......................	Length...	15.5	Width...	19.5 mm.
M^2......................	...	18.0		22.5
M^3......................	...	19.5		22.5

Amer. Mus. No. 29998.

M^3...................... Length... 18.0 mm... Width... 22.0 mm.

Amer. Mus. No. 19355.

M$_1$......................	Length...	17.0 mm...	Width...	12.0 mm.
M$_2$......................	...	17.5		13.0

Amer. Mus. No. 19353.

M$_3$...................... Length... 28.0 mm... Width... 14.0 mm.

Dorcabune hyaemoschoides Pilgrim

Dorcabune hyaemoschoides, Pilgrim, 1915, Rec. Geol. Surv. India, XLV, p. 231, Pl. XXI, fig. 6, Pl. XXII, figs. 2, 3, Pl. XXIII, fig. 1.

Type.—G.S.I. No. B 585, a right mandibular ramus.

Paratypes.—G.S.I. Nos. B 586, a right M_3; B 587, an upper molar; B 589, a right M_2.

Horizon.—Lower Siwaliks, Chinji zone.

Locality.—Near Chinji, Salt Range, Punjab.

Diagnosis.—Similar to *Dorcabune anthracotherioides*, but smaller. See remarks on page 305.

This species is here considered as synonymous with *Dorcabune anthracotherioides;* see the discussion on page 305.

Dorcabune nagrii Pilgrim

Dorcabune nagrii, Pilgrim, 1915, Rec. Geol. Surv. India, XLV, p. 233, Pl. XXI, fig. 5, Pl. XXII, fig. 1.

Type.—(Lectotype.)—G.S.I. No. B 590, a right M^3.

Cotypes.—G.S.I. No. B 591, a left mandibular ramus. Another mandible, not specified.

Horizon.—Middle Siwaliks.

Locality.—Near Nagri, Punjab.

Diagnosis.—Smaller than *D. anthracotherioides* with less developed cingula.

Dorcabune latidens Pilgrim

Dorcabune latidens, Pilgrim, 1915, Rec. Geol. Surv. India, XLV, p. 232, Pl. XXII, figs. 7, 8.

Type.—G.S.I. No. B 106, a mandibular ramus.

Paratypes.—None.

Horizon.—Middle Siwaliks.

Locality.—Near Hasnot, Punjab.

Diagnosis.—A small species with broad molars and a comparatively deep mandibular ramus. See remarks on page 305.

This species is here considered as synonymous with *Dorcabune nagrii;* see the discussion on p. 305.

Dorcabune sindiense Pilgrim

Dorcabune sindiense, Pilgrim, 1915, Rec. Geol. Surv. India, XLV, p. 234, Pl. XXI, figs. 3, 4.

Type.—(Lectotype.)—G.S.I. No. B 598, a left M^3.

Cotype.—G.S.I. No. B 601, a right M_1.

Horizon.—Lower Manchars.

Locality.—Bhagothora, Sind.

Diagnosis.—Comparable in size to *D. nagrii*, but with pronounced brachyodont molars.

A REVISION OF THE GENUS *Dorcabune*

The genus *Dorcabune* was founded on *Dorcabune anthracotherioides*, briefly described by Pilgrim in 1910.

"This interesting species from the Lower Siwaliks of Chenji shows the most extraordinary mingling of Traguloid and Anthracotheroid characters. Its upper molars may be described as like those of a *Dorcatherium*, only of an extreme bunodont and brachyodont type. Parastyle and mesostyle are prominent and isolated. The protoconal crescent is incomplete posteriorly with two marked folds much as in *Telmatodon* and *Hemimeryx*. There is a prominent cingulum and a strong rugose sculpture. M_3 is 26.2 mm. broad and 22.5 mm. long. The same type of structure is displayed in the lower teeth, which, however, differ less, qualitatively, from *Dorcatherium* than the upper ones. The characteristic posterior fold in the protoconid is present. On the whole, the genus may be appropriately placed in the Tragulidae. Lower molars of the species occur in Sind."[76]

In 1915 this same author issued a monograph on the genus, in which he designated five species as belonging to it, namely *D. anthracotherioides* and four new forms, *D. hyaemoschoides*, *D. nagrii*, *D. latidens* and *D. sindiense*. Of these species, two are from the Lower Siwaliks, two from the Middle Siwaliks, and one from the Lower Manchar beds of Sind. The horizons and localities of the species of *Dorcabune*, as recognized by Pilgrim, may be expressed in tabular form.

D. anthracotherioides.... Lower Siwaliks, Chinji zone..... Near Chinji, Punjab.
D. hyaemoschoides.......Lower Siwaliks, Chinji zone.....Near Chinji, Punjab.
D. nagrii ....Middle Siwaliks, Nagri zone.....Near Nagri, Punjab.
D. latidens ....Middle Siwaliks ....Near Hasnot, Punjab.
D. sindiense...........Lower Manchar ....Bhagothoro, Sind.

A careful study of the literature and of material has led me to believe that *Dorcabune hyaemoschoides* is synonymous with *Dorcabune anthracotherioides*, while *Dorcabune latidens* is synonymous with *Dorcabune nagrii*. A criticism of Pilgrim's distinctions for the two species thus dropped will show why they are not considered as valid.

Dorcabune hyaemoschoides

Distinctive characters setting it apart from *D. anthracotherioides*.

Criticisms.

1. Smaller.

1. The differences in size are within the limits of individual variation.

2. Cusps of lower molars less bunodont.

2. Difference very slight.

3. Front arm of protoconid is narrower.

3. A variable character.

4. Talonid narrower.

4. A variable character.

5. Posterior cusp of P_4 extends further in.

5. The supposed P_4 of *D. anthracotherioides* is too fragmentary for comparison.

6. Upper molar distinguished by its more slender cusps and less pronounced parastyle.

6. The upper molar figured by Pilgrim would seem to be a last milk molar, and therefore not directly comparable with a permanent tooth.

Dorcabune latidens

Distinctive characters setting it apart from *D. nagrii*.

Criticisms.

1. Similar in size but the mandible is deeper.

1. A variable character.

[76] Pilgrim, G. E., 1910A, pp. 68–69.

2. Lower molars broader.   2. Indices of length to breadth in M_2:

                                             D. nagrii 73

                                             D. latidens 70

3. Otherwise like *D. anthracotherioides.*

Thus it would seem probable that the two species *Dorcabune hyaemoschoides* and *Dorcabune latidens*, are synonymous with *Dorcabune anthracotherioides* and *Dorcabune nagrii* respectively.

It appears probable that two species of *Dorcabune* are present in the Siwalik beds, a large one, *D. anthracotherioides* of typical Lower Siwalik age, but extending into the lower portion of the Middle Siwaliks, and a smaller form, *D. nagrii* ranging throughout the Middle Siwaliks.

The distinguishing characters of these two species are listed below.

Dorcabune anthracotherioides.

1. A large species.
2. Upper molars characterized by a well developed cingulum, bunodont cusps, heavy barrels on outer cusps, heavy parastyle and mesostyle, a buccal ridge on the posterior side of the protocone, heavy rugose enamel.
3. Lower molars characterized by their bunodonty, marked development of cingulum, double fold on posterior side of protoconid, spurs from external cusps not reaching the internal border of the tooth.

Dorcabune nagrii

1. A small species.
2. Upper molars characterized by the slight development of the cingulum. Otherwise similar to the foregoing species.
3. Lower molars characterized by the slight development of the cingulum. Otherwise similar to the foregoing species.

Dorcabune sindiense is comparable in size to *D. nagrii.* It shows, however, a real difference from the latter form in its pronounced molar brachyodonty and for this reason it may be considered as a separated and a valid species. That it comes from Sind, a region geographically distant from the Punjab would increase the probability of its distinctness.

The brachyodont character of the molars in *D. sindiense* would indicate that this species is in all probability a primitive one, in a stage of phylogenetic development less advanced than the species from the Punjab.

Note: Two specimens, Amer. Mus. No. 19353 and Amer. Mus. No. 29998, which belong quite definitely to the species *D. anthracotherioides*, are recorded from the lower portion of the Middle Siwaliks, indicating that this species persisted beyond the limits of Chinji time.

Dorcatherium Kaup, 1833

Generic type, *Dorcatherium naui* Kaup

Dorcatherium majus Lydekker

Dorcatherium majus, Lydekker, 1876, Pal. Indica (X), I, p. 44, Pl. VII, figs. 4, 6, 9, 10, 11.
    Additional References.—
        Lydekker, R., 1880B, p. 32; 1883C, p. 90; 1885B, p. 34; 1885D, p. 154.
        Pilgrim, G. E., 1910B, p. 203; 1913B, pp. 285, 319.
        Matthew, W. D., 1929, p. 453.
    Type.—(Lectotype.)—G.S.I. No. B 197, two upper molars.
    Cotype.—G.S.I. No. B 198, a maxilla with left M^{2-3}.
    Horizon.—"Siwaliks" (probably Middle Siwaliks) for the type. Lower and Middle Siwaliks for referred specimens.
    Locality.—Kushalgar, near Attock. Also Hasnot and adjacent localities.
    Specimens in the American Museum.—Amer. Mus. No. 19302. Right M^2; Lower Siwaliks, about level of Chinji Rest House, five miles east of Chinji Rest House.
19303. Mandibular ramus with left P_{3-4}, M_1; Lower Siwaliks, five miles east of Chinji Rest House.
19304. Left M^{1-2}; Lower Siwaliks, about level of Chinji Rest House, five miles east of Chinji Rest House.
19354. Left M^3; Lower Siwaliks, about level of Chinji Rest House, five miles east of Chinji Rest House.
19369. Right M_{2-3}; Lower Siwaliks, about level of Chinji Rest House, five miles east of Chinji Rest House.
19520. Right M_{1-2}; Middle Siwaliks, two miles northeast of Hasnot.
19524. Right P_4-M_2; Middle Siwaliks, two miles west of Hasnot.
19939. Left M_3; Middle Siwaliks, four and one half miles west of Hasnot.
    Diagnosis.—A large species of *Dorcatherium*, with strong mesostyle and cingula in the upper molars and well developed accessory pillars in the lower molars.

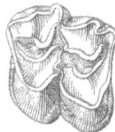

Fig. 138. *Dorcatherium majus* Lydekker. Amer. Mus. No. 19302, right M^2. Lateral view above, crown view below. Natural size.

Dorcatherium majus is a large species of the genus, being equal in size to the largest species of *Dorcabune*. It is characterized by its typical *Dorcatherium* molars, which set it apart from *Dorcabune*. The cheek teeth are relatively less hypsodont than are the molars of *Dorcatherium minus*.

Measurements and figures of some of the specimens in the American Museum collection are given below.

FIG. 139. *Dorcatherium majus* Lydekker. Amer. Mus. No. 19524, right ramus with P_4–M_2. Crown view above, lateral view below. Natural size.

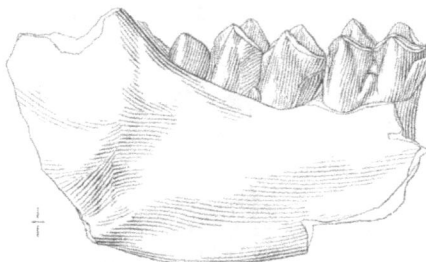

FIG. 140. *Dorcatherium majus* Lydekker. Amer. Mus. No. 19369, right ramus with M_{2-3}. Crown view above, lateral view below. Natural size.

MEASUREMENTS

Dorcatherium majus

	Length	Width	Height
Amer. Mus. No. 19302.			
M²	18.5 mm.	21.5 mm.	
19354.			
M³	20.5	23.5	
19524.			
P₄	14.5	5.0	
M₁	13.5	9.0	
M₂	16.0	11.0	12.0 mm.
19939.			
M₃	25.5	12.0	12.0
19520.			
M₁	14.0	9.0	
M₂	17.0	10.5	13.5

Dorcatherium minus Lydekker

Dorcatherium minus, Lydekker, 1876, Pal. Indica (X), I, p. 46, Pl. VII, figs. 3, 7.

Additional References.—

Lydekker, R., 1880B, p. 32; 1883C, p. 90; 1885B, p. 34; 1885D, p. 154.

Pilgrim, G. E., 1910B, p. 203; 1913B, pp. 317, 319.

Matthew, W. D., 1929, p. 453.

*Type.—*G.S.I. No. B 195, two upper molars, namely right M^{1-2}.

*Paratypes.—*None.

Horizon.—"Siwaliks" (probably Middle Siwaliks) for the type. Lower and Middle Siwaliks for referred specimens.

*Locality.—*Near Hasnot.

*Specimens in the American Museum.—*Amer. Mus. No. 19305. Ramus with left M_{1-2}; Lower Siwaliks, about level of Chinji Rest House, six miles west of Chinji Rest House.

19307. Ramus with left M_{1-2}; Lower Siwaliks, 100 feet above the level of Chinji Rest House, two miles west of Chinji Rest House.

19308. Ramus with right M_{2-3}; Lower Siwaliks, near Chinji Rest House.

19309. Ramus with right M_2; 100 feet above the level of Chinji Rest House, two miles west of Chinji Rest House.

19310. Mandibular ramus with M_{1-3}. Lower Siwaliks, 1600 feet above the level of Chinji Rest House, one and one half miles northeast of Chinji Rest House.

19313. Palatal fragment; 1600 feet above the level of Chinji Rest House, three miles west of Chinji Rest House.

19356. Ramus with right DM_{3-4}; about level of Chinji Rest House, five miles east of Chinji Rest House.

19365. Ramus with left M_{2-3}; about level of Chinji Rest House, ten miles east of Chinji Rest House.

19366. Ramus with left M_{2-3}; Lower Siwaliks, 1600 feet above the level of Chinji Rest House, two miles west of Chinji Rest House.

19367.   Ramus with right M_{2-3}; Lower Siwaliks, 1600 feet above the level of Chinji Rest House, one and one half miles northeast of Chinji Rest House.

19368.   Ramus with left M_{1-3}; near Chinji Rest House.

19374.   Two molar teeth; Lower Siwaliks, 600 feet above the level of Chinji Rest House, two miles west of Chinji Rest House.

19430.   Right M_3; Lower Siwaliks, 1600 feet above the level of Chinji Rest House, twelve miles east of Chinji Rest House.

19517.   Left DM^4-M^1; Middle Siwaliks, lower portion, one mile south of Nathot.

19609.   Ramus with right M_{2-3}; Lower Siwaliks, 1600 feet above the level of Chinji Rest House, one and one half miles west of Chinji Rest House.

19833.   Right M_2; Middle Siwaliks, lower portion, two miles east of Phadial.

29855.   Maxilla with left M^{2-3}; Lower Siwaliks, level of Chinji Rest House, four miles west of Chinji Rest House.

29856.   Right and left maxillary fragments with molars; Lower Siwaliks, level of Chinji Rest House, four miles northeast of that place.

Numerous fragmentary specimens, consisting of upper and lower teeth, without numbers.

Diagnosis.—A small species of the genus with hypsodont molars.

Reference should be made to Lydekkers descriptions and figures of this species. Additional figures are presented here.

Dorcatherium minus is characterized by its small size, relatively hypsodont cheek teeth and the strong development of the cingulum in the upper molars.   In the series of *D. minus* in the American Museum collection there seem to be two variations present.   In most of the upper molars the mesostyle is rather small while the internal cingulum is exceedingly well developed.   In a few molars, however, the mesostyle is comparatively heavy and the internal cingulum is rather slight.

Vestigial internal pillars are frequently present in the lower molars.   In one specimen (Amer. Mus. No. 19309), the pillar is of large size.   This seems to be a retention of a

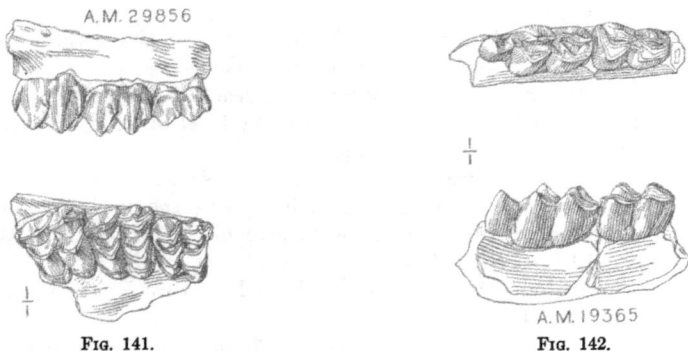

Fig. 141.

Fig. 142.

Fig. 141.   *Dorcatherium minus* Lydekker.   Amer. Mus. No. 29856, maxilla with right M^{1-3}.   Lateral view above, crown view below.   Natural size.

Fig. 142.   *Dorcatherium minus* Lydekker.   Amer. Mus. No. 19365, right M_{2-3}.   Crown view above, lateral view below. Natural size.

primitive character. (Compare "*Hyaemoschus*" of the Orleanais of Baigneaux, in which the external pillar is well developed.)

In an unnumbered specimen, associated with Amer. Mus. No. 19624, the last upper premolars of both sides are preserved. In shape the last premolar is similar to the corresponding tooth in *Hyaemoschus*. It differs from the tooth in the modern species in the heavy antero-external pillar, and the high postero-external wall. On the inner side, there is a posterior cingulum—another distinction. The greatest difference, however, is in the transverse ridge which runs from a point behind the central external cusp, across the median valley, to the middle post of the internal cusp. This structure is entirely lacking in the modern *Hyaemoschus*, one line of evidence pointing to the generic distinctness of the modern and the fossil forms.

MEASUREMENTS

Dorcatherium minus

	Length	Width	Height
Amer. Mus. No. 29856.			
M¹	9.8 mm.	10.0 mm.	7.0 mm.
M²	11.3	12.0	8.5
M³	11.5	13.0	10.0
19517.			
DM⁴	11.0	10.0	6.0
M¹	12.0	11.0	7.5
19365.			
M₂	13.0	7.5	
M₃	18.0	8.0	10.0
19366.			
M₂	12.0	7.5	9.0
M₃	16.0	8.0	10.0

Dorcatherium sp.

Specimens.—Amer. Mus. No. 19306. A palatal fragment with right M¹⁻³; Lower Siwaliks, 1600 feet above the level of Chinji Rest House, twelve miles east of Chinji Rest House.

19613. Ramus with right M₂₋₃; Lower Siwaliks, 600 feet above the level of Chinji Rest House, three miles northwest of Chinji Rest House.

29854. Ramus with right M₂₋₃; Lower Siwaliks, five miles east of Chinji Rest House.

29887. Ramus with left DM₄, M₁; Middle Siwaliks, one mile north of Hasnot.

The specimens listed above are very small, being on the average twenty to thirty per cent smaller than those of *Dorcatherium minus*. The teeth are, on the whole, very much like those of *Dorcatherium minus*, but are relatively less hypsodont. The upper molars are characterized, too, by their strong mesostyle and the slight development of the cingulum, being in these respects like the second variant of *D. minus*, mentioned above.

In view of the fragmentary nature of these specimens, it is not deemed advisable to set them apart with a new specific name, although they undoubtedly do constitute a new species. A species can be created when more adequate material is known.

A. M. 19306

FIG. 143.

A. M. 19613

FIG. 144.

FIG. 143. *Dorcatherium* sp. Amer. Mus. No. 19306, maxilla with right M$^{1-3}$. Lateral view above, crown view below. Natural size.

FIG. 144. *Dorcatherium* sp. Amer. Mus. No. 19613, right ramus with M$_{2-3}$. Crown view above, lateral view below. Natural size.

Measurements and figures are presented here.

Dorcatherium sp. Amer. Mus. No. 19306		*Dorcatherium minus* Amer. Mus. No. 29856	
Antero-posterior dia.... M^1 .... 8.0 mm.		Antero-posterior dia.... M^1 .... 9.8 mm.	
M^2 .... 8.5		M^2 .... 11.3	
M^3 .... 9.5		M^3 .... 11.5	
M$^{1-3}$ 25.2		M$^{1-3}$ 31.5	

Linear difference 20 percent.

Dorcatherium sp. Amer. Mus. No. 19613	29854	*Dorcatherium minus* Amer. Mus. No. 19365
Antero-posterior dia. M$_2$ ..... 8.0 mm.	8.5 mm.	Antero-posterior dia. M$_2$ .... 13.0 mm.
M$_3$ 11.5	13.0	M$_3$ 18.0
M$_{2-3}$ ... 19.0	22.0	M$_{2-3}$ 31.0

Greatest linear difference....... 39 percent.
Least linear difference......... 29 percent.

NOT SPECIFICALLY IDENTIFIED.

Amer. Mus. No. 19602. Tragulid teeth. Lower Siwaliks, one mile southeast of Chinji Rest House.

29953. Tragulid jaw fragment. Lower portion of Middle Siwaliks, one mile south of Nathot.

ON THE GENUS *Dorcatherium*

The name *Dorcatherium* was first used by Kaup in 1833, in reference to some Miocene tragulids from Eppelsheim. Later, in 1883, Rütimeyer concluded that *Hyaemoschus*, a name applied by Gray in 1845 to the modern African tragulid, is synonymous with *Dorcatherium*, and since his time various authors have applied this latter name to the living African animal.

In 1876 Lydekker described certain tragulids from India and these he placed in the genus *Dorcatherium*. Numerous specimens referable to this genus are in the American Museum Siwalik collection.

An examination of the literature and of the specimens in the American Museum would indicate that the Siwalik tragulids are generically quite distinct from the modern South African form. On the other hand, they are seemingly congeneric with the Miocene and Pliocene *Dorcatherium* of Europe.

Dorcatherium would seem to be a specialized tragulid which has developed parallel to *Hyaemoschus*. This postulate is based on certain dental characters, which show the teeth of the fossil form to be different, and in some respects more advanced, than the teeth of the modern African genus.

1. The cheek teeth of *Dorcatherium* are more hypsodont than are the cheek teeth of *Hyaemoschus*.

2. The external styles of the upper molars are strongly developed in *Dorcatherium* and less developed in *Hyaemoschus*.

3. The basal cingulum is well developed in the upper molars of *Dorcatherium*, which is not the case in *Hyaemoschus*.

4. In *Dorcatherium* a vestige of the external median pillar remains in the lower molars. In some teeth this pillar is rather robust.

Unfortunately, the skull of the Siwalik *Dorcatherium* is as yet unknown, so it is impossible at present to discuss the really important skull characters in the fossil form as compared to the recent genus.

Bearing the above listed dental characters of *Dorcatherium* in mind, we see that this genus became specialized, especially in regard to the hypsodonty of the cheek teeth, and separated from the phylogenetic line that eventually led to the African *Hyaemoschus*.

Matthew (1908B, p. 555) set forth the view that *Dorcatherium* is a primitive offshoot of the pecoran stock, quite separate from *Tragulus*. The common characters of the two genera he would attribute to a retention of persistent primitive characters in both of them. Perhaps the phylogenetic relationships of the tragulids might be represented somewhat in the following manner.

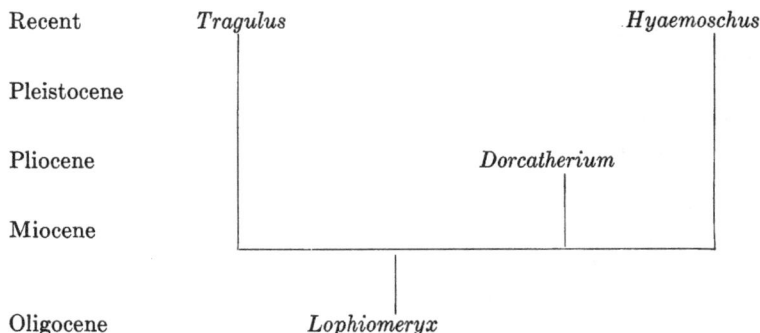

Recent	*Tragulus*		*Hyaemoschus*
Pleistocene			
Pliocene		*Dorcatherium*	
Miocene			
Oligocene	*Lophiomeryx*		

Tragulus Brisson, 1762

Generic type, *Tragulus indicus* Brisson

Tragulus sivalensis Lydekker

Tragulus sivalensis, Lydekker, 1882, Rec. Geol. Surv. India, XV, p. 30.

   Additional References.—
  Lydekker, R., 1884D, pp. 117–118, fig. 6, p. 129; 1885B, p. 35; 1885D, pp. 152–153.
  Pilgrim, G. E., 1910B, p. 203.
  Matthew, W. D., 1929, pp. 459–460.
  Type.—G.S.I. No. B 360, a left M^3.
  Paratypes.—None.
  Horizon.—Middle Siwaliks.
   Locality.—Hasnot, Punjab.
   Diagnosis.—(See Lydekker, R., 1884D, p. 118.)   Upper cheek teeth similar to those of the modern *Tragulus.*

   A species of very doubtful value.   Founded on a single, isolated molar tooth.

CERVOIDEA

CERVIDAE

CERVINAE

Cervus Linnaeus, 1758

Generic type, *Cervus elaphus* Linnaeus

Cervus sivalensis Lydekker

Cervus sivalensis, Lydekker, 1880, Pal. Indica (X), I, p. xvii, Pl. VIII, fig. 5.

   Additional References.—
  Lydekker, R., 1880B, p. 32; 1883C, p. 90; 1885B, p. 33; 1885D, pp. 104–106.
  Pilgrim, G. E., 1910B, p. 203; 1913B, p. 325.
  Brown, B., 1926, p. 6.
  Matthew, W. D., 1929, p. 444.
   Type.—G.S.I. No. B 215, a ramus with right M_{2-3}.
   Paratypes.—None.
   Horizon.—Upper Siwaliks.
   Locality.—Maili, Punjab.
   Specimens in the American Museum.—Amer. Mus. No. 19807.   An antler.   From the
   Upper Siwaliks, one mile west of Chandigarh.
   19829.   A skull.   From the Upper Siwaliks, nine miles west of Kalka.
   Diagnosis.—A large cervid with relatively hypsodont molars.   The skull and antlers resemble these portions in *Cervus duvaucelli*, the skull by virtue of the frontal concavity at the orbits, and the forward swell at the pedicles.   The lacrymal vacuity is smaller than in *C. duvaucelli*.   The brow tine of the antler arises immediately above the burr, and forms an obtuse angle with the beam.

Lydekker has pointed out the resemblances of this species to *Cervus duvaucelli*, both in the characters of the skull and of the antlers.

The skull in the American Museum collection is that of a large cervid. The specimen has suffered from the effects of weathering, so that the surfaces are worn and partially destroyed, and the muzzle anterior to the fourth premolar is missing. The cranium is badly used in the region where the pedicels might have been, making it impossible to determine whether or not this individual possessed antlers.

The brain case is large and the face is deep. The lacrymal fossae are relatively large and deep, though not to the extent that is characteristic of the Barasingha. The palate is wide. The basicranium is well preserved. As is typical of the cervids, the auditory

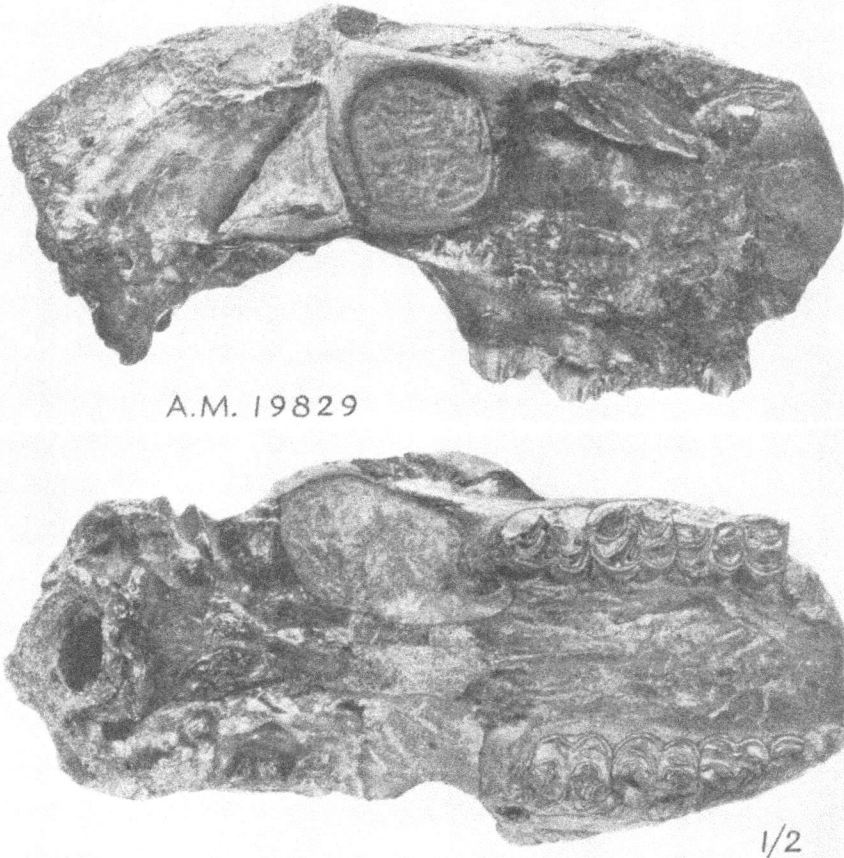

Fig. 145. *Cervus sivalensis* Lydekker. Amer. Mus. No. 19829, skull. Lateral view above, palatal view below. One half natural size.

bullae are quite small and the paroccipital processes are strong.   The basicranial foramina are essentially the same as in *Cervus duvaucelli*, and need not be described here.

The teeth are rather badly damaged.   They are large and quadrate in shape, the folds are open and the enamel rugose.   The internal pillars are very small.

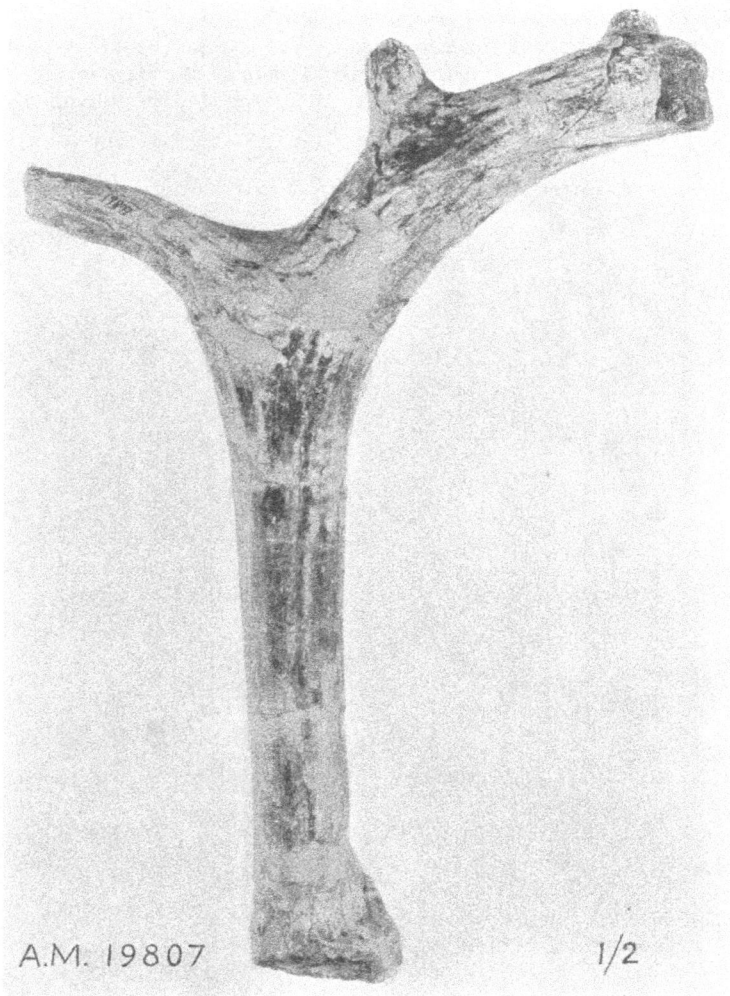

FIG. 146. *Cervus sivalensis* Lydekker.   Amer. Mus. No. 19807, antler.   Lateral view, one half natural size.

Amer. Mus. No. 19829

Width of palate at M¹............................ 56 mm.
Length of molar series........................... 78
P⁴ length....................................... 18
    width....................................... 24
M¹ length....................................... 22
    width.......................................
M² length....................................... 29
    width....................................... 27
M³ length....................................... 28
    width....................................... 25

An antler in the American Museum collection is definitely referable to this species. It is large, heavily built and of round cross section. It is shaped very much like the antler of the Barasingha; that is, there is a brow tine directly above the burr, while further up the beam divides into two branches, a smaller inner one and a larger outer one, the latter bearing three tines.

MEASUREMENTS

Amer. Mus. No. 19807

Anteroposterior diameter of beam above burr................. 37 mm.
Length of beam from burr to branch........................ 270

Cervus triplidens Lydekker

Cervus triplidens, Lydekker, 1876, Pal. Indica (X), I, p. 49, Pl. VIII, figs. 1, 2.
    Additional References.—
        Lydekker, R., 1880B, p. 32; 1883C, p. 90; 1884D, p. 120; 1885B, p. 32.
        Pilgrim, G. E., 1910B, p. 203; 1913B, p. 286.
        Brown, B., 1926, p. 6.
        Matthew, W. D., 1929, pp. 449, 452.
    *Type.—*G.S.I. No. B 204, a right maxilla with the last two molars.
    *Paratypes.—*None.
    Horizon.—"Siwaliks," probably Upper Siwaliks (Brown); Middle Siwaliks (Pilgrim).
    *Locality.—*From the Punjab.
    *Specimens in the American Museum.—*Amer. Mus. No. 19792. Right P³–M³; left M²⁻³, from the Pinjor zone of the Upper Siwaliks, six miles east of Chandigarh.
    *Diagnosis.—*Molars hypsodont, with a large accessory column and rugose enamel.

This species was founded on very meagre material, and consequently it must be regarded as of little value. Lydekker did not designate any definite locality or horizon for it, thereby increasing the difficulties of determining it with any degree of assurance. Pilgrim listed this species as of Middle Siwalik age, while Brown, on the basis of his field observations, suggested that it might be of Upper Siwalik age.

The American Museum specimen is referred to this species on the basis of the large internal pillar, the hypsodonty and the lack of extensions from the enamel walls into the fossettes. This identification must be regarded as somewhat provisional.

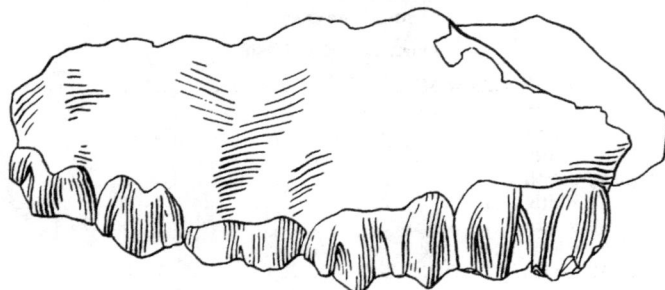

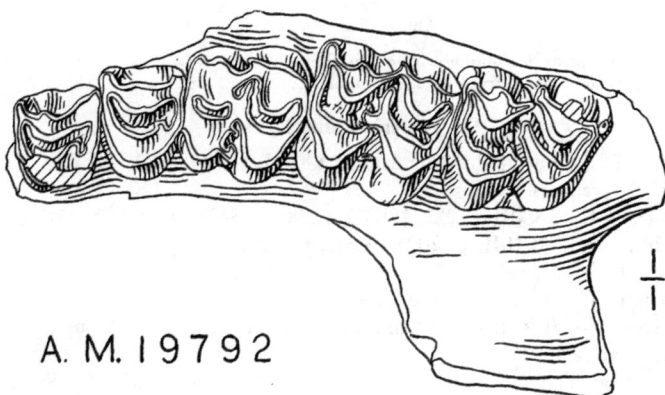

A. M. 19792

FIG. 147. *Cervus triplidens* Lydekker. Amer. Mus. No. 19792, maxilla with left P³–M³. Lateral view above, crown view below. Natural size.

MEASUREMENTS

Amer. Mus. No. 19792

	Length	Width
P³	15.5 mm.	18.0 mm.
P⁴	14.0	19.0
M¹	20.0	24.0
M²	24.0	26.5
M³	26.0	26.0

Cervus simplicidens Lydekker

Cervus simplicidens, Lydekker, 1876, Pal. Indica (X), I, p. 51, Pl. VIII, fig. 3.

Additional References.—

Lydekker, R., 1880B, p. 32; 1883C, p. 90; 1884D, p. 119; 1885B, p. 32.

Pilgrim, G. E., 1910B, p. 203; 1913B, p. 285.

Brown, B., 1926, p. 6.

Matthew, W. D., 1929, pp. 449, 452.

Type.—G.S.I. No. B 204.   A maxillary fragment with left M^{1-2}.

Paratypes.—None.

Horizon.—"Siwaliks," probably Upper Siwaliks (Brown); Middle Siwaliks (Pilgrim).

Locality.—From the Punjab.

Specimens in the American Museum.—Amer. Mus. No. 19811.   A palatal fragment with right DM^{2-4}, M^{1-2}.   From the Upper Siwaliks, one mile south of Mirzapur.

19987.   Some isolated molar teeth.   Upper Siwaliks, near Chandigarh.

Diagnosis.—Molar crowns square, with small accessory pillars, not obliquely set, and with slightly rugose enamel.

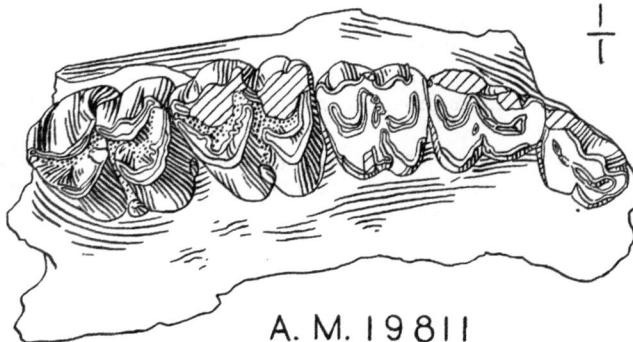

A. M. 19811

FIG. 148. *Cervus simplicidens* Lydekker.   Amer. Mus. No. 19811, maxilla with right DM^{2-4}, M^{1-2}.   Lateral view above, crown view below.   Natural size.

The remarks regarding the status of *Cervus triplidens* apply equally well to this species.

The two specimens in the American Museum collection referred to this species are so placed by virtue of the very small internal pillars, smaller size as compared with *C. triplidens* and the configuration of the enamel folds within the fossettes.

MEASUREMENTS

Amer. Mus. No. 19811

	Length	Width
M^1...............................	17.0 mm.	20.0 mm.
M^2...............................	23.0	24.5
M^3...............................	26.0	23.0
Height of M^3............ 27.0 mm.		

Cervus punjabiensis Brown

Cervus punjabiensis, Brown, 1926, Amer. Mus. Novitates, No. 242.

Type.—Amer. Mus. No. 19911, an incomplete skull with antlers.

Paratypes.—None.

Horizon.—Upper Siwaliks, immediately below the conglomerate.

Locality.—Two miles west of Chandigarh, Ambala District, Punjab.

Specimen in the American Museum.—The type, designated above.

Diagnosis.—"Brain case formed as in *C. axis;* bullae low and rounded; antlers lying in a plane parallel to facial angle, widely expanded, round and comparatively smooth; nasals narrow and muzzle not expanded; teeth hypsodont, with open crescents and enfolded enamel, lacking accessory inner columns; surface smooth." [77]

Portions of Brown's description of this species are quoted below. He has clearly shown that *C. punjabiensis* is intermediate and transitional between *C. axis* and *C. unicolor*.

"This animal was fully mature (apparently about four years old), with permanent teeth erupted and slightly worn. The antlers had probably just passed the 'velvet' stage.

"Compared with the two nearest living deer, *C. axis* and *C. unicolor*, to which it is clearly related, we find a combination of characters that now distinguish these two racial stocks.

"The antlers, supported on short pedicles, are long, round, divergent and moderately smooth, the brow-tine making an obtuse angle with the beam. The rear tine of the terminal fork, which forms the continuation of the beam, is much the longer, the shorter front tine being placed forward and outward. The antlers are comparatively shorter and more slender than in the living chital, but, unlike the chital or any member of the sambur race, they are widely divergent and in profile lie in a plane generally parallel to the facial angle without forward recurve.

"The range of variation in form and size of antlers in the axis deer is considerable, depending chiefly upon the age of the individual, as is shown in a series of antlers developed by an animal during its life in captivity.

"The brain case is formed as in *C. axis*, with a similar development and relationship in each of the skull elements.

"The teeth are hypsodont and higher-crowned than in the several skulls of *C. axis* with which this specimen has been compared, agreeing in this respect with *C. unicolor*.

[77] Brown, B., 1926, p. 1.

They are more oblique to the axis of the skull and to each other than in *C. unicolor*, the anterior premolars approaching the median line as in *C. axis*, thus verifying the narrow muzzle which is indicated by the high-arched narrow nasals. The crescents are more open than in *C. unicolor* or in *C. axis*, with costae on outer faces of teeth less pronounced and inner accessory columns absent. Enamel surface smooth."

"By careful comparison with the figures and with Lydekker's descriptions of these several species, *C. punjabiensis* agrees more closely with specimens referred to *C. simplicidens*, which probably came from the same series of rocks but is distinguished from that species by the more open valleys between the cones, by the enfolded enamel, by the flatter outer face of all teeth, with less pronounced costae, by greater obliquity of teeth to axis of series and total absence of accessory column on last molar, with only a rudimentary accessory column on first molar; enamel perfectly smooth."

MEASUREMENTS

Length of antler on outside curve............................ 695 mm.
Width of antlers tip to tip; double left, normal side............ 670
Circumference of beam above brow-tine...................... 95
Length of molar series.................................... 100

The accompanying figures illustrate this species.

FIG. 149. *Cervus punjabiensis* Brown. Amer. Mus. No. 19911, skull with antlers. Dorsal view, two fifteenths natural size. From Brown, 1926.

FIG. 150. *Cervus punjabiensis* Brown. Amer. Mus. No. 19911, skull with antlers. Right side, two fifteenths natural size. From Brown, 1926.

FIG. 151. *Cervus punjabiensis* Brown. Amer. Mus. No. 19911, skull with antlers. Left side, two fifteenths natural size. From Brown, 1926.

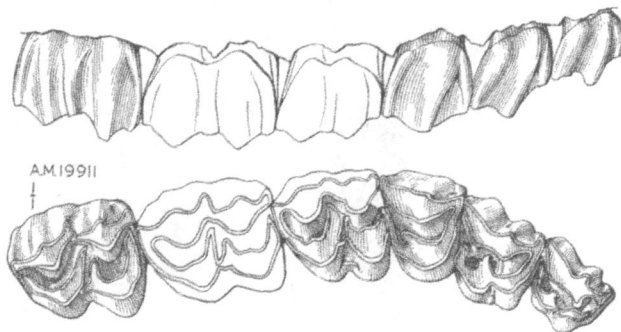

FIG. 152. *Cervus punjabiens's* Brown. Amer. Mus. No. 19911, right P^2–M^3. Lateral view above, crown view below. Natural size. From Brown, 1926.

NOT SPECIFICALLY IDENTIFIED

19510. Cervid mandible. Middle Siwaliks, 1000 feet below the Bhandar bone bed, one mile south of Nathot.

19826. Cervid; left mandibular ramus with fragmentary teeth. Upper Siwaliks, near Siswan.

29815. Cervid; miscellaneous teeth. Middle Siwaliks, one half mile south of Dhok Pathan.

GIRAFFOIDEA

GIRAFFIDAE

THE CLASSIFICATION OF THE GIRAFFIDAE [78]

In 1911 Dr. Pilgrim classified the giraffes (in his monograph entitled "The Fossil Giraffidae of India") as follows:

Family Giraffidae
    Subfamily Palaeotraginae
        Genera *Palaeotragus*
            Samotherium
            Alcicephalus
            Okapia
            Indratherium
            Libytherium
    Subfamily Helladotheriinae
        Genera *Helladotherium*
            Vishnutherium
            Giraffokeryx
    Subfamily Progiraffinae
        Genera *Progiraffa*
    Subfamily Giraffinae
        Genera *Giraffa*
            Orasius
    Subfamily Sivatheriinae
        Genera *Sivatherium*
            Hydaspitherium
            Bramatherium
            Urmiatherium (placed here rather than in the Bovidae)

Pilgrim's classification is marked by the multiplication of subfamilies, of which one, the Progiraffinae, is founded on rather scanty material, and another, the Helladotheriinae, consists of genera that might be placed within two different but well established groups, the Palaeotraginae and the Sivatheriinae.

Pilgrim placed *Giraffokeryx* in the Helladotheriinae, whereas other authors have usually regarded it as belonging in the Palaeotraginae. Bohlin, in 1927, went to the other extreme and reduced the genus *Giraffokeryx* to synonymy with *Palaeotragus*, as will be shown below.

[78] A more detailed treatment of this subject will be found in Colbert, Edwin H., 1935, Amer. Mus. Novitates, No. 800.

Both of these authors based their conclusions on the evidence of teeth alone.   The complete skull of *Giraffokeryx*, described in subsequent pages of this work (pp. 329–333), helps to solve the problem of taxonomic relations of the genus.   It is shown below that *Giraffokeryx* is essentially a *Palaeotragus*-kind of animal with an extra pair of horn cores on the frontals. Therefore the genus properly belongs in the Palaeotraginae, but it is quite separate from the genus *Palaeotragus*.

Bohlin and Matthew have both given conclusive evidence to show that the genera *Helladotherium* and *Vishnutherium* should be included in one subfamily with *Sivatherium*, *Bramatherium*, etc.

Birger Bohlin, in his recent monograph on the Giraffidae (1927), classified the family in the following way.

> Family Giraffidae
> > Subfamily   Palaeotraginae
> > > Genera *Palaeotragus*
> > > > *Giraffokeryx*
> > > > *Achtiaria*
> > > > *Samotherium*
> > > > *Alcicephalus*
> > > > *Chersonotherium*
> > > > *Shanshitherium*
> > Subfamily   Giraffinae
> > > Genera *Honanotherium*
> > > > *Orasius*
> > > > *Giraffa*
> > Subfamily   Okapiinae
> > > Genera *Okapia*
> > Subfamily   Sivatheriinae
> > > Genera *Sivatherium*
> > > > *Indratherium*
> > > > *Bramatherium*
> > > > *Hydaspitherium*
> > > > *Helladotherium*
> > > > *Griquatherium*
> > > > *Vishnutherium*
> > > > *Libytherium*
> > [Subfamily Progiraffinae
> > > Genus *Progiraffa*]

In 1929 Dr. Matthew pointed out the desirability of including *Okapia* among the Palaeotraginae, thereby making three subfamilies for the Giraffidae instead of four, leaving out of account the problematical Progiraffinae.   Matthew's classification (1929, p. 546) was as follows.

> Family Giraffidae
> > Subfamily   Palaeotraginae
> > > Genera *Palaeotragus*
> > > > *Samotherium*

Giraffokeryx
Okapia
Subfamily Giraffinae
Genera *Orasius*
Giraffa
Honanotherium
Subfamily Sivatheriinae
Genera *Helladotherium+Bramatherium*
Hydaspitherium
Sivatherium+Indratherium

A careful consideration of the problem of a classification of the Giraffidae will demonstrate the validity of Dr. Matthew's arrangement, especially as regards the inclusion of *Okapia* in the Palaeotraginae. Bohlin's separation of the okapi into a distinct subfamily is seemingly a flaw in his otherwise admirable classification of this group of mammals. It would seem that he has placed too much emphasis on minute, and for the most part unimportant characters, and in doing this he has disregarded the great preponderance of characters that typify *Okapia* as a truly primitive palaeotragine. *Okapia* is, in all of its essential characters, a structurally primitive Miocene giraffe, more primitive than *Palaeotragus* or *Samotherium*, that has persisted on to the present day in a region conducive to the continuation of such a primitive form.

Bohlin has separated the okapi from the Palaeotraginae because:

1. The frontals are narrow in the modern form;

2. The horns are placed in a slightly different position in *Okapia* from the positions of the horns in *Palaeotragus* or *Samotherium;*

3. The frontals in the okapi tend to develop pneumatic sinuses within them, whereas the sinuses are not pronounced in *Palaeotragus* and *Samotherium;*

4. There are minor differences in the dentition; there is no outer cingulum on DM^2 in the okapi, whereas in *Palaeotragus* and *Samotherium* this cingulum is present;

5. The skeleton of the okapi differs in minor details, especially those of proportions, from the skeleton of *Palaeotragus.*

These are differences of minor importance. Now let us look at *Okapia* and the fossil Palaeotraginae for the purpose of making comparisons between major anatomical characters.

The skull of *Okapia* is in most respects more primitive than the skull of the fossil Palaeotraginae. The canine-premolar diastema of the mandible is much shorter in the okapi than it is in the fossil forms, so we see that the modern genus has retained a short muzzle, a primitive and a diagnostic heritage character. In the okapi the frontals are narrow, which is to be expected in a primitive animal. In *Palaeotragus* the frontals are wide, and this can be considered as a character developed subsequent to the narrow frontal region. An examination of various groups of ungulates will show that the skull tends to elongate first, after which it widens, if there is a tendency to widen at all. That is, elongation precedes lateral expansion. Consequently we may expect a primitive giraffid such as the okapi, to have a narrower frontal region than a more advanced form in which the skull has broadened out.

Of course, as Bohlin has pointed out, the frontals of the okapi contain rather large sinus cavities, which are lacking in *Palaeotragus* and *Samotherium*. It may be quite probable, however, that the development of the frontal sinuses in the okapi are of a secondary nature, and that they have been acquired more or less independently in the long period of time that has elapsed between the Miocene and the present day. But this is no reason for excluding the okapi from a place as a relatively primitive Palaeotragine. It is a primitive form that has developed certain specialized characters during the passage of geologic time.

In *Okapia* the horn cores are rather small, whereas in *Palaeotragus* they are much larger. Thus we may regard the okapi as more primitive in its horn development than is *Palaeotragus*. Of course, one might argue that the small horns in the okapi are degenerate structures, secondarily reduced from larger horns. But in answer to this argument it might be said that the horn cores in the okapi have retained a primitive position over the orbit, and this would favor their being truly primitive structures.

In *Okapia* the dentition is very brachyodont—a primitive character. In *Palaeotragus* and *Samotherium* the teeth are considerably higher than is the case in the modern form, showing that they are more advanced in their phylogenetic development.

The skeleton of the okapi is certainly primitive. It shows little of the elongation of the limbs, or of transverse growth of the skull and skeletal elements that appear variously in the more advanced Giraffidae.

Therefore, considering *Okapia* with regard to its major anatomical characters, without special emphasis on small, single features, we see that it is a very primitive giraffid, more primitive even than *Palaeotragus*, and that it forms quite a satisfactory structural ancestor for the Palaeotraginae. It has the diagnostic characters but not the specializations of the various palaeotragines.

With the foregoing considerations in mind, we may now attempt a classification of the Giraffidae.

GIRAFFIDAE

Large, ruminating artiodactyls, with heavy, rugose cheek teeth. The skull may or may not have horn cores, but if they are present they show a great variety of development. Bones of cranial roof pneumatic. Lateral metapodials and digits atrophied.

PALAEOTRAGINAE

Primitive, medium sized giraffids, having as a rule one pair of supraorbital, frontal horns cores. There may be a second pair of horn cores at the anterior extremities of the frontals. Horn cores in the form of simple tines, well developed in the males, feebly developed or absent in the females. Skull usually elongated, dolichocephalic.

Cheek teeth brachyodont, with moderately coarse sculpture on the enamel. Neck and limbs slightly elongated.

Genera: *Palaeotragus*   *Achtiaria* (synonymous with *Palaeotragus*)
   Giraffokeryx
   Okapia
   Samotherium   *Alcicephalus, Chersonotherium, Shanshitherium*
            (synonymous with *Samotherium*)

   Propalaeomeryx ⎫
   Progiraffa   ⎬ Of doutful status; placed here provisionally.

GIRAFFINAE

Large giraffids, with a moderately brachycephalic skull. Horns variously developed, being on the parietals and frontals, and in *Giraffa* a single median horn is also present, located on the nasals. Horn cores rounded or flattened on the ends and covered with hair. Skull roof with highly developed sinus cavities.

Cheek teeth brachyodont, with heavily rugose enamel. Limbs and neck greatly elongated.

Genera: *Giraffa*
   Orasius
   Honanotherium

SIVATHERIINAE

Gigantic giraffids, with large, heavy, brachycephalic skulls. Horns variously developed, being of frontal and parietal origin. Skull roof with large sinus cavities.

Cheek teeth moderately hypsodont, with heavily rugose enamel. Limbs not elongated but very heavy. Body heavy.

Genera: *Sivatherium* *Indratherium* (synonymous with *Sivatherium*)
   Bramatherium
   Hydaspitherium
   Helladotherium
   Vishnutherium
   Griquatherium
   Libytherium

PALAEOTRAGINAE

Giraffokeryx Pilgrim, 1910

Generic type, *Giraffokeryx punjabiensis* Pilgrim

Giraffokeryx punjabiensis Pilgrim

Giraffokeryx punjabiensis, Pilgrim, 1910, Rec. Geol. Surv. India, XL, p. 69.

Additional References.—

  Pilgrim, G. E., 1910B, pp. 189, 203; 1911, pp. 14–17, Pl. I, figs. 4, 5; Pl. II, figs. 1–16; 1913B, p. 286.

  Bohlin, B., 1927, pp. 41–42.

  Matthew, W. D., 1929, pp. 454, 456, 460, 535–537.

  Colbert, E. H., 1933E.

Type.—(Lectotype.)—G.S.I. No. B 502, a right M³.

Cotypes.—G.S.I. Nos. B 493–B 511 inclusive (with the exception of B 502, selected as the lectotype). Various upper and lower teeth.

Horizon.—Lower Siwaliks, Chinji zone. Also lower portion of the Middle Siwaliks.

Locality.—Chinji and vicinity, Salt Range, Punjab.

Specimens in the American Museum.—Amer. Mus. No. 19311. Miscellaneous teeth. From the Lower Siwaliks, 200 feet above the level of Chinji Rest House, one half mile north of Chinji Rest House.

19317. A right M$_3$. From the base of the Middle Siwaliks, one mile south of Nathot.

19320. Miscellaneous teeth. From the base of the Middle Siwaliks, one mile south of Nathot.

19323. Left and right mandibular rami. Lower Siwaliks, 1600 feet above the level of Chinji Rest House, two miles west of Chinji Rest House.

19324. Portions of mandibular rami. About level of Chinji Rest House, Lower Siwaliks, four miles northeast of Chinji Rest House.

19325. Miscellaneous teeth. Lower Siwaliks, 1600 feet above the level of Chinji Rest House, three miles west of Chinji Rest House.

19326. Fragment of palate. Lower Siwaliks, near Chinji Rest House.

19327. Miscellaneous teeth. Lower Siwaliks, 600 feet above the level of Chinji Rest House, one mile north of Chinji Rest House.

19328. Fragment of mandibular ramus. Lower Siwaliks, 1600 feet above the level of Chinji Rest House, one and one half miles north of that place.

19329. Left mandibular ramus. Lower Siwaliks, about level of Chinji Rest House, four miles west of Chinji Rest House.

19330. Miscellaneous teeth. Lower Siwaliks, about level of Chinji Rest House, four miles northeast of Chinji Rest House.

19331. Right P^4. Lower Siwaliks, about 400 feet above the level of Chinji Rest House, three miles west of Chinji Rest House.

19332. Right mandibular ramus. Lower Siwaliks, 600 feet above the level of Chinji Rest House, five miles west of Chinji Rest House.

19333. A molar tooth. Lower Siwaliks, near Chinji Rest House.

19334. Fragment of a palate. Lower Siwaliks, 600 feet above the level of Chinji Rest House, one mile north of Chinji Rest House.

19335. Miscellaneous teeth and jaw fragments. Lower Siwaliks, about level of Chinji Rest House, four miles west of Chinji Rest House.

19336. Mandibular fragments. Lower Siwaliks, 400 feet above the level of Chinji Rest House, one and one half miles west of Chinji Rest House.

19375. Miscellaneous teeth. Lower Siwaliks, 600 feet above the level of Chinji Rest House, two miles west of Chinji Rest House.

19419. Miscellaneous teeth. Lower Siwaliks, 1600 feet above the level of Chinji Rest House, twelve miles east of Chinji Rest House.

19430. A lower canine. Lower Siwaliks, 1600 feet above the level of Chinji Rest House, twelve miles east of Chinji Rest House.

19472. A right maxilla. Lower portion of the Middle Siwaliks, 1200 feet below the Bhandar bone bed and one mile south of Nathot.

19475. A skull. Lower portion of the middle Siwaliks, 1000 feet below the Bhandar bone bed. One mile south of Nathot.

19587. A right mandibular ramus. Lower Siwaliks, 200 feet above the level of Chinji Rest House, and one and one half miles north of that place.

19593. Miscellaneous teeth. Lower Siwaliks, 100 feet above the level of Chinji Rest House, two miles west of Chinji Rest House.

19596. Right mandibular ramus. Lower Siwaliks, 1600 feet above the level of Chinji Rest House, three miles west of Chinji Rest House.

19611. Unassociated teeth. Lower Siwaliks, 200 feet above the level of Chinji Rest House, four miles west of Chinji Rest House.

19618. Right mandibular ramus, with DM$_4$, M$_1$. Lower Siwaliks, 1600 feet above the level of Chinji Rest House, three miles northwest of Chinji Rest House.

19623. Miscellaneous teeth. Lower Siwaliks, 400 feet above the level of Chinji Rest House, three miles west of Chinji Rest House.

19632. Miscellaneous teeth. Lower Siwaliks, 600 feet above the level of Chinji Rest House, five miles east of Chinji Rest House.

19849. Left mandibular ramus. Lower Siwaliks, 300 feet above the level of Chinji Rest House, one and one half miles west of Chinji Rest House.

19930. Associated teeth. Lower Siwaliks, one half mile east of Rammagar.

Diagnosis.—A medium sized giraffid with four horns, two being at the anterior extremities of the frontals and two on the fronto-parietal region. Posterior horns overhanging the temporal fossae. Basicranium as in other Palaeotraginae. Teeth brachyodont, with rugose enamel as in the other Giraffidae. Limbs and feet presumably of medium length.

A detailed description of an almost complete skull of *Giraffokeryx* has been recently published as one of several papers preliminary to this present study. The reader is referred to this description (Colbert, E. H., 1933E) for the anatomical details of the skull and mandible of the genus under discussion.

A short synopsis of the above cited description is presented below.

Giraffokeryx is a medium sized member of the Giraffidae. It is at once distinguished from other genera of the Giraffidae in general and the Palaeotraginae in particular, by virtue of the fact that it has four horn cores, two immediately in front of the orbits and directed laterally, and two behind the orbits, above the temporal fossae, and directed laterally and posteriorly. The anterior horns are at the anterior ends of the frontal bones, while the posterior horns are either on the frontals, as in *Palaeotragus*, or possibly overspread the frontal-parietal suture as in the recent *Giraffa*. The question of the exact location of the posterior horns with relation to the frontal-parietal suture can not be decided on the basis of the existing material. These posterior horns overhang the temporal fossae.

There is a well developed preorbital vacuity in front of the large orbit. The maxilla is rather deep. This genus is characterized by the large size of its occipital condyles, and by the rather shallow pits in the supraoccipital region, this latter feature being characteristic of the *Palaeotraginae*. The basicranium is wide, as is the palate, and the cranial foramina are arranged much as in the okapi. The auditory bullae are of medium size.

The cheek teeth are brachyodont, with rugose enamel, and on the whole very much like the teeth of *Palaeotragus*.

It would seem logical to suppose that the premaxillaries of *Giraffokeryx* were long, and that there was a correlative elongation in the anterior region of the mandible, between the incisors and canine and the second premolar. Unfortunately the American Museum specimens are incomplete in these very regions, so that it is not possible to be certain about this question. Since *Palaeotragus* has long premaxillaries and a long canine-P$_2$ diastema, it is quite probable that *Giraffokeryx*, a form closely related to *Palaeotragus*, also showed elongation of these regions. The American Museum specimens, illustrated by the accompanying figures, have been thus restored, following the restorations by Bohlin of *Palaeotragus*.

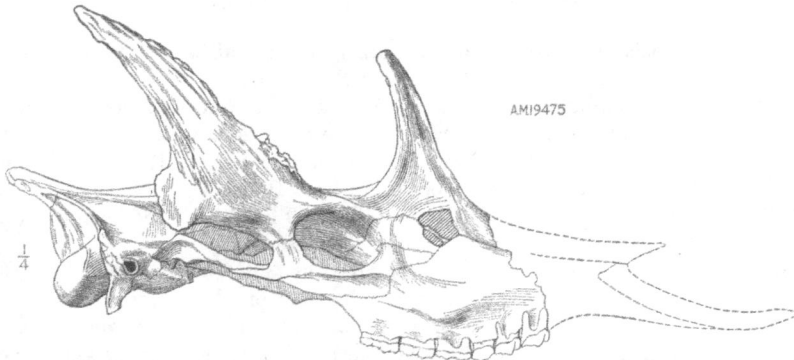

FIG. 153. *Giraffokeryx punjabiensis* Pilgrim. Amer. Mus. No. 19475, skull, right lateral view. Restored muzzle based on comparisons with *Palaeotragus* and *Okapia*. One fourth natural size. From Colbert, 1933.

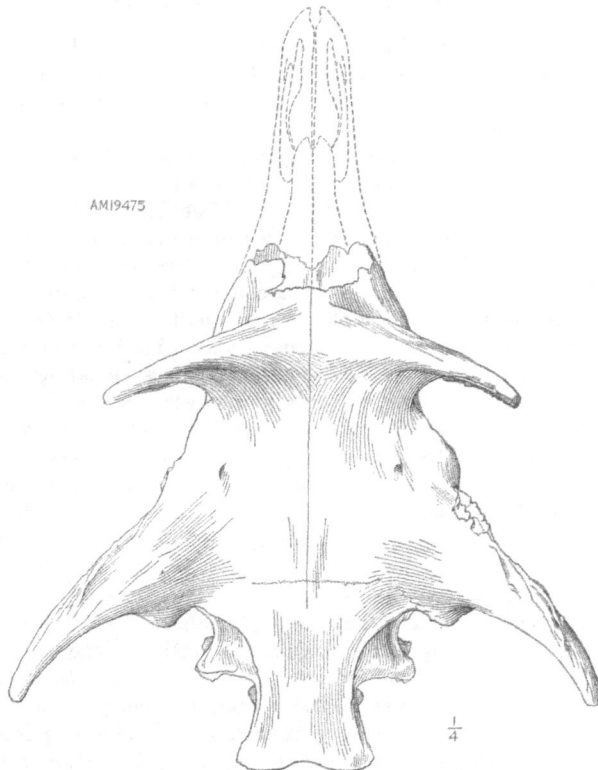

FIG. 154. *Giraffokeryx punjabiensis* Pilgrim. Amer. Mus. No. 19475, skull, dorsal view. Muzzle restored. One fourth natural size. From Colbert, 1933.

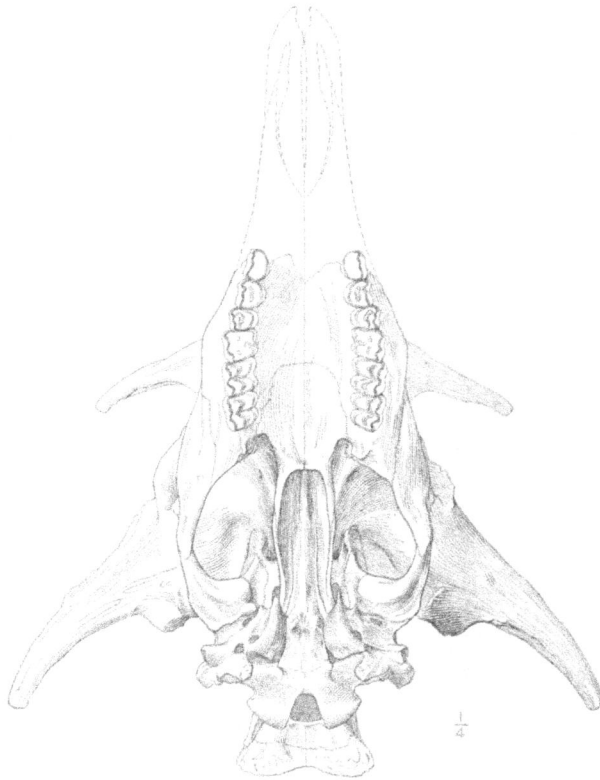

FIG. 155. *Giraffokeryx punjabiensis* Pilgrim. Amer. Mus. No. 19475, skull, palatal view. Muzzle restored. One fourth natural size. From Colbert, 1933.

A.M. 19475
A.M. 19587

FIG. 156. Restoration of the skull and mandible of *Giraffokeryx punjabiensis* Pilgrim, showing the probable appearance of the skull with the crushing removed. One fourth natural size. From Colbert, 1933.

The measurements and figures of certain American Museum specimens of *Giraffokeryx* accompany this description.

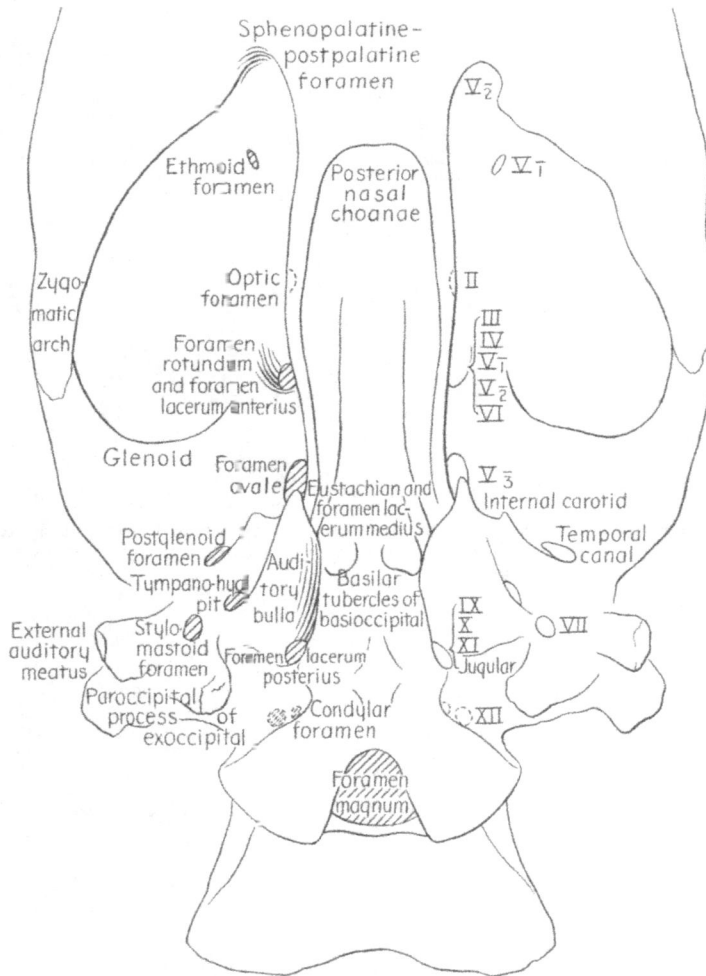

FIG. 157. Diagram of basicranium and adjacent regions of the skull of *Giraffokeryx punjabiensis* Pilgrim. Names of foramina shown on left side of diagram; position of nerves and blood vessels on right. From Colbert, 1933.

MEASUREMENTS

Skull.—Amer. Mus. No. 19475.

Length, P² to condyle	326 mm.
Estimated total length	500
Height above M³ (to superior border of orbit)	85
Restored height	110
Antero-posterior diameter of orbit	66
Postorbital length (front of orbit to condyle)	222
Estimated preorbital length	278
Width between tips of anterior horn cores	277
Width between tips of posterior horn cores	403
Width of confluent base of anterior horn cores	142
Distance between bases of posterior horn cores	123
Greatest width across squamosals	144
Width at narrowest portion of parietals	59
Width across zygomatic arches	179
Width of maxillae above P²	80
Width of palate at M³	66
Width of occipital condyles	76
Distance, anterior border of foramen magnum to border of posterior nasal choanae	150
Width between orbits	134
Width between tips of paroccipital processes	88

Mandible.—Amer. Mus. No. 19587.

Depth of ramus below third molar	43 mm.

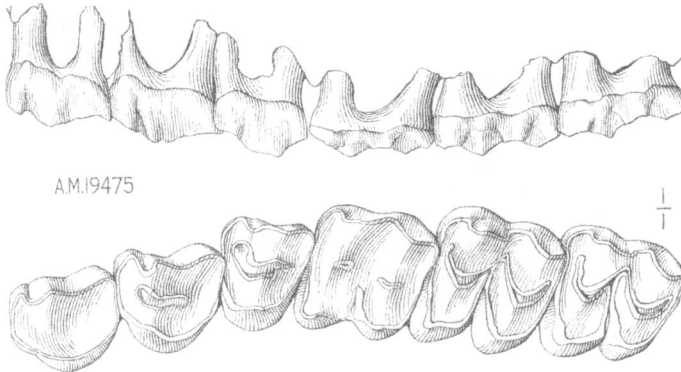

A.M.19475

FIG. 158. *Giraffokeryx punjabiensis* Pilgrim. Amer. Mus. No. 19475, left P²–M³. Lateral view above, crown view below. Natural size.

Upper dentition.—Amer. Mus. No. 19475.

	Antero-posterior	Transverse
P²	22.0 mm.	19.0 mm.
P³	20.5	20.0
P⁴	17.5	21.0
M¹	22.0	24.0
M²	25.0	27.0
M³	24.5	26.0

Lower dentition.—Amer. Mus. No. 19587.

P_2	18.0	9.0
P_3	20.5	12.0
P_4	24.0	15.0
M_1	24.0	16.0
M_2	25.0	17.0
M_3	37.0	17.0

Ratio, premolar to molar length.
Upper dentition.............................. 83
Lower dentition.............................. 71

Matthew in his discussion of Giraffokeryx (1929), indicates a large species, *G. punjabiensis*, and a smaller form which he designates as *Giraffokeryx* sp. indescr. On page 537 of his paper he gives measurements of several specimens in the Calcutta Museum, which he considers to belong to this new species.

As to the type of the smaller species, he says: "No. K 16/ 398 will be the type if the American Museum skull does *not* belong." Further on he states that "The difference from *G. punjabiensis* is about comparable to the difference between *Palaeotragus microdon* and *coelophrys* as figured by Bohlin."

It seems to be rather difficult to maintain two species of this genus in the American Museum material. There are specimens of the larger and the smaller forms, but there are specimens that represent all manner of gradations between the two.

Undoubtedly Dr. Matthew had not seen these various intermediate specimens, because the American Museum collection had not been properly sorted when he made his observations. Thus, in spite of his statement, it would seem best on the basis of the present material to regard the specimens in the American Museum as variants of one species. These variations run—seemingly without order—through the Chinji zone and into the Nagri zone. Large and small specimens occur throughout the vertical range of these beds.

Of course, if there were an association of skeletal material and teeth, differences of real specific value might appear in this series of *Giraffokeryx*. It will be remembered that in the discussion of *Hipparion*, on preceding pages of this work, it was shown that although there is an almost continual gradation in the teeth between the two species, *H. antelopinum* and *H. theobaldi*, the feet are quite distinct in the two forms. Perhaps some such condition may have existed in *Giraffokeryx*, but until more material is assembled it would seem best on the basis of the teeth alone, to consider the Siwalik material as indicative of one variable species.

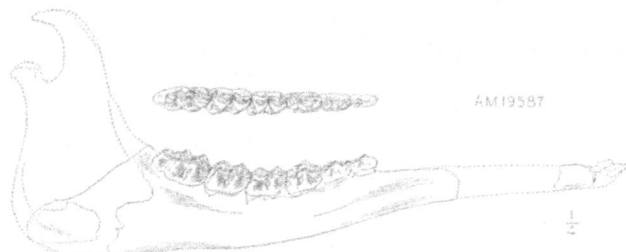

FIG. 159. *Giraffokeryx punjabiensis* Pilgrim. Amer. Mus. No. 19587, right ramus of mandible, lateral view. Crown view of grinding teeth, above. Restored portions indicated by dotted lines. The length of the restored premolar-canine diastema is based on comparisons with *Palaeotragus*. One fourth natural size. From Colbert, 1933.

Giraffokeryx has followed a line of development distinct from the other giraffids in the evolution of the horn cores. Superficially, this genus would appear to resemble *Syndyoceras* of the Miocene of North America, but an examination will soon show the absolute distinctness of these forms. In *Syndyoceras* the anterior horn cores are of premaxillary

A.M. 19472

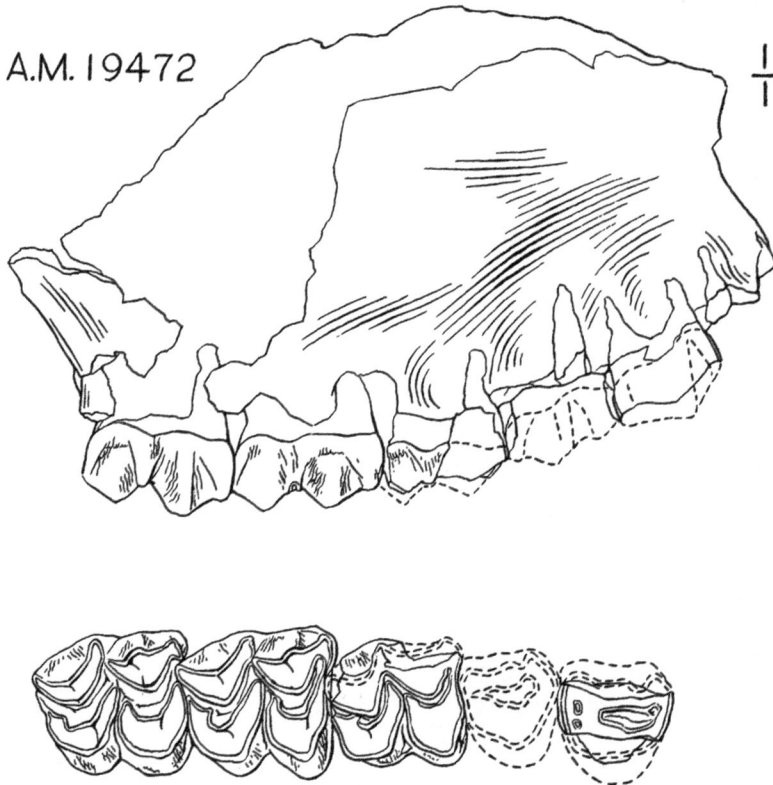

FIG. 160. *Giraffokeryx punjabiensis* Pilgrim. Amer. Mus. No. 19472, right maxilla with P²–M³. Lateral view above, crown view below. Natural size.

A.M. 19320

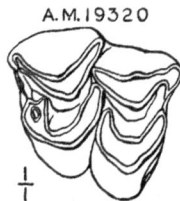

FIG. 161. *Giraffokeryx punjabiensis* Pilgrim. Amer. Mus. No. 19320, right M³. Crown view, natural size.

MEASUREMENTS IN MM. OF THE TEETH OF *Giraffokeryx punjabiensis*
(Anterior-posterior diameter given first in each case)

Specimen	Pm²	Pm³	Pm⁴	M¹	M²	M³	
19475	22.0 × 19.0	20.5 × 20.0	17.5 × 21.0	22.0 × 24.0	25.0 × 27.0	24.5 × 26.0	Lower Middle
19472	....	....	....	23.0 × 22.0	27.0 × 25.5	25.0 × 25.0	
19320	....	20.0 × 18.0	....	....	29.0 × 28.5	....	
W19469*	....	....	....	....	26.5 × 28.0	....	
W19610	....	....	17.0 × 23.0	....	....	....	U. Chinji 2000' A.B.
19325	....	....	18.0 × 24.0	....	29.5 × 27.0	27.5 × 28.0	L. Chinji 1000' A.B.
19632	....	....	....	....	28.0 × 24.0	....	
19334	....	....	....	25.5 × 25.0	27.0 × 27.0	....	
19327	....	....	....	....	24.0 × 26.0	23.0 × 23.5	
							L. Chinji
19623	....	....	....	....	27.0 × 29.0	....	800' A.B.
19311	....	19.0 × 17.5	15.0 × 18.0	23.0 × 22.0	....	....	
19611	....	....	16.5 × 21.0	....	27.0 × 26.0	....	600' A.B.
19593	....	....	....	24.0 × 24.0	....	....	500' A.B.
W19633	....	....	18.0 × 23.0	....	27.0 × —	....	400' A.B.
19330	....	21.5 × 17.5	17.0 × 23.0	....	....	....	
19930	....	22.0 × 20.0	19.5 × 23.5	26.5 × 28.0	....	....	

Specimen	Pm₂	Pm₃	Pm₄	M₁	M₂	M₃	
19317	....	....	....	....	....	37.0 × 18.0	Lower Middle
19320	....	....	....	27.0 × 16.0	27.0 × 15.0	....	
W19508	....	....	....	....	24.5 × 18.5	....	
No. num.	....	....	....	....	27.5 × 18.5	....	
19323	....	20.0 × 12.0	22.0 × 14.5	22.5 × 16.0	25.0 × 18.0	33.0 × 17.0	U. Chinji 2000' A.B.
19419	....	....	....	....	29.0 × 19.0	....	
19332	....	....	....	25.0 × 16.0	26.0 × 18.0	....	L. Chinji 1000' A.B.
							L. Chinji 800' A.B.
W19623	....	22.5 × 15.0	....	....	....	29.0 × 14.5	
19849	16.0 × 9.0	22.0 × 11.0	19.0 × 11.5	20.0 × 14.5	22.0 × 16.0	35.0 × 15.5	
19587	18.0 × 9.0	20.5 × 12.0	24.0 × 15.0	24.0 × 16.0	25.0 × 17.0	37.0 × 17.0	600' A.B.
19335	....	....	....	....	....	39.0 × 20.0	
19324	....	....	22.0 × 15.5	25.5 × 17.5	27.0 × 19.0	38.0 × 17.0	400' A.B.
19329	....	....	23.0 × 15.0	....	....	....	
W19557	....	21.0 × 12.5	....	26.0 × 18.0	....	....	
19593	....	....	....	24.0 × 16.0	?	....	
No. num.	....	....	24.0 × 16.0	....	....	....	

"A.B." means above base. "Base" in this case is a point 400 feet below the level of Chinji Rest House, near the base of the Chinji beds.

* A number preceded by W indicates an odd specimen, found with that number but not considered as of sufficient value to be catalogued.

origin, while in *Giraffokeryx* the anterior horns arise from the frontals. In both genera the posterior horn cores are similar as to their position and origin.

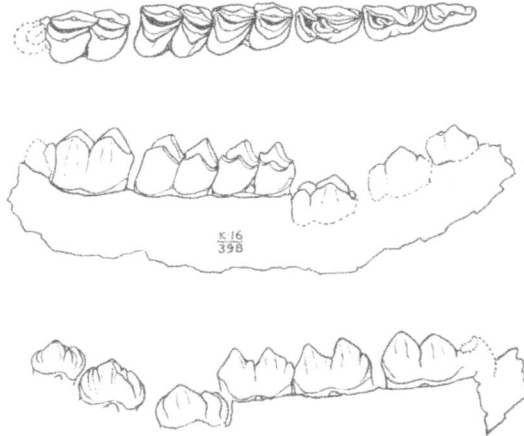

FIG. 162. *Giraffokeryx* sp. Geol. Surv. India No. K 16/398, lower jaw fragment with P₂–M₃. Premolars pre-formed but not wholly emerged; molars very little worn. Crown view above, external lateral view in middle, internal lateral view below. One half natural size. From Matthew, 1929.

Turning to *Bramatherium* and *Sivatherium*, four-horned giraffes from the Middle and Upper Siwaliks, it may be seen that in these genera the horns are frontal and parietal in origin, and they are placed and developed quite differently from the horn cores of *Giraffokeryx*.

When we compare *Giraffokeryx* with *Palaeotragus* and *Samotherium*, we find that the relationships are close among these forms, as was indicated by Bohlin when he placed them in one subfamily. The posterior horns of *Giraffokeryx* may be compared directly with the

FIG. 163.

FIG. 164.

FIG. 163. *Giraffokeryx punjabiensis* Pilgrim. Amer. Mus. No. 19930, lower incisor. Lingual surface, natural size.
FIG. 164. *Giraffokeryx punjabiensis* Pilgrim. Amer. Mus. No. 19430, left inferior canine. Lingual surface, natural size.

horns of *Palaeotragus*, *Samotherium* and *Okapia*. The greatest difference between the Siwalik form and the other genera mentioned is in the position of the posterior horn cores. In the three Palaeotragines listed above the horns are located above the orbit, the posterior edge of the horn core being only slightly back of the posterior edge of the postorbital bar, while in *Giraffokeryx* the horns are well behind the orbit, their anterior edges being above the postorbital process. It would seem that the typical Palaeotragine horn had pushed backward in *Giraffokeryx*, and in the process of this movement had developed a long base. Moreover the backward growth of the horn core in *Giraffokeryx* occasioned the roofing over of the temporal fossa, as mentioned above.

Bohlin has described a so-called four-horned Palaeotragine, namely *Palaeotragus quadricornis*. The anterior horns of this species are merely slight swellings adjacent to and in front of the large supraorbital horns. They might be regarded as incipient horns, rather than as horns of really anatomical significance. In other words, the anterior horns of *Palaeotragus quadricornis* are horns in the early stages of phylogenetic development. They are homologous with the anterior horns of *Giraffokeryx*.

In the modern giraffe there is a bony growth along the median line of the nasals, but it can hardly be compared to the separate anterior horns of *Giraffokeryx*.

There seems to have been some factor in the evolutionary development of the Giraffidae that gave impetus to the appearance of many-horned genera in the late Tertiary and the Pleistocene. In India there were *Giraffokeryx*, *Bramatherium* and *Sivatherium*, all four

FIG. 165. Restoration of the head of *Giraffokeryx punjabiensis*. One sixth natural size.

horned forms. The tendency to develop cranial outgrowths is strong in the giraffes, and has found expression in different ways through successive periods of geologic time.

Variable Characters in the Teeth of *Giraffokeryx*

The characters listed below are individual variations within this genus, and they probably occur within the one species, *G. punjabiensis*. They are found throughout the vertical range of *Giraffokeryx*, namely from the Chinji through the Nagri zone, and would not seem in any way to represent ascending mutations.

1. The enamel loop in the posterior crescent of the last upper premolar and the upper molars may be large or small. Its presence as a "lake" is a function of wear. In well worn teeth it forms a point in the postfossette, or it may disappear entirely.

2. The size of the anterior and posterior crescents as compared to each other may vary.

3. The cingulum on the anterior crescent may be large or small.

4. The internal border of the third lower premolar may be closed or open.

5. The width of the last lower premolar is variable.

6. The size of the median external column in the lower molars is variable, and it may even be absent. When present it is usually largest in the first molar.

7. The development of cingula from the styles of the upper molars is variable.

Foot Bones of *Giraffokeryx*

Certain isolated foot bones from the Lower Siwaliks and the lower portion of the Middle Siwaliks are provisionally referable to *Giraffokeryx*. They are very similar to the corresponding foot bones in *Okapia*.

Sivatheriinae

Sivatherium Falconer and Cautley, 1835

Generic type, *Sivatherium giganteum* Falconer and Cautley

Sivatherium giganteum Falconer and Cautley

Sivatherium giganteum, Falconer and Cautley, 1835, Jour. Asiatic Soc. Bengal, IV, No. 48, p. 706. (Nomen nudum.)

Sivatherium giganteum, Falconer and Cautley, 1836, Asiatic Researches, XIX, pp. 1–24, Pl. I.

Additional References.—

Falconer and Cautley, 1849, Pl. XCI, XCII.

Falconer, H., 1868A, p. 247.

Lydekker, R., 1880B, p. 32; 1882C, pp. 131–140; 1883C, p. 92; 1885B, pp. 21–24; 1885D, pp. 58–69.

Murie, J., 1871, p. 438.

Abel, O., 1904, pp. 629–650.

Pilgrim, G. E., 1910A, p. 69 (*Indratherium majori*); 1910B, p. 203; 1911, p. 22; 1913B, p. 324.

Bohlin, B., 1927, p. 141.

Matthew, W. D., 1929, p. 551.

For references to the female skull of Sivatherium (*Indratherium majori* Pilgrim) see: Falconer, H., 1868A.

Gaudry, A., 1862. (*Helladotherium.*)

Murie, J., 1871.

Rütimeyer, L., 1881. (*Helladotherium.*)

Lydekker, R., 1882C. (*Helladotherium.*)

Major, C. J. F., 1891. (*Hydaspitherium.*)

Schlosser, M., 1903. (*Hydaspitherium.*)

Bohlin, B., 1927.

Type.—Brit. Mus. No. 15283, a skull.

Paratypes.—Brit. Mus. Nos. 40667, a mandible; 39525, fragment of a posterior horn core. A skull in the University of Edinburgh.

Horizon.—Upper Siwaliks.

Locality.—Siwalik Hills.

Specimens in the American Museum.—Amer. Mus. No. 19774. A fragment of a horn core, provisionally referred to this species. From the Upper Siwaliks, six miles west of Kalka.

19797. Fragment of a mandibular ramus with right M_2. From the Upper Siwaliks, near Siswan.

19802. Fragment of a left mandibular ramus with M_{1-3}. From the Upper Siwaliks, near Siswan.

19828. Fragment of a right mandibular ramus with M_3. From the Upper Siwaliks, two and one half miles south of Chandigarh.

19883. A well preserved skull, lacking the nasals, premaxillae and the occipital portion. From the Upper Siwaliks, north of Siswan.

29805. Fragment of a right maxilla with M^{1-3}. From the Upper Siwaliks, six miles west of Kalka.

29835. Fragment of left ramus with M_3. Upper Siwaliks, six miles east of Kalka.

Diagnosis.—A gigantic Pleistocene giraffid, with four horns in the male, an anterior conical pair, arising from the frontals, and a posterior, palmate pair situated on the parietals. As in the other gigantic Siwalik giraffids there are deep pits in the temporal fossa for the temporal muscles, and on the supraoccipital for the neck muscles. The face is very short, the nasals being retracted and strongly curved. The teeth are large, with rugose enamel. Body and limbs heavy, limbs not elongated.

The skull of *Sivatherium* in the American Museum collection is strikingly like the type as regards its preservation. In both specimens the occipital portion is missing, but luckily an occiput collected by Colonel Colvin, now in the Royal Scottish Museum, supplies the needed knowledge of the back portion of the skull.

Little if anything in the way of new knowledge about this genus is supplied by the skull in the American Museum. The information to be obtained from the present specimen is to be considered as supplementary to that set forth by Falconer and Cautley, Lydekker, Murie and others, as the result of their several studies of the type specimens.

The American Museum specimen resembles the type. It is broad and heavy. There are two heavy, forward-projecting horn cores above the orbits. The orbits are relatively small, and do not protude. The nasals are short and arched, and there is a large swelling of the maxillary above the first molar. The infraorbital foramen is over the second premolar.

This skull is broken just behind the front horn cores and the broken edges show that the cranial roof and the horns were filled with large sinus spaces.

The dentition of the skull under discussion is well worn. All of the cheek teeth are heavy and robust, with strongly rugose enamel. In the molars the outer styles are prominent, and there are no cingula. The molars are quadrate, with their transverse axes at right angles to the longitudinal axis. The teeth are hypsodont.

FIG. 166. *Sivatherium giganteum* Falconer and Cautley. Amer. Mus. No. 19883, skull. Lateral view of left side. One third natural size.

Reasoning by analogy with the other ruminants, it is safe to say that there were no upper incisors or canines in *Sivatherium*.

A maxillary fragment, Amer. Mus. No. 29805, shows the character of the molars even better than do the teeth in the skull. Since the teeth in this maxilla are but slightly worn, the degree of their hypsodonty is clearly demonstrated.

Amer. Mus. Nos. 19828 and 29835 show the lower third molar to advantage. The former differs from the latter by reason of the longer posterior talonid, evidently a variable feature in *Sivatherium*.

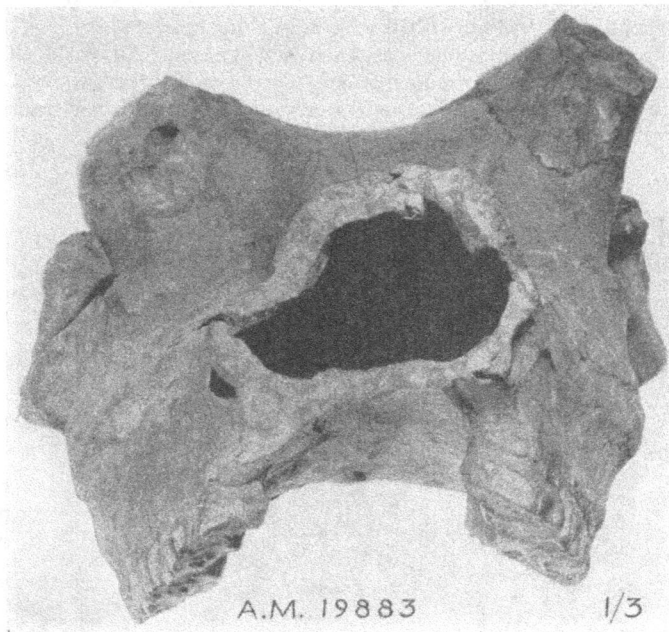

Fɪɢ. 167. *Sivatherium giganteum* Falconer and Cautley. Amer. Mus. No. 19883, skull. Anterior view.
One third natural size.

Tᴀʙʟᴇ

	19828	29835
Length of M_3	60.0 mm.	68.0 mm.
Length of talonid	16.0	23.0
Ratio length of talonid to length of tooth	26.6	34.0

The crown of a left lower incisor associated with 29835 would seem to be referable to this genus and species. It is twenty millimeters wide.

Mᴀʟᴇ ᴀɴᴅ Fᴇᴍᴀʟᴇ Sᴋᴜʟʟꜱ ᴏғ *Sivatherium*

The American Museum skull of *Sivatherium* is that of a male animal. It is quite comparable in size with the skull in the British Museum collection.

Bohlin and Matthew have shown that *Indratherium* is probably the hornless female of *Sivatherium*. Falconer had originally considered the female of *Sivatherium* to be hornless. Matthew likewise indicated the possibility of "*Helladotherium*" being the hornless female of *Hydaspitherium* or *Bramatherium*. If these interpretations of the skulls of the large Siwalik giraffids are correct, as it would seem they are, then these animals are like the primitive Palaeotraginae in the disparity between males and females in the possession of horn cores.

A Supposed Posterior Horn Core of *Sivatherium*

There is a specimen of curious shape and of rather perplexing affinities in the collection made by Mr. Brown in India. The specimen comes from the Upper Siwaliks, near Kalka, and is a fragment of a tremendous horn core or antler. By virtue of its size and shape it

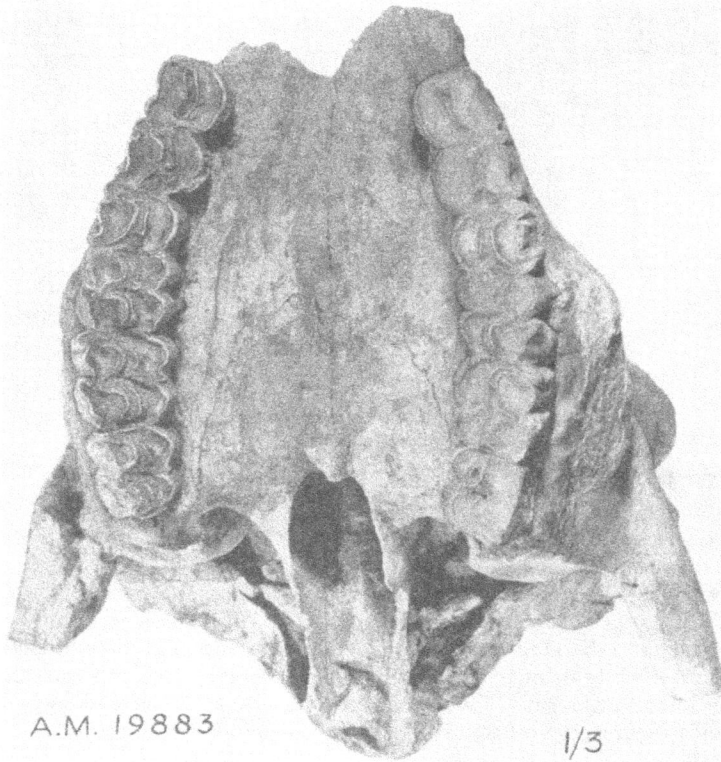

Fig. 168. *Sivatherium giganteum* Falconer and Cautley. Amer. Mus. No. 19883, skull. Palatal view.
One third natural size.

would seem to belong to *Sivatherium*, rather than to any other known fossil form. At the larger end, presumably the proximal one, this horn is somewhat elliptical in cross section. In passing from this to the opposite extremity the long axis of the elliptical cross section twists through an angle of about 120 degrees, so that the two ends of the fragment are directed differently. There are numerous deep nutrient canals or grooves along the surface of this horn core or antler, and these have followed the twisting so that they run in long open spirals. There are two large tuberosities on the sides of the antler or horn core, one at either end of the fragment. They are placed opposite to each other as regards the elliptical cross section of the horn, but due to the twisting of the horn these tuberosities are located topographically one above the other.

In 1904 Abel described a fragmentary horn core or antler from the Pleistocene beds at Adrionopel, and he attributed this specimen to *Sivatherium*. It resembles the fragment

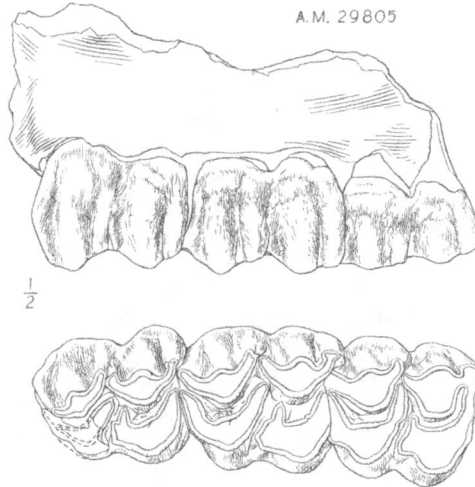

FIG. 169. *Sivatherium giganteum* Falconer and Cautley. Amer. Mus. No. 29805, right maxilla with M^{1-3}. Lateral view above, crown view below. One half natural size.

here being discussed in point of size, in its elliptical cross section and in the very rugose surface, traversed by large blood canals.

The specimen in the American Museum collection is plainly a horn core or an antler, as shown by the nutrient canals on its surface. It can hardly be a cervid antler because of its great size. Likewise, because of its size and shape it would hardly seem referable to any bovid. It does seem to be more easily identified as a giraffid horn core than the horn core of any other known artiodactyl. And because of its size, its shape, and its geologic occurrence, it would seem most probably referable to *Sivatherium*.

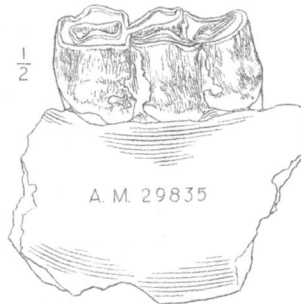

FIG. 170. *Sivatherium giganteum* Falconer and Cautley. Amer. Mus. No. 29835, right third inferior molar. Lateral view. One half natural size.

FIG. 171. Fragment of horn core attributed to *Sivatherium*. Amer. Mus. No. 19774. One third natural size.

MEASUREMENTS OF *Sivatherium giganteum*

A.M. No. 19883. Skull.

Distance P² to anterior border of orbit . . . . 257 mm.
Anterior-posterior diameter of orbit 85
Length P²–M³ 248
Width of maxilla above P² 171
Greatest width of maxilla (above M¹) 310
Width between orbits (anterior border) 283
Anterior-posterior diameter of horn core at base 135
Width of palate at P² 102
Width of palate at M³ 146
Height of skull above M³ 215*
Distance anterior border P²—anterior border nasal choana 215

A.M. No. 19774. *Sivatherium* (?). Fragment of horn core.

Diameter proximal end . . . . 137 mm.
Diameter distal end 107

* Approximate

MEASUREMENTS OF DENTITION OF *Sivatherium* (IN MM.)

(Antero-posterior diameter given first in each case)

A.M. No. 19883.

Right P² . . . . 39 × 44 mm.
P³ . . . . 39 × 47
P⁴ . . . . 38 × 49
M¹ . . . . 45 × 52
M² . . . . 56 × 56
M³ . . . . 53 × 50

A.M. No. 29805.

Right M¹ . . . . 50 × 50
M² . . . . 55 × 52
M³ . . . . 50 × 46

A.M. No. 19802.

Left M₂ . . . . 58 × 38
M₃ . . . . 67 × 33

A.M. No. 19797.

Right M₂ . . . . 54 × 39

A.M. No. 29835.

Left M₃ . . . . 68 × 33

A.M. No. 19828.

Right M₃ . . . . 57 × 29

FIG. 172. Restoration of the skull and mandible of *Sivatherium giganteum*. One sixth natural size.

Indratherium Pilgrim, 1910

Generic type, *Indratherium majori* Pilgrim

Indratherium majori Pilgrim

Indratherium majori, Pilgrim, 1910, Rec. Geol. Surv. India, XL, p. 69.
    Additional References.—
        Pilgrim, G. E., 1910B, p. 203; 1911, pp. 18–19; 1913B, p. 324.
        Bohlin, B., 1927, pp. 141–146.
        Matthew, W. D., 1929, p. 444, 551.
    *Type.—*Brit. Mus. No. 39523, a hornless skull.
    *Paratypes.—*None.
    *Horizon.—*Upper Siwaliks.
    *Locality.—*Siwalik Hills, Markanda Valley.

FIG. 173. Restoration of the head of *Sivatherium giganteum*. One sixth natural size.

Diagnosis.—"This new name is proposed for the reception of the Giraffoid hornless skull from the Markanda valley, now in the British Museum. Dr. Forsyth Major showed its distinctness from *Helladotherium*. It differs from *Hydaspitherium* by the greater proportionate breadth of the teeth, by the enormous size of the premolars, and by the presence of frontal horn swellings, which lead one to suppose that if it be a hornless female, then the male would have frontal horns instead of the parietal ones of *Hydaspitherium megacephalum*." [78]

This genus and species are probably synonymous with *Sivatherium giganteum*, the skull designated by Pilgrim as the type being a hornless female of the genus described by Falconer and Cautley. In fact, this skull was originally described as a female *Sivatherium* by Falconer and Cautley. Lydekker later referred this specimen to *Helladotherium* and to *Hydaspitherium*. Dr. Matthew (1929, p. 552) discusses the affinities of "*Indratherium*."

Bramatherium Falconer, 1845

Generic type, *Bramatherium perimense* Falconer

Bramatherium perimense Falconer

Bramatherium perimense, Falconer, 1845, Quar. Jour. Geol. Soc., London, I, p. 363.

Additional References.—

Bettington, A., 1845, p. 340.

Falconer, H., 1868A, p. 399.

Lydekker, R., 1876A, p. 42; 1880B, p. 32; 1882C, p. 130; 1883C, p. 89; 1885B, p. 24; 1885D, p. 69.

Pilgrim, G. E., 1910B, p. 203; 1911, p. 19; 1913B, p. 301.

Bohlin, B., 1927, p. 162.

Matthew, W. D., 1929, p. 550.

Type.—(Lectotype.)—Brit. Mus. No. 48933, a maxilla with left P^4–M^3.

Cotype.—Maxilla with left P^2–M^2, figured in Falconer's "Palaeontological Memoirs," I, Pl. XXXIII, figs. 1, 2.

Neotype.—Royal College of Surgeons. No. 1436, a skull. This specimen is hereby designated as a neotype because it shows the distinctive features of the genus and species, not shown by the type.

Horizon.—Middle Siwaliks of Perim Island, for type and neotype. Also Dhok Pathan zone of Middle Siwaliks, Punjab.

Locality.—Perim Island and the Punjab.

Specimen in the American Museum.—Amer. Mus. No. 19771, a partial skull from the Middle Siwaliks near Dhok Pathan, Punjab.

Diagnosis.—A gigantic Upper Tertiary giraffe having four horns, two of which grow up from the fronto-parietal region, and two of which extend laterally from the parietals. Face short, with nasals considerably retracted. A large groove occupies the parietal region, just below the horn core bases as an accomodation for the temporalis muscles. Deep pits are located in the supraoccipital for the heavy neck muscles. Teeth large and heavy, with rugose enamel. Limbs and body presumably heavy and massive.

[78] Pilgrim, G. E., 1910A, pp. 69–70.

There is no doubt as to the generic identity of the American Museum skull of *Bramatherium* with the skull figured by Bettington. Unfortunately the skull found by Mr. Brown had been subjected to considerable weathering so that the points of the horn cores and the face anterior to the orbit were lost, thus making a detailed specific identification by means of the teeth impossible. Since the cranium is, however, quite similar in both of the skulls, the American Museum specimen from the Salt Range is assigned to the same species as that from Perim Island.

The skull in the American Museum is approximately a fifth larger than the Perim Island skull, and since the former has large horn cores, is probably that of a mature adult (as contrasted with a young adult).

As in the other Siwalik giraffoids the "horns" form the most striking feature of this specimen. They are four in number, a large anterior pair and a much smaller posterior pair. The anterior horns are placed entirely on the frontals, and due to the posterior position of these horns the frontals have been carried back to cover the top of the skull. These large frontal horns are confluent at their bases, and this confluence of the horn cores extends well up towards their distal extremities, so that they are actually united for about half of their length. As seen from in front, the horns diverge upward and outward, and as seen from the side they are inclined somewhat posteriorly. Due to the fact that the tips have been weathered away in the American Museum specimen, the interior structure is plainly visible. As in the other large giraffes, the horns are hollow and divided into numerous chambers by bony partitions.

The posterior horns are placed laterally on the parietals, and they project out at about right angles to the median axis of the skull. They are much smaller than the other horns, and whereas the front horns are roughly rectangular in cross section, these parietal protuberances are round in cross section and conical in shape.

The frontals would seem to be very large, if the determination of the fronto-parietal suture *behind* the large horn cores is correct, and it would seem that they have extended back to accomodate the large horn cores. If such is the case we have here a growth in the skull roof similar to that characteristic of the Bovidae, in which the frontals often extend far back, carrying the horns to the "top" of the skull. Of course, there is the possibility that these large horn cores in *Bramatherium* rest on both frontals and parietals, as in the modern giraffe, in which case the frontals would be of less extent. It might be well to say, however, that the evidence for the position of the fronto-parietal suture behind the large horns, as shown by the skull in the American Museum collection, is fairly conclusive. The orbits project strongly, in contrast to *Sivatherium*.

As a corollary to the large frontals, the parietals are restricted antero-posteriorly, so that they form but a small portion of the top of the skull. There is a lateral extension of the parietal on either side of the skull as in the okapi.

A very distinctive and peculiar feature of the parietal is the large, deep groove which traverses it along the side of the skull, running from the lambdoidal crest, just beneath the lateral horn cores, in a forward and downward direction, terminating anteriorly on the edge of the alisphenoid just behind the optic foramen. It would seem probable that this temporal groove, as it may be called, is an adaptation for the accomodation of a large temporalis muscle. In a normal mammal the temporalis muscle runs well up over the top of the skull, often being attached to a strong sagittal crest. In *Bramatherium*, however, the growth of the transverse horn cores would seem to have limited the upward growth of this muscle, so that out of necessity it was forced to establish a strong surface of origin on the side of the cranium. Evidently the proper space for the origin of the temporalis was obtained by the development of a deep groove, thus affording a muscle attachment below the lateral horn and medially to the masseter muscle.

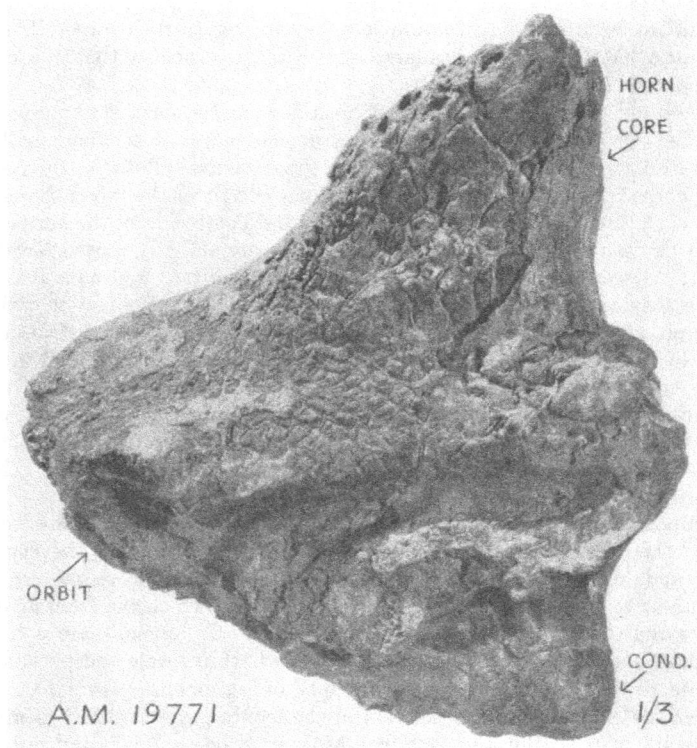

FIG. 174. *Bramatherium perimense* Falconer. Amer. Mus. No. 19771, posterior portion of cranium. Lateral view of left side. One third natural size.

The parietal-squamosal suture runs parallel to the temporal groove, and just below it. As seen from the side the squamosal is triangular in shape, with the long base of the triangle formed by the above mentioned suture. The attachment for the zygomatic process is small, showing that the arch was relatively weak, a common character among the ruminants. The arch was evidently short, for the distance between the squamosal attachment and the back of the orbit is quite short. In this respect *Bramatherium* resembled the modern *Giraffa*. There are two small indentations on the side of the squamosal, in front of the arch.

The junction of the parietal and the supraoccipital is marked by a heavy lambdoidal ridge. Just below this ridge are two large pits in the supraoccipital, which together with the lambdoidal crest afforded attachments for the ligamentum nuchae and for the semi-spinalis capitis muscles. The heavy character of the crest and the large size of the supra-orbital pits indicates the great strength of the muscles that supported the head.

The occipital condyles are large, as is the foramen magnum, a condition similar to that of *Sivatherium* and *Hydaspitherium*. As seen from behind, the paroccipital process is very wide, and combined with the lateral ridge of the squamosal forms a broad, transverse plate of bone. The same adaptation is found in *Hydaspitherium* and *Sivatherium*.

FIG. 175. *Bramatherium perimense* Falconer. Amer. Mus. No. 19771, posterior portion of cranium. Occipital view. One third natural size.

Bettington, in his original description of the *Bramatherium* skull, failed to present any details concerning the basicranium. Nor did Lydekker or Pilgrim, subsequently writing about Siwalik giraffes, see fit to discuss this region of the skull. Thus the following de-scription must be offered as the first account of the important basicranial features in a species that has been known for almost a century.

The basicranium will be considered under three heads, namely;
1. The cranial floor and its foramina.
2. The auditory apparatus.
3. The attachments for the lower jaw.

1. *The Cranial Floor and its Foramina.*

The basicranium of *Bramatherium* is wide, as might be expected. The basioccipital is broad with large tuberosities for the attachment of the rectus capitis ventralis major muscles.

In the distribution of the basicranial foramina, the genus under consideration resembles *Giraffa*. The optic foramen is opposite the posterior border of the orbit. Behind it is the large foramen formed by the confluence of the foramen lacerum anterius and the foramen rotundum. The combined posterior exit of the alisphenoid canal and the foramen ovale form a round opening placed at some distance in front of the auditory bulla, a resemblance to the giraffe. This last feature offers a nice contrast to *Giraffokeryx* and *Okapia*, in which genera the posterior exit of the alisphenoid canal and the foramen ovale form a long opening immediately in front of and partly above the bulla. Indeed the disposition of these foramina indicates the close relationship between *Bramatherium* and *Giraffa*, and their distinctness from the Palaeotraginae. *Okapia* and *Giraffokeryx* represent the more primitive stage in giraffoid evolution, in which the foramina have not yet become as completely joined as in the more advanced forms. In fact, in an *Okapia* skull at hand, there is a small bridge of bone marking the boundary between the alisphenoid canal and the foramen ovale, a remnant of an earlier condition in which these two foramina were entirely separated from each other.

As to the foramen lacerum medius, *Bramatherium* may again be compared to the giraffe. In both, the opening is just anterior to the bulla, and the alisphenoid is grooved above, as a course for the branch of the carotid artery. The foramen lacerum medius is confluent with the bony opening of the eustachian tube, as in the other giraffids.

The stylomastoid foramen is seemingly large as an accomodation for a heavy facialis nerve. Likewise the postglenoid foramen is large. The foramen lacerum posterius is similar to the corresponding foramen in *Giraffa*.

A heavy covering of matrix conceals the condylar foramen, but it is probably single as in the giraffe, rather than double as in *Okapia*.

2. *The Auditory Apparatus.*

When viewed as to its relative size, the auditory bulla of *Bramatherium* is intermediate between that of *Giraffa* and that of *Okapia*. That is, it is inflated but not to the degree characteristic of the okapi. In the giraffe the bulla is much less inflated, comparable in this respect with the more primitive forms of giraffids. Structurally, however, the bullae are similar in *Bramatherium* and *Giraffa*. Since the walls of the bulla are thick in *Bramatherium*, no attempt has been made to open the bulla for an examination of its interior structure.

In both *Bramatherium* and *Giraffa* the external auditory meatus is relatively short and directed somewhat posteriorly.

3. The Attachments for the Lower Jaw.

Only the inner portions of the glenoids are preserved in *Bramatherium*. There is a deep pit occupying the space between the postglenoid process and the foramen ovale, and from the size of this pit it may be presumed that the articular meniscus was quite large. A

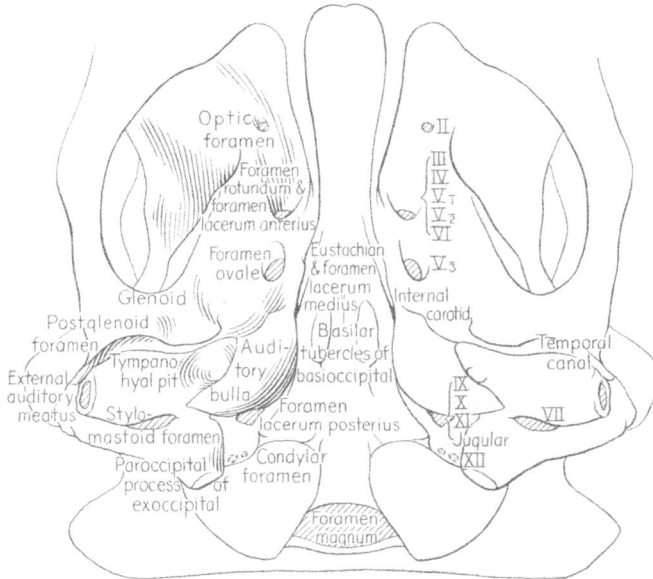

Fig. 176. *Bramatherium perimense* Falconer. Diagram of basicranium. Names of foramina on left, positions of nerves and blood vessels on right.

large pit for the meniscus is characteristic of the Giraffidae. The pterygoids are broken away, but they seem to have been rather heavy. As remarked above, the zygomatic arches were short and relatively weak.

The mandible in *Bramatherium* must have been fairly slender, as in *Giraffa*, and was attached by strong temporalis and pterygoid muscles. The masseter was probably relatively weak.

REMARKS CONCERNING THE HORN CORES OF *Bramatherium*

The question of the homologies of the horns in *Bramatherium* is indeed perplexing, as it is in the various other Siwalik giraffes. Lydekker considered the large conjoined anterior horns of *Bramatherium* to be homologous with the common horn base of *Hydaspitherium*, and with the posterior horns of *Sivatherium* and *Giraffa*. If these homologies are valid, the horns of the various genera mentioned would needs be fronto-parietal in origin, that is, they would straddle the fronto-parietal suture as in the modern giraffe. It is very probable, as pointed out above, that the large horns of *Bramatherium* are entirely frontal but this point can not be absolutely settled on the basis of present evidence.

Of course, in these fossil giraffes, as in the modern forms, the horn cores are probably secondary fusions of dermal structures with the roof bones of the skull, rather than primary outgrowths of the cranial bones. Therefore they might vary somewhat in position and still be homologous in origin.

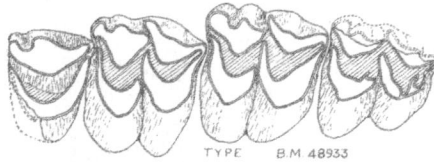

FIG. 177. *Bramatherium perimense* Falconer. Brit. Mus. No. 48933, type, left P⁴–M³. Crown view, one half natural size. From Matthew, 1929.

The small posterior-lateral horns of *Bramatherium* are certainly of parietal origin. Both Lydekker and Pilgrim speak of them as being occipital, but on this point these two authorities are seemingly mistaken.

MEASUREMENTS

Bramatherium. Amer. Mus. No. 19771

Antero-posterior diameter, base of horn cores (approximate)...... 180 mm.
Transverse diameter, base of horn cores...................... 185
Width between bases of posterior horns...................... 180
Greatest width of skull at squamosals........................ 272
Width of occipital condyles................................. 142
Distance, anterior border of orbit to condyle.................. 310
Diameter of posterior horn................................. 90

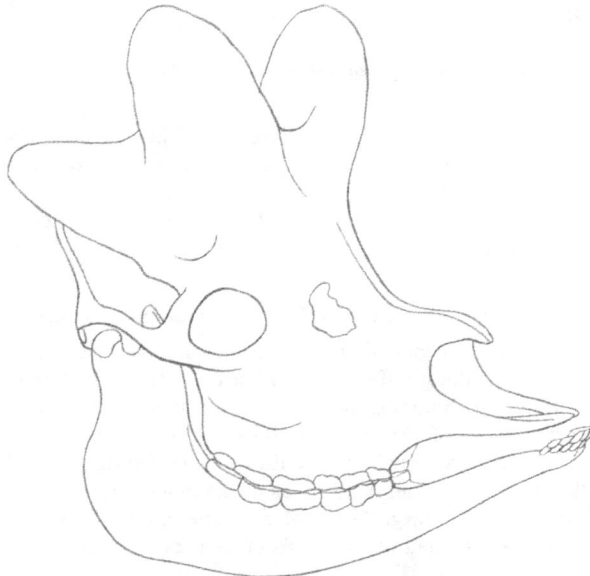

FIG. 178. Restoration of the skull of *Bramatherium perimense*. One sixth natural size.

FURTHER NOTES ON THE SKULL OF *Bramatherium*, AS DERIVED FROM A STUDY OF THE
COMPLETE SKULL IN THE ROYAL COLLEGE OF SURGEONS

The original description of *Bramatherium* was based on maxillary fragments, containing
various cheek teeth. The finest specimen of the genus and species, the skull figured and
described by Bettington (but not named by that author), and now housed in the Royal

FIG. 179. Restoration of the head of *Bramatherium perimense*. One sixth natural size.

College of Surgeons in London, was not described by Falconer nor by subsequent authors. A new cast of this specimen, recently acquired by the American Museum, gives certain anatomical characters not included in the description of the specimen in the Siwalik Collection made by Mr. Brown. Additional information given by the London skull will be presented below.

This skull shows the close relationship existing between the genus under consideration and *Sivatherium*. In both genera the preorbital portion of the skull is short, while the postorbital region is relatively long. In both *Bramatherium* and *Sivatherium* there are four large horn cores on the frontals and parietals, structures that add materially to the size of the cranium. In both the orbit is relatively small and round and the palate is wide. Both are characterized by the excavations in the occiput for the attachments of neck muscles.

Although the premaxillaries are missing in the skull of *Bramatherium*, it would seem likely that they were comparatively short, as in *Sivatherium*. Accompanying these short premaxillaries the nasals also must have terminated at no great distance in front of the orbit.

The front of the orbit is located above the anterior edge of the second molar. There is a distinct ridge running from the anterior surface of the lateral horn to the back of the orbit. This ridge defines the upper boundary of the temporal groove, and in life it must have served as a demarcation for the upper limits of the temporalis muscles.

There is a rather deep depression in the parietal beneath the anterior horn core and behind and above the orbit. This depression was probably developed negatively, in contrast to the strong expansion of the horns, and it is doubtful as to whether it served in any functional manner.

The zygomatic arch is quite short.

Other detailed characters, especially those of the basicranium, are treated in the preceding section having to do with the skull in the American Museum collection. Descriptions of the teeth are to be found in the publications of Falconer, Lydekker and Pilgrim.

Measurements

Bramatherium, cast of skull. Amer. Mus. No. 27016

Estimated length of skull, condyles to tips of premaxillaries	525 mm.
Preorbital length	210
Postorbital length	315
Diameter of orbit	68
Antero-posterior diameter of anterior horn cores, at base	175
Transverse diameter of anterior horn cores at tips	340
Transverse diameter of anterior horn cores at base	150
Width between bases of posterior horn cores	185
Distance between anterior border of orbit and condyles	315
Diameter of posterior horn cores	90
Transverse diameter of condyles	133
Height of maxilla above first molar	270
Width across lambdoidal crest	260
Height of occiput	158
Width of palate at first molar	103
Length of molar series	129

Hydaspidotherium Lydekker, 1876

Hydaspitherium Lydekker, emend., 1878

Generic type, *Hydaspitherium megacephalum* Lydekker

Hydaspitherium megacephalum Lydekker

Hydaspidotherium megacephalum, Lydekker, 1876, Rec. Geol. Surv. India, IX, p. 154.
Hydaspitherium megacephalum, Lydekker, 1878, Pal. Ind. (X), I, p. 159.

Additional References.—

Lydekker, R., 1880B, p. 32; 1882C, pp. 118–126, Pl. XVII, figs. 7, 10, 11, Pl. XVIII, fig. 3, Pl. XIX; 1883C, p. 91; 1885B, p. 25.

Pilgrim, G. E., 1910B, p. 203; 1911, p. 20; 1913B, p. 304.

Bohlin, B., 1927, pp. 141, 145, 155, 162.

Matthew, W. D., 1929, pp. 541, 551.

Type.—G.S.I. No. D 150, a skull.

Paratypes.—None.

Horizon.—Middle Siwaliks, Dhok Pathan zone.

Locality.—Salt Range, Punjab, from near Hasnot.

Specimens in the American Museum.—Amer. Mus. No. 19488. A fragmentary palate with right P^4–M^3. From the Middle Siwaliks, four miles east of Dhok Pathan.

19669. Fragment of a ramus with right P_2–M_3. Middle Siwaliks, one half mile south of Dhok Pathan.

19319. A canine tooth doubtfully referred to this genus. From the Middle Siwaliks, one mile north of Hasnot.

19297. Upper premolar. Middle Siwaliks, four miles east of Dhok Pathan.

19298. Right mandibular ramus. Middle Siwaliks, four miles east of Dhok Pathan.

19301. Fragment of palate with left DM^{2-4}. Middle Siwaliks, one mile northeast of Hasnot.

19315. Upper premolar. Middle Siwaliks, one half mile northeast of Hasnot.

19322. Upper molar. Middle Siwaliks, one and one half miles northeast of Hasnot.

19691. Foot bones. Middle Siwaliks, one half mile southwest of Dhok Pathan.

19731. Lower milk teeth. Middle Siwaliks, one mile northeast of Hasnot.

19938. Palate with right M^{1-3}. Middle Siwaliks, one mile north of Hasnot.

Diagnosis.—A gigantic giraffid with two horn cores, fused at their bases into one solid mass, on the fronto-parietal region. The face is short, the nasals retracted. There is a large parietal or temporal groove below the horn core, for the accomodation of the temporal muscles. Teeth large, quadrate, with rugose enamel. Limbs massive and not extraordinarily elongated.

A great deal of confusion exists concerning the various genera and species of Middle Siwalik giraffes. This is especially so in the case of the genera *Hydaspitherium* and *Vishnutherium*, the latter being founded on somewhat tenuous evidence.

The specimens in the collection made by Mr. Brown are somewhat smaller than the typical *Hydaspitherium megacephalum*, and are about equal to *Vishnutherium iravadicum* in size. The last two upper molars of the maxillary fragment (Amer. Mus. No. 19488) are almost identical to the corresponding teeth in *Vishnutherium*, and if this were a well established genus, the American Museum material would be classified with it. In view, however, of the possibility of the *Vishnutherium* teeth representing a small species of *Hydaspitherium*, or even a small variety of *H. megacephalum* (as Matthew has suggested), the specimens in the American Museum collection are classified as the latter species. It seems advisable to identify them in this way, because of their fragmentary nature.

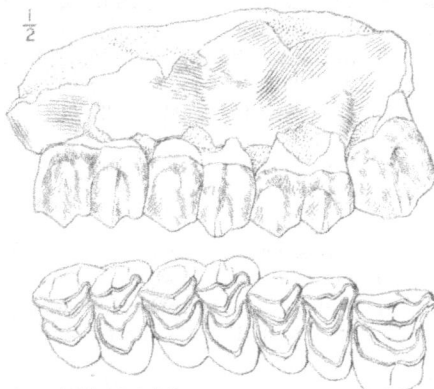

FIG. 180. *Hydaspitherium megacephalum* Lydekker. Amer. Mus. No. 19488, right maxilla with P⁴–M³. Lateral view above, crown view below. One half natural size.

Of course, these teeth might belong to *Bramatherium*.

There is little need for discussion of the specimens in the American Museum identified as *Hydaspitherium megacephalum*. The teeth are similar to the teeth of *Hydaspitherium* and *Vishnutherium* figured by Lydekker and by Pilgrim. Suffice it to say that the teeth are robust, with heavily rugose enamel and with strong outer styles in the upper molars.

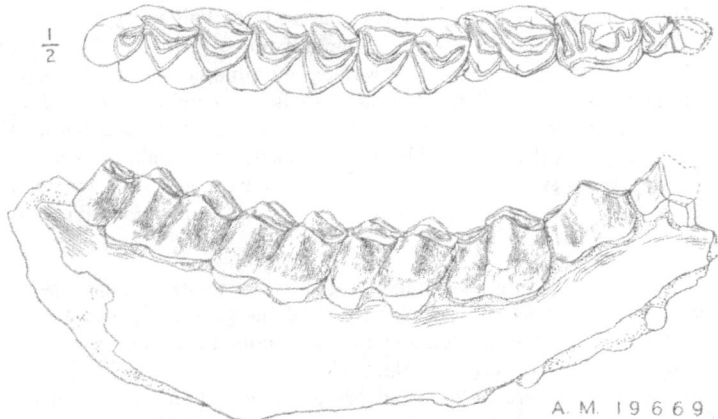

FIG. 181. *Hydaspitherium megacephalum* Lydekker. Amer. Mus. No. 19669, right mandibular ramus with P₂–M₃. Crown view above, lateral view below. One half natural size.

The one canine is of interest. It is quite large, of the proper size to belong to this genus, and on its unworn edges are numerous tubercles.

Figures and measurements of the specimens accompany these remarks.

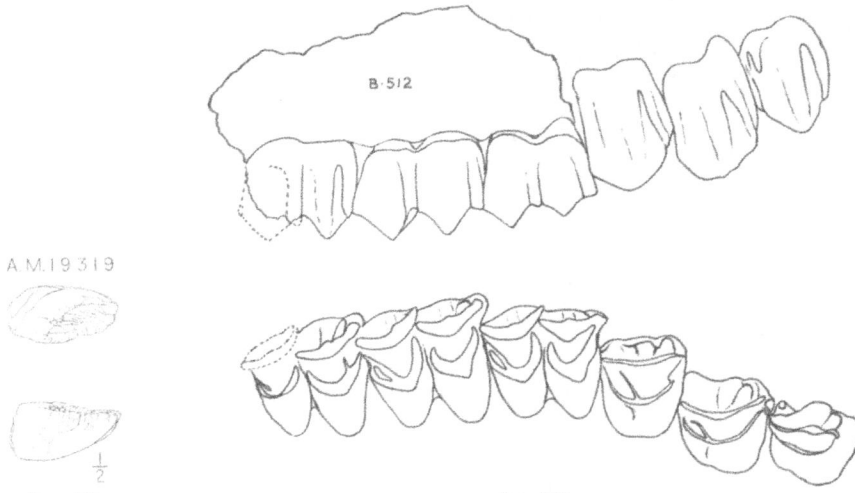

FIG. 182.   FIG. 183.

FIG. 182. Right inferior canine referred to *Hydaspitherium megacephalum* Lydekker. Amer. Mus. No. 19319. Dorsal view above, external lateral view below. One half natural size.

FIG. 183. *Hydaspitherium megacephalum* Lydekker. G.S.I. No. B 512, right maxilla with P²-M³. Lateral view above, crown view below. One half natural size. From Matthew, 1929.

"This is the immature (and incomplete) skull figured by Pilgrim as *Helladotherium grande* (Pilgrim, 1911, Pal. Ind. IV, Pl. III) except for the p² which is omitted from his figure." (Matthew, W. D., 1929, p. 541, fig. 45.)

MEASUREMENTS

Amer. Mus. No. 19488, palate of *Hydaspitherium megacephalum*

	Length	Width
Right P⁴	29.0 mm.	35.0 mm.
M¹	36.0	35.0
M²	37.5	40.0
M³	38.5	39.0

Amer. Mus. No. 19669, mandible of *Hydaspitherium megacephalum*

	Length	Width
Right P₂	21.0 mm.	13.0 mm.
P₃	30.0	21.0
P₄	31.0	24.0
M₁	38.0	27.0
M₂	38.0	28.0
M₃	50.0	28.5

Hydaspitherium grande Lydekker

Hydaspitherium grande, Lydekker, 1878, Rec. Geol. Surv. India, XI, p. 93.
Helladotherium grande, Pilgrim, 1910, Rec. Geol. Surv. India, XL, p. 69.

Additional References.—

Lydekker, R., 1880B, p. 32; 1882C, pp. 126–129, Pl. XX, XXI, fig. 2; 1883C, pp. 84, 91; 1885B, pp. 28–29.

Pilgrim, G. E., 1910B, p. 203; 1911, pp. 11–13, Pl. III, figs. 1–4.

Matthew, W. D., 1929, pp. 449, 542–543, fig. 46.

Type.—G.S.I. No. B 155 a left upper molar.

Paratypes.—None.

Horizon.—Middle Siwaliks.

Locality.—From the Punjab.

Diagnosis.—A large species, considerably larger than *H. megacephalum*, but otherwise similar to it.

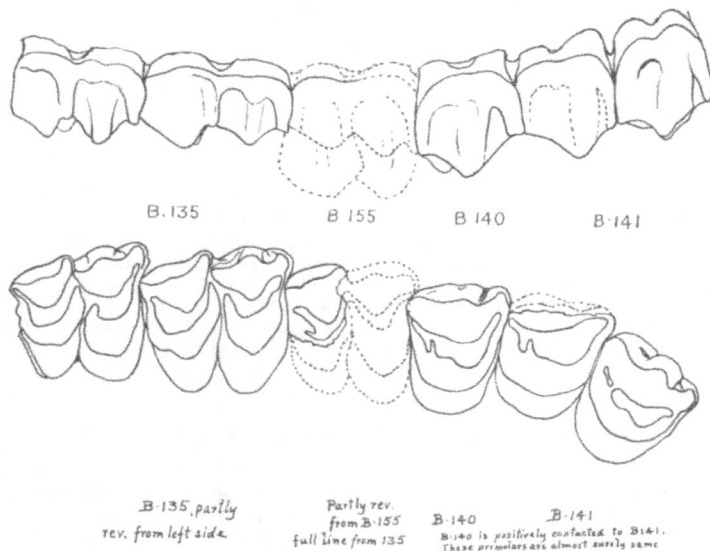

Fig. 184. *Hydaspitherium grande* Lydekker. G.S.I. Nos. B 135, B 155, B 140, B 141, right upper cheek teeth. Lateral view above, crown view below. One half natural size. From Matthew, 1929.

"The dotted outlines are those of the little-worn type molar, Ind. Mus. B 155, modified to show the change in form resulting from wear to a corresponding degree with the remainder of the series. The molars No. B 135 were figured by Lydekker as *H. megacephalum;* the premolars Nos. B 140, B 141, although separately catalogued, fit together by unmistakable contact as parts of the same individual, and are almost certainly the same individual as the molars, though no contacts can be made." (Matthew, W. D., 1929, p. 542, fig. 46.)

The generic name *Helladotherium* was applied by Pilgrim in 1910 and again in 1911 to *Hydaspitherium grande* Lydekker. It certainly was Pilgrim's intention to transfer the species named to a genus different from that used by the original author, and not to create a new species of the genus *Helladotherium*.

Dr. Matthew (1929) has shown that the genus *Helladotherium* never reached the Siwalik region. Therefore the original designation as given by Lydekker is applied to the species under discussion.

Hydaspitherium magnum Pilgrim

Hydaspitherium magnum, Pilgrim, 1910, Rec. Geol. Surv. India, XL, p. 70.

Additional References.—

Pilgrim, G. E., 1910B, p. 203; 1911, pp. 20–21, Pl. IV, fig. 1, Pl. V.

Matthew, W. D., 1929, pp. 449, 544, fig. 47.

Type.—(Lectotype.)—G.S.I. No. B 514, maxilla with left P^2–M^3; G.S.I. No. B 515, mandibular ramus with left M_{1-3}; G.S.I. No. B 516, left P_3 from mandible. (All specimens represent one individual.)

Cotypes.—None.

Horizon.—Middle Siwaliks, Dhok Pathan zone.

Locality.—Hasnot, Punjab.

Diagnosis.—(After Matthew, 1929, p. 544.) "Molars larger and more robust than *H. grande*, and p^4 more trihedral in form. Lower teeth broader and somewhat longer; the difference in depth of mandible I do not take seriously."

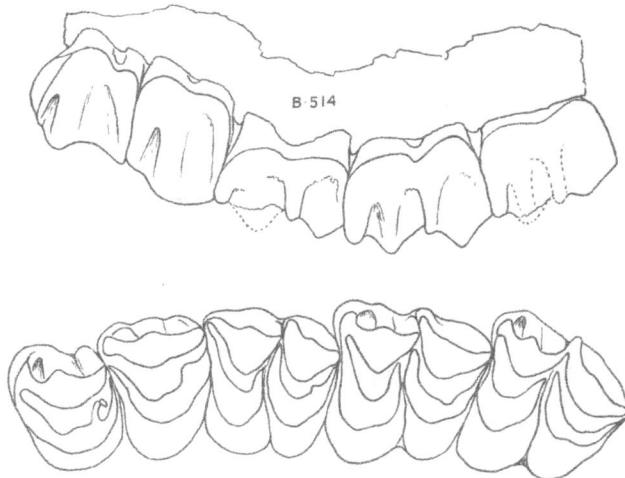

FIG. 185. *Hydaspitherium magnum* Pilgrim. G.S.I. No. B 514, type, left maxilla with P^3–M^3. Lateral view above, crown view below. One half natural size. From Matthew, 1929.

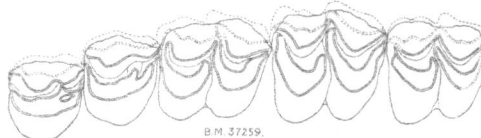

FIG. 186. *?Hydaspitherium* sp. Brit. Mus. No. 37259, left P^3–M^3. Crown view, one half natural size. From Matthew, 1929.

Hydaspitherium birmanicum Pilgrim

Hydaspitherium birmanicum, Pilgrim, 1910, Rec. Geol. Surv. India, XL, p. 70.
  Additional References.—
    Pilgrim, G. E., 1910B, p. 203; 1911, pp. 21–22, Pl. IV, fig. 2.
  *Type.—*G.S.I. No. B 517, a right upper molar.
  *Paratypes.—*None.
  *Horizon.—*Irrawaddy Series, equivalent to the Middle Siwaliks.
  *Locality.—*Near Singa, Upper Burma.

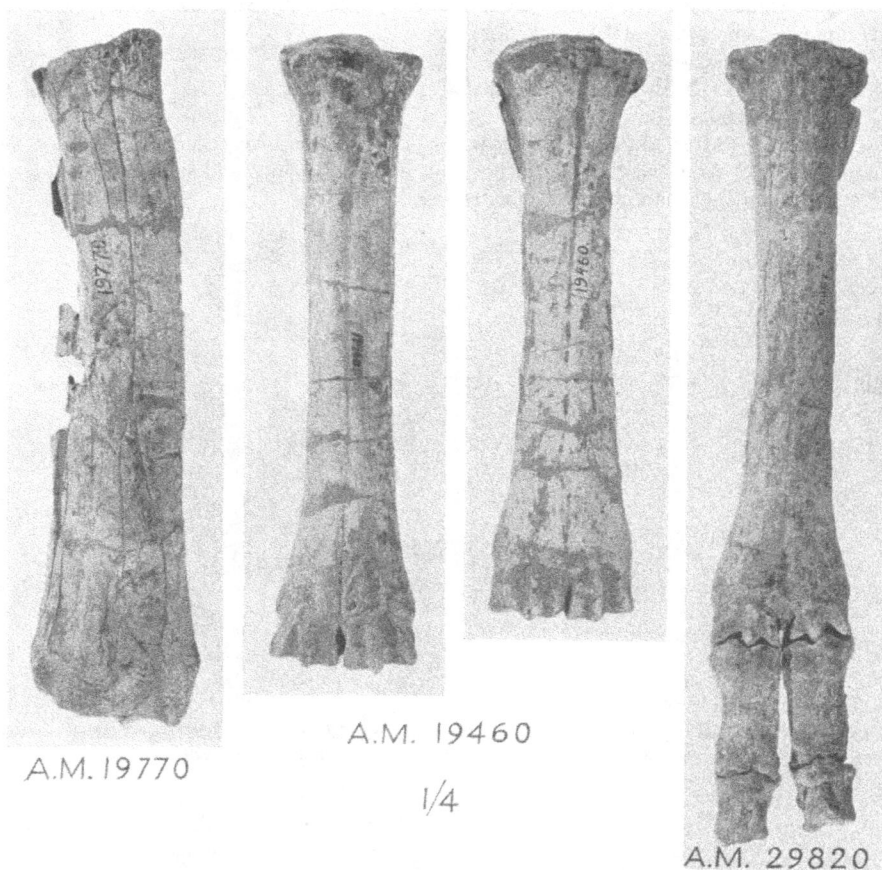

FIG. 187. *Hydaspitherium* or *Bramatherium.* Amer. Mus. Nos. 19770, 19460, 29820, elements from the fore limb. Left to right; right ulna-radius, left metacarpals, right metacarpals with proximal and median phalanges. Anterior views. One fourth natural size.

Diagnosis.—"A single upper molar, found near Singu in Upper Burma, differs little from *H. megacephalum* except by its much smaller size. The front horn of the second crescent is not so long and the enamel folds penetrate the crown less deeply." [79]

This species, founded on a single fragmentary tooth, cannot be considered as of much value.

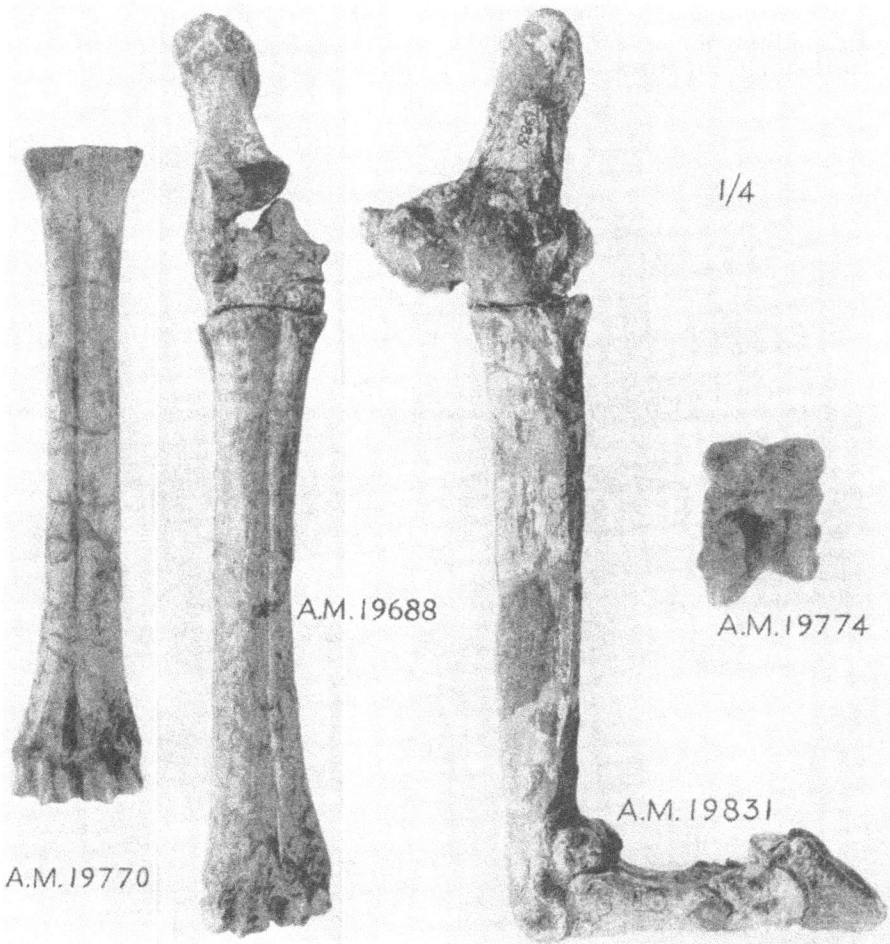

FIG. 188. *Hydaspitherium* or *Bramatherium*. Amer. Mus. Nos. 19770, 19688, 19831, elements from the hind limb. Left to right; right metatarsals, right tarsus and metatarsals, left pes (lateral view).
Sivatherium. Amer. Mus. No. 19774, astragalus. Anterior view. All figures one fourth natural size.

[79] Pilgrim, G. E., 1910A, p. 70.

Vishnutherium Lydekker, 1876

Generic type, *Vishnutherium iravaticum* Lydekker

Vishnutherium iravaticum, Lydekker, 1876, Rec. Geol. Surv. India, IX, p. 173.
   Additional References.—
      Lydekker, R., 1880B, p. 32; 1882C, pp. 112–116, Pl. XVI, fig. 7; 1883C, p. 93;
         1885B, p. 29.
      Pilgrim, G. E., 1910B, p. 203; 1911, pp. 17–18.
      Matthew, W. D., 1929, pp. 544–545.
   Type.—G.S.I. No. B 168, a left mandibular ramus.

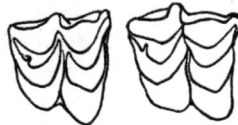

FIG. 189.   *?Vishnutherium* sp.   G.S.I. No. K 16/483, upper molars and lower premolars.   Crown views, one half natural size.   From Matthew, 1929.

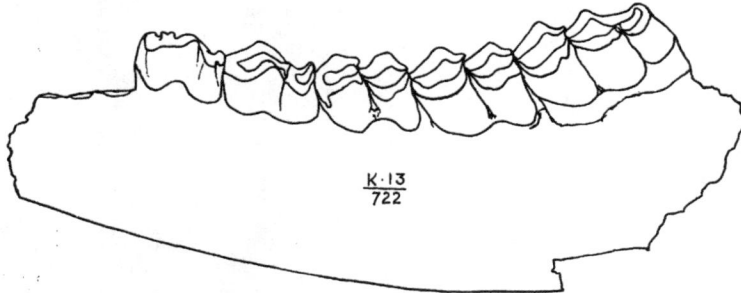

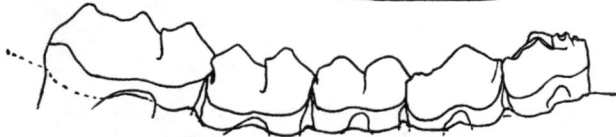

FIG. 190.   *?Vishnutherium* sp.   G.S.I. No. K 13/722, left mandibular ramus.   Crown view above, external lateral view in middle, internal lateral view below.   One half natural size.   From Matthew, 1929.

Paratypes.—None.

Horizon.—Irrawaddy beds, equivalent to the Middle Siwaliks.

Locality.—Burma. Referred specimens are from the Punjab.

Diagnosis.—Similar to *Hydaspitherium*, but smaller.

"The type of the genus . . . appears very doubtfully separable from *H. megacephalum*. In absence of adequate topotypes the genus and species are practically indeterminate." [80]

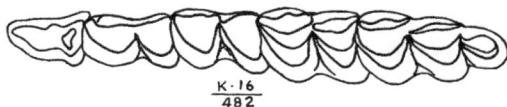

FIG. 191. *?Vishnutherium* sp. G.S.I. No. K 16/482, left P_3–M_3. Crown view, one half natural size. From Matthew, 1929.

LIMB AND FOOT BONES REFERABLE TO *Hydaspitherium* OR *Bramatherium*

Amer. Mus. No. 19460. Left metacarpals. From the Middle Siwaliks, one half mile south of Dhok Pathan.

19461. Left astragalus. Middle Siwaliks, one half mile south of Dhok Pathan.

19464. Right tarsus, metatarsus and proximal phalanges of a young animal. Middle Siwaliks, one half mile south of Dhok Pathan.

19688. Right metatarsals and calcaneum. Middle Siwaliks, one half mile south of Dhok Pathan.

19770. Right radius-ulna and metatarsals. Middle Siwaliks, one half mile south of Dhok Pathan.

19831. Left tarsus, metatarsus and phalanges, articulated. Middle Siwaliks, one half mile south of Dhok Pathan.

29820. Right metacarpals with proximal and medial phalanges. Middle Siwaliks, one half mile south of Dhok Pathan.

The excellent series of giraffid foot bones in the American Museum collection was found in a quarry near Dhok Pathan. The several bones show considerable differences, one from another, differences that may be of specific or even of generic value. On the other hand, since they were associated together in a quarry, it is quite probable that they represent size and age variations in one species.

Among the metacarpals, the two specimens numbered 19460 are different from each other. One is considerably shorter and heavier than the other. No. 29820 is about intermediate between the two.

The metacarpals are characterized by the deep groove along the plantar surface, probably an adaptation for the accomodation of the heavy flexor digitorum ligaments.

The carpal articulations are broad, but otherwise similar to those of the modern giraffe. The combined trapezium-magnum facet occupies a greater portion of the articular surface; a low ridge separates it from the unciform facet. The phalanges are heavy.

As to the hind foot, No. 19464 is definitely from a young animal, in which the epiphyses were not fully ankylosed to the long bones. As in the fore foot, the metatarsals are characterized by a deep plantar groove. The phalanges are heavy and large.

[80] Matthew, W. D., 1929, p. 544.

DIMENSIONS OF LIMB AND FOOT BONES OF MIDDLE SIWALIK GIRAFFIDAE

Hydaspitherium (?) OR *Bramatherium* (?)

A.M. 19770.	Right radius-ulna.	
	Approximate length	500 mm.
	Width at distal end	115
A.M. 19460.	Left metacarpals (two specimens).	
	Length	432
	Width at distal end	100
	Width at middle of shaft	58
	Width at proximal end	95
	Length	398
	Width at distal end	98
	Width at middle of shaft	61
	Width at proximal end	103
A.M. 29820.	Right metacarpals.	
	Length	427
	Width at distal end	96
	Width at middle of shaft	56
	Width at proximal end	96
	Length of proximal phalanx	98
	Length of median phalanx	50
A.M. 19831.	Left metatarsals.	
	Length	434
	Width at distal end	83
	Width at middle of shaft	43
	Width at proximal end	78
	Length of proximal phalanx	102
	Length of median phalanx	60
	Length of distal phalanx (along inferior surface)	80
A.M. 19688.	Right metatarsals.	
	Length	431
	Width at distal end	89
	Width at middle of shaft	52
	Width at proximal end	82
A.M. 19770.	Right metatarsals.	
	Length	450
	Width at distal end	89
	Width at middle of shaft	48
	Width at proximal end	81
A.M. 19464.	Right metatarsals.	
	Length	369
	Width at distal end	72
	Width at middle of shaft	36
	Width at proximal end	68
	Length of proximal phalanx	89
A.M. 19831.	Calcaneum and astragalus.	
	Total length of calcaneum	183
	Greatest width of astragalus	65

The tarsal articulations may be compared to those of the modern giraffe; they consist of two broad facets, one for the cuboid and one for the fused cuneiforms. The calcaneum is very large, the distal end of the tuber being somewhat expanded. A groove is present to serve as an attachment for the tendon of Achilles. The astragalus is broad, with a shallow trochlea, an adaptation to the support of great weight.[81]

GIRAFFINAE

Giraffa Brisson 1762

Generic type, *Giraffa giraffa* Brisson = *Cervus camelopardalis* Linnaeus

Giraffa punjabiensis Pilgrim

Giraffa punjabiensis, Pilgrim, 1910, Rec. Geol. Surv. India, XL, p. 69.

Additional References.—

Pilgrim, G. E., 1910B, p. 203; 1911, p. 8; 1913B, p. 286.

Bohlin, B., 1927, p. 122.

Matthew, W. D., 1929, pp. 539–541, figs. 43, 44.

Type.—(Lectotype.)—G.S.I. No. B 184, maxillary fragments.

Cotype.—G.S.I. No. B 173, mandibular ramus.

Horizon.—Middle Siwaliks, Dhok Pathan zone.

Locality.—Punjab.

Specimens in the American Museum.—Amer. Mus. No. 19318. Fragment of a mandible with right P_2–M_2. Also a right M^1, possibly from a different individual. From the bone bed at Bhandar, Middle Siwaliks, one and one half miles northeast of Hasnot, Punjab.

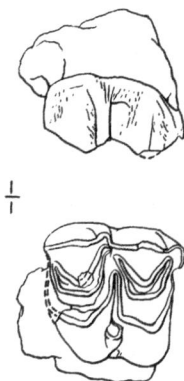

A. M. 19318

FIG. 192. *Giraffa punjabiensis* Pilgrim. Amer. Mus. No. 19318, right M^1. Lateral view above, crown view below. Natural size.

19300. Palate with DM^{3-4}, M^1. Middle Siwaliks, four miles east of Dhok Pathan.

Diagnosis.—Smaller than *Giraffa sivalensis* or *Giraffa camelopardalis*. Upper premolars relatively small and narrow; upper molars long with seldom any tubercles in the median valleys; lower molars long; lobes of molars set less obliquely to axis of jaw than in the recent giraffe.

[81] For a study of the increase in width of the astragalus accompanying an increase in size and weight, see Osborn, H. F., 1929, "Titanotheres of Ancient Wyoming, Dakota and Nebraska," U. S. Geological Survey, Monograph 55.

The American Museum specimen referred to this species represents an advanced form of a giraffine, closely related to the modern *Giraffa*. The lower molars are quite comparable in size to those of *Giraffa*, but the premolars are somewhat narrower than they are in the recent species. The enamel is heavily rugose. The greatest differences between the lower teeth of the Siwalik and of the recent forms are to be found in the presence of heavy median external pillars and antero-external cingula on the molars of the fossil.

The upper molar associated with the jaw would seem to belong to a smaller individual of this species. It is somewhat different from the corresponding tooth in the modern giraffe in that it has a strong median internal pillar as contrasted with the weakly developed pillar in *Giraffa*, and also because along the ectoloph the barrels and styles are feeble, while in the modern form they are strong.

Measurements and figures of the American Museum specimens are given below.

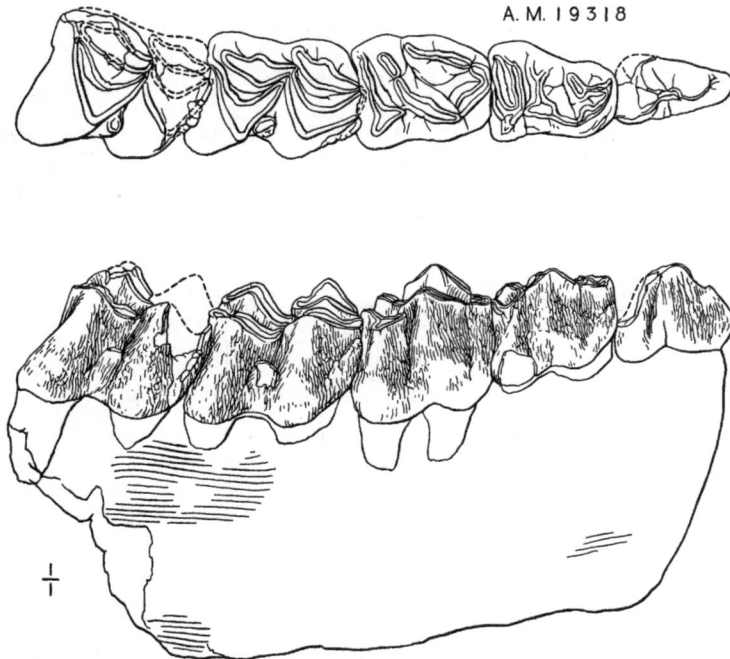

FIG. 193. *Giraffa punjabiensis* Pilgrim. Amer. Mus. No. 19318, right mandibular ramus with P_2–M_2. Crown view above, lateral view below. Natural size.

MEASUREMENTS

Amer. Mus. No. 19318, *Giraffa punjabiensis*

	Length	Width
R M^1	23.5 mm.	23.5 mm.
R P_2	20.0	11.0
P_3	21.0	17.0
P_4	24.0	20.0
M_1	27.0	22.0
M_2	26.0	25.0

Giraffa punjabiensis is referable to the genus *Orasius* of Wagner, typified by Pontian giraffes from Pikermi and Samos. Matthew has shown that the name *Orasius* is synonymous with *Giraffa*, and on the other hand Bohlin has clearly demonstrated that the Pliocene *Orasius* is generically distinct from the recent *Giraffa*. Matthew (1929, p. 546) suggested the name *Bohlinia* for the various species of *Orasius*, but as he purposely did not specify a generic type the name did not become valid. Matthew's reluctance to assign the name

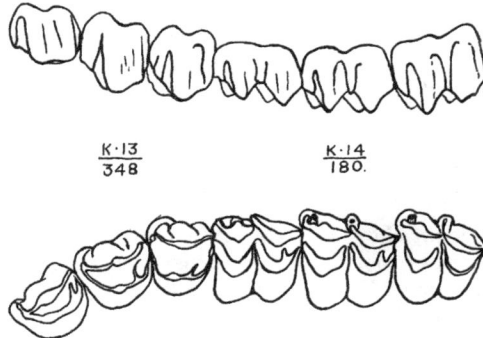

FIG. 194. *Giraffa punjabiensis* Pilgrim. G.S.I. Nos. K 13/348, K 14/180, left P^2–M^3. Lateral view above, crown view below. One half natural size. From Matthew, 1929.

"The teeth differ considerably from those of the type of *G. punjabiensis* in the direction of "*Vishnutherium*" sp. (infra) and may very likely belong to a distinct unnamed species." (Matthew, W. D., 1929, p. 539, fig. 43.)

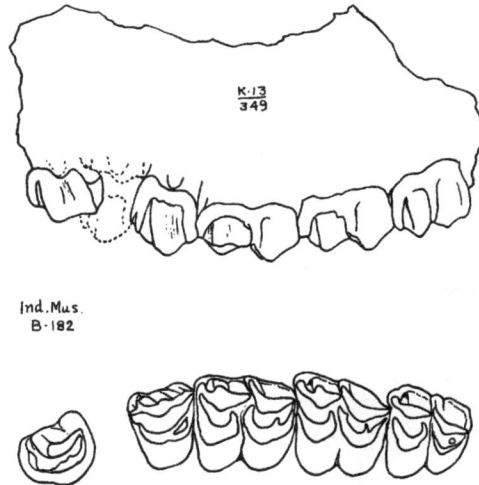

FIG. 195. *Giraffa punjabiensis* Pilgrim. G.S.I. No. K 13/349, left maxilla with P$_2$–M$_3$. Lateral view above, crown view below. One half natural size. From Matthew, 1929.

"Drawn from Ind. Mus. No. K 13/349 which is probably the same individual as No. B 182. This specimen represents the same species as Fig. [40] and comes from the same horizon and locality." (Matthew, W. D., 1929, p. 540, fig. 44.)

definitely to the several species of *Orasius* was due to his clear recognition of the fact that the tooth from the Siwalik Hills, described by Lydekker as *Propalaeomeryx*, might be generically identical with *Orasius*, in which case *Propalaeomeryx* would become the proper generic term.

"The only way of clearing it up would be a more definite determination of the locality of the '*Propalaeomeryx*' type upper molar and obtaining adequate topotypes to determine its character."[82]

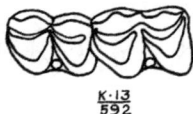

FIG. 196. *Giraffa punjabiensis* Pilgrim. G.S.I. No. K 13/592, lower molars. Crown view, one half natural size. From Matthew, 1929.

Since there is this doubt about the relation of the specimens identified as *Orasius* to *Propalaeomeryx*, it seems best not to propose a new generic term, but rather to retain Pilgrim's name of *Giraffa punjabiensis*. At the same time it must be remembered that there are distinctions, probably of generic importance between *Giraffa punjabiensis* and the recent *Giraffa*.

Giraffa sivalensis (Falconer and Cautley)

Camelopardalis sivalensis, Falconer and Cautley, 1843, Proc. Geol. Soc., London, IV, p. 244.
Giraffa sivalensis, Lydekker, 1885, Cat. Foss. Mam., Brit. Mus., Pt. 2, p. 71.
Honanotherium sivalense, Bohlin, 1927, Pal. Sinica (C), IV, Fasc. I, p. 121.

    Additional References.—

        Falconer, H., 1845, p. 356; 1859, p. 197; 1868A, pp. 197–206, Pl. XVI (copy of original description).

        Gaudry, A., 1862, p. 250.

        Lydekker, R., 1876A, p. 58, Pl. VII, figs. 14, 15; 1876B, p. 104; 1878A, p. 83; 1880B, p. 32; 1882C, pp. 103–112; 1883C, pp. 84, 90; 1885B, pp. 30–31.

        Pilgrim, G. E., 1910B, p. 203; 1911, pp. 10–11.

        Matthew, W. D., 1929, pp. 444, 546–549.

    Type.—Brit. Mus. No. 39747, a cervical vertebra.

    Paratypes.—The type and cotypes of *G. affinis*. Brit. Mus. Nos. 39756, 39756a, 39755, 39757. Upper and lower teeth.

    Horizon.—Upper Siwaliks.

    Locality.—Siwalik Hills.

    Diagnosis.—A large species, but smaller than the modern giraffe, and differing from it by certain details of the cheek teeth. The posterior half of the last upper molar in the fossil is reduced.

[82] Matthew, W. D., 1929, p. 546.

Camelopardalis affinis Falconer and Cautley

Camelopardalis affinis, Falconer and Cautley, 1843, Proc. Geol. Soc., London, IV, p. 246.

Additional References.—
> Falconer, H., 1868A, pp. 201–203, Pl. XVI, figs. 5–10.
> Lydekker, R., 1876B, p. 104; 1882C, pp. 103–104; 1883C, p. 90; 1885D, p. 71.
> Pilgrim, G. E., 1911, p. 10.
> Matthew, W. D., 1929, p. 549.

Type.—(Lectotype.)—Brit. Mus. No. 39756a, left M^{1-3}.

Cotypes.—Brit. Mus. Nos. 39756, right M^3; 39755, left M_3; 39757, right P_4.

Horizon.—Upper Siwaliks.

Locality.—Siwalik Hills.

Diagnosis.—Like *Giraffa sivalensis*, but larger, being comparable in size to the modern giraffe.

Synonymous with *G. sivalensis*.

Giraffa priscilla Matthew

Giraffa priscilla, Matthew, 1929, Bull. Amer. Mus. Nat. Hist., LVI, pp. 537–538.

Type.—G.S.I. No. B 511, a left M^3.

Paratype.—G.S.I. No. B 492, fragment of mandible.   Also left M^2 and right M^3, field number K 14/25.

Horizon.—Lower Siwaliks, Chinji zone.

Locality.—Vicinity of Chinji, Salt Range, Attock District, Punjab.

Diagnosis.—(Matthew, 1929, p. 538.) "Distinguished from *Giraffokeryx* by the broader and more brachyodont teeth, prominent styles (especially note metastyle), prominent anterior rib; in m_3 the more oblique-set inner crescents, broader third lobe with strong accessory basal cusp in front of it, as well as shorter crown."

INCERTAE SEDIS

Propalaeomeryx Lydekker, 1883

Generic type, Propalaeomeryx sivalensis Lydekker

Propalaeomeryx sivalensis Lydekker

Propalaeomeryx sivalensis, Lydekker, 1883, Pal. Indica (X), II, pp. 173–174, fig. 2.
Palaeomeryx sivalensis, Lydekker, 1885, Cat. Foss. Mam., Brit. Mus., Pt. 2, pp. 119–120.
Progiraffa sivalensis, Pilgrim, 1910, Rec. Geol. Surv. India, XL, p. 69.

Additional References.—
> Lydekker, R., 1885B, pp. 33–34.
> Pilgrim, G. E., 1910B, p. 203; 1911, pp. 5–6, Pl. I, fig. 3; 1913B, p. 317.
> Matthew, W. D., 1929, pp. 454, 459, 535.

Type.—G.S.I. No. B 337, a left upper molar.

Paratypes.—None.

Horizon.—Lower Siwaliks.

Locality.—Near Rurki.

Diagnosis.—Upper molar selenodont and low crowned; enamel faintly rugose; cingulum imperfectly developed.

This species is based on a single tooth, and consequently is of little value.

Homologies of the Horn Cores in Various Giraffids

The tendency for the formation of horn cores is exceedingly strong among the various kinds of Giraffidae. The presence and the form of the horn cores in the giraffes is as characteristic, and almost as varied as in their near relatives, the Bovidae. In fact, the horn cores are definitive evolutionary features of the giraffes, and a proper understanding of the method of formation of these structures, and of their homologies, is a distinct aid to any proper understanding of the evolution of the group.

In two papers appearing in the Proceedings of the Zoological Society of London in 1907, E. Ray Lankester has given a helpful discussion of the method of formation of the horn cores in the modern giraffe. He has shown that the horn cores begin to develop in the skin of the forehead, as separate elements which he calls ossicones. As the animal approaches maturity, these ossicones become coalesced with the roof of the skull, and finally in the fully adult giraffe the frontal sinuses of the skull extend up into the horn cores. It is an interesting fact that whereas in the Bovidae the horn cores are always developed from the frontal bones, in the Giraffidae these structures may be attached to various elements of the skull roof.

Lankester shows that in the okapi, a primitive form that has persisted on to modern times, the ossicones of the young animal coalesce with the frontal bones and develop as frontal structures, but in the giraffe, a specialized genus, the ossicones in the young animal develop above the parietal bones and later, due to factors of growth, they spread forward over the frontals so that in the adult animal they rest partly on the frontals and partly on the parietals.

It would seem as if the development of a supra-orbital, frontal horn core is a primitive character among the Giraffidae, whereas the growth of horn cores on other skull elements or on regions of the cranial roof other than above the eye, is a character indicative of specializations towards more advanced or aberrant types.

If we examine the various genera of fossil giraffes we find that they bear out this contention. Among the Palaeotraginae, which constitute the most primitive of the Giraffidae, the horn cores are located on the frontals above the orbits. In *Giraffokeryx punjabiensis* and *Palaeotragus quadricornis*, two rather aberrant forms of palaeotragines, there are extra horn cores anterior to the supra-orbital horn cores. In *Palaeotragus quadricornis* these anterior horn cores, if so they may be called, are quite small swellings of the frontal bones; in *Giraffokeryx* they are fully developed horn cores.

It may be mentioned here that the horn cores of the palaeotragine genera, including *Okapia*, are not directly over the orbit, but at a slight distance behind the eye, the anterior edge of the horn cores in each case extending forward at the base to a point above the middle of the eye.

Palaeotragus quadricornis may represent the beginnings of a specialization towards an aberrant type, and *Giraffokeryx* may represent the culmination of such a specialization. These two forms are departures from the primitive palaeotragine type, and the anterior horn cores developed in them are to be regarded as neomorphic structures.

Coming to the subfamilies of advanced giraffids, the Sivatheriinae and the Giraffinae, we see that the horn cores vary considerably from the primitive supra-orbital, frontal type. In the Sivatheriinae there was a tendency towards the development of very large and elaborate horn cores. In *Hydaspitherium*, as pointed out on a preceding page, there is a pair of gigantic horn cores, so large that they are fused at their bases into one solid structure, covering the cranial roof. These horns are seemingly homologous with the primitive supra-orbital horns of the Palaeotraginae.

In *Bramatherium* there is a similar pair of large horns, fused at the bases and covering the top of the head. In addition there is a pair of smaller horns extending laterally from the parietals, at the posterior edge of the base of the large horn cores. These lateral horn cores are neomorphs. They are probably homologous with the large palmate horns of *Sivatherium*. In *Sivatherium* the relatively small, conical horn cores above and slightly in front of the orbits may be homologous with the primitive supraorbital horn cores of the

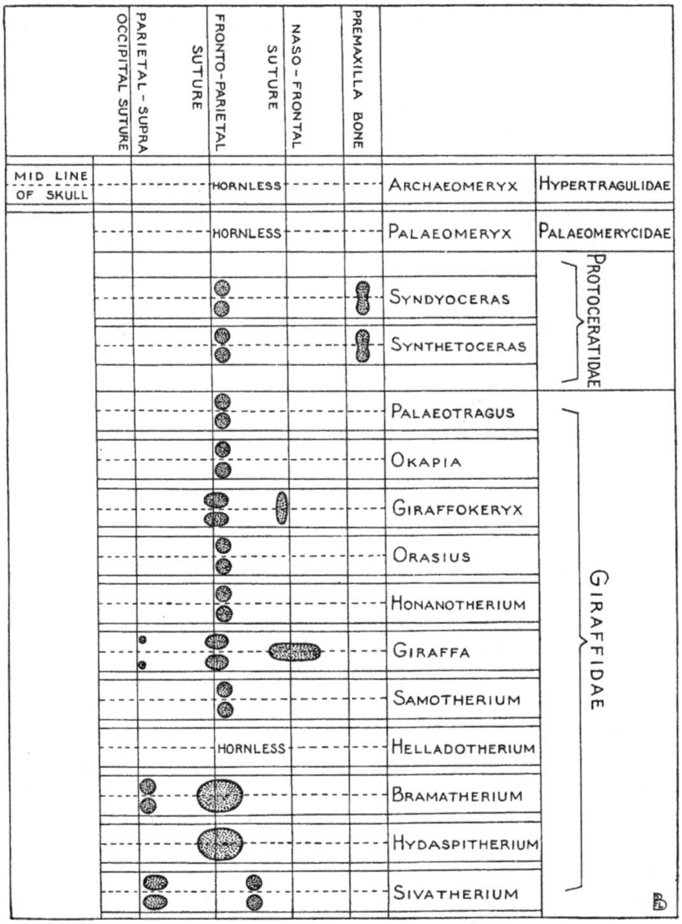

Fig. 197. Homologies of the horn cores in the Giraffidae, Protoceratidae, Palaeomerycidae and Hypertragulidae.

Palaeotraginae. These are the homologies suggested by Oldfield Thomas (1901, p. 483). On the other hand, Birger Bohlin (1927, p. 166) has suggested the possibility of the posterior horn cores of *Sivatherium* being homologous with the primitive supra-orbital frontal horn cores of the Palaeotraginae.

Among the Giraffinae, *Giraffa* is distinguished by five horn cores, or protuberances. The largest pair, resting on the frontal and parietals and straddling the fronto-parietal suture, is probably homologous with the primitive supraorbital horn cores of the Palaeotraginae. The single median protuberance located partly on the frontals and partly on the nasals, has no homologies among the other Giraffidae. Often in *Giraffa*, there are two protuberances on the parietal crests, and these may be homologous with the posterior horns of *Bramatherium* and *Sivatherium*. *Orasius* and *Honanotherium*, fossil genera belonging to the Giraffinae, are characterized by a single pair of horns, homologous with the horns of *Palaeotragus*.

The above conclusions as to the homologies of the horn cores in the Giraffidae are graphically represented in the accompanying chart (Fig. 197).

Were the horn cores in the fossil Giraffidae covered with horny sheaths, as in the Bovidae, or with hair, as in the modern giraffes? This is a question that at present can not be answered. Lankester (1907) thought that he had discovered the remnants of a horny sheath on the tip of the horn cores in an okapi. He therefore concluded that the horn cores in the Giraffidae were originally covered by horny sheaths, and that their present form—that is, bony upgrowths with a covering of hairy skin, is secondary and due to retrogressive evolution. On the other hand, it is quite possible that the supposed horny tip on the okapi horn, considered by Lankester as the remnant of a sheath, is in reality heavy tough skin, hardened and worn free of hair by constant rubbing against branches and tree trunks.

Consequently it may be likely that the hairy horn core of the modern giraffe is a primitive character that typified the various fossil forms, and that the Giraffidae never developed a horny sheath during the history of their evolutionary history.

THE PHYLOGENY OF THE GIRAFFIDAE

In the accompanying chart an attempt has been made to represent the phylogeny of the Giraffidae in a graphic form. On this chart the geologic distribution of the family is shown along the vertical axis, while the geographical extent is represented along the horizontal axis. The three subfamilies of the Giraffidae are indicated by parallel shading; the Palaeotraginae being indicated by vertical lines, the Giraffinae by oblique lines and the Sivatheriinae by horizontal lines. The primitive Palaeomerycidae, from which the Giraffidae might have been derived, are shown also.

It will be seen that the Giraffidae are of Miocene derivation, having their origin in the Holarctic region. The family is characterized by the rapidity of evolution of its subfamilies and genera, all of the great variety of giraffid form and structure having been established since late Miocene times.

The evolutionary development of the family took place in Europe and Asia. The okapi and the giraffe, the one a persistent primitive genus and the other a genus that specialized early in the evolutionary history of the group, migrated to Africa from the

Holarctic center of origin. The survival of these two genera in Africa, far from the center of origin for the family, is what might be expected. Matthew has shown, in his "Climate and Evolution," that persistent primitive species migrate away from the center of origin and that more specialized forms take their place. Or, to put it in a different way, the primitive and inadaptive species are pushed out by the specialized, adaptive species, so that they must needs find refuge in peripheral regions, far distant from their original range.

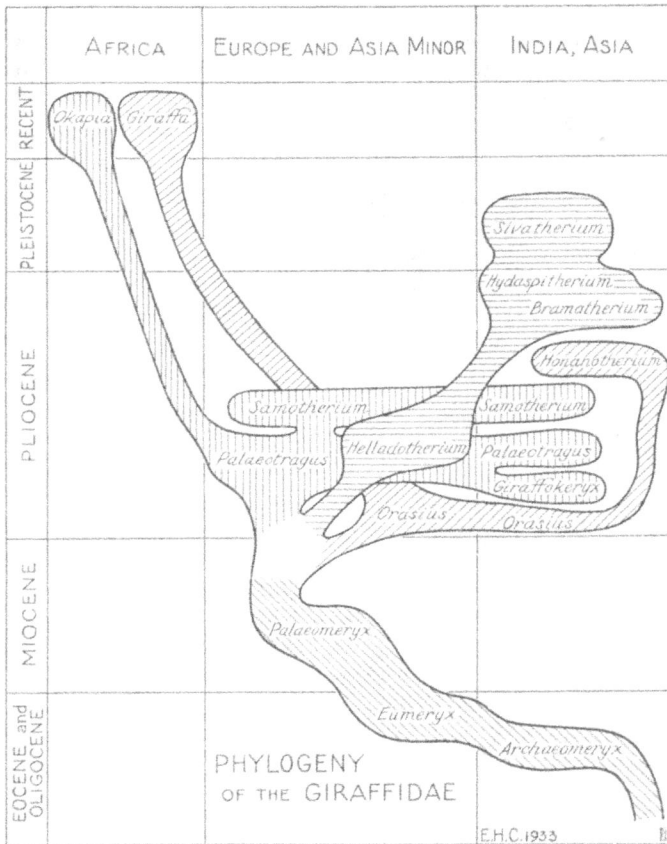

FIG. 198. Phylogeny of the Giraffidae. Development through geologic time is represented along the vertical axis of this chart; geographic distribution is shown along the horizontal axis. The shading indicates taxonomic divisions as follows:

Giraffidae
    Palaeotraginae...vertical lines
    Sivatheriinae...horizontal lines
    Giraffinae...oblique lines
Palaeomerycidae...oblique lines.

BOVOIDEA

BOVIDAE

The Siwalik Bovidae in the American Museum have been systematically studied by Dr. Guy E. Pilgrim, formerly Superintendent of the Geological Survey of India. In order to avoid any conflict with Dr. Pilgrim's studies, the bovids have been purposely left out of the present work. A list of the Siwalik Bovidae, as known to date, is given in the faunal tabulations on page 36.

Dr. Pilgrim has named several new genera and species of Siwalik bovids, and these will be published in a forthcoming Bulletin of the American Museum of Natural History. Dr. Pilgrim's names, now in manuscript, are not included in the faunal list in this paper, because it is thought advisable not to risk any possibility of antedating his publication.

UNIDENTIFIED SPECIMENS

The specimens, listed by number below, in the American Museum Siwalik Collection, have not been definitely identified. They consist for the most part of vertebrae or of fragmentary specimens not readily identifiable.

Amer. Mus. Nos. 19291, 19293, 19294, 19295, 19296, 19299, 19312, 19321, 19351, 19379, 19384, 19439, 19445, 19486, 19489, 19561, 19575, 19649, 19768, 19850, 19897, 19938, 19975, 19988, 29839.

THE MIGRATIONS OF CERTAIN MAMMALS TO AND FROM THE SIWALIKS

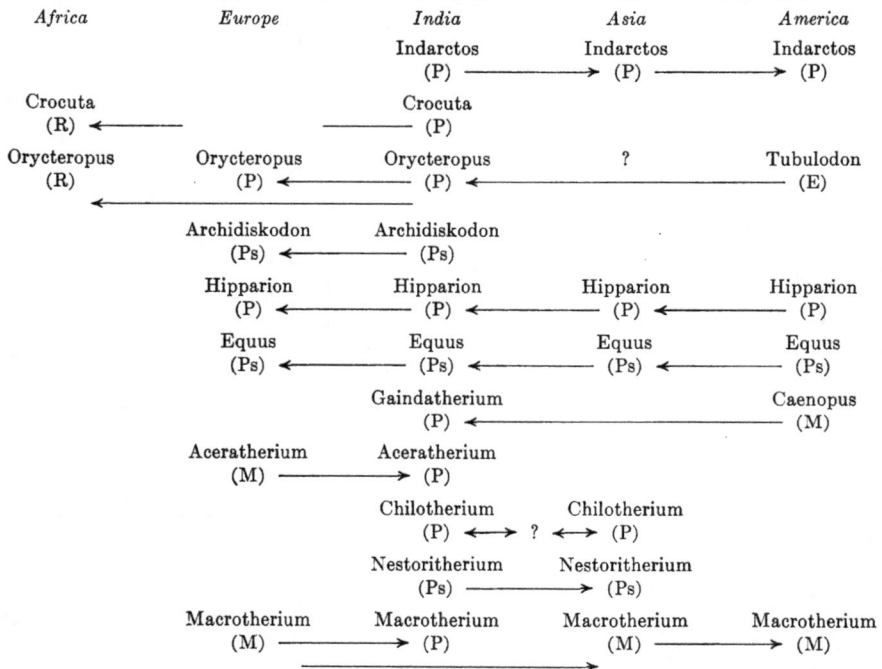

Africa	Europe	India	Asia	America
		Indarctos (P) ⟶	Indarctos (P) ⟶	Indarctos (P)
Crocuta (R) ⟵	⟵	Crocuta (P)		
Orycteropus (R) ⟵	Orycteropus (P) ⟵	Orycteropus (P) ⟵	?	Tubulodon (E)
	Archidiskodon (Ps) ⟵	Archidiskodon (Ps)		
	Hipparion (P) ⟵	Hipparion (P) ⟵	Hipparion (P) ⟵	Hipparion (P)
	Equus (Ps) ⟵	Equus (Ps) ⟵	Equus (Ps) ⟵	Equus (Ps)
		Gaindatherium (P) ⟵		Caenopus (M)
Aceratherium (M) ⟶	Aceratherium (P)			
		Chilotherium (P) ⟷ ? ⟷	Chilotherium (P)	
		Nestoritherium (Ps) ⟶	Nestoritherium (Ps)	
Macrotherium (M) ⟶	Macrotherium (P)	Macrotherium (M) ⟶	Macrotherium (M)	

THE MIGRATIONS OF CERTAIN MAMMALS TO AND FROM THE SIWALIKS

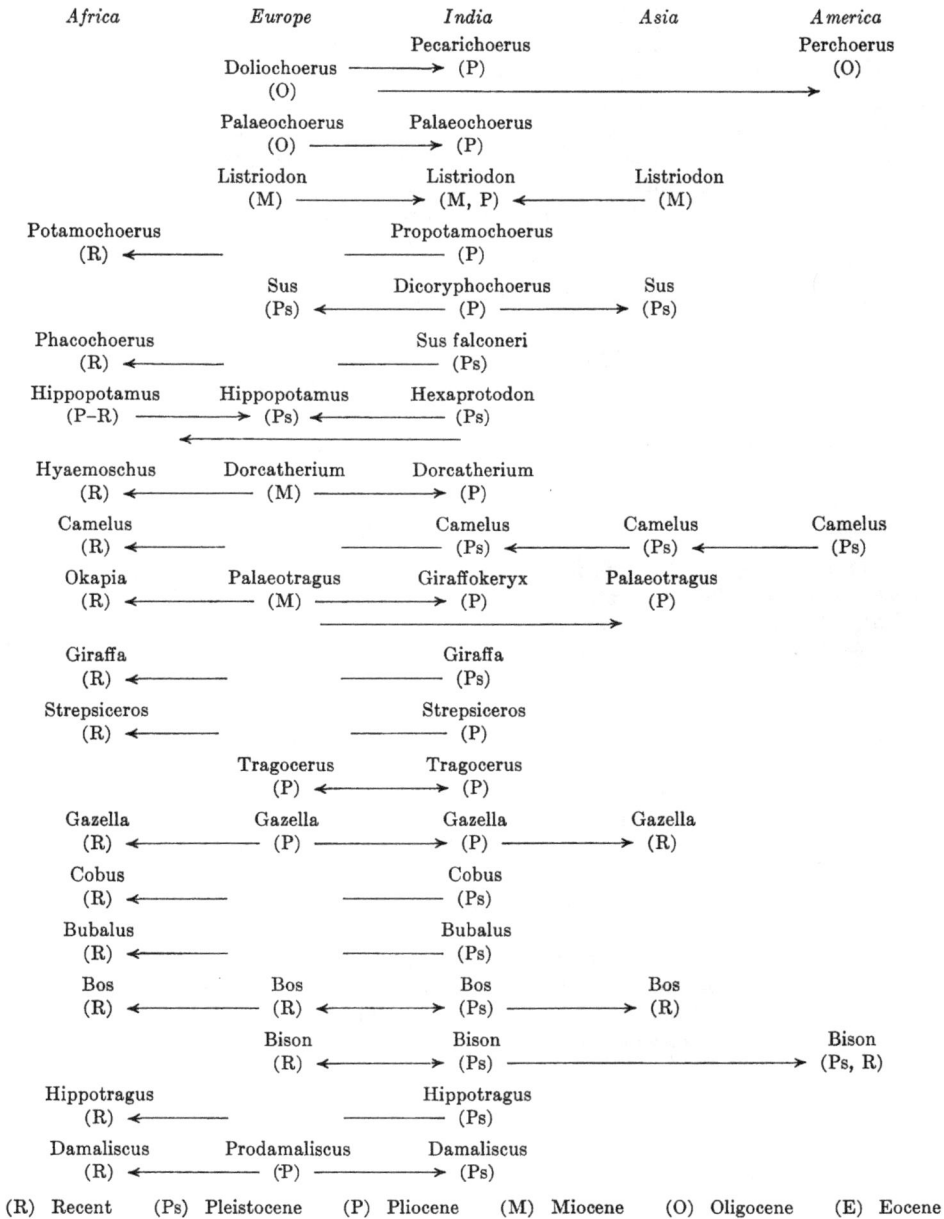

Africa	*Europe*	*India*	*Asia*	*America*
		Pecarichoerus		Perchoerus
	Doliochoerus ⟶	(P)		(O)
	(O)			
	Palaeochoerus	Palaeochoerus		
	(O) ⟶	(P)		
	Listriodon	Listriodon	Listriodon	
	(M) ⟶	(M, P) ⟵	(M)	
Potamochoerus		Propotamochoerus		
(R) ⟵		(P)		
	Sus	Dicoryphochoerus	Sus	
	(Ps) ⟵	(P) ⟶	(Ps)	
Phacochoerus		Sus falconeri		
(R) ⟵		(Ps)		
Hippopotamus	Hippopotamus	Hexaprotodon		
(P–R) ⟶	(Ps) ⟵	(Ps)		
	⟵			
Hyaemoschus	Dorcatherium	Dorcatherium		
(R) ⟵	(M) ⟶	(P)		
Camelus		Camelus	Camelus	Camelus
(R) ⟵		(Ps) ⟵	(Ps) ⟵	(Ps)
Okapia	Palaeotragus	Giraffokeryx	Palaeotragus	
(R) ⟵	(M) ⟶	(P) ⟶	(P)	
		⟶		
Giraffa		Giraffa		
(R) ⟵		(Ps)		
Strepsiceros		Strepsiceros		
(R) ⟵		(P)		
	Tragocerus	Tragocerus		
	(P) ⟵	(P)		
Gazella	Gazella	Gazella	Gazella	
(R) ⟵	(P) ⟶	(P) ⟶	(R)	
Cobus		Cobus		
(R) ⟵		(Ps)		
Bubalus		Bubalus		
(R) ⟵		(Ps)		
Bos	Bos	Bos	Bos	
(R) ⟵	(R) ⟵	(Ps) ⟶	(R)	
	Bison	Bison		Bison
	(R) ⟵	(Ps) ⟶		(Ps, R)
Hippotragus		Hippotragus		
(R) ⟵		(Ps)		
Damaliscus	Prodamaliscus	Damaliscus		
(R) ⟵	(P) ⟶	(Ps)		

(R) Recent (Ps) Pleistocene (P) Pliocene (M) Miocene (O) Oligocene (E) Eocene

PART VI

GENERAL SUMMARY AND CONCLUSIONS

The collection of Siwalik mammals in the American Museum of Natural History was made by Dr. Barnum Brown in two regions in the Punjab. One of these areas was in the Siwalik Hills, between the Sutlej and the Jumna Rivers, and the other was in the Salt Range between the Indus and the Jhelum Rivers.

The Siwalik Series consists of about twenty thousand feet of continental deposits, forming a continuous succession of beds, with virtually no stratigraphic breaks, from the Upper Miocene through the Lower Pleistocene. Progressing from the lower to the upper parts of the Siwalik Series the beds become increasingly coarser, an indication that they were deposited contiguous to an uplifting mountain mass. There is a great amount of variation in lithology in the Siwalik beds.

The Siwaliks may be divided in the following manner.

	Upper Siwaliks	Boulder Conglomerate zone
		Pinjor zone
		Tatrot zone
Siwalik Series	Middle Siwaliks	Dhok Pathan zone
		Nagri zone
	Lower Siwaliks	Chinji zone
		Kamlial zone

The present detailed study of the Siwalik problem would seem to indicate that the Siwalik faunas are (for the most part) relict assemblages, stratigraphically more advanced than their homotaxial affinities would indicate them to be. A true conception of the age of the Siwalik deposits is based, not on the general resemblances of their faunas, but rather on the appearances within these faunas of immigrant forms from outside regions. This was the argument first advanced by Dr. Matthew in 1929.

On these grounds the Kamlial zone of the Lower Siwaliks may be regarded as of Upper Miocene age, while the Chinji zone would seem to be of Lower Pliocene age. The Middle Siwaliks may be considered as higher up in the Pliocene, while the Tatrot zone of the Upper Siwaliks would seem to represent a transition from the Pliocene to the Pleistocene. The Pinjor zone and the Boulder Conglomerate zone of the Upper Siwaliks are of Pleistocene age.

The several Siwalik faunas constitute an abundant assemblage of mammalian genera and species.

From the field studies of Dr. Brown it has been possible to make a careful record of the geographic position and the stratigraphic level of nearly all of the fossils in the American Museum Siwalik collection. This record is of great aid in the study of faunal sequences and in the formulation of views as to the phylogenetic relationships of the various Siwalik mammals.

378

Brief remarks concerning the several orders of mammals in the American Museum Siwalik collection are given in the following paragraphs.

Primates.

A great many genera and species of primates have been described from the Siwalik deposits. Undoubtedly they represent a considerable amount of synonymy, which, owing to the fragmentary nature of the specimens and to their importance, is to be expected. India was a great center of evolution for the advanced primates during Upper Tertiary times, and some of the Siwalik forms have progressed far towards the hominid type in their structure.

Rodentia.

The remains of rodents in the Siwaliks are rather rare. Two new rodents have been described from the American Museum collection.

Carnivora.

Dissopsalis is an hyaenodont occurring in the Lower Siwaliks, and it is therefore the last known survivor of the creodont line. It is structurally but slightly more advanced than *Sinopa*, of the Eocene of North America.

The remains of Fissipedia are scanty in the American Museum collection. Of especial interest is the genus *Indarctos*, which is found not only in the Pliocene of India, but also in Pliocene beds of China and of California. *Arctotherium*, of the Pleistocene of South America would seem to be derived directly from *Indarctos*. Here we have a remarkable example of the migration of a genus from India to North America, and on (by a derived genus) to South America.

The genus *Ictitherium* in the Siwaliks is intermediate in position between the Viverridae and the Hyaenidae. It would seem to be a type ancestral to the later hyaenas, and it would seem to indicate that the hyaenas are of relatively late origin as an offshoot from the viverrids.

Tubulidentata.

Some teeth of *Orycteropus* in the American Museum collection are the first records of this mammalian order in the Siwalik deposits. They constitute the most eastwardly record yet known of the order in the Old World.

Proboscidea.

The Siwalik proboscideans in the American Museum have been thoroughly studied by Professor Osborn. The genus *Synconolophus*, a specialized short jawed mastodont, is especially interesting among the Siwalik proboscideans. *Archidiskodon planifrons*, a primitive elephantine proboscidean first appears in the uppermost Pliocene in India, and almost simultaneously it is found in the uppermost Pliocene of England. Evidently this form migrated northward from India to northern Europe with the advent of the Pleistocene.

Perissodactyla.

Equoidea.

A study of the Siwalik Equidae affords the most reliable criteria for the correlation of the Upper Tertiary deposits of India. Two genera are present in the Siwalik Series, *Hipparion* in the Lower and Middle Siwaliks, and *Equus* in the Upper Siwaliks. Both of these forms undoubtedly migrated from North America to the Asiatic region. The two species of *Hipparion* in the Siwaliks are large forms, with specialized cheek teeth. Since they are clearly much more advanced than the primitive species of *Hipparion* first appearing in North America, we are justified in thinking that a certain amount of time must have elapsed between the origin of the genus in the New World and its subsequent migration to the Old World. The presence of *Equus* in the Upper Siwaliks is a strong argument for regarding these beds as of Pleistocene age.

Chalicotherioidea.

Two genera of chalicotheres are found in the Siwalik Series, namely *Macrotherium* and *Nestoritherium*. These genera are found in Europe and in Asia, showing that the chalicotheres were widely spread throughout the Old World in Upper Tertiary and Lower Quaternary times.

The chalicotheres would seem to have had their origin in North America, in the genus *Eomoropus*, of Middle Eocene age. By Upper Eocene times the group had reached Eurasia, where it enjoyed a long period of wide adaptive radiation.

The later Tertiary chalicotheres may be divided into two groups, one characterized by elongated, comparatively hypsodont cheek teeth and long feet, and the other typified by quadrate, brachyodont cheek teeth and short feet.

Rhinocerotoidea.

The Siwalik rhinoceroses are noteworthy because of their variety; they represent several distinct lines of phylogenetic development. In the Lower Siwaliks is the new genus *Gaindatherium*, which would seem to be directly ancestral to the genus *Rhinoceros* occurring in the Upper Siwaliks. In the Middle Siwaliks are *Aceratherium* and *Chilotherium*. *Aceratherium perimense* would seem to be a gigantic and specialized member of the *Aceratherium* branch of rhinocerine development.

Artiodactyla.
Suoidea.
Tayassuidae.

Primitive peccaries were seemingly present in the Siwaliks, and were represented by the new genus, *Pecarichoerus*, a form closely related to *Doliochoerus* of Europe. Both of these genera were probably derived from the primitive peccary stock, which in turn was closely related to the primitive suid stock.

Suidae.

India was a center for the adaptive radiation of the pigs during the Upper Tertiary, and the Siwalik deposits are replete with numerous and varied genera, representative of most of the branches of suilline evolution. The most primitive suid type, *Palaeochoerus*, is found in the lower portions of the Siwaliks, and it is a survivor of the central stem from which the more specialized kinds of pigs developed. Some especially significant lines of suid phylogeny in the Siwaliks are exemplified by the *Conohyus-Tetraconodon* development, in which the premolars became extraordinarily large and the animals attained great size; by *Listriodon*, a lophodont form; by *Dicoryphochoerus*, a genus leading to *Sus*, and correlatively by *Sus falconeri*, which would seem to be more or less in the direct line of the wart hogs; by *Sivahyus* and *Hippohyus*, in which genera the teeth became quite hypsodont with complexly folded enamel; by *Propotamochoerus*, a form ancestral to *Potamochoerus;* and by certain aberrantly specialized types such as *Sanitherium*.

Anthracotherioidea.

There are several genera of anthracotheres in the Siwaliks, of which *Merycopotamus* of the Upper Siwaliks is the most advanced. *Merycopotamus* is a large anthracothere with a strong *Hippopotamus*-like habitus. The question arises as to whether the numerous *Hippopotamus*-like characters present in *Merycopotamus* are due to a real relationship between the two genera, or whether they are to be attributed to convergence in evolution.

Hippopotamoidea.

A detailed comparison of *Hexaprotodon* of the Siwaliks (which is slightly more primitive than *Hippopotamus*) with *Merycopotamus*, with more primitive anthracotheres, and with certain genera of the Suidae, would seem to offer some evidence in favor of the derivation of the Hippopotamidae from the Anthracotheriidae. It is quite possible that the Hippopotamidae diverged from the Anthracotheriidae at a comparatively late date in the Tertiary, that is, during Upper Miocene times. Consequently it is likely that early Tertiary Hippopotamidae of primitive form will never be found because they never existed.

Cameloidea.

Camelus sivalensis of the Upper Siwaliks is closely comparable with the modern *Camelus*.

Traguloidea.

Dorcabune, from the Lower and Middle Siwaliks, is a primitive tragulid with bunodont cheek teeth. It is seemingly structurally antecedent to *Dorcatherium*, also found in the Siwaliks. *Dorcatherium* is seemingly related to the modern *Hyaemoschus* of Africa, and is on a branch of phylogenetic development separate from *Tragulus* of the Oriental region.

Cervoidea.

Cervus was present in the Middle and Upper Siwaliks. *Cervus punjabiensis* would seem to be a species rather intermediate between the chital and the sambar, while *Cervus sivalensis* is closely related to the barasingha.

Giraffoidea.

India was a center for the adaptive radiation of the giraffes during the period in which the Siwalik deposits were accumulating. In the Lower Siwaliks is *Giraffokeryx*, a primitive but rather aberrant four-horned giraffid, closely related to *Palaeotragus*. In the Middle Siwaliks are *Bramatherium* and *Hydaspitherium*, and in the Upper Siwaliks is *Sivatherium*, all of them giraffids of gigantic size, having highly developed frontal and parietal horn cores.

The family Giraffidae may be divided into three subfamilies, the Palaeotraginae the Sivatheriinae and the Giraffinae. The Palaeotraginae consist of primitive genera, of which *Giraffokeryx* is one, and of which Okapia is a persisting form. The Sivatheriinae are made up of gigantic giraffids having highly developed horn cores. *Bramatherium*, *Hydaspitherium* and *Sivatherium* belong in this group. The Giraffinae consist of the modern *Giraffa* and its Pleistocene and Tertiary relatives. *Orasius* of the Middle Siwaliks belongs in this subfamily.

Bovoidea.

The Siwalik bovids are closely comparable to the Upper Tertiary and Quaternary bovids of Europe and Asia. Many of the genera of bovids which were present in the Siwaliks are now to be found in Africa, and this would indicate that there was a general migration of the Eurasiatic Upper Tertiary bovids into the Ethiopean region, where they persist on at the present time.

BIBLIOGRAPHY

ABEL, O.   1904.   "Über eine Fund von *Sivatherium giganteum* bei Adrianopel."   Sitz. der Kaiserl. Akad. der Wissenschaf. in Wien, CXIII, Abt. I, 23 pp.

   1910.   "Kritische Untersuchengen über die Paläogenen Rhinocerotiden Europas."   Abhandl. der K. K. Geol. Reichenst. Wien, XX, Heft 3, pp. 1–52, Pls. I, II.

   1911.   "Die Palaeontologische Excursion nach Pikermi am 17.   April, 1911."   Druck von Paul Gerin, Wien.

   1920.   "Studien über die Lebensweise von *Chalicotherium*."   Acta Zoologica, I, Heft 1–2, pp. 21–60.

   1926.   "Die Geschichte der Equidae auf dem Boden Nordamerikas."   Verh. Zool.-Bot. Gesellsch., Wien, 74/75 Bd. Jahrgang 1924/1925, pp. 159–164.

   1928.   "Ein Beitrag zur Stammesgeschichte der Pferde: die Phylogenetische Stellung von *Hipparion* und *Neohipparion*."   Sonderabdruck, Akad. Anzeiger Wiss., LXV, No. 5, pp. 28–30.

ADAMS, LEITH.   1867.   "Wandering of a Naturalist in India."   Edinburgh.

ALEXEJEW, A.   1916.   "Animaux fossiles du village Novo-Elisavetovka."   Odessa.

ANDERSON, R. V. V.   1927.   "Tertiary Stratigraphy and Orogeny of the Northern Punjab."   Bull. Geol. Soc. America, XXXVIII, pp. 665–720.

ANDREWS, C. W.   1896.   "On a Skull of *Orycteropus gaudryi* from Samos."   Proc. Zool. Soc., London, pp. 196–199.

   1906.   "A Descriptive Catalogue of the Tertiary Vertebrata of the Fayûm, Egypt."   London, pp. i–xxxvii, 1–324, Plates.

   1923.   "An African Chalicothere."   Nature, CXII, No. 2819, p. 696.

ARAMBOURG, C.   1934.   "Mammifères miocènes du Turkana (Afrique Orientale)."   Ann. de Paléontologie, XXII, pp. 123–146, Pls. XII, XIII.

ARAMBOURG, C., AND PIVETEAU, J.   1929.   "Les Vertébrés du Pontien de Salonique."   Ann. de Paléontologie, XVIII, pp. 59–138, Pls. III, XIV.

ASHLEY, G. H., AND OTHERS.   1933.   "Classification and Nomenclature of Rock Units."   Bull. Geol. Soc. America, XLIV, pp. 423–459.

BAKER, W. E.   1835A.   "On the Fossil Elk of the Himálaya."   Jour. Asiatic Soc. Bengal, IV, pp. 506–507.

   1835B.   "Selected Specimens of the Sub-Himálayan Fossils in the Dádupur Collection."   Jour. Asiatic Soc. Bengal, IV, pp. 565–570.

   1835C.   "Note on the Fossil Camel of the Sub-Himálayas."   Jour. Asiatic Soc. Bengal, IV, pp. 694–695.

BAKER, W. E., AND DURAND, H. M.   1836A.   "Fossil Remains of the Smaller Carnivora from the Sub-Himálayas."   Jour. Asiatic Soc. Bengal, V, pp. 579–584, Pl. XXVII.

   1836B.   "Table of Sub-Himálayan Fossil Genera in the Dádúpur Collection."   Jour. Asiatic Soc. Bengal, V, pp. 291–293, 661–669, 739–740.

BARBOUR, E. H.   1905.   "Notice of a New Fossil Mammal from Sioux County, Nebraska."   Neb. Geol. Surv., II, Pt. 3, pp. 1–4, 1 plate.

   1906A.   "The Skull of *Syndyoceras*."   Science (n.s.), XXIII, pp. 288–289.

   1906B.   "The Skulls of *Syndyoceras* and *Protoceras*."   Science (n.s.), XXIII, p. 623.

BATE, D. M. A.   1906.   "The Pigmy Hippopotamus of Cyprus."   Geol. Mag., n.s., III, pp. 241–245, Pl. XV.

BETTINGTON, ALBEMARLE.   1846.   "Memorandum on Certain Fossils, more Particularly a New Ruminant found at the Island of Perim, in the Gulf of Cambay."   Jour. Roy. Asiatic Soc. of Gt. Brit. and Ireland, XVI, T. 2, pp. 340–348.

DE BLAINVILLE, H. M. D.   1864.   "Ostéographie ou Description Iconographique Comparée du Squelette et du Systéme Dentaire des Mammiféres Récent et Fossiles pour Servir de Base à la Zoologie et à la Geologie."   I–IV.   (Issued in parts from 1839 to 1864.)

BLANFORD, W. T.   1876.   "The African Element in the Fauna of India; a Criticism of Mr. Wallace's Views."   Ann. Mag. Nat. Hist. (4), XVIII, pp. 277–294.

1879. "Geology of Western Sind." Mem. Geol. Surv. India, XVII, Pt. 1.

1885. "Homotaxis as Illustrated from Indian Formations." Rec. Geol. Surv. India, XVIII, Pt. 1, pp. 32–57.

1891. "The Fauna of British India, Ceylon and Burma." London.

BOHLIN, BIRGER. 1927. "Die Familie Giraffidae." Pal. Sinica, Ser. C, IV, Fas. 1, pp. 1–179, Pls. I–XII.

BORISSIAK, A. 1914. "Mammifères Fossiles de Sebastopol." Part I. Mem. Comité Geologique St. Pétersbourg, Nouvelle Serie, Liv. 87, pp. 1–104, Pls. I–X.

1915. "Mammifères Fossiles de Sebastopol." Part II. Mem. Comité Geologique St. Pétersbourg, Nouvelle Serie, Liv. 137, pp. 1–24, Pls. I–III.

1918. "The Remains of Chalicotherioidea from the Oligocene Deposits of Turgai." Ann. Soc. Pal. Russ., III, pp. 43–51, Pl. VII.

1923A. "Sur un Nouveau Représentant des Rhinocéros gigantesques de l'Oligocene d'Asie." Mem. de la Soc. Geol. de France, XXV, Fas. 3, pp. 5–15, Pl. XI–XIII.

1923B. "On the Remains of Anthracotheriidae from the Indricotherium Beds." Bull. Acad. Sci. Russ., pp. 103–110.

1924A. "On the Indricotheriinae." Bull. Acad. Sci. Russ., pp. 127–150.

1924B. "Uber die Unterfamilie Indricotheriinae, Boriss.-Baluchitheriinae, Osb." Centralblatt für Min., No. 118, pp. 571–575.

1927A. "*Aceratherium depereti*, n. sp. from the Jilancik Beds." Bull. Acad. Sci. U. R. S. S., pp. 769–785.

1927B. "*Brachypotherium aurelianense* Nouel var. nov. *cailiti*, from the Miocene Deposits of the Turgai Region." Bull. Acad. Sci. U. R. S. S., pp. 273–281, Pl. I.

1927C. "On the Brachypotherium from the Jilancik Beds of Turgai." C. R., Acad. Sci. U. R. S. S., pp. 93–94.

1927D. "On the Paraceratherium." C. R., Acad. Sci. U. R. S. S., pp. 1–2.

BOSE, P. N. 1880. "Undescribed Fossil Carnivora from the Sivalik Hills." Quar. Jour. Geol. Soc., London, XXXVI, pp. 119–136.

BOULE, M. 1890. "Sur la Limite entre le Pliocène et le Quaternaire." Bull. Soc. Geol. de France, 3rd Ser., XVIII, pp. 945–947.

BREUNING, STEPHAN. 1924. "Beiträge zur Stammesgeschichte der Rhinocerotidae." Verhandl. der Zool.-Bot. Gesellsch., Wien, LXXIII, pp. 5–46.

BRONN, H. G. 1838. "Lethaea Geognostica, oder Abbildungen und Beschreiben der für die Gebirgs-Formation bezeichendsten Versteinerungen." II.

BROOM. R. 1909. "On the Milk Dentition of *Orycteropus*." Ann. So. African Mus., V, Pt. 7, pp. 381–384.

1930. "A New Extinct Pig from the Diamond Gravels of Windsarton, South Africa." Rec. Albany Mus., IV, pp. 167–168.

BROWN, BARNUM. 1925. "Glimpses of India." Natural History, New York, XXV, pp. 109–125.

1926. "A New Deer from the Siwaliks." Amer. Mus. Novitates, No. 242.

BROWN, BARNUM, GREGORY, WILLIAM KING, AND HELLMAN, MILO. 1924. "On Three Incomplete Anthropoid Jaws from the Siwaliks, India." Amer. Mus. Novitates, No. 130.

BUCKLAND, WILLIAM. 1823. "Reliquae Diluvianae; or Observations on the Organic Remains Contained in Caves, Fissures and Diluvial Gravel, and on other Geological Phenomena, attesting the Action of an Universal Deluge." London.

1828. "Geological Account of a Series of Animal and Vegetable Remains, and of Rocks, collected by J. Crawfurd, Esq., on a Voyage up the Irawadi to Ava." Trans. Geol. Soc., London, Second Ser., II, Pt. 3, pp. 377–392.

CAUTLEY, P. T. 1868. "On the Structure of the Sewalik Hills, and the Organic Remains found in them." Pal. Memoirs of Hugh Falconer, London, I, pp. 30–43, Pl. III.

CAUTLEY, P. T., AND FALCONER, H. 1835. "Synopsis of Fossil Genera and Species from the upper deposits of the tertiary strata of the Siválik Hills, in the collection of the authors." Jour. Asiatic Soc. Bengal, IV, pp. 706–707.

1868. "Notice on the Remains of a Fossil Monkey from the Tertiary Strata of the Sewalik Hills." Pal. Memoirs of Hugh Falconer, London, I, pp. 292–297.

See also Falconer, Hugh and Cautley, P. T.

CLIFT, WILLIAM. 1828. "On the Fossil Remains of two new Species of Mastodon, and of other Verte-brated Animals found on the Left Bank of the Irawadi." Trans. Geol. Soc., London, Second Ser., II, Pt. 3, pp. 369–376.

COLBERT, EDWIN H. 1933A. "The Skull of *Dissopsalis carnifex* Pilgrim, a Miocene Creodont from India." Amer. Mus. Novitates, No. 603.

1933B. "The Presence of Tubulidentates in the Middle Siwalik Beds of Northern India." Amer. Mus. Novitates, No. 604.

1933C. "A New Mustelid from the Lower Siwalik Beds of India." Amer. Mus. Novitates, No. 605.

1933D. "The Skull and Mandible of *Conohyus*, a Primitive Suid from the Siwalik Beds of India." Amer. Mus. Novitates, No. 621.

1933E. "A Skull and Mandible of *Giraffokeryx punjabiensis* Pilgrim." Amer. Mus. Novitates, No. 632.

1933F. "Two New Rodents from the Lower Siwalik Beds of India." Amer. Mus. Novitates, No. 633.

1933G. "An Upper Tertiary Peccary from India." Amer. Mus. Novitates, No. 635.

1934A. "An Upper Miocene Suid from the Gobi Desert." Amer. Mus. Novitates, No. 690.

1934B. "A New Rhinoceros from the Siwalik Beds of India." Amer. Mus. Novitates, No. 749.

1934C. "Chalicotheres from Mongolia and China in the American Museum." Bull. Amer. Mus. Nat. Hist., LXVII, Art. VIII, pp. 353–387.

1935A. "Distributional and Phylogenetic Studies on Indian Fossil Mammals. I. American Museum Collecting Localities in Northern India." Amer. Mus. Novitates, No. 796.

1935B. "Distributional and Phylogenetic Studies on Indian Fossil Mammals. II. The Correlation of the Siwaliks of India as Inferred by the Migrations of *Hipparion* and *Equus*." Amer. Mus. Novitates, No. 797.

1935C. "Distributional and Phylogenetic Studies on Indian Fossil Mammals. III. A Classification of the Chalicotherioidea." Amer. Mus. Novitates, No. 798.

1935D. "Distributional and Phylogenetic Studies on Indian Fossil Mammals. IV. The Phylogeny of the Indian Suidae and the Origin of the Hippopotamidae." Amer. Mus. Novitates, No. 799.

1935E. "Distributional and Phylogenetic Studies on Indian Fossil Mammals. V. The Classification and Phylogeny of the Giraffidae." Amer. Mus. Novitates, No. 800.

1935F. "The Proper Use of the Generic Name, *Nestoritherium*." Jour. Mammalogy, XVI, pp. 233–234.

COOPER, C. FORSTER. 1913. "New Anthracotheres and Allied Forms from Baluchistan." Ann. Mag. Nat. Hist., Ser. 8, XII, pp. 514–522.

1920. "Chalicotheroidea from Baluchistan." Proc. Zool. Soc., London, pp. 357–366.

1922. "*Macrotherium salinum* sp. n., a new Chalicothere from India." Ann. Mag. Nat. Hist., Ser. 9, X, pp. 542–544.

1923. "On the Skull and Dentition of *Paraceratherium bugtiense*." Phil. Trans. Roy. Soc. London, Ser. B, CCXII, pp. 369–394.

1924. "The Anthracotheriidae of the Dera Bugti Deposits in Baluchistan." Pal. Indica, N.S., VIII, No. 2, pp. 1–56, Pls. I–VII.

1934. "The Extinct Rhinoceroses of Baluchistan." Phil. Trans. Roy. Soc. London, Ser. B, CCXXIII, pp. 564–616, Pls. LXIV–LXVII.

CROIZET, L'ABBE, AND JOBERT, AINDE. 1828. "Recherches sur les Ossemens Fossiles du Département du Puy-de-Dome."

CUVIER, G. 1812. "Recherches sur les Ossemens Fossiles." Paris. First Edition.

1822. Ibid. Second Edition.

1825. Ibid. Third Edition.

COPE, E. D. 1880. "On the Extinct Cats of America." Amer. Naturalist, XIV, pp. 833–838.

1881. "The Systematic Arrangement of the Order Perissodactyla." Proc. Amer. Phil. Soc., XIX, pp. 377–401.

1884. "Lydekker on the Extinct Mammalia of India." Amer. Naturalist, XVIII, pp. 617–618.

1887. "The Perissodactyla." Amer. Naturalist, XXI, pp. 1014–1019.

1888A. "Lydekker's Catalogue of Fossil Mammalia in the British Museum, Part V." Amer. Naturalist, XXII, pp. 164–165.

1888B. "The Phylogeny of the Horses." Amer. Naturalist, XXII, pp. 345–346.

DAMES, W. 1883. "Über das Vorkommen von Hyaenarctos in den Pliocan von Pikermi bei Athen." Gesellsch. Naturfor. Freunde, Sitzung 16 Oct., pp. 132–139.

DEPÉRET, CHARLES. 1892. "La Faune de Mammifères Miocènes de la Grive-Saint-Alban." Archives du Mus. d'Hist. Nat. de Lyon, V, pp. 1–95.

    1893. "Note sur la Succession Stratigraphique des Faunes Mammifères Pliocènes d'Europe et du Plateau Antra, en Particulier." Bull. Soc. Geol. de France, Ser. 3, XXI, pp. 524–540.

    1908. "L'Histoire Geologique et la Phylogenie des Anthracotherides." C. R. Acad. Sci., CXLVI; pp. 158–162.

DEPÉRET, CHARLES, AND LLUECA, GOMEZ. 1928. "Sur l'Indarctos arctoides." Bull. Soc. Geol. de France, 4 Ser., XXVIII, pp. 149–160.

DIETRICH, W. O. 1927. "Ueber einem Schädel von Ictitherium (Fam. Viverridae)." Neues Jahrbuch für Min. Geol. und Pal., Stuttgart, LVII, Abt. B, pp. 364–371.

DURAND, H. M. 1836. "Specimens of the Hippopotamus and the Fossil Genera of the Sub-Himalayas in the Dadapur Collection." Asiatic Researches, XIX, pp. 54–59.

DUVERNOY, G. L. 1853. "Memoire sur les Orycteropes du Cap, du Nil ou d'Abyssinie, et du Sénégal." Ann. des Sci. Nat., Zoologie, Paris, 3 Ser., XIX, pp. 181–203.

    1854. "Nouvelles Études sur les Rhinoceros Fossiles." Arch. du Mus., Paris, VII, pp. 1–144, Pls. I–VIII.

DUBOIS, EUGENE. 1897. "Ueber drei Ausgestorbene Menschenaffen." Neues Jahrbuch für Min. Geol. und Pal., Stuttgart, I, p. 84.

FALCONER, HUGH. 1831. "Note on Certain Specimens of Animal Remains from Ava." Gleanings in Science, Calcutta, III, p. 167.

    1837. "Note on the Occurrence of Fossil Bones in the Sewalik Range, eastward of Hardwar." Jour. Asiatic Soc. Bengal, VI, p. 233.

    1845. "Description of some Fossil Remains of Dinotherium, Giraffe, and other Mammalia, from Perim Island, Gulf of Cambay, Western Coast of India." Jour. Geol. Soc. London, I, pp. 356–372.

    1859. "Descriptive Catalogue of the Fossil Remains of Vertebrata from the Siwalik Hills, the Nerbudda, Perim Island, Etc., in the Museum of the Asiatic Society of Bengal." Calcutta.

    1868A. "Palaeontological Memoirs." Volume I. Edited by R. I. Murchison. Contains the following:

    1. "Introductory Observations on the Geography, Geological Structure, and Fossil Remains of the Sewalik Hills." Pp. 1–29, Pl. II.

    2. "On the Fossil Hippopotamus of the Sewalik Hills." Pp. 130–148, Pls. XI–XIII. (With P. T. Cautley.)

    3. "Description of a Fragment of a Jaw of an Unknown Extinct Pachydermatous Animal, from the Valley of the Murkanda." Pp. 149–156.

    4. "On the Species of Fossil Rhinoceros found in the Sewalik Hills." Pp. 157–172, Pl. XIV.

    5. "On the Fossil Rhinoceros of Central Tibet, and its Relation to the Recent Upheaval of the Himalayahs." Pp. 173–185. Pl. XV.

    6. "On the Fossil Equidae of the Sewalik Hills." Pp. 186–189.

    7. "On some Fossil Remains of *Anoplotherium* and Giraffe from the Sewalik Hills." Pp. 190–207, Pl. XVI. (With P. T. Cautley.)

    8. "On *Chalicotherium sivalense*." Pp. 208–226, Pl. XVII.

    9. "On the Fossil Camel of the Sewalik Hills." Pp. 227–246, Pl. XVIII. (With P. T. Cautley.)

    10. "*Sivatherium giganteum;* a New Fossil Ruminant Genus from the Valley of the Murkunda." Pp. 247–279, Pls. XIX–XXI. (With P. T. Cautley.)

    11. "On the Fossil Quadrumana of the Sewalik Hills." Pp. 292–313, Pl. XXIV. (With P. T. Cautley.)

    12. "On *Felis cristata*, a New Fossil Tiger from the Sewalik Hills." Pp. 315–320, Pl. XXV. (With P. T. Cautley.)

    13. "On *Ursus (Hyaenarctos) Sivalensis;* a New Fossil Species from the Sewalik Hills." Pp. 321–330, Pl. XXVI. (With P. T. Cautley.)

    14. "On *Enhydriodon (Amyxodon)*, a Fossil Genus allied to *Lutra*, from the Tertiary Strata of the Sewalik Hills." Pp. 331–338, Pl. XXVII.

15. "On the Fossil Carnivora of the Sewalik Hills." Pp. 339–343. (Including notes by W. E. Baker and H. M. Durand.)

16. "Description of some Fossil Remains of *Dinotherium*, Giraffe, *Bramatherium*, and other Mammalia from Perim Island, Gulf of Cambay, Western Coast of India." Pp. 391–411, Pl. XXXIII.

17. "Note on Certain Specimens of Fossil Animal Remains from Ava." Pp. 412–413.

18. "Note on Fossil Remains found in the Valley of the Indus, below Attock, and at Jubbulpoor." Pp. 414–419, Pl. XXXIV.

19. "A Description of the Plates in the Fauna Antiqua Sivalensis." Pp. 421–556.

1868B. "Palaeontological Memoirs." Volume II. Contains the following:

1. "Note on the Existing *Hippopotamus Liberiensis* (Morton), with a Synopsis of the Hippopotamidae, Fossil and Recent." Pp. 404–407.

FALCONER, HUGH, AND CAUTLEY, P. T. 1836A. "*Sivatherium giganteum*, a New Fossil Ruminant Genus from the Valley of the Markanda in the Siwalik Branch of the Sub-Himalayan Mountains." Asiatic Researches, XIX, pp. 1–24.

1836B. "Note on the Fossil Hippopotamus of the Sivàlik Hills." Asiatic Researches, XIX, pp. 39–53.

1836C. "Note on the Fossil Camel of the Sivàlik Hills." Asiatic Researches, XIX, pp. 115–134.

1836D. "Note on the Felis Cristata, a New Fossil Tiger from the Sivàlik Hills." Asiatic Researches, XIX, pp. 135–142.

1836E. "Note on the Ursus Sivalensis, a New Fossil Species, from the Sivàlik Hills." Asiatic Researches, XIX, pp. 193–200.

1837. "On additional fossil species of the Order Quadrumana from the Sewalik Hills." Jour. Asiatic Soc. Bengal, VI, pp. 354–360.

1843. "On some fossil remains of *Anoplotherium* and *Giraffe*, from the Siwalik Hills, in the north of India." Proc. Geol. Soc., London, IV, pp. 235–249.

1846. "Fauna Antiqua Sivalensis." Pt. 1, Proboscidea. London.

1847A. "Fauna Antiqua Sivalensis." Pt. 2, Proboscidea.

1847B. Ibid. Pt. 3, Proboscidea.

1847C. Ibid. Pt. 4, Proboscidea.

1847D. Ibid. Pt. 5, Proboscidea.

1847E. Ibid. Pt. 6, Proboscidea.

1847F. Ibid. Pt. 7, Hippopotamus.

1847G. Ibid. Pt. 8, Suidae and Rhinocerotidae.

1849. "Fauna Antiqua Sivalensis." Pt. 9, Equidae, Camelidae and *Sivatherium*.

FILHOL, M. H. 1876. "Étude du Squelette du Cynohyaenodon." Ann. des Sci. Nat., pp. 179–192, Pls. XVII, XVIII.

1880. "Mammifères fossiles de Saint Gerand le Puy." Bib. l'Ecole des Hautes Etudes, Sec. des Sci. Nat., XX, Art. 1, pp. 6–40, Pls. IV–IX.

1882A. "Memoir sur Quelques Mammiferes Fossiles des Phosphorites du Quercy." Toulouse.

1882B. "Etudes des Mammiferes Fossiles de Ronzon." Ann. Sci. Geol., XII.

1890. "Etudes sur les Mammifères Fossiles de Sansan." Bib. de l'Ecole des Hautes Etudes, Sec. Sci. Nat., XXXVII, pp. 294–305, Pls. XLIII–XLVI.

1894. "Quelques Mammifères fossiles nouveaux du Quercy." Ann. des Sci. Nat., Zoologie, Ser. 7, XVI, pp. 129–150.

1895A. "Observations concernant la restauration d'un squelette d'*Hippopotamus lemerlei*." Bull. Mus. d'Hist. Nat., No. 1, pp. 88–91.

1895B. "Observations concernant les Mammifères contemporains des Aepyornis à Madagascar." Bull. Mus. d'Hist. Nat., No. 1, p. 14.

FERMOR, L. I. 1932. "General Report of the Geological Survey of India for the year 1931." Rec. Geol. Surv. India, LXVI, Pt. 1, pp. 118–120.

FOOTE, R. B. 1880. "*Rhinoceros deccanensis*." Pal. Indica (X), I, Pt. 1, pp. 1–19, Pls. I–III.

FORBES, EDWARD. 1868. "Note on the Fossil Freshwater Shells from the Sewalik Hills." Palaeontological Memoirs of Hugh Falconer, I, pp. 389–390.

FRAIPONT, J. 1908. "*L'Okapi*—Ses affinités avec les Giraffidés vivants et fossiles." Acad. Roy. de Belgique. Bull. de la Classe des Sci., No. 12, Bruxelles, pp. 1099–1130, Pls. I–IV.

FRICK, CHILDS. 1921. "Extinct Vertebrate Faunas of the Badlands of Bautista Creek and San Timoteo Canyon, Southern California." Bull. Dept. Geol. Sci., Univ. California, XII, No. 5, pp. 277–424, Pls. XLIII–L.

1926. "The Hemicyoninae and an American Tertiary Bear." Bull. Amer. Mus. Nat. Hist., LVI, pp. 1–119.

GAUDRY, A. 1861. "Note sur la Giraffe et l'*Helladotherium* trouvés à Pikermi (Grece)." Bull. Soc. Geol. de France, 2 ser., XVIII, p. 587.

1862–1867. "Animaux fossiles et Geologie de l'Attique." Paris.

1873. "Animaux fossiles du Mont Leberon." Paris.

1876. "Sur une Hippopotamus fossile decouvert a Bone (Algerie)." Bull. Soc. Geol. de France, 3 Ser., IV, p. 501.

1886. "Sur l'Age de la faune de Pikermi, du Léberon et de Maragha." Bull. Soc. Geol. de France, 3 Ser., XIII, p. 287.

GRANDIDIER, A., AND FILHOL, M. H. 1894. "Observations relatives aux ossements d'Hippopotames trouvés dans le Marais d'Ambolisatra a Madagascar." Ann. des Sci. Nat., XVI, pp. 131–176.

GREGORY, W. K. 1910. "The Orders of Mammals." Bull. Amer. Mus. Nat. Hist., XXVII, pp. 1–524.

1922. "The Origin and Evolution of the Human Dentition." Baltimore.

1934. "A Half Century of Trituberculy, the Cope-Osborn Theory of Dental Evolution." Proc. Amer. Phil. Soc., LXXIII, pp. 169–317.

GREGORY, W. K., AND HELLMAN, MILO. 1926. "The Dentition of *Dryopithecus* and the Origin of Man." Anthrop. Papers, Amer. Mus. Nat. Hist., XXVIII, pp. 1–123. See also Brown, Gregory and Hellman, 1924.

GILL, T. 1872. "Arrangement of the Families of Mammals and Synoptical Tables of Characters of the Subdivisions of Mammals." Smithsonian Miscell. Coll., No. 230, pp. i–vi, 1–98.

HALL, E. RAYMOND. 1933. (See Stock, Chester and Hall, E. Raymond.)

HAUGHTON, S. H. 1921. "A Note on some Fossils from the Vaal River Gravels." Trans. Geol. Soc. South Africa, XXIV, pp. 11–16, Pl. I.

HELBING, H. 1925. "Das Genus *Hyaenaelurus* Biedermann." Sonderabdruck Eclogae Geologicae Helvetiae, XIX, pp. 214–245, Pl. VI.

1932. "Ueber einen Indarctos Schädel aus dem Pontian der Insel Samos." Abhandl. Schweizerischen Pal. Gesellsch., LII, pp. 1–18, Pl. I.

HELLMAN, MILO. See Brown, Gregory and Hellman, 1924; and Gregory and Hellman, 1926.

HEUSER, PAUL. 1913. "Ueber die Entwicklung des Milchzahngebisses des Afrikanischen Erdferkels." Zeitschr. Wiss. Zool., Leipzig, CIV, pp. 622–691.

VAN HOEPEN, E. C. N., AND VAN HOEPEN, H. E. 1932. "Vrystaatse Wilde Varke." Pal. Navorsing vase die Nasionale Museum Bloemfontein, II, Vierde Stuck, pp. 39–62.

HOLLAND, T. H. 1926. "Indian Geological Terminology." Mem. Geol. Surv. India, LI, Pt. 1.

HOLLAND, W. J., AND PETERSON, O. A. 1913. "The Osteology of the Chalicotheroidea." Mem. Carnegie Mus., III, No. 2, pp. 189–406, Pls. XLVIII–LXXVII.

JEPSEN, G. L. 1932. "*Tubulodon taylori*, a Wind River Eocene Tubulidentate from Wyoming." Proc. Amer. Phil. Soc., LXXI, No. 5, pp. 255–274.

JOBERT, AINDE. See Croizet, l'Abbe and Jobert, Ainde, 1828.

VAN KAMPEN, P. N. 1905. "Die Tympanalgegend des Saugetierschädels." Morph. Jahrbuch, XXXIV, pp. 321–722.

VAN DER KLAAUW, C. J. 1931. "The Auditory Bulla in some Fossil Mammals." Bull. Amer. Mus. Nat. Hist., LXII, pp. 1–352.

KAUP, J. J. 1832–1835. "Description d'ossements fossiles de Mammiféres inconnus jusqu'à présent qui se trouvent au Muséum grand-ducal de Darmstadt." Darmstadt.

1859. "Beiträge zur näheren Kentniss der urweltlichen Säugethiere." Heft IV, p. 3.

KILLGUS, H. 1922. "Unterpliozäne Säuger aus China." Pal. Zeitschr., V, Heft 3, pp. 251–257.

KITTL, E. 1887. "Beiträge zur Kentniss der fossilen Säugethiere von Maragha in Persine, I. Carnivoren." Ann. Naturhist. Hofmus. Wien, II, pp. 317–338, Pls. XIV–XVIII.

1889. "Reste von Listriodon aus dem Miocän Niederösterreichs." Beitr. Pal. Ost. Ung., VII, p. 233, Pl. XIV.

1901. "Die Fossile Saugethiere von Maragha." Ann. k. k. Naturhist. Hofmus., Wien, I.

VON KOENIGSWALD, G. H. R. 1932. "*Metaschizotherium fraasi*, N. G., N. Sp., ein Neuer Chalicotheriide aus dem Obermiocän von Steinheim a Albuch." Palaeontographica, Suppl. Band VIII, I, pp. 1–24, Pls. I–III.

KOWALESKY, W. 1873A. "Monographie der Gattung *Anthracotherium* Cuv. und Versuch einer natürlichen Classification der fossilen Hufthiere." Palaeontographica, XXII, pp. 131–347, Pls. VII–XV.
1873B. "On the Osteology of the Hyopotamidae." Phil. Trans. Roy. Soc. London, CLXIII, pp. 19–94, Pls. XXXV–XL.

KOKEN, E. 1885. "Ueber fossile Säugethiere aus China." Pal. Abhandl., Berlin, III, Heft 2, pp. 31–114, Pls. VI–XII.

KRETZOI, N. 1929A. "Materialen zur Phylogenetischen Klassification der Aeluroideen." Proc. Internat. Congr. Zool., Budapest, 1927, pp. 1293–1355, Pls. XLIII–XLIV.
1929B. "Feliden Studien." Mag. Kir. Foldt. Int. Haziny., Budapest, V, XXIV, pp. 1–22.

LANKESTER, E. RAY. 1901. "On *Okapia*, a new Genus of Giraffidae from Central Africa." Trans. Zool. Soc. London, XVI, Pt. 6, pp. 279–307, Pls. XXX–XXXIII.
1907A. "The Origin of the Lateral Horns of the Giraffe in Foetal Life on the Area of the Parietal Bones." Proc. Zool. Soc. London, pp. 100–115.
1907B. "The Existence of Rudimentary Antlers in the Okapi." Proc. Zool. Soc. London, pp. 126–135.
1910. "Monograph of the Okapi." London.

LEWIS, G. E. 1933A. "Notice of the Discovery of *Plesiogulo brachygnathus* in the Siwalik Measures of India." Amer. Jour. Sci. (5), XXVI, p. 80.
1933B. "Preliminary Notice of a new Genus of Lemuroid from the Siwaliks." Amer. Jour. Sci. (5), XXVI, pp. 134–138.
1934A. "Preliminary Notice of New Man-Like Apes from India." Amer. Jour. Sci. (5), XXVII, pp. 161–179, Pls. I, II.
1934B. "Preliminary Notice of a New Species of *Hippohyus* from India." Amer. Jour. Sci. (5), XXVII, pp. 457–459.

LÖNNBERG, EINAR. 1906. "On a New *Orycteropus* from the Northern Congo and some Remarks on the Dentition of the Tubulidentata." Arkiv. für Zool. k. Svensk. Vet. Akad., Stockholm, III, No. 3, pp. 1–35.
1924. "On a New Fossil Porcupine from Honan, with some Remarks about the Development of the Hystricidae." Pal. Sinica, Ser. C, I., Fas. 3, pp. 1–15, 1 plate.

LYDEKKER, R. 1876A. "Molar Teeth and other Remains of Mammalia." Pal. Indica (X), I, Pt. 2, pp. 19–87, Pls. IV–X.
1876B. "Notes on the Osteology of *Merycopotamus dissimilis*." Rec. Geol. Surv. India, IX, pp. 144–154.
1877A. "Notices of New and Other Vertebrata from Indian Tertiary and Secondary Rocks." Rec. Geol. Surv. India, X, pp. 30–43.
1877B. "Notices of New or Rare Mammals from the Siwaliks." Rec. Geol. Surv. India, X, pp. 76–83.
1877C. "Note on the Genera *Choeromeryx* and *Rhagatherium*." Rec. Geol. Surv. India, X, p. 225.
1878A. "Notices of Siwalik Mammals." Rec. Geol. Surv. India, XI, pp. 64–104.
1878B. "Crania of Ruminants." Pal. Indica (X), I, Pt. 3, pp. 88–171, Pls. XI–XXVIII.
1879. "Notices of Siwalik Mammals." Rec. Geol. Surv. India, XII, pp. 33–52.
1880A. "Supplement to Crania of Ruminants." Pal. Indica (X), I, Pt. 4, pp. 172–181, Pls. XXIA, B, XXIIIA.
1880B. "A Sketch of the History of the Fossil Vertebrata of India." Jour. Asiatic Soc. Bengal, XLIX, Pt. 2, pp. 8–40.
1880C. "Preface" to Volume I of Pal. Indica. Pal. Indica (X), I, pp. vii–xix.
1881A. "Siwalik Rhinocerotidae." Pal. Indica (X), II, Pt. 1, pp. 1–62, Pls. I–X.
1881B. "Note on Some Siwalik Carnivora." Rec. Geol. Surv. India, XIV, pp. 57–66.
1882A. "Siwalik and Narbada Equidae." Pal. Indica (X), II, Pt. 3, pp. 67–98, Pls. XI–XV.
1882B. "Note on some Siwalik and Jamna Mammals." Rec. Geol. Surv. India, XV, pp. 28–33.
1882C. "Siwalik Camelopardalidae." Pal. Indica (X), II, Pt. 4, pp. 99–142, Pls. XVI–XXII.
1883A. "Siwalik Selenodont Suina." Pal. Indica (X), II, Pt. 5, pp. 143–177, Pls. XXIII–XXV.

1883B. "Note on the Probable Occurrence of Siwalik Strata in China and Japan." Rec. Geol. Surv. India, XVI, p. 158.

1883C. "Synopsis of the Fossil Vertebrata of India." Rec. Geol. Surv. India, XVI, pp. 61–93.

1884A. "Siwalik and Narbada Carnivora." Pal. Indica (X), II, Pt. 6, pp. 178–363, Pls. XXVI–XLV.

1884B. "Additional Siwalik Perissodactyla and Proboscidea." Pal. Indica (X), III, Pt. 1, pp. 1–34, Pls. I–V.

1884C. "Siwalik and Narbada Bunodont Suina." Pal. Indica (X), III, Pt. 2, pp. 35–104, Pls. VI–XII.

1884D. "Rodents, Ruminants and Synopsis of Mammalia." Pal. Indica (X), III, Pt. 3, pp. 105–134, Pl. XIII.

1884E. "Catalogue of Vertebrate Fossils from the Siwaliks of India, in the Science and Art Museum, Dublin." Sci. Trans. Roy. Dublin Soc., III (Ser. II), pp. 69–86, Pl. III.

1884F. "Introductory Observations." Pal. Indica (X), II, pp. ix–xv.

1885A. "Note on a Third Species of *Merycopotamus*." Rec. Geol. Surv. India, XVIII, Pt. 3, pp. 145–146.

1885B. "Catalogue of Siwalik Vertebrata in the Indian Museum." Part 1, Mammalia. Calcutta.

1885C. "Catalogue of Fossil Mammalia in the British Museum." Part 1, London.

1885D. "Catalogue of Fossil Mammalia in the British Museum." Part 2, London.

1885E. "Note on Three Genera of Fossil Artiodactyla, with Description of a New Species." Geol. Mag., Dec. III, II, pp. 63–72.

1886A. "Catalogue of Fossil Mammalia in the British Museum." Part 3, London.

1886B. "Catalogue of Fossil Mammalia in the British Museum." Part 4, London.

1886C. "Siwalik Mammalia—Supplement I." Pal. Indica (X), IV, Pt. 1, pp. 1–18, Pls. I–VI.

1886D. "The Fauna of the Karnul Caves." Pal. Indica (X), IV, Pt. 2.

1886E. "On the Fossil Mammalia of Maragha, in Northwestern Persia." Quar. Jour. Geol. Soc., London, XLII, pp. 173–176a.

1886F. "Introductory Observations." Pal. Indica (X), III, pp. xi–xxiv.

1887. "Description of a Jaw of *Hyotherium* from the Pliocene of India." Quar. Jour. Geol. Soc., London, XLIII, pp. 19–23.

1888. "Notes on Indian Fossil Vertebrata; the Ulna of *Hyaenarctos*." Rec. Geol. Surv. India, XXI, pp. 145–146.

1889. "Note on the Pelvis of a Ruminant from the Siwaliks." Rec. Geol. Surv. India, XXII, Pt. 4, pp. 212–214.

1890. "On the Occurrence of the Striped Hyaena in the Tertiary of the Val d'Arno." Quar. Jour. Geol. Soc. London, XLVI, pp. 62–65.

1891. "On a Collection of Mammalian Bones from Mongolia." Rec. Geol. Surv. India, XXIV, Pt. 4, pp. 207–211.

MAJOR, C. J. Forsyth. 1885. "On the Mammalian Fauna of the Val d'Arno." Quar. Jour. Geol. Soc., London, XLI, p. 2.

1888. "Sur un gisement d'ossements fossiles dans l'île de Samos, contemporains de l'âge de Pikermi." C. R. Acad. Sci., CVII, pp. 1178–1182.

1891A. "Considérations nouvelles sur la faune des vertébrés du miocène supérieur dans l'île de Samos." C. R. Acad. Sci., CXIII, pp. 608–610.

1891B. "Sur l'âge de la faune de Samos." C. R. Acad. Sci., CXIII, pp. 708–710.

1891C. "On the Fossil Remains of Species of the Family Giraffidae." Proc. Zool. Soc. London, pp. 315–326.

1893. "Exhibition of, and remarks upon, a tooth of the Ant-bear (*Orycteropus*) from the Upper Miocene of Maragha (Persia)." Proc. Zool. Soc. London, p. 239.

1897A. "On the Species of *Potamochoerus*, the Bush-Pigs of the Ethiopian Region." Proc. Zool. Soc. London, pp. 359–370, Pls. XXV–XXVI.

1897B. "On the Malagasy Rodent Genus *Brachyuromys;* and on the Mutual Relations of some Groups of the Muridae (Hesperomyinae, Microtinae Murinae, and 'Spalacidae') with each other and with the Malagasy Nesomyinae." Proc. Zool. Soc. London, pp. 695–720, Pls. XXXVII–XL.

1897C. "On Sus verrucosus Müller et Schlegel and allies from the Eastern Archipelago." Ann. and Mag. Nat. Hist. (6), XIX, pp. 521–542.

1899. "On Fossil and Recent Lagomorpha." Trans. Linn. Soc., 2nd ser., VII, Pt. 9, pp. 433–520, Pls. XXXVI–XXXIX.

1902A. "On *Mustela palaeattica* from the Miocene of Pikermi and Samos." Proc. Zool. Soc. London, pp. 109–114, Pl. VII.

1902B. "Exhibition of and remarks upon, some remains of a pigmy Hippopotamus from Cyprus." Proc. Zool. Soc. London, pp. 238–239.

1902C. "On the remains of the Okapi received by the Congo Museum in Brussels." Proc. Zool. Soc. London, pp. 73–79.

1902D. "On the pigmy Hippopotamus from the Pleistocene of Cyprus." Proc. Zool. Soc. London, pp. 107–112, Pls. IX, X.

1902E. "On a Specimen of the Okapi lately received at Brussels." Proc. Zool. Soc. London, pp. 339–350.

1902F. "Some account of a nearly complete skeleton of *Hippopotamus madagascariensis* Guld. from Sirabé, Madagascar, obtained in 1895." Geol. Mag., IX, No. 455, pp. 193–199, Pl. XII.

MARSH, O. C. 1874. "On the Structure and Affinities of the Brontotheriidae." Amer. Jour. Sci. (3), VII, p. 82.

MATTHEW, W. D. 1902. "New Canidae from the Miocene of Colorado." Bull. Amer. Mus. Nat. Hist., XVI, pp. 281–290.

1904. "New or Little Known Mammals from the Miocene of South Dakota." Pt. II, Carnivora and Rodentia. Bull. Amer. Mus. Nat. Hist., XX, pp. 246–265, figs. 1–13.

1908A. "Mammalian Migrations between Europe and North America." Amer. Jour. Sci. (4), XXV, pp. 68–70.

1908B. "Osteology of *Blastomeryx* and Phylogeny of the American Cervidae." Bull. Amer. Mus. Nat. Hist., XXIV, pp. 535–562.

1909A. "Observations upon the Genus *Ancodon*." Bull. Amer. Mus. Nat. Hist., XXVI, pp. 1–7.

1909B. "Faunal Lists of the Tertiary Mammalia of the West." Bull. U. S. Geol. Surv., CCCLXI, pp. 91–138.

1909C. "The Carnivora and Insectivora of the Bridger Basin, Middle Eocene." Mem. Amer. Mus. Nat. Hist., IX, pp. 291–567. Pls. XLII–LII.

1915A. "Climate and Evolution." Ann. N. Y. Acad. Sci., XXIV, pp. 171–318.

1915B. "The Tertiary sedimentary record and its problems." Dana Commem. Lectures. Problems of American Geology, pp. 377–478.

1918. "Contributions to the Snake Creek Fauna; with notes upon the Pleistocene of western Nebraska; American Museum Expedition of 1916." Bull. Amer. Mus. Nat. Hist., XXXVIII, pp. 183–229, Pls. IV–X.

1924A. "Third Contribution to the Snake Creek Fauna." Bull. Amer. Mus. Nat. Hist., L, pp. 59–210.

1924B. "A New Link in the Ancestry of the Horse." Amer. Mus. Novitates, No. 131.

1924 C. "Correlation of the Tertiary Formations of the Great Plains." Bull. Geol. Soc. Amer., XXXV, pp. 743–754.

1926. "The Evolution of the Horse. A Record and its Interpretation." Quar. Rev. Biol., I, pp. 139–185.

1929. "Critical Observations upon Siwalik Mammals." Bull. Amer. Mus. Nat. Hist., LVI, pp. 437–560.

1931. "Critical Observations on the Phylogeny of the Rhinoceroses." Univ. of Calif., Bull. Dept. Geol., XX, No. 1, pp. 1–9.

1932. "A Review of the Rhinoceroses with a description of Aphelops material from the Pliocene of Texas." Univ. of Calif., Bull. Dept. Geol., XX, No. 12, pp. 411–480, Pls. LXI–LXXIX.

MATTHEW, W. D., AND COOK, H. J. 1909. "A Pliocene Fauna from Western Nebraska." Bull. Amer. Mus. Nat. Hist., XXVI, pp. 361–414.

MATTHEW, W. D., AND GRANGER, WALTER. 1923. "New Fossil Mammals from the Pliocene of Sze-Chuan, China." Bull. Amer. Mus. Nat. Hist., XLVIII, pp. 563–598.

MATTHEW, W. D. (WITH E. H. COLBERT). 1934. "A Phylogenetic Chart of the Artiodactyla." Jour. Mammalogy, XV, pp. 207–209.

M'CLELAND, J. 1838. "On the Genus *Hexaprotodon*." Jour. Asiatic Soc. Bengal, VII, p. 1038.

DE MECQUENEM, R. 1925. "Contribution à l'étude des fossiles de Maragha." Ann. Pal., Paris, XIV, pp. 1–36, Pls. V–IX.

MEDLICOTT, H. B. 1873. "The Plains of the United Provinces." Rec. Geol. Surv. India, VI, Pt. 1, pp. 9–17.

MERRIAM, J. C. 1911. "Tertiary Mammals Beds of Virgin Valley and Thousand Creek in Northwestern Nevada." Pt. II. Vertebrate Faunas. Bull. Dept. Geol. Sci., Univ. Calif., VI, pp. 199–304, Pls. XXXII–XXXIII.

MERRIAM, J. C., STOCK, CHESTER AND MOODY, C. L. 1916. "An American Pliocene Bear." Bull. Dept. Geol. Sci., Univ. Calif., X, pp. 87–109.

MERRIAM, J. C., AND STOCK, CHESTER. 1927. "A Hyaenarctid Bear from the Later Tertiary of the John Day Basin of Oregon." Carnegie Inst. of Washington, Publ. 346, pp. 39–44.

VON MEYER, H. 1848. "Index Palaeontologicus."

1865. "Ueber die fossilen Reste von Wirbelthieren, welche die Herren von Schlagentweit von ihren Reisen in Indien und Hochasien mitgebracht haben." Palaeontographica, XV, pp. 1–40.

MIDDLEMISS, C. S. 1890. "Geology of the Sub-Himalayas." Mem. Geol. Surv. India, XXIV, Pt. 2.

MILLER, G. S. 1906. "Notes on Malayan Pigs." U. S. Nat. Mus., XXX, No. 1466, pp. 737–758.

1912. "Catalogue of the Mammals of Western Europe in the Collection of the British Museum." London.

1927. "Revised Determinations of some Tertiary Mammals from Mongolia." Pal. Sinica, Ser. C, V, Fasc. 2, pp. 5–20.

MILNE-EDWARDS, A. 1864. "Recherches Anatomiques, Zoologiques et Palaeontologiques sur la famille de Chevrotains." Ann. de Sci. Nat., 5 ser., Zool., II, pp. 49–167, Pl. II–XII.

MIVART, ST. GEORGE. 1882. "On the Classification and Distribution of the Aeluroidea." Proc. Zool. Soc. London, pp. 135–208.

1885. "On the Anatomy, Classification and Distribution of the Arctoidea." Proc. Zool. Soc. London, pp. 340–404.

MOOK, C. C. 1932. "A New Species of Fossil Gavial from the Siwalik Beds." Amer. Mus. Novitates, No. 514.

1933. "A Skull with Jaws of *Crocodilus sivalensis* Lydekker." Amer. Mus. Novitates, No. 670.

MURIE, J. 1871. "On the Systematic Position of the *Sivatherium giganteum*." Geol. Mag., VIII, pp. 438–448.

1872. "The Horns, Viscera and Muscles of the Giraffe," Ann. Mag. Nat. Hist., Fourth Ser., IX, pp. 177–195.

NOETLING, F. 1897. "Note on a Worn Femur of *Hippopotamus iravadicus* Caut. and Falc., from the Lower Pliocene of Burma." Rec. Geol. Surv. India, XXX, Pt. 4, pp. 242–248, Pl. XIX–XX.

1901. "Fauna of the Miocene Beds of Burma." Pal. Indica, N.S., I, Pt. 3, p. 378, Pl. XXV, figs. 24–25.

OLDHAM, R. D. 1893. "Geology of India." Calcutta.

OSBORN, H. F. 1893. "The Ancylopoda, *Chalicotherium* and *Artionyx*." Amer. Naturalist, XXVII, pp. 118–133.

1898. "The Extinct Rhinoceroses." Mem. Amer. Mus. Nat. Hist., I, pp. 75–164, Pls. XIIA–XX.

1900. "Correlation between Tertiary Mammal Horizons of Europe and America." Ann. N. Y. Acad. Sci., XIII, pp. 1–72.

1901. "The Phylogeny of the Rhinoceroses of Europe." Ann. N. Y. Acad. Sci., XIII, p. 505.

1909. "Cenozoic Mammal Horizons of Western North America." Bull. U. S. Geol. Survey, CCCLXI, pp. 1–90.

1910A. "Correlation of the Cenozoic through its Mammalian Life." Jour. Geol., XVIII, pp. 201–215.

1910B. "The Age of Mammals in Europe, Asia and North America." New York.

1912. "Craniometry of the Equidae." Mem. Amer. Mus. Nat. Hist. (n.s.) I, Pt. III, pp. 55–100.

1913. "*Eomoropus*, an American Eocene Chalicothere." Bull. Amer. Mus. Nat. Hist., XXXII, pp. 261–274.

1915. "Review of the Pleistocene of Europe, Asia and Africa." Ann. N. Y. Acad. Sci., XXVI, pp. 215–315.

1918. "Equidae of the Oligocene, Miocene and Pliocene of North America; Iconographic Type Revision." Mem. Amer. Mus. Nat. Hist. (n.s.), II, pp. 1–330, Pls. I–XLIV.

1926. "Additional New Genera and Species of the Mastodontoid Proboscidea." Amer. Mus. Novitates, No. 238.

1929A. "New Eurasiatic and American Proboscideans." Amer. Mus. Novitates, No. 393.

1929B. "The Titanotheres of Ancient Wyoming, Dakota and Nebraska." U. S. Geol. Survey, Monograph 55, Pts. 1 and 2.

OWEN, R. 1839. "Notes on the Anatomy of the Nubian Giraffe." Trans. Zool. Soc., London, III, pp. 217–248, Pls. XL–XLV.

1840–1845. "Odontography." London.

1846. "History of British Fossil Mammals." London.

1850. "On the Development and Homologies of the Molar Teeth of the Wart Hog." Phil. Trans. Roy. Soc., London, CXL, Pt. 1, pp. 481–498, Pls. XXXIII–XXXIV.

1852. "On the Anatomy of the Indian Rhinoceros." Trans. Zool. Soc., London, IV, Pt. 2, pp. 31–58.

1870. "On Fossil Remains of Mammals found in China." Quar. Jour. Geol. Soc., London, XXVI, pp. 417–434, Pls. XXVII–XXIX.

PASCOE, E. H. 1920. "Petroleum in the Punjab and Northwest Frontier Province." Mem. Geol. Surv. India, XL, Pt. 3.

1923. "General Report for 1922." Rec. Geol. Surv. India, LV, Pt. 1, pp. 40–42.

PEARSON, H. S. 1923. "Some Skulls of Perchoerus from the White River and John Day Formations." Bull. Amer. Mus. Nat. Hist., XLVIII, pp. 61–96.

1927. "On the Skulls of Early Tertiary Suidae, together with an Account of the Otic Region in some other Primitive Artiodactyla." Phil. Trans. Roy. Soc., London, Ser. B, CCXV, pp. 389–460.

1928. "Chinese Fossil Suidae." Pal. Sinica, Ser. C, V, Fasc. 5, pp. 1–75, Pls. I–IV.

1929 "The Hinder End of the Skull in Merycopotamus and in Hippopotamus minutus." Jour. Anatomy, LXIII, pp. 237–241.

PENTLAND, J. B. 1828. "Description of Fossil Remains of Some Animals from the Northeast Border of Bengal." Trans. Geol. Soc., London, 2nd Ser., II, pp. 394–395.

PETERSON, O. A. 1919. "Report on the Material Discovered in the Upper Eocene of the Uinta Basin by Earl Douglass in the Years 1908–1909, and by O. A. Peterson in 1912." Ann. Carnegie Mus., XII, Nos. 2–4, pp. 40–168.

DAL PIAZ, G. 1932. "I Mammiferi dell'Oligocene veneta. Anthracotherium monsvialense." Mem. dell'Inst. Geol. della R. Univ. di Padova, X, pp. 1–63, Pls. I–XVI.

PILGRIM, G. E. 1905. "On the Occurrence of Elephas antiquus (namadicus) in the Godavari Alluvium, with remarks on the Species, its Distribution and the Age of the Associated Indian Deposits." Rec. Geol. Surv. India, XXXII, Pt. 3, pp. 199–218, Pls. VIII–XII.

1906. "Fossils of the Irrawaddy Series from Rangoon." Rec. Geol. Surv. India, XXXIII, Pt. 2, pp. 157–158.

1908. "The Tertiary and Post-Tertiary Deposits of Baluchistan and Sind, with Notices of New Vertebrates." Rec. Geol. Surv. India, XXXVII, Pt. 2, pp. 139–168.

1910A. "Notices of New Mammalian Genera and Species from the Tertiaries of India." Rec. Geol. Surv. India, XL, Pt. 1, pp. 63–71.

1910B. "Preliminary Note on a Revised Classification of the Tertiary Freshwater Deposits of India." Rec. Geol. Surv. India, XL, Pt. 3, pp. 185–205.

1910C. "On the Changes of Climate in India during the Post-Glacial Portion of the Pleistocene." Postglaziale Klimaveranderungen, Stockholm, pp. 441–442.

1911. "The Fossil Giraffidae of India." Pal. Indica, N.S., IV, No. 1, pp. 1–29, Pls. I–V.

1912. "The Vertebrate Fauna of the Gaj Series in the Bugti Hills and the Punjab." Pal. Indica, N.S., IV, No. 2, pp. 1–83, Pls. I–XXXI.

1913A. "Correction in Generic Nomenclature of Bugti Fossil Mammals." Rec. Geol. Surv. India, XLIII, Pt. 1, pp. 75–76.

1913B. "Correlation of the Siwaliks with Mammal Horizons of Europe." Rec. Geol. Surv. India, XLIII, Pt. 4, pp. 264–326, Pls. XXVI–XXVIII.

1914A. "Further Description of *Indarctos salmontanus* Pilgrim, the New Genus of Bear from the Middle Siwaliks, with some Remarks on the Fossil Indian Ursidae." Rec. Geol. Surv. India, XLIV, Pt. 3, pp. 225–233.

1914B. "Description of Teeth Referable to the Lower Siwalik Creodont Genus *Dissopsalis* Pilgrim." Rec. Geol. Surv. India, XLIV, Pt. 4, pp. 265–279.

1915A. "New Siwalik Primates and their Bearing on the Question of the Evolution of Man and the Anthropoids." Rec. Geol. Surv. India, XLV, Pt. 1, pp. 1–74.

1915B. "Note on the New Feline Genera *Sivaelurus* and *Paramachaerodus* and on the Possible Survival of the Subphylum in Modern Times." Rec. Geol. Surv. India, XLV, Pt. 2, pp. 138–155.

1915C. "The Dentition of the Tragulid Genus *Dorcabune*." Rec. Geol. Surv. India, XLV, Pt. 3, pp. 226–238.

1918. "Preliminary Note on Some Recent Collections from the Basal Beds of the Siwaliks." Rec. Geol. Surv. India, XLVIII, pp. 98–101.

1919. "Suggestions Concerning the History of the Drainage of Northern India, arising out of a Study of the Siwalik Boulder Conglomerate." Jour. Asiatic Soc. Bengal, N.S., XV, No. 2, pp. 81–99, Pls. I, II.

1925. "The Migrations of Indian Mammals." Presidential Address, Section of Geology, Twelfth Indian Science Congress.

1926A. "The Fossil Suidae of India." Pal. Indica, N.S., VIII, No. 4, pp. 1–65, Pls. I–XX.

1926B. "The Tertiary Formations of India and the Interrelation of Marine and Terrestrial Deposits." Proc. Pan-Pacific Sci. Cong., Australia, 1926, pp. 896–931.

1927A. "The Lower Canine of *Tetraconodon*." Rec. Geol. Surv. India, LX, Pt. 2, pp. 160–163, Pl. XIV.

1927B. "A *Sivapithecus* Palate and other Primate Fossils from India." Pal. Indica, N.S., XIV, pp. 1–24, 1 plate.

1928. "The Lower Canine of an Indian Species of *Conohyus*." Rec. Geol. Surv. India, LXI, Pt. 2, pp. 196–205.

1931. "Catalogue of the Pontian Carnivora of Europe." Cat. British Museum, London.

1932. "The Fossil Carnivora of India." Pal. Indica, N.S., XVIII, pp. 1–232, Pls. I–X.

1932. "The Genera *Trochictis*, *Enhydrictis* and *Trocharion*, with Remarks on the Taxonomy of the Mustelidae." Proc. Zool. Soc. London, Pt. 4, 1932, pp. 845–867, Pls. I, II.

1934. "Correlation of Ossiferous Sections in the Upper Cenozoic of India." Amer. Mus. Novitates, No. 704.

PILGRIM, G. E., AND HOPWOOD, A. T. 1928. "Catalogue of the Pontian Bovidae of Europe." Cat. British Museum, London.

PINFOLD, E. S. 1918. "Notes on Structure and Stratigraphy in the Northwest Punjab." Rec. Geol. Surv. India, XLIX, Pt. 3, pp. 149–159.

PEI, W. C. 1931. "Mammalian Remains from Locality 5 at Chouk'outien." Pal. Sinica, Ser. C, VII, Fasc. 2, pp. 1–18, 1 plate.

POCOCK, R. I. 1916A. "The Tympanic Bulla in Hyaenas." Proc. Zool. Soc. London, Pt. 1, pp. 303–307.

1916B. "The Alisphenoid Canal in Civets and Hyaenas." Proc. Zool. Soc. London, Pt. 2, pp. 442–445.

1921A. "The External Characters and Classification of the Procyonidae." Proc. Zool. Soc. London, Pt. 2, pp. 389–442.

1921B. "The Auditory Bulla in the Mustelidae." Proc. Zool. Soc. London, Pt. 3, pp. 473–486.

1921C. "The External Characters of some Species of Lutrinae." Proc. Zool. Soc. London, Pt. 3, pp. 535–546.

1921D. "The External Characters and Classification of the Mustelidae." Proc. Zool. Soc. London, Pt. 4, pp. 803–837.

POHLE, H. 1919. "Die Unterfamilie der Lutrinae." Archiv. für Naturgeschichte, Abt. A, Heft IX, 85, pp. 1–247, Pls. I–IX.

1928. "Die Raubtiere von Oldoway." Wiss. Ergib. Oldoway Exped., p. 50.

POHLIG, H. 1886. "On the Pliocene of Maragha, Persia, and its Resemblance to that of Pikermi in Greece." Quar. Jour. Geol. Soc., London, XLII, No. 166, pp. 177–282.

POMEL, A. 1848. "Recherches sur les caractères et les rapports entre eux des divers genres vivants et fossiles des Mammiféres ongulés." C. R., Acad. Sci., Paris, XXVI, pp. 686–688.

—— 1853. "Catalogue Méthodique et descriptif des vertébrés fossiles découvert dans le bassin hydrographique supérieur de la Loire, et surtout dans la vallée de son affluent principal, l'Allier." Auvergne Ann. Sci., XXVI, pp. 81–229.

—— 1896. "Les Hippopotames." Pal. Monograph, No. VIII, Carte Geol. de l'Algerie.

—— 1897. "Les Suilliens." Pal. Monograph, No. X, Carte Geol. de l'Algerie.

RINGSTRÖM, T. 1924. "Näshorner der Hipparion-Fauna Nord-Chinas." Pal. Sinica, Ser. C, I, Fasc. 4, pp. 1–156, Pls. I–XII.

—— 1927. "Ueber Quartäre und Jungterteäre Rhinocerotiden aus China und der Mongolei." Pal. Sinica, Ser. C, IV, Fasc. 3, pp. 1–21, Pls. I–II.

RODLER, A., AND WEITHOFER, K. A. 1890. "Die Wiederkäuer der Fauna von Maragha." Besonders Abgedr. Denkschr. Math.-Naturwiss. Classe, K. Akad. der Wissensch. LVII, pp. 753–772, Pls. I–VI.

ROMAN, F. 1914. "Sur les Rhinocerides du Bassin de Mayence." C. R. Acad. Sci., Paris, CLVIII, p. 1224.

ROSE, C. 1892. "Beiträge zur Zahnentwicklung der Edentaten." Anat. Anz., 7, Jahrgang Nos. 16–17, p. 495.

—— 1894. "Zur Phylogenie des Säugetiergebisses." Biol. Zentralbl., XII, p. 624.

RUSSEL, L. S. 1934. "Revision of the Lower Oligocene Vertebrate Fauna of the Cypress Hills, Saskatchewan." Royal Canadian Inst., XX, pp. 63–67.

RÜTIMEYER, L. 1857. "Ueber Lebende und Fossile Schweine." Verhandl. naturf. Gesellsch., Basal, I, pp. 517–554.

—— 1857. "Ueber *Anthracotherium magnum* und *hippoideum*." Neue Denkschr. Schweizer Gesellsch. Naturwiss., XV, Art. VIII, pp 1–32, Pls. I, II.

—— 1881. "Beiträge zur einer Natürlichen Geschichte der Hirsche." Abhandl. der Schweizerischen Pal. Gesellsch., VIII, pp. 1–124, Pls. I–X.

SCHWALBE, G. 1915. "Ueber den fossilen Affen *Oreopithecus bamboli*." Zeitschr. für Morph. und Anthropologie, XIX, Heft I, pp. 150–254.

SCOTT, W. B. 1892. "The Evolution of the Premolar Teeth in the Mammals." Proc. Acad. Nat. Sci., Philadelphia, pp. 405–444.

—— 1894. "The Structure and Relationships of *Ancodus*." Jour. Acad. Nat. Sci., Philadelphia, IX, pp. 461–497, Pls. XXIII, XXIV.

—— 1895. "The Osteology and Relationships of *Protoceras*." Jour. Morphology, XI, pp. 303–374, Pls. XX–XXII.

—— 1913. "A History of Land Mammals in the Western Hemisphere." New York.

SCLATER, W. L. 1900. "The Mammals of South Africa." Vol. I. London.

SCHLOSSER, MAX. 1890. "Die Differenzierung des Saugetiergebisses." Biol. Centralblatt, X, No. 8, pp. 238–277.

—— 1903. "Die Fossilen Saugethiere Chinas, nebst einer Odontographie der recenten Antilopen." Abhandl. der Bayer. Akad. der Wiss., II, Cl. 22, Bd. I, pp. 193–217.

—— 1904. "Fossilen Cavicornia von Samos." Beitrage Pal. und Geol. Osterreich-Ongarns und des Orients, Band XVII, pp. 21–118, Pls. VII–XVI.

—— 1912. "Beiträge zur Kentniss der Saugethierreste aus Süddeutschen Bemergen." Geol. und Pal. Abhandl., IX, p. 83.

—— 1921. "Die Hipparionen Fauna von Veles in Mazedonien." Abhandl. Bayer. Akad. der Wiss. Math.-Phys., Klasse XXXIX, Band 4, pp. 1–55, Pls. I–II.

—— 1924A. "Tertiary Vertebrates from Mongolia." Pal. Sinica, Ser. C, I, Fasc. 1, pp. 1–119, Pls. I–VI.

—— 1924B. "Fossil Primates from China." Pal. Sinica, Ser. C, I, Fasc. 2, pp. 1–14, Pl. I.

SEFVE, IVAR. 1927. "Die Hipparionen Nord-Chinas." Pal. Sinica, Ser. C, IV, Fasc. 2, pp. 1–91, Pls. I–VII.

SIMONESCU, J. 1932. "Tertiäre und Pleistocäne Camelidae in Rümänien." Bull. Sec. Sci. Acad. Roumaine, XV, pp. 1–8.

SIMPSON, G. G. 1931. "A New Classification of Mammals." Bull. Amer. Mus. Nat. Hist., LIX, Art. V, pp. 259–293.

1933. "Glossary and Correlation Charts of North American Tertiary Mammal-Bearing Formations." Bull. Amer. Mus. Nat. Hist.. LXVII, Art. III, pp. 79–121.

SONNTAG, C. F., WOOLARD, H. H., AND CLARK, W. E. leGros 1925–1926. "A Monograph of *Orycteropus afer*." Proc. Zool. Soc. London. (1925, pp. 331–437, 1185–1233; 1926, pp. 445–485.)

STEHLIN, H. G. 1899. "Geschichte des Suiden Gebisses." Abhandl. der Schw. Pal. Gesellsch., XXVI, pp. 1–527, Pls. I–X.

1904. "Une Faune à Hipparion à Perrier." Bull. de la Soc. Geol. de France, 4 Ser., IV, pp. 432–444.

1910. "Zur Revision der Europaischen Anthracotherien." Verh. Naturforschenden Gesellsch., Basel, XXI, pp. 165–185.

1927. "Ueber einen Baluchitheriumfund aus dem Punjab." Sonderabdruck Eclogae Geologicae Helvetiae, XX, pp. 297–301.

1929A. "Artiodactylen mit Fünffingriger Vorderextremität aus den Europäischen Oligocän." Verh. Naturforschenden Gesellsch., Basel, XL, pp. 599–625.

1929B. "Bemerkungen zu der Frage nach der Unmittelbaren Ascendenz der Genus Equus." Sonderabdruck Eclogae Geologicae Helvetiae, XXII, pp. 187–201.

1931. "Une Mandibule de Giraffidé de Tokoum (Perse)." Sonderabdruck Eclogae Geologicae Helvetiae, XXIV, pp. 275–279.

STIRTON, R. A. 1932. "A New Genus of Artiodactyla from the Clarendon Lower Pliocene of Texas." Bull. Dept. Geol. Sci., Univ. of Calif., XXI, No. 6, pp. 147–168, Pls. VI–XI.

STIRTON, R. A., AND McGREW, P. O. 1935. "A Preliminary Notice on the Miocene and Pliocene Mammalian Faunas near Valentine, Nebraska." Amer. Jour. Sci. (5), XXIX, pp. 125–132.

STROMER, E. 1926. "Reste Land und Susswasser Bewohnender Wirbeltiere aus den Diamentfeldern Deutsch-Sudwestafrikas." Die Diamentenwuste Sudwest Afrikas, II, pp. 107–153.

STOCK, CHESTER, AND HALL, E. R. 1933. "The Asiatic Genus *Eomellivora* in the Pliocene of California." Jour. Mammalogy, XIV, No. 1, pp. 63–65.

TEILHARD DE CHARDIN, P. 1915. "Les Carnassiers des Phosphorites du Quercy." Ann. de Pal., Paris, IX, pp. 103–191, Pls. XII–XX.

1922. "Sur une Faune de Mammifères Pontiens Provenant de la Chine Septentrionale." C. R. Acad. Sci, Paris, Seance de 20 Nov., 1922, pp. 979–981.

1927. "Observations sur la Lenteur d'Evolution des Faunes de Mammifères Continentales." Palaeobiologica, I, pp. 55–60.

TEILHARD DE CHARDIN, P., AND YOUNG, C. C. 1931. "Fossil Mammals from the Late Cenozoic of Northern China." Pal. Sinica, Ser. C, IX, Fasc. 1, pp. 1–66, Pls. I–X.

TEILHARD DE CHARDIN, P., AND STIRTON, R. A. 1934. "A Correlation of some Miocene and Pliocene Mammalian Assemblages in North America and Asia with a Discussion of the Mio-Pliocene Boundary." Univ. of Calif., Bull. Dept. Geol. Sci., XXIII, No. 8, pp. 277–290.

THEOBALD, W. 1874. "Geology of Pegu." Mem. Geol. Surv. India, X, Pt. 2.

1881. "The Siwalik Group of the Sub-Himalayan Region." Rec. Geol. Surv. India, XIV, Pt. 1, pp. 66–125.

THOMAS, OLDFIELD. 1890. "A Milk Dentition in *Orycteropus*." Proc. Roy. Soc. London, XLVII, pp. 246–248.

1905. "On *Hylochoerus*, the Forest Pig of Central Africa." Proc. Zool. Soc. London, pp. 193–199.

TOMES, C. S. 1914. "A Manual of Dental Anatomy." London.

TROUESSART, E. L. 1897. "Catalogus Mammalium tam Viventium quam Fossilium." Berlin.

1905. "Catalogus Mammalium tam Viventium quam Fossilium." Supplement, Berlin.

VAUGHAN, T. W. 1924. "Criteria and Status of Correlation and Classification of Tertiary Deposits." Bull. Geol. Soc. America, XXXV, pp. 677–742.

VREDENBURG, E. W. 1906. "The Classification of the Tertiary System in Sind with Reference to Zonal Distribution of Eocene Echinoidea." Rec. Geol. Surv. India, XXXIV, Pt. 3, pp. 172–198.

WADIA, D. N. 1919. "Geology of India." London.

1928. "The Geology of the Poonch State and Adjacent Portions of the Punjab." Mem. Geol. Surv. India, LI, Pt. 2.

1929. "Note on the Joya Dome Fold near Chakwal, Punjab." Rec. Geol. Surv. India, LXI, pp. 358–362.

1932. "The Tertiary Geosyncline of Northwest Punjab and the History of Quaternary Earth Movements and Drainage of the Gangetic Trough." Quar. Jour. Geol. Min. and Met. Soc. of India, IV, No. 3, pp. 70–95.

WAGNER, A. 1857. "Neue Beiträge zur Kenntniss der Fossilen Säugethier-Ueberreste von Pikermi." Abhandl. Bayer. Akad. Wissenschaf. Math., CL, VIII Band, I Abth., pp. 3–50.

WEBER, MAX. 1927. "Die Saugetiere." Revised Edition. Jena.

WEITHOFER, K. A. 1888. "Beitrage zur Kentniss der Fauna von Pikermi bei Athen." Beitr. Pal. Ost.-Ung., Wien, VI, p. 233.

WINDLE, B. C. A., AND PARSONS, F. G. 1901. "On the Muscles of the Ungulata." Proc. Zool. Soc. London, Pt. 1, pp. 656–704.

1903. "On the Muscles of the Ungulata." Proc. Zool. Soc. London, Pt. 2, pp. 262–298.

WANG, K. M. 1928. "Die Obermiozänen Rhinoceratiden von Bayern." Pal. Zeitschr., Berlin, X, pp. 184–212.

WYNNE, A. B. 1874. "Notes on the Geology of the Neighbourhood of Mari Hills Station in the Punjab." Rec. Geol. Surv. India, VII, pp. 64–74.

WOOD, H. E. 1927. "Some Early Tertiary Rhinoceroses and Hyracodonts." Bull. Amer. Palaeontology, XIII (No. 50), pp. 3–89, Pls. I–VII.

YOUNG, C. C. 1927. "Fossile Nagetiere aus Nord-China." Pal. Sinica, Ser. C, V, Fasc. 3, pp. 1–83, Pls. I–III.

1932. "On the Artiodactyla from the Sinanthropus Site at Chouk'outien." Pal. Sinica, Ser. C, VIII, Fasc. 2, pp. 1–101, Pls. I–XXIX.

ZDANSKY, O. 1923. "Fundorte der Hipparion-Fauna um Pao-Te-Hsien in N. W. Shanshi." Bull. Geol. Surv. China, No. 5, pp. 69–81, Pls. I–V.

1924. "Jungtertiare Carnivoren Chinas." Pal. Sinica, Ser. C, II, Fasc. 1, pp. 1–149, Pls. I–XXXIII.

1925A. "Quartäre Carnivoren aus Nord-China." Pal. Sinica, Ser. C, II, Fasc. 2, pp. 1–27, Pls. I–IV.

1925B. "Die Fossile Hirsche Chinas." Pal. Sinica, Ser. C, II, Fasc. 3, pp. 1–94, Pls. I–XVI.

1926. "*Paracamelus gigas* Schlosser." Pal. Sinica, Ser. C, II, Fasc. 4, pp. 1–44, Pls. I–IV.

1927. "Weitere Bemerkungen über fossile Carnivoren aus China." Pal. Sinica, Ser. C, IV, Fasc. 4, pp. 1–30, Pls. I–II.

1930. "Die Alttertiaren Saugetiere Chinas, nebst Stratigraphischen Bemerkungen." Pal. Sinica, Ser. C, VI, Fasc. 2, pp. 1–87, Pls. I–V.

VON ZITTEL, K. A. 1923. "Grundzüge der Paläontologie, Mammalia." Revised. Berlin.

Specimens from various museums or collections are listed by number in the following manner.

American Museum of Natural History, New York . . . . Amer. Mus. or A. M.
British Museum (Natural History), London . . . . Brit. Mus. or B. M.
Geological Survey of India Collection, Calcutta . . . . G. S. I.
Yale Peabody Museum, New Haven . . . . Y. P. M.
Science and Art Museum, Dublin . . . . Sci. and Art Museum, Dublin
Royal College of Surgeons Museum, London . . . . Royal College of Surgeons

---

ADDENDUM

Plesiogulo brachygnathus (Schlosser) has been recorded by G. E. Lewis (1933) from the upper portion of the Middle Siwaliks, near Bhandar. This species was inadvertently omitted from the faunal list and from the systematic discussion of the Carnivora.

INDEX

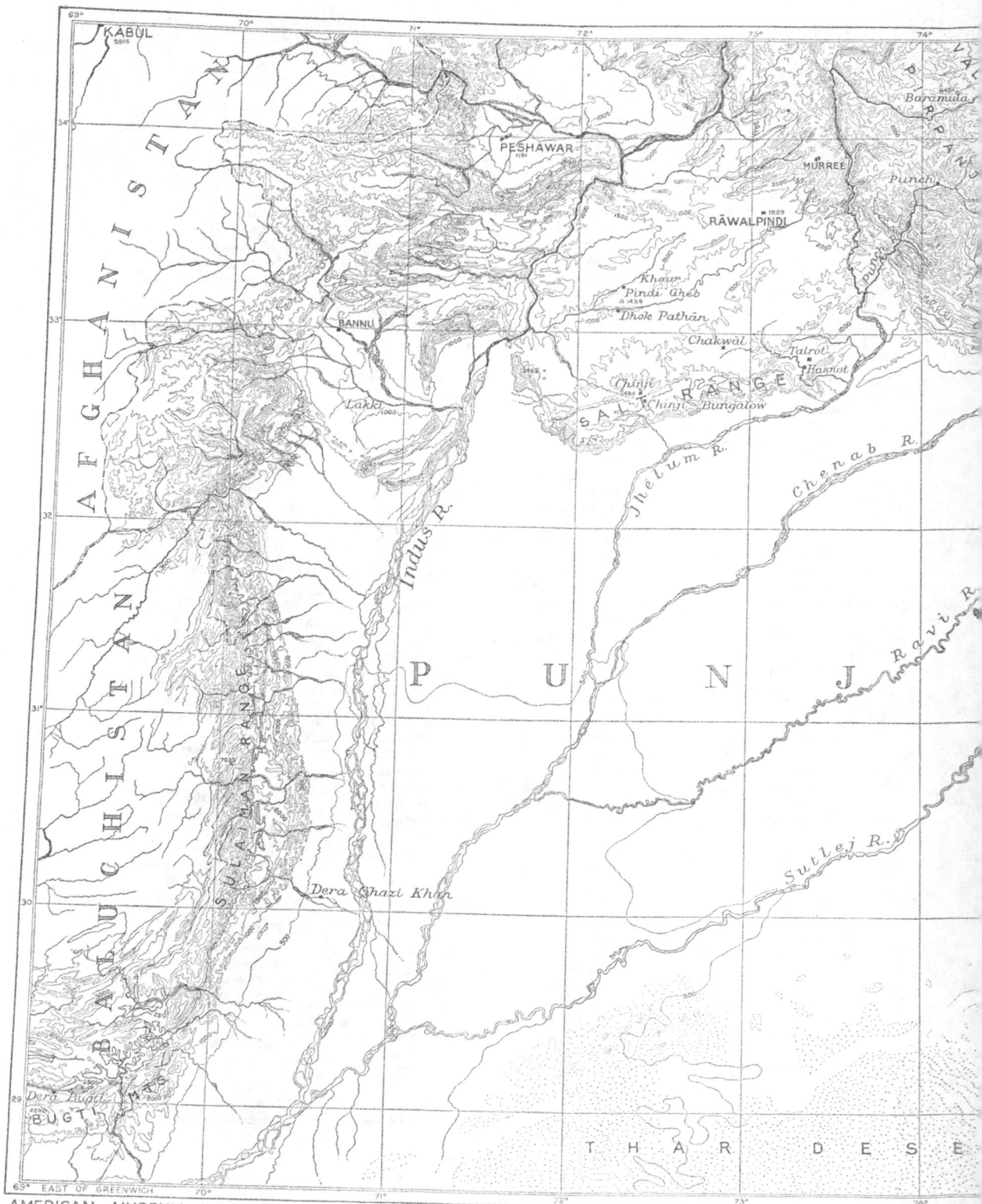

Map of Northwestern India and adjacent territory, showing

AMERICAN MUSEUM OF NATURAL HISTORY 1927.
Compiled from the Survey of India 1:1000000 sheets Nrs. 38, 39, 43, 44, 52, 53.

KĀBUL

A F G H A N I S T A N

PESHĀWAR

MURREE

Punch

RĀWALPINDI

Baramula

VAL P P AN

BANNU

Kheur
Pindi Gheb

Dhole Pathān

Chakwāl

Talrot
Hasnot

Chinji
Chinji Bungalow

S A L T   R A N G E

Lakki

Indus R.

Jhelum R.

Chenab R.

P U N J

Ravi R.

B A L U C H I S T A N

S U L A I M A N   R A N G E

Sutlej R.

Dera Ghazi Khan

Dera Bugti
BUGTI HILLS

T H A R   D E S E

63° EAST OF GREENWICH

MAP OF
NORTHWESTERN INDIA
AND
ADJACENT TERRITORY.

SCALE
10 0 10 20 30 40 50 60 70 80 90 100 MILES

G R E A T

H I M A L A Y A

T I B E T

T I B B A   R A N G E

R A N G E

K A S H M I R

L A D H A R   R A N G E

S I W A L I K

N A G

B

SRINAGAR

JAMMU

LKOT

AMRITSAR

Nurpur

Kangra

Bilaspur

Haritalyangar

SIMLA

Mirzapur
Siswan
Pinjaur
Chandigarh
Moginand
Kasauli
Dagshai
Kalka

DELHI

Ramnagar

Indus R.

Sutlej R.

Beas R.

Sutlej R.

Ghaggar R.

Jumna R.

Ganges R.

△ 23410
△ 21275
△ 16529
△ 16500
△ 20040
△ 21380
△ 20000
△ 18490
△ 17040
△ 21760
△ 20500
△ 20650
△ 22250
△ 23050
△ 20200
△ 22210
△ 20800
△ 20720
△ 21760
△ 22170
△ 23190
△ 22953
△ 25447
△ 25645
△ 23360

Erwin J. Raisz A.M.

Museum collecting localities in the Siwalik Hills and the Salt Range. Scale, 1/40 inch to mile. Drawn by Erwin J. Raisz, 1927.